Lecture Notes in Computer Science 6112

Commenced Publication in 1973
Founding and Former Series Editors:
Gerhard Goos, Juris Hartmanis, and Jan van Leeuwen

Aurélio Campilho   Mohamed Kamel (Eds.)

Image Analysis and Recognition

7th International Conference, ICIAR 2010
Póvoa de Varzim, Portugal, June 21-23, 2010
Proceedings, Part II

 Springer

Volume Editors

Aurélio Campilho
University of Porto, Faculty of Engineering
Institute of Biomedical Engineering
4200-465 Porto, Portugal
E-mail: campilho@fe.up.pt

Mohamed Kamel
University of Waterloo, Department of Electrical
and Computer Engineering
Waterloo, Ontario, N2L 3G1, Canada
E-mail: mkamel@uwaterloo.ca

Library of Congress Control Number: 2010928206

CR Subject Classification (1998): I.4, I.5, I.2.10, I.2, I.3.5, F.2.2

LNCS Sublibrary: SL 6 – Image Processing, Computer Vision, Pattern Recognition,
and Graphics

ISSN   0302-9743
ISBN-10   3-642-13774-1 Springer Berlin Heidelberg New York
ISBN-13   978-3-642-13774-7 Springer Berlin Heidelberg New York

Typesetting: Camera-ready by author, data conversion by Scientific Publishing Services, Chennai, India
Printed on acid-free paper   06/3180

Preface

ICIAR 2010, the International Conference on Image Analysis and Recognition, held in Póvoa do Varzim, Portugal, June 21-23, was seventh in the ICIAR series of annual conferences alternating between Europe and North America. The idea of organizing these conferences was to foster the collaboration and exchange between researchers and scientists in the broad fields of image analysis and pattern recognition, addressing recent advances in theory, methodology and applications. During the years the conferences have become a forum with a strong participation from many countries. This year, ICIAR was organized along with AIS 2010, the International Conference on Autonomous and Intelligent Systems. Both conferences were organized by AIMI—Association for Image and Machine Intelligence.

For ICIAR 2010, we received a total of 164 full papers from 37 countries. The review process was carried out by members of the Program Committee and other reviewers; all are experts in various image analysis and pattern recognition areas. Each paper was reviewed by at least two reviewers, and checked by the Conference Chairs. A total of 89 papers were finally accepted and appear in the two volumes of these proceedings. The high quality of the papers is attributed first to the authors, and second to the quality of the reviews provided by the experts. We would like to sincerely thank the authors for responding to our call, and to thank the reviewers for their careful evaluation and feedback provided to the authors. It is this collective effort that resulted in the strong conference program and high-quality proceedings.

This year included a competition on "Fingerprint Singular Points Detection" and a challenge on "Arabidopsis Thaliana Root Cell Segmentation Challenge," which attracted the attention of ICIAR participants.

We were very pleased to be able to include in the conference program keynote talks by three well-known experts: Alberto Sanfeliu, Universitat Politècnica de Catalunya, Spain; Edwin Hancock University of York, UK and José Santos-Victor, Institute for Systems and Robotics, Instituto Superior Técnico, Portugal. We would like to express our sincere gratitude to the keynote speakers for accepting our invitation to share their vision and recent advances in their specialized areas.

We would like to thank Khaled Hammouda, the webmaster of the conference, for maintaining the Website, interacting with the authors and preparing the proceedings. Special thanks are also due to the members of the local Organizing Committee for their advice and help. We are also grateful to Springer's editorial staff, for supporting this publication in the LNCS series. We would like to acknowledge the professional service of Viagens Abreu in taking care of the registration process and the special events of the conference.

Finally, we were very pleased to welcome all the participants to ICIAR 2010. For those who did not attend, we hope this publication provides a good view into the research presented at the conference, and we look forward to meeting you at the next ICIAR conference.

June 2010   Aurélio Campilho
                                                                   Mohamed Kamel

ICIAR 2010 – International Conference on Image Analysis and Recognition

General Chair

Aurélio Campilho
University of Porto, Portugal
campilho@fe.up.pt

General Co-chair

Mohamed Kamel
University of Waterloo, Canada
mkamel@uwaterloo.ca

Local Organizing Committee

Ana Maria Mendonça
University of Porto
Portugal
amendon@fe.up.pt

Jorge Alves Silva
University of Porto
Portugal
jsilva@fe.up.pt

António Pimenta Monteiro
University of Porto
Portugal
apm@fe.up.pt

Pedro Quelhas
Biomedical Engineering Institute
Portugal

Gabriela Afonso
Biomedical Engineering Institute
Portugal
iciar10@fe.up.pt

Conference Secretariat

Viagens Abreu SA
Porto, Portugal
congresses.porto@viagensabreu.pt

Webmaster

Khaled Hammouda
Waterloo, Ontario, Canada
hammouda@pami.uwaterloo.ca

Advisory Committee

M. Ahmadi	University of Windsor, Canada
P. Bhattacharya	Concordia University, Canada
T.D. Bui	Concordia University, Canada
M. Cheriet	University of Quebec, Canada
E. Dubois	University of Ottawa, Canada
Z. Duric	George Mason University, USA
G. Granlund	Linköping University, Sweden
L. Guan	Ryerson University, Canada
M. Haindl	Institute of Information Theory and Automation, Czech Republic
E. Hancock	The University of York, UK
J. Kovacevic	Carnegie Mellon University, USA
M. Kunt	Swiss Federal Institute of Technology (EPFL), Switzerland
J. Padilha	University of Porto, Portugal
K.N. Plataniotis	University of Toronto, Canada
A. Sanfeliu	Technical University of Catalonia, Spain
M. Shah	University of Central Florida, USA
M. Sid-Ahmed	University of Windsor, Canada
C.Y. Suen	Concordia University, Canada
A.N. Venetsanopoulos	University of Toronto, Canada
M. Viergever	University of Utrecht, The Netherlands
B. Vijayakumar	Carnegie Mellon University, USA
J. Villanueva	Autonomous University of Barcelona, Spain
R. Ward	University of British Columbia, Canada
D. Zhang	The Hong Kong Polytechnic University, Hong Kong

Program Committee

A. Abate	University of Salerno, Italy
P. Aguiar	Institute for Systems and Robotics, Portugal
M. Ahmed	Wilfrid Laurier University, Canada
N. Alajlan	King Saud University, Saudi Arabia
J. Alirezaie	Ryerson University, Canada
H. Araújo	University of Coimbra, Portugal
N. Arica	Turkish Naval Academy, Turkey
J. Barbosa	University of Porto, Portugal
J. Barron	University of Western Ontario, Canada
J. Batista	University of Coimbra, Portugal
C. Bauckhage	York University, Canada
A. Bernardino	Technical University of Lisbon, Portugal
G. Bilodeau	École Polytechnique de Montréal, Canada
J. Bioucas	Technical University of Lisbon, Portugal

B. Boufama	University of Windsor, Canada
T.D. Bui	Concordia University, Canada
J. Cardoso	University of Porto, Portugal
E. Cernadas	University of Vigo, Spain
F. Cheriet	École Polytechnique de Montréal, Canada
M. Cheriet	University of Quebec, Canada
M. Coimbra	University of Porto, Portugal
M. Correia	University of Porto, Portugal
L. Corte-Real	University of Porto, Portugal
J. Costeira	Technical University of Lisbon, Portugal
A. Dawoud	University of South Alabama, USA
M. De Gregorio	Istituto di Cibernetica "E. Caianiello" - CNR, Italy
Z. Duric	George Mason University, USA
N. El Gayar	Nile University, Egypt
M. El-Sakka	University of Western Ontario, Canada
P. Fieguth	University of Waterloo, Canada
M. Figueiredo	Technical University of Lisbon, Portugal
G. Freeman	University of Waterloo, Canada
V. Grau	University of Oxford, UK
M. Greenspan	Queen's University, Canada
L. Guan	Ryerson University, Canada
F. Guibault	École Polytechnique de Montréal, Canada
M. Haindl	Institute of Information Theory and Automation, Czech Republic
E. Hancock	University of York, UK
C. Hong	Hong Kong Polytechnic, Hong Kong
K. Huang	Chinese Academy of Sciences, China
J. Jiang	University of Bradford, UK
B. Kamel	University of Sidi Bel Abbès, Algeria
G. Khan	Ryerson University, Canada
M. Khan	Saudi Arabia
Y. Kita	National Institute AIST, Japan
A. Kong	Nanyang Technological University, Singapore
M. Kyan	Ryerson University, Canada
J. Laaksonen	Helsinki University of Technology, Finland
Q. Li	Western Kentucky University, USA
X. Li	University of London, UK
R. Lins	Universidade Federal de Pernambuco, Brazil
J. Lorenzo-Ginori	Universidad Central "Marta Abreu" de Las Villas, Cuba
G. Lu	Harbin Institute, China
R. Lukac	University of Toronto, Canada
A. Mansouri	Université de Bourgogne, France
A. Marçal	University of Porto, Portugal
J. Marques	Technical University of Lisbon, Portugal

M. Melkemi   Univeriste de Haute Alsace, France
A. Mendonça   University of Porto, Portugal
J. Meunier   University of Montreal, Canada
M. Mignotte   University of Montreal, Canada
A. Monteiro   University of Porto, Portugal
M. Nappi   University of Salerno, Italy
A. Padilha   University of Porto, Portugal
F. Perales   University of the Balearic Islands, Spain
F. Pereira   Technical University of Lisbon, Portugal
E. Petrakis   Technical University of Crete, Greece
P. Pina   Technical University of Lisbon, Portugal
A. Pinho   University of Aveiro, Portugal
J. Pinto   Technical University of Lisbon, Portugal
F. Pla   Universitat Jaume I, Spain
P. Quelhas   Biomedical Engineering Institute, Portugal
M. Queluz   Technical University of Lisbon, Portugal
P. Radeva   Autonomous University of Barcelona, Spain
B. Raducanu   Autonomous University of Barcelona, Spain
S. Rahnamayan   University of Ontario Institute of Technology
                           (UOIT), Canada
E. Ribeiro   Florida Institute of Technology, USA
J. Sanches   Technical University of Lisbon, Portugal
J. Sánchez   University of Las Palmas de Gran Canaria,
                           Spain
B. Santos   University of Aveiro, Portugal
A. Sappa   Computer Vision Center, Spain
G. Schaefer   Nottingham Trent University, UK
P. Scheunders   University of Antwerp, Belgium
J. Sequeira   Ecole Supérieure d'Ingénieurs de Luminy,
                           France
J. Shen   Singapore Management University, Singapore
J. Silva   University of Porto, Portugal
B. Smolka   Silesian University of Technology, Poland
M. Song   Hong Kong Polytechnical University,
                           Hong Kong
J. Sousa   Technical University of Lisbon, Portugal
H. Suesse   Friedrich Schiller University Jena, Germany
S. Sural   Indian Institute of Technology, India
S. Suthaharan   USA
A. Taboada-Crispí   Universidad Central "Marta Abreu"
                           de las Villas, Cuba
M. Vento   University of Salerno, Italy
J. Vitria   Computer Vision Center, Spain
Y. Voisin   Université de Bourgogne, France
E. Vrscay   University of Waterloo, Canada
L. Wang   University of Melbourne, Australia

Z. Wang	University of Waterloo, Canada
M. Wirth	University of Guelph, Canada
J. Wu	University of Windsor, Canada
F. Yarman-Vural	Middle East Technical University, Turkey
J. Zelek	University of Waterloo, Canada
L. Zhang	The Hong Kong Polytechnic University, Hong Kong
L. Zhang	Wuhan University, China
G. Zheng	University of Bern, Switzerland
H. Zhou	Queen Mary College, UK
D. Ziou	University of Sherbrooke, Canada

Reviewers

A. Abdel-Dayem	Laurentian University, Canada
D. Frejlichowski	West Pomeranian University of Technology, Poland
A. Mohebi	University of Waterloo, Canada
Y. Ou	University of Pennsylvania, USA
R. Rocha	Biomedical Engineering Institute, Portugal
F. Sahba	University of Toronto, Canada

Supported by

 AIMI – Association for Image and Machine Intelligence

 Department of Electrical and Computer Engineering Faculty of Engineering University of Porto Portugal

 INEB – Instituto de Engenharia Biomédica Portugal

 PAMI – Pattern Analysis and Machine Intelligence Group University of Waterloo Canada

Table of Contents – Part II

Biomedical Image Analysis

Biometrics

Applications

Table of Contents – Part I

Feature Extraction and Pattern Recognition

Computer Vision

Shape, Texture and Motion Analysis

Coding, Indexing and Retrieval

Face Detection and Recognition

Automated Vertebra Identification from X-Ray Images

Xiao Dong and Guoyan Zheng

Institute for Surgical Technology and Biomecnahics, University of Bern,
Stauffacherstrasse 78, CH-3014 Bern, Switzerland
guoyan.zheng@ieee.org

Abstract. Automated identification of vertebra bodies from medical images is important for further image processing tasks. This paper presents a graphical model based solution for the vertebra identification from X-ray images. Compared with the existing graphical model based methods, the proposed method does not ask for a training process using training data and it also has the capability to automatically determine the number of vertebrae visible in the image. Experiments on digitially reconstructed radiographs of twenty-one cadaver spine segments verified its performance.

1 Introduction

Automated identification of vertebra bodies from medical images is important for further image processing tasks such as segmentation, registration, reconstruction and inter-vertebra disk identification. Due to the complexity of the spine structure, simple feature (for example, landmarks or edges) based solutions are not reliable and researchers are paying more attention on graphical model based solutions[1][2]. The current vertebra or intervertebral disk identification approaches usually face the following difficulties:

Unknown object number. Detecting an unknown number of vertebrae or intervertebral disks invokes a model selection problem. In [1][2], they focus on either the lumbar or the whole spine so that the number of intervertebral disks is taken as fixed and they can thus build their graphical models with a fixed number of nodes and avoid the model selection problem. In [3], the number of vertebrae is detected by a generalized Hough transformation (GHT) along the detected spinal cord. The robustness of the vertebra number detection is highly dependent on the image quality.

Off-line training. Due to the complexity of the spine structure, most of the existing work on the spine area ask for the involvement of prior knowledge which is usually obtained by off-line training. In [1][2], both the low level image observation models and the high level disk context potentials need to be trained using training data. In [3], statistical surface models for each vertebra, the sacrum, the vertebra coordinate system and generalized Hough transform models are obtained from the training data. Besides the fact that

A. Campilho and M. Kamel (Eds.): ICIAR 2010, Part II, LNCS 6112, pp. 1–9, 2010.

the model training and model building are complex problems themselves, the dependency on training data makes these approaches only applicable to the data with similar characteristics with the training data.

Our contributions to overcome the difficulties are: (1) firstly we designed a graphical model of the spine, which can determine the number of visible vertebrae during the inference procedure; (2) secondly, in the graphical model, both the low level image observation model and the high level vertebra context potentials need not to be learned from training data. Instead they are designed so that they are capable of self-learning from the image data during the inference procedure.

2 Method

2.1 Graphical Model

Similar to [2], we build a graphical model $G = \{V, E\}$ with N nodes for the spine structure as shown in Fig. 1. Each node $V_i, i = 0, 1, ..., N - 1$ represents a connected disk-vertebra-disk component of the spine, in which both the disks and the vertebral body are modelled as rectangular shapes. We assign a parameter set $\mathbf{X}_i = \{x_i, y_i, r_i, h_i, \theta_i, h_i^u, \theta_i^u, h_i^l, \theta_i^l\}$ to V_i to describe the positions and the shapes of V_i as shown in Fig. 2. $E = \{e_{i,j}\}, i, j = 0, 1, 2, ..., N - 1$ defines a connection matrix of the graph G. On this graphical model, we define the component observation model $p(\mathbf{I}|X_i), i = 0, 1, ..., N - 1$ of a single component and potentials $p(X_i, X_j), i, j = 0, 1, ...N-1, e_{i,j} = 1$ among neighboring components. $\{p(\mathbf{I}|\mathbf{X}_i), i = 0, 1, ..., N - 1\}$ represents the probabilities that the configurations of the nodes match the observed images I and the potentials $\{p(\mathbf{X}_i, \mathbf{X}_j)\}$ encode the geometrical constraints between components. The identification of the spinal structure is then to find the configurations of $\{V_i\}, \mathbf{X} = \{\mathbf{X}_0, \mathbf{X}_i, ..., \mathbf{X}_{N-1}\}$, that maximizes

$$P(X|I) \propto \prod_i p(I|\mathbf{X}_i) \prod_{e_{i,j}=1} p(\mathbf{X}_i, \mathbf{X}_j) \qquad (1)$$

2.2 Component Observation Model

The component observation model $p(\mathbf{I}|\mathbf{X}_i)$ is to match a template, which is determined by $\mathbf{X}_i$, with the observed image $\mathbf{I}$ defined as

$$p(\mathbf{I}|\mathbf{X}_i) = p_I(\mathbf{I}|\mathbf{X}_i)p_G(\mathbf{I}|\mathbf{X}_i)p_V(\mathbf{I}|\mathbf{X}_i) \qquad (2)$$

The three items in (2) come from the intensity, gradient and local variance of the template as described below:

Intensity observation model $p_I(\mathbf{I}|\mathbf{X}_i)$**:** Given $\mathbf{X}_i$, it determines a disk-vertebra -disk template on the 2D X-ray image plane as shown in Fig. 2. We assume that the interior area of the vertebra body has a homogeneous intensity distribution, a Gaussian model $\mathcal{N}(\mu_i, \sigma_i)$, which is different from the intensity

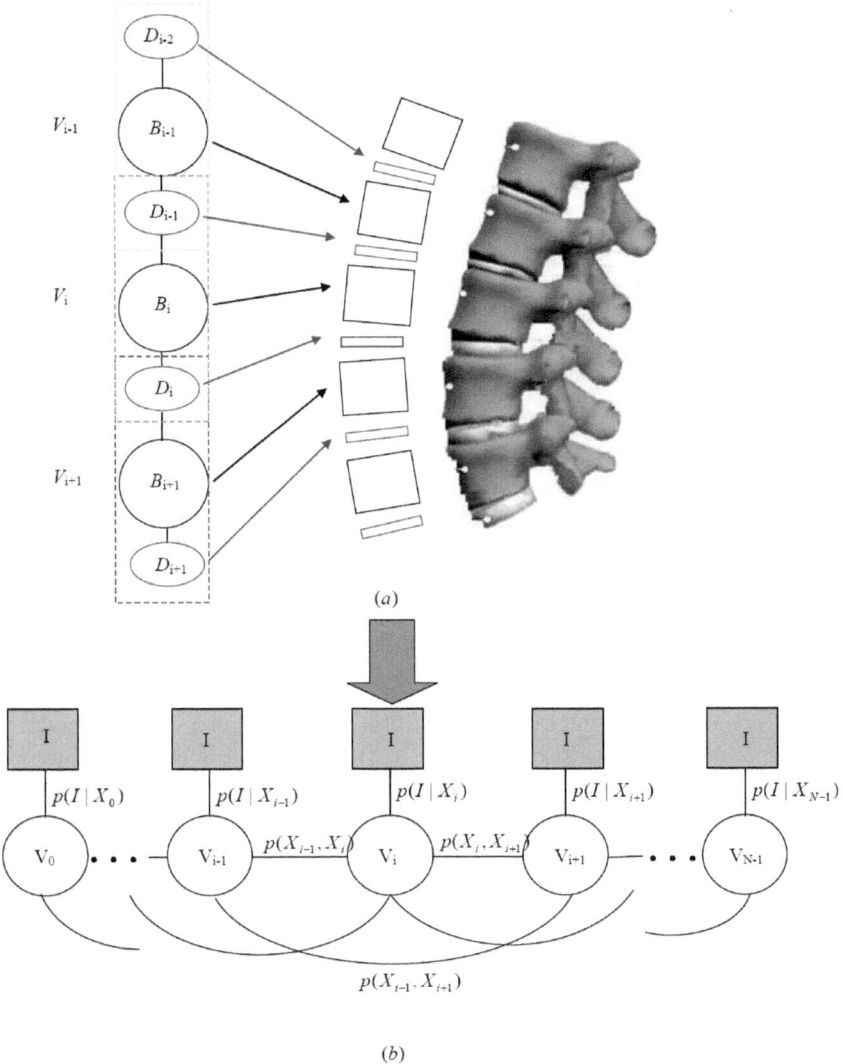

(a)

(b)

Fig. 1. Graphical model of the spine

distribution of the border region, which is defined as a small neighbourhood outside the vertebra body. For each pixel s that falls in the interior and border region of the template as shown in Fig. 2, the image appearance value of s is defined as

$$p(s|\mathbf{X}_i) = e^{-\frac{(I(s)-\mu_i)^2}{2\sigma_i^2}}$$

(3)

We define $p_I(\mathbf{I}|\mathbf{X}_i) = e^{\omega_I c_I^i}$, where c_I^i is the cross-correlation between the image appearance values $p(s|\mathbf{X}_i)$ and a binary template which sets value 1 to the interior area of the template and 0 to the border region. $\omega_I > 0$ is

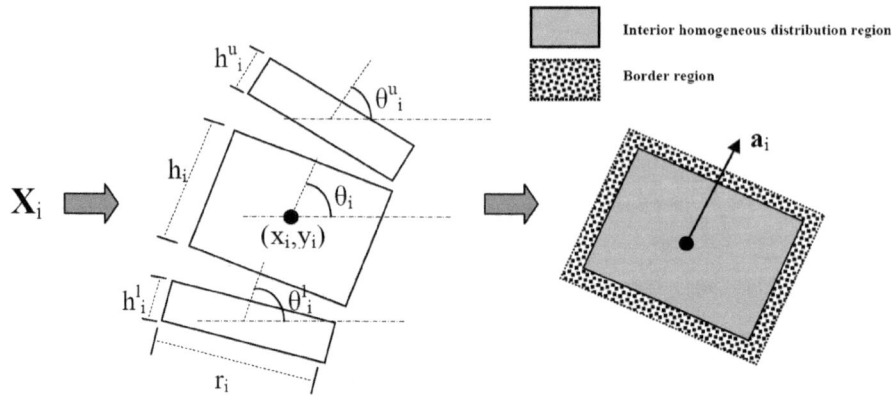

Fig. 2. Vertebra body template for the component observation model

a weighting factor. Intuitively this means that we assume that the interior region of the template should obey the Gaussian distribution and the border area should have a different intensity distribution. The Gaussian model $\mathcal{N}(\mu_i, \sigma_i)$ can be learned from the observed image once $\mathbf{X}_i$ is given.

Gradient observation model $p_G(\mathbf{I}|\mathbf{X}_i)$**:** Similar to $p_I(\mathbf{I}|\mathbf{X}_i)$, we can define $p_G(\mathbf{I}|\mathbf{X}_i) = e^{\omega_G c_G^i}$, where c_G^i is the cross-correlation between the gradient image values of the observed image in the template area and a binary gradient template, which sets 0 in the interior area and 1 in the border region. This means strong gradient values should only happen on the border of the vertebra template.

Local variance observation model $p_V(\mathbf{I}|\mathbf{X}_i)$**:** We compute the local variance image I_V of the image I, which is defined as the intensity variance in a small window centered at each pixel. We define $p_V(\mathbf{I}|\mathbf{X}_i) = e^{\omega_V c_V^i}$, where c_V^i is the cross-correlation between the local variance values and a binary template identical to the gradient template.

We only consider the image observation model of the vertebra bodies but ignore the observation model of the disks. This is due to the fact that for X-ray images with different view directions, a unified observation model for the disks is difficult to design.

It can also be observed that three components in the component observation model do not need to be trained with training data as in [1][2]. Instead they can adjust their parameters by a self-learning from the X-ray images.

2.3   Potentials between Components

We define inter-node potentials to set constraints on the geometries of the nodes $\{\mathbf{V}_i\}$ so that all the nodes will be assembled to a meaningful spine structure.

We define

$$p(\mathbf{V}_i, \mathbf{V}_j) = p_S(\mathbf{V}_i, \mathbf{V}_j) p_O(\mathbf{V}_i, \mathbf{V}_j) p_D(\mathbf{V}_i, \mathbf{V}_j) \qquad (4)$$

Size constraints. $p_S(\mathbf{V}_i, \mathbf{V}_j)$ is used to set constraints on the sizes of the neighbouring components defined as

$$p_S(\mathbf{V}_i, \mathbf{V}_j) = e^{-(\omega_r \frac{|r_i - r_j|}{|r_i + r_j|} + \omega_h \frac{|h_i - h_j|}{|h_i + h_j|})/|i - j|} \qquad (5)$$

which means that neighbouring components should have similar sizes and the strength of the constraints should decay with the distance between components.

Orientation constraints. We define

$$p_O(\mathbf{V}_i, \mathbf{V}_j) = e^{-\omega_o \mathbf{a}_i \bullet \mathbf{a}_i / |i - j|} \qquad (6)$$

to ensure that neighbouring vertebra bodies should have similar orientations, in which $\mathbf{a}_i$ is the 2-dimensional axis of the vertebra body template as shown in Fig. 2.

Distance constraints. For direct neighboring nodes $\mathbf{V}_i, \mathbf{V}_j$, i.e. $|i - j| = 1$, we also define constraints on the spatial distance between their vertebra body centers. Without losing any generality, for the case $j = i + 1$ we define $p_D(\mathbf{V}_i, \mathbf{V}_j)$ as

$$p_D(\mathbf{V}_i, \mathbf{V}_j) = \begin{cases} e^{-\omega_D \frac{d_{C,ij}}{d_{h,ij}}} & , d_{C,ij} < \frac{d_{h,ij}}{4} \\ e^{-\omega_D \frac{d_{C,ij} - (d_{h,ij} + h_i^l + h_j^u)/2}{d_{h,ij}}} & , \frac{5}{4} d_{h,ij} > d_{C,ij} > \frac{3}{4} d_{h,ij} \\ 0 & , \text{elsewhere} \end{cases} \qquad (7)$$

where $d_{C,ij} = \|C_i - C_j\|, d_{h,ij} = h_i + h_j$. Intuitively this constraint means that we ask direct neighboring components should either be closely connected side-by-side or merge to one object. This makes our graphical model capable of adjusting the configuration of the nodes in the component chain to find the number of vertebrae during the inference procedure.

2.4 Optimization

In [1], the optimization is achieved by a generalized EM algorithm given the known disk number and a proper initialization. In [2] the candidate configuration for each object can be detected by searching the whole data volume using trained random classification trees and the inference is achieved by the A^* algorithm. Both of their optimization methods are not suitable for our graphical model. Firstly we do not have a proper initialization of the configuration of our components as in [1]. Secondly, the configuration of each object in our case is high dimensional so that the complete search for candidate configurations of each object as presented in [2] is computational costly. Our optimization procedure to find the solution of Eq. (1) consists of two levels:

1. The optimization to find the configuration(s) of each individual component $\mathbf{V}_i$ by a particle filtering.
2. The optimization to find the joint configuration of the component set by a belief propagation based inference [4].

For an object $\mathbf{V}_i$ and its configuration parameter set $\mathbf{X}_i$

1. Randomly generate a set of K configuration of $\mathbf{V}_i$, $\mathbf{X}_i^k$, $k = 0, 1, ...K - 1$.
2. Compute the *believe* of each configuration as $w_i^k \propto p(\mathbf{I}|\mathbf{X}_i^k)$.
3. Taking the randomly generated configurations $\mathbf{X}_i^k$ as candidate configurations of each component and the believes w_i^k as local believes, run a (loopy) belief propagation on the graphical model to approximate the joint distribution of all the components.
4. Update the configuration of $\mathbf{X}_i^k$, $k = 0, 1, ...K - 1$ using the marginal distribution of each component $\overline{w}_i^k$, which can be easily obtained from the belief propagation procedure, The basic idea of the configuration update is to generate new configurations near the configurations with higher believes $\overline{w}_i^k$.
5. Repeat 2-4 till the procedure converges.

Algorithm 1. Optimization of the joint configuration of the component set

The basic concept of our optimization algorithm is that

- The particle filtering part (step 1,2 and 4) is used to find probable candidates for each individual object.
- The BP part (step 3) is used to set regularization on the components so that only the candidates that can fulfill the inter-component constraints will be selected.

2.5 The Determination of the Number of Vertebrae

Vertebra number determination is a key factor for a correct vertebra detection. In our approach it is solved in a semiautomatic method as described as follows:

- Users click two landmarks on the X-ray image to indicate the first and the last visible vertebra.
- From the landmarks and the projection parameters of the X-ray image, we can estimate an upper bound of the number of vertebrae N^{upper} and construct a graph model with N^{upper} nodes.
- Carry out the inference procedure described in Algorithm 1. Due to the distance potential between neighboring components, neighboring components will either be located side-by-side or overlap. Therefore the extra vertebrae will merge with their neighbours, i.e., multiple nodes may be located at the same vertebra.
- After the optimization, a simple mean-shift based clustering on the center positions of the components using the mean height of the vertebra bodies as its bandwidth can easily merge overlapping components and therefore find the number of vertebrae [5].

3   Experimental Results

We validated the present approach on digitally reconstructed radiographs (DRRs) of twenty-one cadaver spine segments, where eight of them were from cervical region, six of them were from thoracic region and the rest were from lumbar region. The DRRS were constructed from the CT volumes of the associated spine segments, resulting in a detection of totally 132 vertebrae from the DRRs (45 cervical vertebrae, 56 thoracic vertebrae, and 31 lumbar vertebrae). For each CT volume, a pair of DRRs consisting of an anterior-posterior (AP) image and a lateral-medial (LM) image were generated.

For each pairs of DRRs, we started the detection from the LM image due to the observation that the vertebra bodies in the LM image were more homogeneous than those in the AP image. As soon as all the vertebrae were detected from the LM image, we could apply the same approach to the AP image but with a fixed number of the vertebrae that is determined from the LM image. For each detection, the user specified two points as the input to our approach with one picked around the center of the top vertebra and the other around the center of the bottom vertebra. Our approach was then used to detect all vertebrae from the input image pair. The outputs from our approach include the number of vertebrae in the image, as well as the three-dimensional (3D) location and orientation of each vertebra, which are reconstructed from the associated two-dimensional (2D) detection results in both images. Figure 3 shows three examples of the automated detection of vertebrae in three different anatomical regions.

The automated vertebra body detection results are presented in Table 1. Although our approach had false/miss detection on four images, the false/miss vertebra detection rate was low. From the totally 132 vertebrae, our approach could successfully detect 122 vertebrae, which results in a 92.4% success rate.

Table 1. Automated vertebra body detection results

Spine Regions	Detection Results	Image Number	Vertebra Number
Cervical Vertebrae	Correct	6	38
	False/miss	2	7
Thoracic Vertebrae	Correct	4	53
	False/miss	2	3
Lumbar Vertebrae	Correct	7	31
	False/miss	0	0

4   Discussion and Conclusion

In this paper we proposed a graphical model based method for automated detection of vertebra bodies from X-ray images. We validated our method on DRRs of twenty-one cadaver spine segments of different regions. Compared to previously introduced approach, our approach has the following advantages: (1) It need not to be trained using training data, (2) It does not ask for the prior information

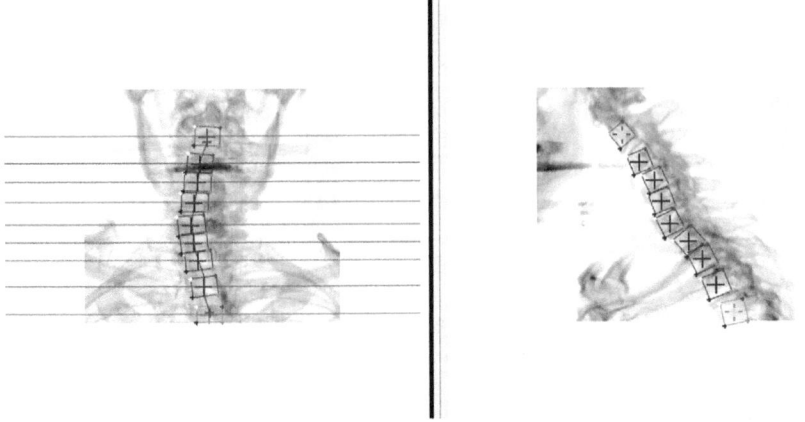

(a) Cervical vertebra detection example

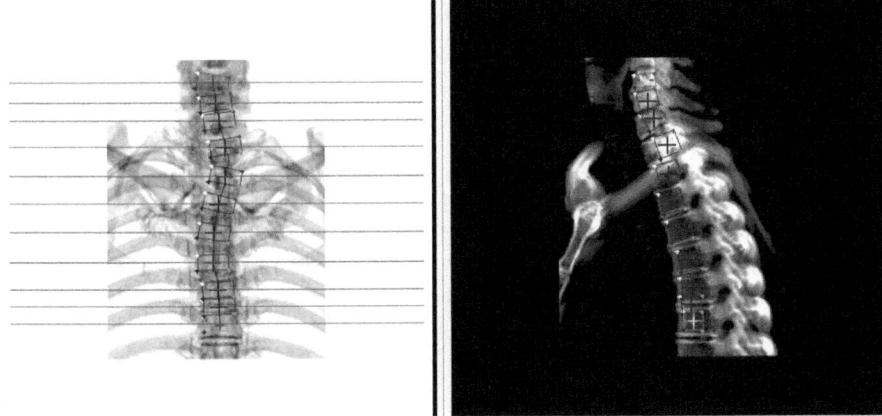

(b) Thoracic vertebra detection example

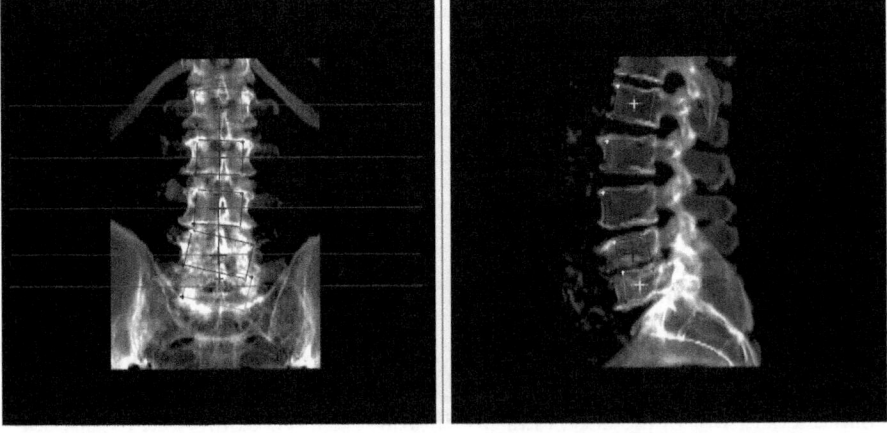

(c) Lumbar vertebra detection example

Fig. 3. Examples of detection vertebrae in different spine regions

of the examined anatomical region and (3) It can automatically identify the number of vertebrae visible in the image and therefore does not ask for a prior information of the vertebra number to be identified. Our future work focuses on investigating the performance of the proposed approach on clinical x-ray images.

References

1. Corso, J., Alomari, R., Chaudhary, V.: Lumbar disc localization and labeling with a probabilistic model on both pixel and object features. In: Metaxas, D., Axel, L., Fichtinger, G., Székely, G. (eds.) MICCAI 2008, Part I. LNCS, vol. 5241, pp. 202–210. Springer, Heidelberg (2008)
2. Schmidt, S., Kappes, J., Bergtholdt, M., Pekar, V., Dries, S., Bystrov, D., Schnoerr, C.: Spine detection and labeling using a parts-based graphical model. In: Karssemeijer, N., Lelieveldt, B. (eds.) IPMI 2007. LNCS, vol. 4584, pp. 122–133. Springer, Heidelberg (2007)
3. Klinder, T., Ostermann, J., Ehm, M., Franz, A., Kneser, R., Lorenz, C.: Automated model-based vertebra detection, identification, and segmentation in ct images. Medical Image Analysis 13, 471–482 (2009)
4. Murphy, K.P., Weiss, Y., Jordan, M.I.: Loopy belief propagation for approximate inference: An empirical study. In: Proceedings of Uncertainty in AI, pp. 467–475 (1999)
5. Comaniciu, D., Meer, P.: Mean shift: a robust approach toward feature space analysis. IEEE Transactions on Pattern Analysis and Machine Intelligence 24, 603–619 (2002)

Towards Non Invasive Diagnosis of Scoliosis Using Semi-supervised Learning Approach

Lama Seoud[1], Mathias M. Adankon[1], Hubert Labelle[2],
Jean Dansereau[1], and Farida Cheriet[1]

[1] École Polytechnique de Montréal
lama.seoud@polymtl.ca, mathias-mahouzonsou.adankon@polymtl.ca,
jean.dansereau@polymtl.ca, farida.cheriet@polymtl.ca
[2] Sainte Justine Hospital Research Center, Montréal, Québec, Canada
hubert.labelle@recherche.ste-justine.qc.ca

Abstract. In this paper, a new methodology for the prediction of scoliosis curve types from non invasive acquisitions of the back surface of the trunk is proposed. One hundred and fifty-nine scoliosis patients had their back surface acquired in 3D using an optical digitizer. Each surface is then characterized by 45 local measurements of the back surface rotation. Using a semi-supervised algorithm, the classifier is trained with only 32 labeled and 58 unlabeled data. Tested on 69 new samples, the classifier succeeded in classifying correctly 87.0% of the data. After reducing the number of labeled training samples to 12, the behavior of the resulting classifier tends to be similar to the reference case where the classifier is trained only with the maximum number of available labeled data. Moreover, the addition of unlabeled data guided the classifier towards more generalizable boundaries between the classes. Those results provide a proof of feasibility for using a semi-supervised learning algorithm to train a classifier for the prediction of a scoliosis curve type, when only a few training data are labeled. This constitutes a promising clinical finding since it will allow the diagnosis and the follow-up of scoliotic deformities without exposing the patient to X-ray radiations.

1 Introduction

Scoliosis is a three-dimensional deformity of the spine and the ribcage that affects the general appearance of the trunk. In general, one of the first symptoms of scoliosis is the manifestation of a hump on the back, called the rib hump. It constitutes one of the most disturbing aspects of the deformity for the patients.

The management of scoliosis depends essentially on the severity, the type and the risk of progression of the curve. Those parameters are commonly evaluated on standard frontal and lateral X-rays of the patient's trunk in upright position. However, there are several limitations attributed to the radiographic evaluation of scoliosis. First of all, it provides only bi-dimensional information that is not sufficient to fully evaluate a complex three-dimensional pathology like scoliosis. Second, only the internal deformities can be evaluated in the radiographs while the patients' main concern is their external appearance. And last but not least, as X-ray acquisition is invasive, and considering the risks associated with radiation exposure from repeated radiographs, its frequency

A. Campilho and M. Kamel (Eds.): ICIAR 2010, Part II, LNCS 6112, pp. 10–19, 2010.

is limited to every 6 months which represents a long interval for the follow-up of a progressive scoliosis.

In order to evaluate the scoliosis on a more frequent basis, non invasive imaging techniques that provide a three-dimensional reconstruction of the trunk surface have been proposed in the literature. The main challenge currently is to relate the topographic measurements with the radiographic ones. In this context, several authors have tried to predict the severity of scoliosis [1,2,3,4] or the 3D shape of the spine [5] from metric evaluations on the surface of the back or of the trunk, using statistical methods [4] or machine learning techniques like neural networks [2] and supervised support vector machines [3,5]. To build such classifiers, a large set of labeled samples, called the training data, is necessary. As the labeling is based on radiographic measurements, the size of the training database is thus limited by the X-rays acquisition frequency. This affects negatively the performance of the classifiers.

Since the back surface acquisition is totally non-invasive, it would be advantageous to complement the training database with topographic data that is not necessarily labeled, which means that no X-ray information is available for this data. Training using both labeled and unlabeled data is called semi-supervised learning. Recently, one such approach has been proposed in the literature [6] and its effectiveness has been tested on artificial data as well as on real problems such as the character recognition. The results have shown the usefulness and the good performance of this method even when the number of labeled data is too small.

The aim of the current study is to prove the feasibility of using a semi-supervised learning approach for the prediction of scoliosis curve types by analysing the back surface of the trunk.

2   3D Back Surface Analysis

2.1   Data Acquisition

Currently, at Sainte-Justine Hospital Research Centre (SJHRC) in Montreal (Canada), the back surface of the trunk is acquired using an optical digitizer (InSpeck Inc., Montreal, Canada), comprised of a color CCD camera and a structured light projector. The acquisition process consists of projecting successively four fringe patterns, obtained by phase-shifting a set of light fringes, onto the surface. Based on the four resulting images, the system computes, by interferometry and triangulation, the depth of each surface point relative to the reference plane of the digitizer. A fifth image, with no fringes, acquires the texture of the surface which is then mapped onto the 3D model (figure 1).

The resulting mesh consists of more than 10,000 nodes, depending on the size of the patient. The accuracy of this system was evaluated in [7], using markers placed on a mannequin whose coordinates were previously recorded by a computer measuring machine. The results showed a reconstruction accuracy of 0.56 mm over the back.

During the acquisition, the patient is in upright position, with his arms slightly abducted on the sides. Prior to the acquisition, a nurse locates by palpation and places markers over several anatomic landmarks on the trunk, such as the center of the posterior superior iliac spines (CPSIS) and the vertebral prominence (VP). These markers are used for clinical measurements and registration.

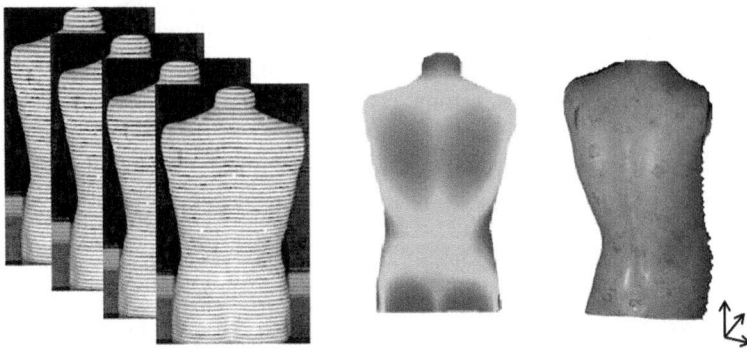

Fig. 1. On the left: four fringe patterns are projected on the back surface of a mannequin. In the center: color representation of depth. On the right: 3D reconstruction of the surface with the mapped texture.

2.2 Features Extraction

To evaluate the scoliosis deformities on the surface of the trunk, local measurements are computed on horizontal cross-sections of the trunk. More specifically, 50 sections, equally spaced along the vertical axis of the trunk, are automatically extracted starting from the CPSIS and going up to the VP (Figure 2). On each of the 50 sections, an automatic algorithm computes the back surface rotation (BSR). This clinical measurement is related to the amplitude of the rib hump and is defined as the angle, projected onto the axial plane, between the dual tangent to the section and the X-axis of the acquisition reference frame. The BSR can be negative or positive depending on the side of the hump. The accuracy of this measurement computed on a 3D reconstruction of the trunk surface was previously evaluated at 1.4mm [8].

Thus, each back is characterized by 50 BSR values. In order to filter outliers and obtain smoother value sets, an averaging window was applied to each set. Moreover, for each patient, the angle values were normalized between -100 and 100 degrees to compensate for differences between the patients in term of severity. Finally, the BSR values corresponding to the upper 5 sections were not considered because they were too noisy and the values of the BSR are not relevant in this area in the context of scoliosis assessment.

3   Semi-supervised Learning

Pattern recognition problems are solved with classifiers which are designed using prototypes of the data to be recognized. This data, called the training set, consists of the patterns and their labels (the category of the pattern). This is the supervised learning where the features extracted from the patterns and their labels are used for modeling the classifier parameters [9,10]. However, the labeling process can become extremely expensive and cumbersome. For instance, the labeling of handwritten documents, images, or web pages requires both human expertise and insight, whereas in the field of

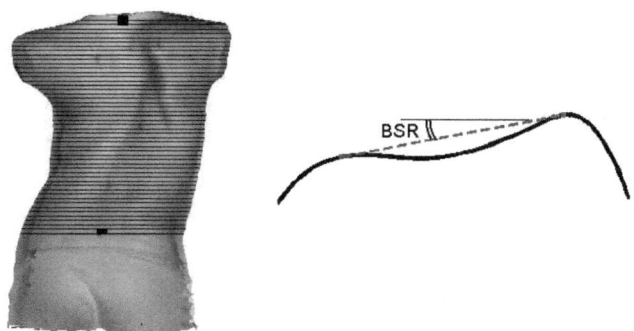

Fig. 2. On the left: 50 horizontal sections (black lines) of the trunk are extracted between the CPSIS (black squared marker on the bottom) and the VP (black squared marker on the top). On the right: the BSR is the angle defined by the dual tangent (red dashed line) to the back section (black curve) and the projection of the lateral axis onto the axial plane (black line).

medicine or biology, the labeling process may require some data acquisitions that are limited for ethical reasons. Thus, it may be very difficult, or even impossible, to label all the available data. The alternative is the semi-supervised learning [11,12,13], where both labeled and unlabeled data are used to train the classifier. Hence, it is not necessary to label all the data collected in order to build the classifier.

In this work, we propose using the least squares support vector machine (LS-SVM) in order to classify scoliosis curve types. To this end, we have collected labeled and unlabeled data in order to train our machine in semi-supervised mode.

The LS-SVM is an interesting variant of the SVM proposed by Suykens et al.[14,15]. The standard Vapnik SVM classifier [16] is modified to transform the QP problem into a linear one. These modifications are formulated in the LS-SVM definition as follows, when we consider a binary classification problem with $\{(x_1, y_1), ..., (x_\ell, y_\ell)\}$ the training dataset, $x_i \in \mathbb{R}$ and $y_i \in \{-1, 1\}$.

$$\min_{w,b,\xi} \frac{1}{2} w'w + \frac{1}{2} C \sum_{i=1}^{\ell} \xi_i^2 \qquad (1)$$

$$\text{s.t.} \quad \xi_i = y_i - [w'\phi(x_i) + b] \quad \forall i = 1, ..., \ell \qquad (2)$$

where w' denotes the transpose of w, ϕ is the mapping function used implicitly via the kernel function $k(x_i, x_j) = \phi(x_i).\phi(x_j)$ for non linear problems, C is used to balance the trade-off between maximizing the margin and minimizing the training error quantified by the variable ξ.

The original SVM formulation is modified at two points. First, the inequality constraints with the slack variable ξ_i expressed in SVM formulation are replaced by equality constraints. Second, a squared loss function is considered in the objective function. These two essential modifications simplify the problem, which becomes linear.

The Lagrangian of problem (2) is expressed by :

$$\mathcal{L}(w, b, \xi, \alpha) = \frac{1}{2}w'w$$

$$+ \frac{1}{2}C\sum_{i=1}^{\ell}\xi_i^2 - \sum_{i=1}^{\ell}\alpha_i\{y_i - [w'\phi(x) + b] - \xi_i\}$$

where α_i are Lagrange multipliers.

The Karush−Kuhn−Tucker (KKT)[1] conditions for optimality yield

$$\begin{cases} \frac{\partial \mathcal{L}}{\partial w} = 0 \Rightarrow w = \sum_{i=1}^{\ell}\alpha_i\phi(x_i) \\ \frac{\partial \mathcal{L}}{\partial b} = 0 \Rightarrow \sum_{i=1}^{\ell}\alpha_i = 0 \\ \frac{\partial \mathcal{L}}{\partial \xi_i} = 0 \Rightarrow \alpha_i = C\xi_i, \quad \forall i = 1, ..., \ell \\ \frac{\partial \mathcal{L}}{\partial \alpha_i} = 0 \Rightarrow \xi_i = y_i - [w'\phi(x_i) + b] \quad \forall i = 1, ..., \ell \end{cases}$$

We note that the system coming from the KKT conditions is linear, and that its solution is found by solving the system of linear equations expressed in the following matrix form :

$$\begin{pmatrix} K + C^{-1}I & 1' \\ 1 & 0 \end{pmatrix} \begin{pmatrix} \alpha \\ b \end{pmatrix} = \begin{pmatrix} Y \\ 0 \end{pmatrix} \tag{3}$$

where :

$$K_{ij} = k(x_i, x_j)$$
$$Y = (y_1, ..., y_\ell)'$$
$$\alpha = (\alpha_1, ..., \alpha_\ell)'$$
$$1 = (1, ..., 1)$$

In our previous work [6], we have proposed the semi-supervised LS-SVM ($S^2LS - SVM$) using the following expressions:

$$\min_{w,b,\xi,\xi^*,y_1^*,...,y_n^*} \frac{1}{2}w'w + \frac{1}{2}C\sum_{i=1}^{\ell}\xi_i^2 + \frac{1}{2}C^*\sum_{j=1}^{n}(\xi_j^*)^2 \tag{4}$$

$$\text{s.t.} \quad \xi_i = y_i - [w'\phi(x_i) + b], \quad \forall i = 1, ..., \ell \tag{5}$$

$$\xi_j^* = y_j^* - [w'\phi(x_j^*) + b], \quad \forall j = 1, ..., n \tag{6}$$

$$y_j^* \in \{-1, 1\} \quad \forall j = 1, ..., n \tag{7}$$

In this equation, the parameters C and C^* balance the error between the labeled $\{(x_1, y_1), ..., (x_\ell, y_\ell)\}$ and unlabeled data $\{x_1^*, ..., x_n^*\}$.

Considering the combinatorial view of the optimization problem (4), the variables w, b, ξ, ξ^* and y^* are optimized at different levels. Then, for a given fixed set $y_1^*, ..., y_n^*$,

---

[1] KKT conditions are necessary conditions for optimality obtained from first derivative.

the optimization over (w, b) is standard LS-SVM training, and we obtain a linear system in dual space expressed in matrix form by:

$$\begin{pmatrix} K + \Gamma & \mathbf{1}' \\ \mathbf{1} & 0 \end{pmatrix} \begin{pmatrix} \alpha \\ b \end{pmatrix} = \begin{pmatrix} Y \\ 0 \end{pmatrix} \tag{8}$$

where:

$K_{ij} = k(x_i, x_j)$

$Y = (y_1, ..., y_\ell, y_1^*, ..., y_n^*)'$

$\alpha = (\alpha_1, ..., \alpha_\ell, \alpha_1^*, ..., \alpha_n^*)'$

$\mathbf{1} = (1, ..., 1)$

Γ is a diagonal matrix with $\Gamma_{ii} = 1/C$ for $i = 1, ..., \ell$ and $\Gamma_{ii} = 1/C^*$ for $i = \ell + 1, ..., \ell + n$

Two methods are proposed in [6] for solving the semi-supervised problem expressed in (4). In this paper, we used the second approach which is described as follows.

The unlabeled examples are labeled gradually during the learning process: one sample is labeled and added to the labeled set. The added sample is chosen in order to obtain the smallest increase in the objective function. The criterion we use to select this point is based on the value of α_j^*; because considering the equation $\alpha_j^* = C^* \xi_j^*$, it is clear that the error is proportional to the value of α^*.

First, we identify the label of the point to be labeled according to the objective function. Next, for each remaining unlabeled sample, we compute $a_j^{(1)} = \alpha_j^*$ if the identified label is 1 and $a_j^{(-1)}$ for the opposite label. As the goal is to find the unlabeled sample x^* with the smallest increase in the objective function, we select, at each step, the unlabeled sample, the corresponding α_j^* of which will be the smallest if it is added to the previous solution. We repeat this procedure until all unlabeled samples are labeled.

4   Dataset and Experimental Setup

This study was conducted on a cohort of 159 adolescents with scoliosis who were candidates for surgery. Among the cohort, 101 patients had their topographic and radiographic acquisitions done at the same visit. Based on the radiographic measurements of each of those patient, the scoliosis curve type was determined according to the common clinical classification that distinguishes between 4 types of curves: thoracic major curves, thoracolumbar major curves, lumbar major curves and double major curves. The number of lumbar major curves being too small (4/101), we mixed them with the thoracolumbar major curves (18/101), being quite similar. The distribution of the patients among the three considered classes is illustrated in Table 1. For the remaining 58 patients, their radiographs were not acquired at the same date as the trunk topography, thus they were considered unlabeled.

For all the patients, the 45 features are automatically extracted according to subsection 2.2, and four classifiers are trained distinctly. The classifiers' performance is evaluated on the same testing dataset composed of 69 labeled samples chosen quasi-randomly among the database, under the condition of having a class distribution as similar as possible to the one of the whole cohort.

Table 1. Distribution of the patients among the classes

Classes	Curve type	Number of patients
Class 1	Thoracic major curve	45
Class 2	Double major curve	34
Class 3	Lumbar or thoracolumbar major curve	22
Unlabeled	Unknown	58
Total		159

A first classifier (C-SSL1) is built using the semi-supervised learning algorithm as described in section 3 and using a total of 90 training data: the 58 unlabeled samples and the remaining 32 labeled ones. A second classifier (C-SL1) is trained using a supervised LS-SVM and a training database made of the same 32 labeled data as for the C-SSL1. C-SL1 constitutes the reference case since it is trained with all the available labeled data.

In order to compare the supervised and the semi-supervised learning when only a small amount of training data are labeled, a third classifier (C-SSL2) is built using the semi-supervised learning as described in section 3 and using, for training, only 12 labeled data (5 of class 1, 4 of class 2, 3 of class 3) chosen pseudo-randomly (under the condition of having a class distribution among the retained data as similar as possible to the one of the whole cohort) and the remaining 78 training samples considered unlabeled. A fourth classifier (C-SL2) is trained using a supervised LS-SVM algorithm and the same 12 labeled data as in C-SSL2 are considered for training.

Since, we have a multi-class problem, each classifier is built by training three machines using the *one-against-all* strategy. We used a radial basis function (RBF) kernel and performed model selection with cross validation procedure [17].

5   Results and Discussion

Table 2 presents the prediction rates obtained in testing each of the four classifiers. These results show first that the performance of the classifiers trained using the semi-supervised algorithm (C-SSL1 and C-SSL2), compared respectively to C-SL1 and C-SL2, is significantly improved by the addition of unlabeled data. Second, the performance of C-SSL2, trained with only 12 labeled samples tends to be equal to the one of the ideal classifier C-SL1, that is built using the maximum number of available labeled samples. The latter outcome answers to the main goal of the semi-supervised learning: even with a few labeled data (12 among 90 training samples), the generalisation capacity of the classifier C-SSL2 is similar to the reference case.

Table 3 illustrates the confusion matrices obtained in testing the four classifiers. It shows first that all the classifiers clearly distinguish between patterns of class 3 and class 1. The major confusion is between classes 1 and 2 and between classes 2 and 3. This is also illustrated by the plot of the mean BSR values for each class (figure 3). In fact, for some double major curves (class 2) the thoracic hump is more accentuated than the lumbar hump which results in a pattern that is quiet similar to the thoracic major curves (class 1). The same logic follows in the case of some double major curves (class 2) where the lumbar hump is more prominent than the thoracic hump which results

Table 2. Learning and testing databases of the four classifiers and the prediction rates in testing

Classifiers	Training data		Testing data	Prediction rate
	Lab.	Unlab.		
CSSL1	32	58	69	87.0%
CSL1	32	-	69	82.6%
CSSL2	12	78	69	79.7%
CSL2	12	-	69	75.4%

in a pattern that is quiet similar to the lumbar major curves (class 3). Nevertheless, it seems that, with the semi-supervised learning, the unlabeled samples guide the classifier towards more generalizable boundaries for class 2 (prediction rate of 65.2% for C-SSL2 versus 56.5% for C-SL2 and 73.9% for C-SSL1 versus 65.2% for C-SL1) and for class 3 (prediction rate of 80.0% for C-SSL2 versus 66.7% for C-SL2 and 100% for C-SSL1 versus 93.3% for C-SL1). Furthermore, table 3 illustrates once again how the behavior of C-SSL2, trained with only 12 labeled samples out of the 90 training samples, tends to be similar to the reference classifier (C-SL1).

Table 3. Confusion matrices in testing the four classifiers

	C-SSL1			C-SL1			C-SSL2			C-SL2		
	Target Class			Target Class			Target Class			Target Class		
Predicted class	1	2	3	1	2	3	1	2	3	1	2	3
1	28	4	0	28	6	0	29	8	0	29	9	0
2	3	17	0	3	15	1	2	14	3	2	13	5
3	0	2	15	0	2	14	0	1	12	0	1	10
Prediction rate per class (%)	90.3	73.9	100	90.3	65.2	93.3	93.6	65.2	80.0	93.6	56.5	66.7

In this study, we considered the BSR as the only clinical index to describe the back surface deformity. This choice is based on the clinical observation that the rib hump generally appears on the convex side of each spinal curve. Our results show that the BSR, computed on 45 cross-sections of the back, is a good discriminant feature. However, in future works, other characteristics of the scoliosis deformity will be considered in order to reduce even more the misclassification rate.

Furthermore, in this work, only the back surface of the trunk is acquired. As demonstrated in the literature, measurements made on the back surface are sensitive to the patient's posture during the acquisition. To overcome this imprecision, the entire trunk's surface can be reconstructed using four optical digitizers placed all around the patient and a registration process [7]. The measurements can thus be computed in a patient specific coordinates system. Moreover, as the whole shape of the trunk is deformed in 3D, it could be more interesting to consider inclined cross-sections that follows the general shape of the trunk to compute local measurements [8].

Finally, in the present paper, the back surfaces are classified in three different classes. Due to the small number of lumbar curves in the cohort, no distinction is made between

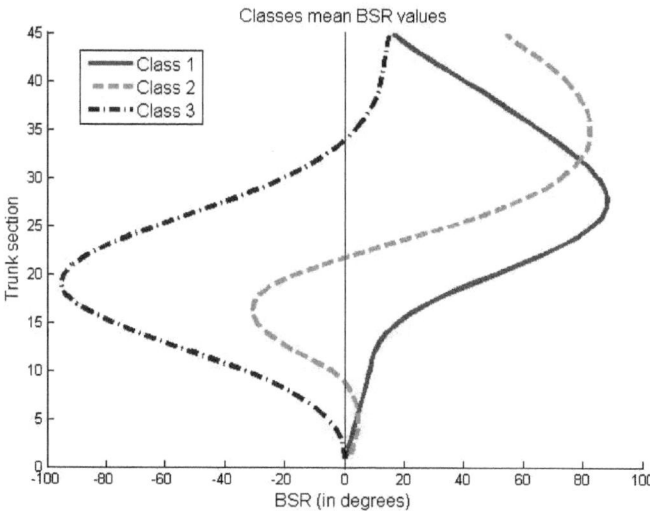

Fig. 3. Mean BSR values for each of the 45 sections of the trunk, computed for each class

lumbar and thoracolumbar major curves. However, in the presence of more lumbar curves, we could consider a fourth class in our classification. Furthermore, the current classification can be considered a high level classification and in future works, we can refine the classification's resolution by identifying features on the trunk's surface that could define distinct clusters within each class. This would be valuable in clinic since the actual scoliosis classification systems relies only on the spinal deformity and do not take into account the general appearance of the trunk which constitutes the patient's major preoccupation.

6   Conclusion

In conclusion, this preliminary study constitutes a proof of feasibility of the semi-supervised learning in the clinical context of classifying the scoliosis curve types based on the analysis of the back surface of the trunk. With only 12 labeled samples out of 90 training data, it is possible to predict the scoliosis curve type with a success rate similar to the reference case where 32 labeled data are used for supervised learning. Moreover, the unlabeled samples improve significantly the definition of the boundaries between the classes. Unlike supervised learning, there is no need to consider a large set of labeled data to build a consistent classifier with high generalization performance.

On a more clinical aspect, these results are valuable since it demonstrates that it is possible to identify the scoliosis curve type without exposing patients to ionizing radiation. This is an important finding since it could reduce the frequency of X-ray acquisitions during the scoliosis progression follow-up.

Acknowledgments

We would like to thank the GRSTB (Groupe de Recherche en Sciences et Technologies Biomédicales), the MENTOR program and the CIHR (Canadian Institutes of Health Research) for their financial support.

References

1. Ajemba, P., Durdle, N., Hill, D., Raso, J.: Classifying torso deformity in scoliosis using orthogonal maps of the torso. Med. Biol. Eng. Comput. 45, 575–584 (2007)
2. Jaremko, J., Poncet, P., Ronsky, J., Harder, J., Dansereau, J., Labelle, H., Zernicke, R.: Genetic algorithm-neural network estimation of Cobb angle from torso asymmetry in scoliosis. J. Biomech. Eng. 124(5), 496–503 (2002)
3. Ramirez, L., Durdle, N., Raso, J., Hill, G.: A support vector machines classifier to assess the severity of idiopathic scoliosis from surface topography. IEEE Trans. Inf. Technol. Biomed. 10, 84–91 (2006)
4. Stokes, I., Moreland, M.: Concordance of back surface asymmetry and spine shape in idiopathic scoliosis. Spine 14, 73–78 (1989)
5. Bergeron, C., Cheriet, F., Ronsky, J., Zernicke, R., Labelle, H.: Prediction of anterior scoliotic spinal curve from trunk surface using support vector regression. Eng. Appl. Artificial Intell. 18, 973–983 (2005)
6. Adankon, M., Cheriet, M., Biem, A.: Semisupervised least squares support vector machine. IEEE Trans. Neural Net. 20(12), 1858–1870 (2009)
7. Pazos, V., Cheriet, F., Song, L., Labelle, H., Dansereau, J.: Accuracy assessment of human trunk surface 3D reconstructions from an optical digitising system. Med. Biol. Eng. Comput. 43, 11–15 (2005)
8. Pazos, V., Miled, F., Debanne, P., Cheriet, F.: Analysis of trunk external asymmetry in side-bending. IRSSD, Liverpool (2008)
9. Duda, R.O., Hart, P.E., Stork, D.G.: Pattern Classification, 2nd edn. John Wiley and Sons, New York (2001)
10. Bishop, C.M.: Pattern Recognition and Machine Learning. Springer, Heidelberg (2006)
11. Seeger, M.: Learning with labeled and unlabeled data. Inst. Adapt. Neural Comput., Univ. Edinburgh, Edinburgh, U.K., Tech. Rep. (2001)
12. Zhu, X.: Semi-supervised learning literature survey. Comput. Sci. Univ. Wisconsin-Madison, Madison, WI, Tech. Rep. 1530 (2007),
    http://www.cs.wisc.edu/~jerryzhu/pub/sslsurvey.pdf
13. Adankon, M., Cheriet, M.: Help-Training for semi-supervised discrimininative classifier. Application to SVM. In: 19th Internationale Conference on Pattern Recognition, Tampa, pp. 1–4 (2008)
14. Suykens, J.A.K., Van Gestel, T., De Brabanter, J., De Moor, B., Vandewalle, J.: Least Squares Support Vector Machines. World Scientific, Singapore (2002)
15. Suykens, J.A.K., Vandewalle, J.: Least squares support vector machine classifiers. Neural Process. Lett. 9(3), 293–300 (1999)
16. Vapnik, V.N.: Statistical Learning Theory. Wiley, New York (1998)
17. Adankon, M., Cheriet, M.: Model selection for LS-SVM. Application to handwriting recognition. Pattern Recognit. 42(12), 3264–3270 (2009)

Articulated Model Registration of MRI/X-Ray Spine Data

Rola Harmouche[1], Farida Cheriet[1], Hubert Labelle[2], and Jean Dansereau[1]

[1] École Polytechnique de Montréal,
2500, chemin de Polytechnique, Montréal, H3T 1J4
[2] Hôpital Ste-Justine,
3175, Chemin de la Côte-Sainte-Catherine, Montréal H3T 1C5

Abstract. This paper presents a method based on articulated models for the registration of spine data extracted from multimodal medical images of patients with scoliosis. With the ultimate aim being the development of a complete geometrical model of the torso of a scoliotic patient, this work presents a method for the registration of vertebral column data using 3D magnetic resonance images (MRI) acquired in prone position and X-ray data acquired in standing position for five patients with scoliosis. The 3D shape of the vertebrae is estimated from both image modalities for each patient, and an articulated model is used in order to calculate intervertebral transformations required in order to align the vertebrae between both postures. Euclidean distances between anatomical landmarks are calculated in order to assess multimodal registration error. Results show a decrease in the Euclidean distance using the proposed method compared to rigid registration and more physically realistic vertebrae deformations compared to thin-plate-spline (TPS) registration thus improving alignment.

1 Introduction

Idiopathic scoliosis is a disease of unknown cause characterized by a complex three-dimensional curvature of the spine with onset most often discovered during puberty [1]. 5.1% of adolescent girls and 3.5% of adolescent boys are affected by scoliosis, with the more severe cases requiring treatment being girls [2]. Curvature measures are usually obtained from standing X-rays on which the skeletal structures are visible. This spinal deviation in turn affects the external appearance of a scoliotic patient, which is usually characterized by a lateral trunk asymmetry and or a rib hump. Such external deformations are often aesthetically undesirable for patients and can cause psychological problems. In more severe cases, the spinal curvature can affect the physical functioning of the patient with symptoms such as chronic back problems or pulmonary problems [3]. When the spinal curvature is very pronounced surgery is necessary in order to correct some of the undesirable deformation. Surgery is most often undertaken in prone position, where a rod and screws are used to fuse the vertebrae causing the deformation. Surgeons rely on their experience and intuition in order to

A. Campilho and M. Kamel (Eds.): ICIAR 2010, Part II, LNCS 6112, pp. 20–29, 2010.

establish the adequate instrumentation that would lead to the desirable post-operative external trunk appearance. However, changes to the external shape of the trunk due to surgery are not directly related to the changes in the shape of the vertebral column. This might be due to several factors including the possible non-rigid deformation of muscles following changes in the structure of the vertebral column. As a result, the surgeon cannot predict the effects of the instrumentation on the shape of the trunk prior to completion of surgery. A geometric model combining information from the vertebral column and the surrounding soft tissue can aid in surgical planning. This additional information about soft tissue can be obtained using magnetic resonance images (MRI), which are most often acquired in prone position. Combining the X-ray and MRI data requires multimodal image registration where the images are acquired in different postures. The aim of the present work is to register MRI and X-ray data of the spine as a first step towards full MRI/X-ray registration.

So far, little work has been done on MRI/X-ray registration. Van de Kraats et al. [4] register MRI to X-ray data using fiducials manually placed on cadaveric data. The placement of fiducials is obviously not realistic in real patient data. Tomazevic et al. [5] use a novel similarity measure in order to rigidly register a series of 2D X-ray images to CT and MRI data. 11 X-ray images are required per patient for accurate results; which is not possible to acquire in normal clinical settings due to radiation issues. Registration of vertebral information from these two image modalities is difficult for two main reasons. First, obtaining 3D information of the spine from the 2D X-ray images, which is the case of the majority of X-ray systems, has only been feasible using manual intervention so far [6]. Thus, registration methods that rely on intensity information, such as mutual information for example, are not adequate for 2D X-ray to 3D MRI registration due to the lack of intensity and spatial correspondences between the two modalities. Second, with the knowledge that vertebrae are rigid structures, traditional non-rigid registration algorithms are not appropriate for the task at hand. Vertebral structures extracted from MRI data have been modeled as rigid bodies for registration purposes [7] but only using unimodal 2D MRI data. Our team has recently developed an articulated model representation for the spine using X-ray data but did not using for registration purposes [8]. This model was used by Kadoury et al. [9] in order to register a preoperative reconstructed X-ray personalized model to the intraoperative CT data of a scoliotic surgical patient. The work in [9] optimizes using Markov random fields which requires significant computation time. In addition, such a model, consisting exclusively of vertebral information, does not provide complementary information to the CT data. This exclusivity limits the application's clinical benefits. Preoperative MR images on the other hand contain soft tissue information which can be useful for surgical planning.

This paper proposes the use of articulated models for the 3D semi-rigid registration between X-ray and MRI image reconstructions. Taking into account the vertebrae's physical characteristics, they are modeled as rigid bodies, and inter-vertebral rigid transformations are calculated using local vertebral coordinate systems constructed using manually extracted landmarks. The overall transformation between the vertebrae extracted from the two image modalities is calculated from the composition of local and global transformations. In order to assess registration error, Euclidean distances between registered corresponding points using this method are compared to those obtained using rigid and thin-plate-spline registration.

This article is organized as follows: Section 2 describes the articulated model, the experimental setup, and the methods used to validate our work. The results of the proposed method are shown in section 3, followed by a conclusion in section 4.

2   Proposed Method

In this work, we propose the use of articulated models in order to register a 3D reconstruction of the spine obtained from X-rays of patients with one obtained from MRIs. Vertebrae are considered as rigid bodies and inter-vertebral rigid transformations are calculated using correspondence points located on the vertebrae on each of the image modalities. In this section, the preprocessing work consisting of the 3D reconstruction of the vertebrae from medical images and the extraction of the correspondence points used for the registration will first be explained. This is followed by a description of the proposed articulated model used in order to align the vertebrae reconstructed from the MRI and X-ray data. Finally the method used to validate the multimodal 3D alignment and to asses the whether the shape of the column is consistent in both postures will be explained.

2.1   3D Reconstruction of Vertebrae and Point Extraction

Prior to the registration, we gathered MRI and X-ray data available at Ste-Justine hospital in Montreal from five patients with scoliosis for this study. In order to generate a 3D model of the spine from MRI data, T1-weighted MRI images are acquired using a Siemens Symphony system (1.5 Tesla, TR/TE = 771/15, 704x704, 350 FOV). Sagittal slices of 0.5mm by 0.5mm in-plane resolution and 3mm thickness are acquired with a 3.6mm separation between slices. The 3D shape of the seventeen thoracic and lumbar vertebrae is manually segmented from these images (figure 1(a)) and eight landmark points are manually labeled using TomoVision's SliceOmatic software so that they can be used to generate the articulated model. Those landmarks are placed on the left and right edges of the posterior, anterior, inferior and superior ends of the vertebral body for all thoracic and lumbar vertebrae (figure 1(a)).

In order to generate a 3D model of the spine from X-ray data, landmarks manually labeled on Postero-anterior and lateral radiographs are used to generate 3D landmark points representing the vertebral column. The 3D position of the points is obtained using an explicit calibration method and optimizing the calibration parameters with the Levenberg Marquardt method [6]. The obtained

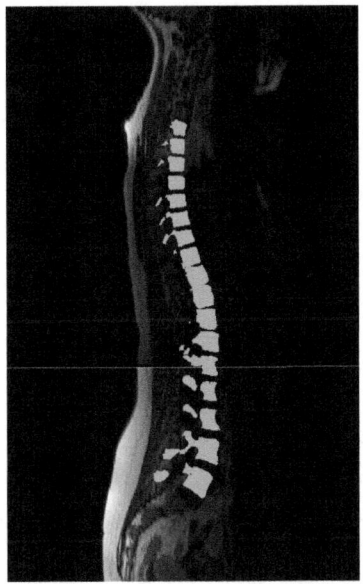 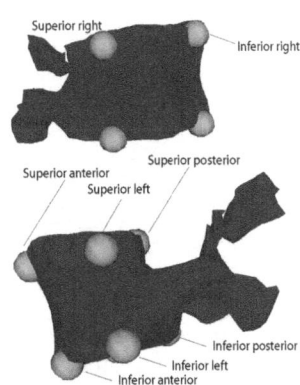

(a) MRI image and extracted vertebrae   (b) Correspondences on vertebrae

Fig. 1. 3D reconstruction of vertebrae from MRI sagittal slices along with manually labeled correspondences on each of the vertebrae

landmarks are used to map a generic vertebral dictionary onto the patient space. Of those landmarks, the same eight as in the case of the MRI data are used in order to generate the articulated model.

2.2 Articulated Model Deformations

The vertebrae reconstructed from the MRI data are aligned with those of the Xray data using the articulated model proposed by Boisvert et al. [8], which models the spine as a series of local inter-vertebral rigid transformations (figure 2). Inter-vertebral transformations are first calculated on each of the image modalities separately. In order to calculate the inter-vertebral transformation $T_{i,i+1}$ from vertebra V_i to the consecutive vertebra V_{i+1}, a local coordinate system is defined for each vertebra using the landmarks described above in the following manner: The 3D coordinates of landmarks on each vertebra are averaged to find the center of the coordinate system. The z-axis is defined as passing through the center from inferior to the superior end of the vertebra, the y-axis from left to right, and the x-axis from posterior to anterior. The Gram Schmidt algorithm[10] is then used to construct an orthogonal basis from these axis forming the local coordinate systems. The position and orientation of the first vertebra is defined using the transformation between the absolute world coordinate system and the first vertebra's local coordinate system ($T_{0,1}$). The intervertebral transformation

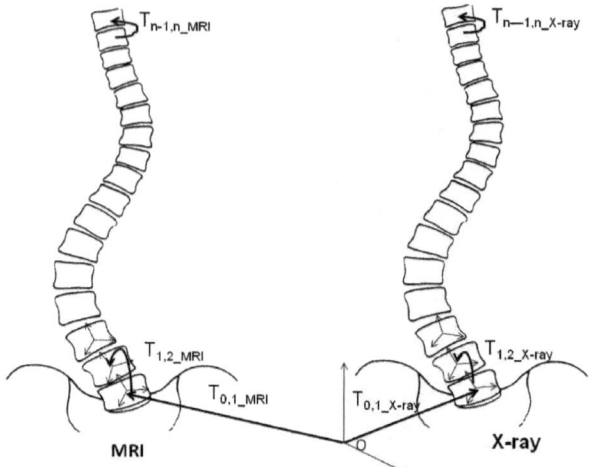

Fig. 2. Local and global transformations forming the articulated model required to align MRI onto X-ray vertebrae

matrices are then calculated as rigid transformations between the local coordinate systems of two consecutive vertebrae.

The global transformation $T_{0,i}$ between the world coordinates and the ith vertebra on each of the two image modalities is obtained by composing the local inter-vertebral transformations with the global transformation of the first vertebra in the following manner:

$$T_{0,i} = T_{i-1,i} \circ T_{i-2,i-1} \circ T_{i-3,i-2} \circ \ldots \circ T_{1,2} \circ T_{0,1}. \tag{1}$$

Finally, in order to register any vertebra i on the MRI image V_{i-MRI} to its corresponding vertebra on the X-ray data $V_{i-X-ray}$, the inverse of the transformation from the absolute world coordinates $T_{0,i_M RI}$ is first applied, followed by the transformation from absolute world coordinates to $V_{i-X-ray}$:

$$T_{i-MRI-X-ray} = T_{0,i_X-ray} \circ T_{0,i_M RI}^{-1}. \tag{2}$$

It must be noted that the point correspondences only serve to create the articulated model and are not required during the registration process itself. The inter-vertebral transformations are obtained in the local coordinates of the vertebrae without directly relying on absolute landmark positions. The method does not require point correspondences provided that another method of obtaining the local coordinate system is used.

3   Results

In order to test the validity of the articulated model registration technique, this method was compared to rigid registration and thin-plate-spline registration.

Rigid registration is carried out by minimizing the least squares distance between the source and the target landmarks. In order to calculate the error for these methods, half of the landmarks were randomly selected and used for registration and the remaining ones were used for validation purposes. The registration error is then defined as the Euclidean distance between the corresponding vertebral landmarks of the registered MRI/X-ray data. This is done in order to verify whether this error is decreased using the proposed method thus signifying better alignment. Errors are reported for the thoracic and the lumbar parts of the spine separately in order to assess in which part our proposed method brings the greatest improvement.

This section will first show some qualitative results of the registered spine between MRI and X-ray using rigid, thin-plate-spline, and articulated model registration. Then, quantitative results showing registration error are shown.

3.1 Qualitative Results

Registration results between the MRI and X-ray reconstructions are shown in figure 3(a) using rigid registration, in figure 3(b) using thin-plate-spline registration, and in figure 3(c) using our proposed articulated model registration. The landmarks used for registration are shown using spheres for the MRI data and cubes for the X-ray data. The rigid registration results show that the curvature is less pronounced at the bottom of the column in the case of the MRI (light gray) as opposed to the X-ray (dark gray) data, which can be explained by the fact that the patient is lying on a flatbed. It can be seen that our proposed method leads to more accurate alignment of the vertebrae when compared to rigid registration. The difference between the registration accuracy in the lower part of the spine can be seen in figures 3(d) and 3(f). Thin-plate-spline registration gives good alignment between the two modalities (figures 3(b) and 3(e)). However, it must be noted that non-rigid deformations such as thin-plate-splines applied to rigid structures like vertebrae misleadingly lead to smaller error values as they do not preserve their physical characteristics. In order to illustrate with an extreme case, figures 3(g) and (h) show an X-ray vertebra and its corresponding MRI vertebra, respectively. Even though the corresponding points are very well aligned, a misplaced landmark at the inferior end of the X-ray target vertebra caused a highly non-rigid deformation of the MRI vertebra and changed it's true physical form thus leading to erroneous results.

3.2 Quantitative Results

Table 1 shows the quantitative results for this study. The Euclidean distance between the MRI and X-ray landmarks is first calculated using our proposed articulated model method and compared to both rigid and thin-plate-spline registration for the lumbar and thoracic vertebrae together. A significant decrease in the registration error can be seen when our proposed method is used compared to rigid registration. The proposed method only provides a slight decrease in error compared to the thin-plate-spline method. Second, the results

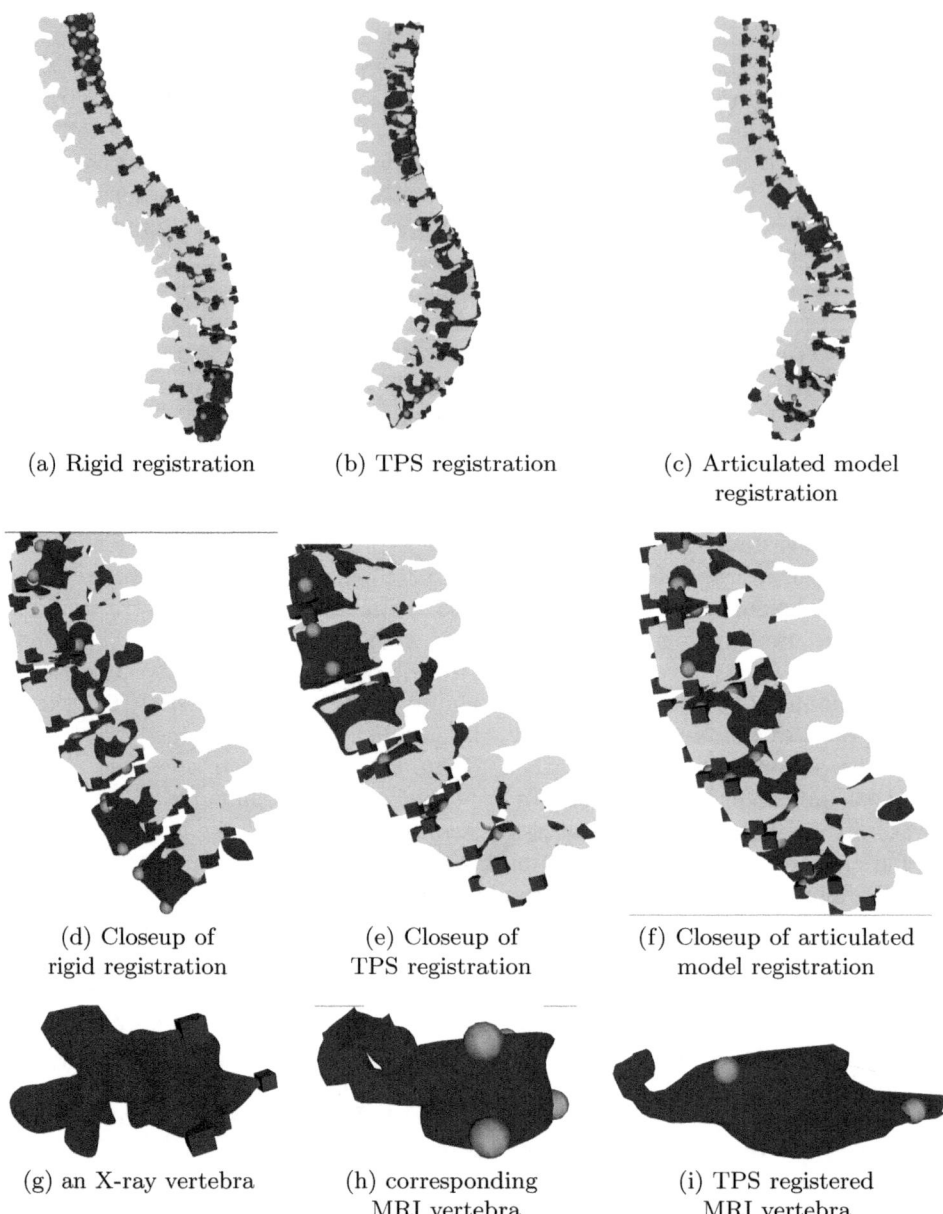

(a) Rigid registration   (b) TPS registration   (c) Articulated model
                                                          registration

(d) Closeup of   (e) Closeup of   (f) Closeup of articulated
rigid registration   TPS registration   model registration

(g) an X-ray vertebra   (h) corresponding   (i) TPS registered
                            MRI vertebra   MRI vertebra

Fig. 3. Registration between vertebrae extracted from MRI (dark grey) and X-ray data
(light grey). The correspondences used for the articulated model are represented by
cubes (X-ray data) and by spheres (MRI data). The alignment between the vertebral
column appears to be more precise using the proposed method (c) when compared
to simple rigid registration (a). (d), (e) and (f) focus on the lower part of the spine
where our proposed method in (f) gives particularly better results. Registration using
TPS seems to yield good landmark alignment (b and e), however at the expense of
non-rigidly deforming MRI vertebrae, which are rigid structures (i).

are assessed for the thoracic and lumbar areas separately. Registration errors are generally higher in the lumbar area for all methods, and, when compared with rigid registration, the improvement ratio upon use of the proposed method $(rigid - proposed)/proposed$ is about 0.97 in both areas of the spine. When compared to thin-plate-spline registration however, the improvement ratio upon use of the proposed method drops to 0.07 overall, 0.13 in the thoracic area, and performs slightly worse in the lumbar area. However, as mentioned in the previous section, thin-plate-splines non-rigidly deform the vertebrae, which is not physically sound, as vertebrae are rigid structures. Thus, preliminary results for multimodal image registration of the spine using an articulated model to represent the deformation of vertebrae are promising when compared to simple rigid registration. This proposed method is also more physically appropriate than thin-plate-spline registration.

Table 1. Registration errors in mm for rigid registration (RR), thin-plate-spline registration (TPS) and for the proposed method (PM).Case five has smaller size vertebrae explaining the overall smaller error values.

Case	Overall RR	Overall TPS	Overall PM	thoracic RR	thoracic TPS	thoracic PM	lumbar RR	lumbar TPS	lumbar PM
01	12.08	7.18	9.08	10.79	6.52	7.21	15.18	6.52	13.57
02	9.04	5.77	4.60	8.03	4.83	3.95	11.46	6.72	6.17
03	8.12	3.70	3.17	6.54	3.22	3.00	11.91	4.19	3.58
04	13.63	5.88	4.63	12.79	6.23	4.81	15.63	5.50	4.20
05	4.45	3.35	2.52	3.78	3.15	2.30	6.09	3.56	3.07
mean	9.46	5.18	4.80	8.39	4.79	4.25	12.05	5.30	6.12

4   Conclusion

This paper described a method in order to register two 3D reconstructions of the spine of patients with scoliosis - one obtained from X-rays and the other one obtained from MRIs. The proposed method uses an articulated model consisting of a series of rigid transformations, taking into account inter-vertebral transformations and thus providing a more accurate representation of the movement of the vertebral column when compared to rigid registration. The method also takes into account bone rigidity providing a more realistic deformation model when compared to non-rigid registration techniques. Corresponding points on the vertebrae of each image modality are extracted and used in order to build the articulated model, and the transformation between the vertebrae of the two images are obtained using a composition of transformations. Results show a decrease in the overall registration error when using the proposed method compared to simple rigid registration, and no difference in the improvement ratio between the thoracic and the lumbar area. When compared to thin-plate-spline registration, the proposed method gives a more physically accurate registration by preserving the shape of the vertebrae, which in actuality are rigid structures.

Since the landmark extraction is done somewhat manually and the extraction process is different in the two modalities, landmark localization errors are to be expected. In the case of the X-ray reconstruction, the localization error has been previously calculated to be around $2.1 \pm 1.5mm$ [11]. The landmark localization error in the case of the MRIs is yet to be established. The automation of landmark extraction would greatly improve the multimodal registration in addition to increasing consistency and reproducibility.

The registration of the vertebral bodies serves as a preliminary step towards the multimodal image registration of MRI and X-ray data. Little work has been done in this field so far, and no other method registers MRI to X-ray data that can be acquired in a clinical setting. This registration can be used for the purpose of evaluating the effects of posture differences on the shape of the spine. It will also allow the construction of a geometric model of the torso of patients with scoliosis combining musculo-skeletal information allowing further studies in treatment techniques which would best benefit patients and improve their quality of life.

References

1. Cottalorda, Kohler, R.: Recueil terminologique de la scoliose idiopathique. In: Sauramps Medical (ed.) La scoliose idiopathique (sous la direction de J. Bérard et R. Kohler), pp. 33–40 (1997)
2. Asher, M., Beringer, G., Orrick, J., Halverhout, N.: The current status of scoliosis screening in north america, 1986: Results of a survey by mailed questionnaire. Spine 14, 652–662 (1989)
3. Roach, J.: Adolescent idiopathic scoliosis. Orthop. Clin. North Am. 30, 353–365 (1999)
4. van de Kraats, E.B., van Walsum, T., Verlaan, J.J., Oner, F.C., Viergever, M.A., Niessen, W.J.: Noninvasive magnetic resonance to three-dimensional rotational x-ray registration of vertebral bodies for image-guided spine surgery. Spine 29(3), 293–297 (2004)
5. Tomazevic, D., Likar, B., Pernus, F.: 3-D/2-D registration by integrating 2-D information in 3-D. IEEE Transactions on Medical Imaging 25(1), 17–27 (2006)
6. Cheriet, F., Dansereau, J., Petit, Y., Labelle, H., de Guise, J.A.: Towards the self-calibration of a multi-view radiographic imaging system for the 3D reconstruction of the human spine and rib cage. International Journal of Pattern Recognition and Artificial Intelligence 13(5), 761–779 (1999)
7. Little, J.A., Hill, D.L.G., Hawkes, D.J.: Deformations incorporating rigid structures. In: Proceedings of the Workshop on Mathematical Methods in Bio- medical Image Analysis, San Francisco, CA, pp. 104–113. IEEE Comput. Soc. Press, Los Alamitos (1996)
8. Boisvert, J., Pennec, X., Labelle, H., Cheriet, F., Ayache, N.: Principal Spine Shape Deformation Modes Using Riemannian Geometry and Articulated Models. In: Perales, F.J., Fisher, R.B. (eds.) AMDO 2006. LNCS, vol. 4069, pp. 346–355. Springer, Heidelberg (2006)

9. Kadoury, S., Paragios, N.: Surface/Volume-Based Articulated 3D Spine Inference through Markov Random Fields. In: Yang, G.-Z., Hawkes, D., Rueckert, D., Noble, A., Taylor, C. (eds.) MICCAI 2009. LNCS, vol. 5762, pp. 92–99. Springer, Heidelberg (2009)
10. Golub, G.H., Van Loan, C.F.: Matrix Computations, 3rd edn. Johns Hopkins, Baltimore (1996)
11. Delorme, S., Labelle, H., Poitras, B., Rivard, C.-H., Coillard, C., Dansereau, J.: Pre-, Intra-, and Postoperative Three-Dimensional Evaluation of Adolescent Idiopathic Scoliosis. Journal of spinal disorders 13(2), 93–101 (2000)

Multimodality Image Alignment Using Information-Theoretic Approach

Mohammed Khader[1], A. Ben Hamza[1], and Prabir Bhattacharya[2]

[1] Concordia Institute for Information Systems Engineering
Concordia University, Montréal, QC, Canada
[2] Department of Computer Science
University of Cincinnati, OH, USA

Abstract. In this paper, an entropic approach for multimodal image registration is presented. In the proposed approach, image registration is carried out by maximizing a Tsallis entopy-based divergence using a modified simultaneous perturbation stochastic approximation algorithm. This divergence measure achieves its maximum value when the conditional intensity probabilities of the transformed target image given the reference image are degenerate distributions. Experimental results are provided to demonstrate the registration accuracy of the proposed approach in comparison to existing entropic image alignment techniques. The feasibility of the proposed algorithm is demonstrated on medical images from magnetic resonance imaging, computer tomography, and positron emission tomography.

Keywords: Image registration; Tsallis entropy; stochastic optimization.

1 Introduction

Multimodality imaging is widely considered to involve the incorporation of two or more imaging modalities that are acquired by different scanners. The goal of multimodal image registration is to align intermodal images created by different medical diagnostic modalities in order to improve diagnosis accuracy [1, 2]. Intermodal images display complementary and shared information about the object in images with different intensity maps. Therefore, distance measures used for multimodal image registration must be insensitive to differing intensity maps. Recently, much attention has been paid to the multimodal image registration problem using information-theoretic measures [4–6]. The latter will be the focus of this paper. The most popular approach in multimodal image registration maximizes the mutual information (MI) between the reference and target images [3–5]. This approach involves maximizing the information (entropy) contained in each image while minimizing the information (joint entropy) contained in the overlayed images. Although MI has been successfully applied to multimodality image registration, it is worth noting that the MI-based registration methods might have the limited performances, once the initial misalignment of the two images is large or equally the overlay region of the two images is

A. Campilho and M. Kamel (Eds.): ICIAR 2010, Part II, LNCS 6112, pp. 30–39, 2010.

relatively small [7]. Moreover, MI is sensitive to the changes that occur in the distributions (overlap statistics) as a result of changes in the region of overlap. To circumvent these limitations, various methods have been proposed to improve the robustness of MI-based registration, including normalized mutual information (NMI) and Rényi entropy based approaches [8–10]. The NMI approach, proposed by Studholme *et al.* [8], is a robust similarity measure that allows for fully automated intermodal image registration algorithms. Furthermore, NMI-based registration is less sensitive to the changes in the overlap of two images. Cahill *et al.* [9] introduced a modified NMF-based approach that is invariant to changes in overlap size. Inspired by the successful application of mutual information, He *et al.* proposed in [10] a generalized information-theoretic approach to ISAR image registration by estimating the target motion during the imaging time using a Rényi entropy based divergence. Sabuncu *et al.* [11] proposed a minimal spanning graph for multimodal image registration using Jensen-Rényi divergence [10] by joint determination of both the alignment measure and a descent direction with respect to alignment parameters.

In recent years, there has been a concerted research effort in statistical physics to explore the properties of Tsallis entropy, leading to a statistical mechanics that satisfies many of the properties of the standard theory [12]. In [13], a Tsallis entropy-based image mutual information approach, combined with the simultaneous perturbation stochastic approximation (SPSA) algorithm [14], was proposed leading to accurate image registration results compared to the classical mutual information [4, 5]. In this paper, we propose a multimodal entropic image registration approach by maximizing the Jensen-Tsallis divergence using a modified SPSA algorithm [15]. To increase the accuracy of multimodal image alignment, we apply a histogram-based modality transformation [16] to the target image prior to maximizing the Jensen-Tsallis divergence measure between the reference and the transformed target images.

The outline of this paper is as follows. In Section 2, we formulate the image alignment problem. In Section 3, we describe the proposed multimodal image alignment method and provide its most important algorithmic steps. In Section 4, we provide experimental results to show the effectiveness and the registration accuracy of the proposed approach. And finally, we conclude in Section 5.

2   Problem Formulation

In the continuous domain, an image is defined as a real-valued function $I : \Omega \rightarrow \mathbb{R}$, and Ω is a nonempty, bounded, open set in $\mathbb{R}^2$ (usually Ω is a rectangle in $\mathbb{R}^2$). We denote by $\boldsymbol{x} = (x, y)$ a pixel location in Ω. Given two misaligned images, the reference image I and the target image J, the image alignment or registration problem may be formulated as an optimization problem

$$\ell^* = \arg\max_{\ell} D\left(I(\boldsymbol{x}), J(\Phi_{\ell}(\boldsymbol{x}))\right), \tag{1}$$

where $D(\cdot, \cdot)$ is a dissimilarity measure that quantifies the discrepancy between the reference image and the transformed target image; and $\Phi_{\boldsymbol{\ell}} : \Omega \leftarrow \Omega$ is a spatial transformation mapping parameterized by a parameter vector $\boldsymbol{\ell}$. An example of such a mapping is a Euclidean transformation with a parameter vector $\boldsymbol{\ell} = (\boldsymbol{t}, \theta, \boldsymbol{s})$, where $\boldsymbol{t} = (t_x, t_y)$ is a translational parameter vector, θ is a rotational parameter, and $\boldsymbol{s} = (s_x, s_y)$ is a scaling parameter vector.

The goal of image registration is to align the target image to the reference image by maximizing the dissimilarity measure $D(I(\boldsymbol{x}), J(\Phi_{\boldsymbol{\ell}}(\boldsymbol{x})))$ using an optimization scheme in order to find the optimal spatial transformation parameters. Note that since the image pixel values are integers, a bilinear interpolation may be used to determine the values of $J(\Phi_{\boldsymbol{\ell}}(\boldsymbol{x}))$ when $\Phi_{\boldsymbol{\ell}}(\boldsymbol{x})$ is not an integer.

3   Proposed Multimodal Image Registration Approach

Our proposed approach may now be described as follows: Given two images that need to be registered, we first compute their conditional intensity probabilities and the Jensen-Tsallis divergence between them. Then we optimize this divergence measure using the modified SPSA algorithm.

Without loss of generality, we consider a Euclidean transformation $\Phi_{\boldsymbol{\ell}}$ with a parameter vector $\boldsymbol{\ell} = (\boldsymbol{t}, \theta)$, i.e. a transformation with translation parameter vector $\boldsymbol{t} = (t_x, t_y)$, and a rotation parameter θ. In other words, for an image pixel location $\boldsymbol{x} = (x, y)$ the Euclidean transformation is defined as $\Phi_{\boldsymbol{\ell}}(\boldsymbol{x}) = R\boldsymbol{x} + \boldsymbol{t}$, where R is a rotation matrix given by

$$R = \begin{pmatrix} \cos\theta & \sin\theta \\ -\sin\theta & \cos\theta \end{pmatrix}.$$

Denote by $\mathcal{X} = \{x_1, x_2, \ldots, x_n\}$ and $\mathcal{Y} = \{y_1, y_2, \ldots, y_n\}$ the sets of pixel intensity values of the reference image $I(\boldsymbol{x})$ and the transformed target image $J(\Phi_{\boldsymbol{\ell}}(\boldsymbol{x}))$ respectively. Let X and Y be two random variables taking values in $\mathcal{X}$ and $\mathcal{Y}$.

3.1   Jensen-Tsallis Divergence

Shannon's entropy of a probability distribution $\boldsymbol{p} = (p_1, p_2, \ldots, p_k)$ is defined as $H(\boldsymbol{p}) = -\sum_{j=1}^{k} p_j \log(p_j)$. A generalization of Shannon entropy is Tsallis entropy [12] given by

$$H_\alpha(\boldsymbol{p}) = \frac{1}{1-\alpha}\left(\sum_{j=1}^{k} p_j^\alpha - 1\right) = -\sum_{j=1}^{k} p_j^\alpha \log_\alpha(p_j), \quad \alpha \in (0,1) \cup (1,\infty). \quad (2)$$

where $\log_\alpha$ is the α-logarithm function defined as $\log_\alpha(x) = (1-\alpha)^{-1}(x^{1-\alpha} - 1)$ for $x > 0$, and α is an exponential order also referred to as entopic index. This generalized entropy is widely used in statistical physics applications [12].

Definition 1. *Let* $p_1, p_2, \ldots, p_n$ *be* n *probability distributions. The Jensen-Tsallis divergence is defined as*

$$D_\alpha^\omega(p_1, \ldots, p_n) = H_\alpha\left(\sum_{i=1}^n \omega_i p_i\right) - \sum_{i=1}^n \omega_i H_\alpha(p_i),$$

where $H_\alpha(p)$ *is Tsallis entropy, and* $\omega = (\omega_1, \omega_2, \ldots, \omega_n)$ *is a weight vector such that* $\sum_{i=1}^n \omega_i = 1$ *and* $\omega_i \geq 0$.

Using the Jensen inequality, it is easy to check that the Jensen-Tsallis divergence is nonnegative for $\alpha > 0$. It is also symmetric and vanishes if and only if all the probability distributions are equal, for all $\alpha > 0$. Moreover, the Jensen-Tsallis divergence is a convex function and achieves its maximum value when $p_1, p_2, \ldots, p_n$ are degenerate distributions, that is $p_i = (\delta_{ij})$, where $\delta_{ij} = 1$ if $i = j$ and 0 otherwise [15].

3.2 Modality Transformation

Assume that I and J are normalized images, that is $I(x), J(x) \in [0, 1]$. By looping through all pixel locations, the joint histogram $h(I, J)$ of the images I and J may be written as:

$$h\big(\lfloor I(x)N \rfloor, \lfloor J(x)N \rfloor\big) = h\big(\lfloor I(x)N \rfloor, \lfloor J(x)N \rfloor\big) + 1, \tag{3}$$

where N is the number of bins, and $\lfloor \alpha \rfloor$ denotes the floor function (largest integer not greater than α).
Using the following optimization scheme

$$J_T(x) = \arg\max_i h\big(\lfloor iN \rfloor, \lfloor J(x)N \rfloor\big), \tag{4}$$

we can transform the target image J into another image J_T that has a similar modality representation as I. In other words, the iterative scheme given by Eq. (4) finds every pixel value i in the image I that overlaps most often with a pixel value in the image J. It is worth pointing out that in medical images two regions may have the same intensity value in one modality. However, in another modality both regions may have completely different intensity values.

3.3 Proposed Algorithm

The proposed algorithm consists of the following main steps:

(i) Find the conditional intensity probabilities

$$p_i = p_i\big(J(\Phi_\ell(x))|I(x)\big) = (p_{ij})_{j=1,\ldots,n}, \quad \forall i = 1, \ldots, n,$$

where $p_{ij} = P(Y = y_j | X = x_i), \; j = 1, \ldots, n.$

(ii) Find the optimal parameter vector $\boldsymbol{\ell}^{\star} = (\boldsymbol{t}^{\star}, \theta^{\star})$ of the Jensen-Tsallis objective function

$$\boldsymbol{\ell}^{\star} = \arg\max_{\boldsymbol{\ell}} \, D_{\alpha}^{\omega}(\boldsymbol{p}_1, \ldots, \boldsymbol{p}_n) \qquad (5)$$

using the modified SPSA optimization algorithm [15].

Note that if the images I and J are exactly matched, then $\boldsymbol{p}_i = (\delta_{ij})$ and the Jensen-Tsallis divergence is therefore maximized. The conditional probability p_{ij} is estimated using the normalized conditional histogram. In other words, p_{ij} estimates the probability that a pixel has intensity j in the transformed target image J, given that is has an intensity i in the reference image I.

4   Experimental Results

The results of the proposed multimodal image registration algorithm are presented in this section. We tested the performance of the proposed approach on a medical imaging dataset. Fig. 1 shows the medical images that were used to validate the performance of the proposed algorithm. These multimodal images were obtained from the Vanderbilt Retrospective Image Registration Evaluation (RIRE) database [19], which contains magnetic resonance (MR), computer tomography (CT), and positron emission tomography (PET) images for various patients. Each patient dataset contains MR images from several protocols, including T1-weighted, T2-weighted, PD-weighted, etc. Most of these datasets contain MR, CT, and PET images. Fig. 1 shows the images from the patient 5 dataset, where the MR-T1, CT, and PET images are shown in Fig. 1(a), (b), and (c), respectively. These images are available online at [20]. It is worth pointing out that multimodal medical images are used to provide as much information about the patient as possible. MR and CT images provide complementary information, with CT proving a good visual description of bone tissue, whereas soft tissues are better visualized by MR images. Moreover, MR and CT are anatomical modalities that display geometric features of the object. On the other hand, PET is a functional modality that displays a metabolic map of the object and captures very reliably the metabolic activity. For example, in radiation treatment for cancer therapy, CT and PET are commonly used modalities to define cancerous lesions and plan treatment strategies. CT and PET modalities display different, but complementary information and involve different acquisition processes. These differences make registering CT and PET data one of the most challenging medical image registration problems.

4.1   Modality Transformation

To increase the accuracy of the proposed multimodal image registration, we apply the histogram-based modality transformation to the target image prior to maximizing the Jensen-Tsallis divergence between the reference and transformed target images. This modality transformation involves finding the maxima of the

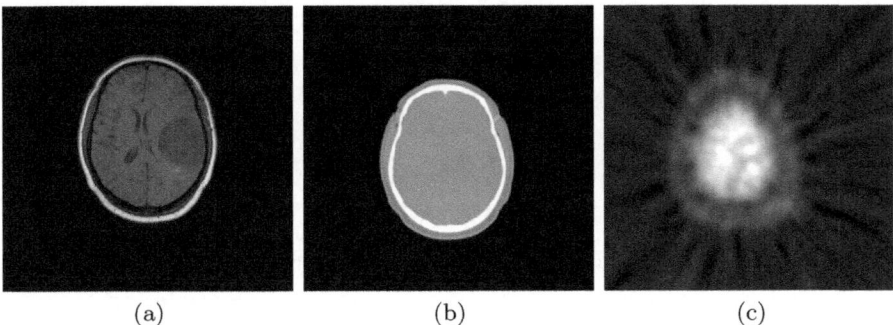

(a)   (b)   (c)

Fig. 1. Multimodal images from the patient 5 dataset: (a) MR-T1 image, (b) CT image, (c) PET image

joint histogram of both images to transform one image modality representation into another modality, allowing more accurate image registration results for MI, NMI, and the proposed approach.

4.2 Registration Functions

In all the experiments we used an entropic index $\alpha = 2$ and the normalized histogram of the reference image as the weight vector $\boldsymbol{\omega}$ for the Jensen-Tsallis divergence. For fair comparison, it is worth mentioning that we also used modality transformation to align MR-T1 to CT and MR-T1 to PET when comparing the performance of the proposed algorithm to MI and NMI-based approaches.

To validate the proposed approach, we first applied a Euclidean transformation $\Phi_{\boldsymbol{\ell}}$ with different values of the parameter vector $\boldsymbol{\ell} = (t_x, t_y, \theta)$ to the references images shown in Fig. 1. Then, we run iteratively the modified SPSA algorithm to find the optimal parameter vector $\boldsymbol{\ell}^* = (t_x^\star, t_y^\star, \theta^\star)$. We also compared the image alignment results of the proposed approach to MI and NMI-based registration methods. An ideal registration function that measures the dissimilarity between two images should be smooth and concave with respect to different transformation parameters. Also, the global maximum of the registration function should be close to the correct transformation parameters that align two images perfectly [6]. Moreover, the capture range around the global maximum should be as large as possible, and the number of local maxima of the registration function should be as small as possible. These criteria will be used to evaluate the registration functions generated by MI, NMI, and Jensen-Tsallis respectively. The registration function of the proposed algorithm can be generated by computing the Jensen-Tsallis of two images under all possible transformations. For the medical images used in our experiments, their relative transformation parameters can be determined with the aid of four fiducial markers implanted in the patients. In other words, we can first align all testing images into the same space by using the four fiducial markers, and then compute the Jensen-Tsallis measure between two testing images under different rigid-body transformations, thereby obtaining a registration function of

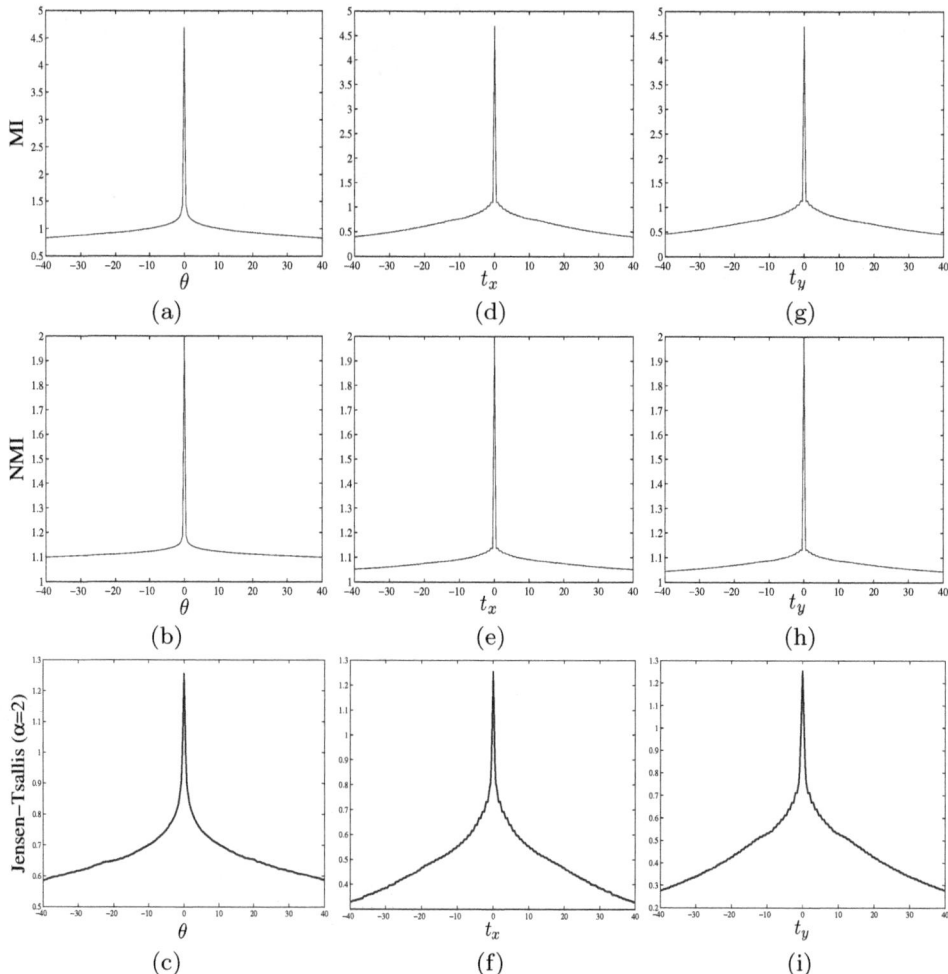

Fig. 2. Registration functions of MI, NMI, and proposed approach in aligning MRI-T1 images. From top to bottom: (a), (b) and (c) rotation only; (d), (e) and (f) translation along x-axis; (g), (h) and (i) translation along y-axis.

Jensen-Tsallis. Similarly, the registration functions of MI and NMI can be obtained. The registration functions of MI, NMI, and Jensen-Tsallis with respect to different rotation and translation parameters are depicted in Fig. 2. It is evident from Fig. 2 that the registration functions of Jensen-Tsallis are much smoother than those of MI and NMI. Moreover, the capture range in the registration function of Jensen-Tsallis is considerably large. In particular, the change of registration function with respect to rotations is smoothly extended relatively far from the global maximum, indicating a better performance of the proposed approach.

4.3   Robustness and Accuracy

We tested the accuracy and robustness of the proposed method by comparing the conditional probabilities of the target image given the reference image after applying various alignment methods. The more linear conditional probability indicates more accurate registration method. Note that if the reference image I and the target image J are exactly matched, then the conditional probability plots a straight diagonal line. The conditional probabilities before and after aligning different images modalities with the MR-T1 image using MI, NMI, and Jensen-Tsallis respectively, are presented in Fig. 3 through Fig. 5. Fig. 3 shows that the proposed approach is more robust to noise than MI and NMI. Note that the presence of noise causes the joint distribution to be less correlated, and thus increases the joint entropy estimate, resulting in a lower registration accuracy for MI and NMI-based approaches. Fig. 4 and Fig. 5 depict the conditional probabilities for aligning MRI-T1 with CT and PET images respectively, after applying modality transformation to CT and PET images. From these figures, it is evident that the linearity of the conditional probabilities indicates the effectiveness and a much improved registration accuracy of the proposed algorithm in comparison to MI and NMI.

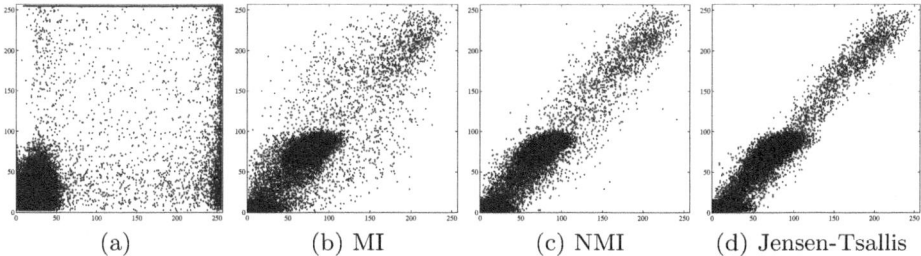

| (a) | (b) MI | (c) NMI | (d) Jensen-Tsallis |

Fig. 3. 2D plots of conditional probabilities for various alignment methods of MR-T1 to noisy MR-T1 images: (a) after applying Euclidean transformation $\Phi_{\boldsymbol{\ell}}$ to reference image with parameter vector $\boldsymbol{\ell} = (5, 5, 5)$; (b)-(d) after aligning the images by MI, NMI, and proposed approach respectively

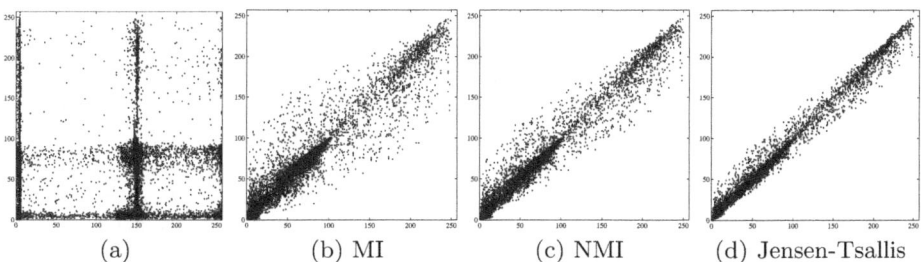

| (a) | (b) MI | (c) NMI | (d) Jensen-Tsallis |

Fig. 4. 2D plots of conditional probabilities for various alignment methods of MR-T1 to CT images: (a) after applying Euclidean transformation $\Phi_{\boldsymbol{\ell}}$ to reference image with parameter vector $\boldsymbol{\ell} = (5, 5, 5)$; (b)-(d) after applying modality transformation and then aligning the images by MI, NMI, and proposed approach respectively

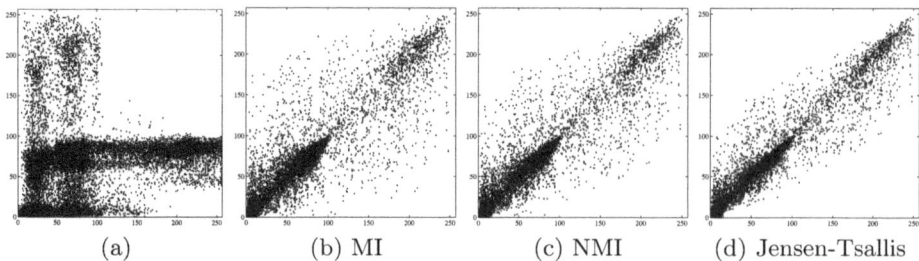

| (a) | (b) MI | (c) NMI | (d) Jensen-Tsallis |

Fig. 5. 2D plots of conditional probabilities for various alignments methods of MR-T1 to PET images: (a) after applying Euclidean transformation Φ_{ℓ} to reference image with parameter vector $\ell = (5, 5, 5)$; (b)-(d) after applying modality transformation and then aligning the images by MI, NMI, and proposed approach respectively

5 Conclusions

An entropic framework for multimodal image registration is proposed in this paper. The proposed algorithm was applied to medical images of different modalities. The experimental results on MR to CT and MR to PET registrations indicate the feasibility of the proposed approach and a much better performance compared to MI and NMI-based methods, not only in terms of wider capture range of registration functions but also in terms of robustness to noise. Future work will focus on non-rigid multimodal image registration.

References

1. Bankman, I.: Handbook of Medical Image Processing and Analysis. Academic Press, London (2008)
2. Hajnal, J.V., Hill, D.L.G., Hawkes, D.J.: Medical Image Registration. CRC Press, Boca Raton (2001)
3. Wells, W.M., Viola, P., Atsumi, H., Nakajima, S., Kikinis, R.: Multi-modal volume registration by maximization of mutual information. Medical Image Analysis 1(1), 35–51 (1996)
4. Viola, P., Wells, W.M.: Alignment by maximization of mutual information. International Journal of Computer Vision 24(2), 154–173 (1997)
5. Maes, F., Collignon, A., Vandermeulen, D., Marchal, G., Suetens, P.: Multimodality image registration by maximization of mutual information. IEEE Trans. on Medical Imaging 16(2), 187–198 (1997)
6. Luan, H., Qi, F., Xue, Z., Chen, L., Shen, D.: Multimodality image registration by maximization of quantitative-qualitative measure of mutual information. Pattern Recognition 41, 285–298 (2008)
7. Pluim, J.P.W., Maintz, J.B., Viergever, M.A.: Mutual-information-based registration of medical images: a survey. IEEE Trans. on Medical Imaging 22(8), 986–1004 (2003)
8. Studholme, C., Hill, D.L.G., Hawkes, D.J.: An overlap invariant entropy measure of 3D medical image alignment. Pattern Recognition 32, 71–86 (1999)

9. Cahill, N.D., Schnabel, J.A., Noble, J.A., Hawkes, D.J.: Revisiting overlap invariance in medical image alignment. In: Proc. IEEE Computer Vision and Pattern Recognition Workshops, pp. 1–8 (2008)
10. He, Y., Ben Hamza, A., Krim, H.: A generalized divergence measure for robust image registration. IEEE Trans. on Signal Processing 51(5), 1211–1220 (2003)
11. Sabuncu, M.R., Ramadge, P.: Using spanning graphs for efficient image registration. IEEE Trans. on Image Processing 17(5), 788–797 (2008)
12. Tsallis, C.: Possible generalization of Boltzmann-Gibbs statistics. Journal of Statistical Physics 52, 479–487 (1988)
13. Martin, S., Morison, G., Nailon, W., Durrani, T.: Fast and accurate image registration using Tsallis entropy and simultaneous perturbation stochastic approximation. Electronic Letters 40(10), 595–597 (2004)
14. Spall, J.C.: Multivariate stochastic approximation using a simultaneous perturbation gradient approximation. IEEE Trans. on Automatic Control 37(3), 332–341 (1992)
15. Mohamed, W., Ben Hamza, A.: Nonextensive entropic image registration. In: Kamel, M., Campilho, A. (eds.) ICIAR 2009. LNCS, vol. 5627, pp. 116–125. Springer, Heidelberg (2009)
16. Kroon, D.-J., Slump, C.H.: MRI modality transformation in demon registration. In: IEEE Inter. Symposium on Biomedical Imaging: From Nano to Macro, pp. 963–966 (2009)
17. Burbea, J., Rao, C.R.: On the convexity of some divergence measures based on entropy functions. IEEE Trans. on Information Theory 28(3), 489–495 (1982)
18. Spall, J.C.: Implementation of the simultaneous perturbation algorithm for stochastic optimization. IEEE Trans. on Aerospace and Electronic Systems 34(3), 817–823 (1998)
19. West, J., Fitzpatrick, F., Wang, M., Dawant, B., Maurer, C., Kessler, R., Maciunas, R.: Comparison and evaluation of retrospective intermodality image registration techniques. In: Proc. SPIE Conference on Medical Imaging (1996)
20. http://www.insight-journal.org/rire/

Retinal Images: Optic Disk Localization and Detection

M. Usman Akram, Aftab Khan, Khalid Iqbal, and Wasi Haider Butt

Department of Computer Engineering,
College of Electrical & Mechanical Engineering,
National University of Sciences & Technology, Pakistan
usmakram@gmail.com, aftabkhan.nust@googlemail.com,
kiqbal@ceme.nust.edu.pk, butt.wasi@gmail.com

Abstract. Automated localization and detection of the optic disc (OD) is an essential step in the analysis of digital diabetic retinopathy systems. Accurate localization and detection of optic disc boundary is very useful in proliferative diabetic retinopathy where fragile vessels develop in the retina. In this paper, we propose an automated system for optic disk localization and detection. Our method localizes optic disk using average filter and thresholding, extracts the region of interest (ROI) containing optic disk to save time and detects the optic disk boundary using Hough transform. This method can be used in computerized analysis of retinal images, e.g., in automated screening for diabetic retinopathy. The technique is tested on publicly available DRIVE, STARE, diaretdb0 and diaretdb1 databases of manually labeled images which have been established to facilitate comparative studies on localization and detection of optic disk in retinal images. The proposed method achieves an average accuracy of 96.7% for localization and an average area under the receiver operating characteristic curve of 0.958 for optic detection.

1 Introduction

Diabetes has associated complications such as vision loss, heart failure and stroke. Patients with diabetes are more likely to develop eye problems such as cataracts and glaucoma, but the disease's affect on the retina is the main threat to vision [1]. Complication of diabetes, causing abnormalities in the retina and in the worst case severe vision loss, is called diabetic retinopathy [1].

To determine if a person suffers from diabetic retinopathy, fundus or retina image is used. Performing the mass screening of diabetes patients will result in a large number of images that need to be examined. The cost of manual examination and screening is prohibiting the implementation of screening on a large scale. A possible solution could be the development of an automated screening system for diabetic retinopathy [2]. Such a system should be able to distinguish between affected eye and normal eye. This will significantly reduce the workload for the ophthalmologists as they have to examine only those images classified as possibly abnormal by the system. Accurate optic disk localization

A. Campilho and M. Kamel (Eds.): ICIAR 2010, Part II, LNCS 6112, pp. 40–49, 2010.

and detection of its boundary is main and basic step for automated diagnosis systems [3]. OD is a bright yellowish disk in human retina from where the blood vessels and optic nerves emerge [1].

Figure 1 shows a healthy retinal image including main retinal features i.e. optic disk, blood vessels and macula. The shape, color and size of optic disk help in localization and detection. However, these properties show a large variance

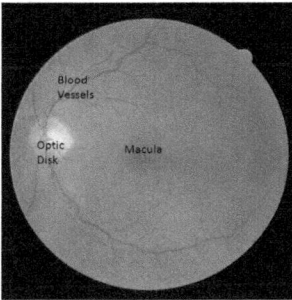

Fig. 1. Healthy retinal image with features

Figure 2 shows examples of a swollen optic nerve, where the circular shape and size are distorted. It also shows an example with a bright circular lesion that appears similar to optic disk.

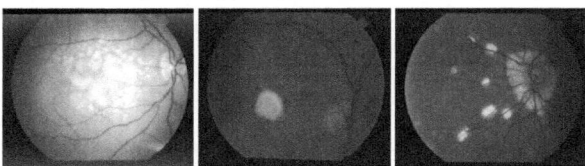

Fig. 2. Retinal images with lesions and distorted shape of optic disk

Localizing the centre and rim of the optic disk is necessary to differentiate the optic disk from other features of the retina and as an important landmark. Techniques described in the literature for optic disk localization are typically aimed at either identifying the approximate centre of the optic disk or placing the disk within a specific region such as a circle or square. In [4], an approximate location of the optic disk is estimated where the location of the optic disk is hypothesized by searching for regions of high intensity, diversity of gradient directions, and convergence of blood vessels. Sinthanayothin [5] located the position of the optic disk by finding the region with the highest local variation in the intensity. Hoover [6] utilized the geometric relationship between the optic disk and main blood vessels to identify the disk location. He described a method based on a fuzzy voting mechanism to find the optic disc location. Mendels et al. [7] and

Osareh et al. [8] introduced a method for the disk boundary identification using free-form deformable model technique. Li and Chutatape [9][10] used a PCA method to locate the optic disk and a modified active shape model (ASM) to refine the optic disk boundary based on point distribution model (PDM) from the training sets. A method based on pyramidal decomposition and Hausdorff distance based template matching was proposed by Lalonde et al. [11].

In this paper, we propose a Hough Transformation based technique for OD localization and detection. In our proposed method, firstly, optic disk localization is done by averaging and then detecting the maximum gray values from an image histogram. Secondly, ROI is extracted and optic disk detection is done by taking Circular Hough Transform of an image which is followed by clinical validation of our Hough based technique. We test the validity of our method on four different publicly available databases i.e DRIVE [12], STARE [13], diaratdb0 and diaretdb1 [14].

This paper is organized in four sections. In Section 2, systematic overview of our methodology is explained. Section 2 also presents the step by step techniques required for an automated optic disk localization and detection system. Experimental results of the tests on the images of the different databases and their analysis are given in Section 3 followed by conclusion in Section 4.

2   System Overview

A systematic overview of the proposed technique is shown in figure 3. In summary, given a pair of color retinal images, the first step localizes the optic disk, the second step extracts ROI and in the third step optic disk detection takes place. The results are utilized for clinical validation of optic disk detected images.

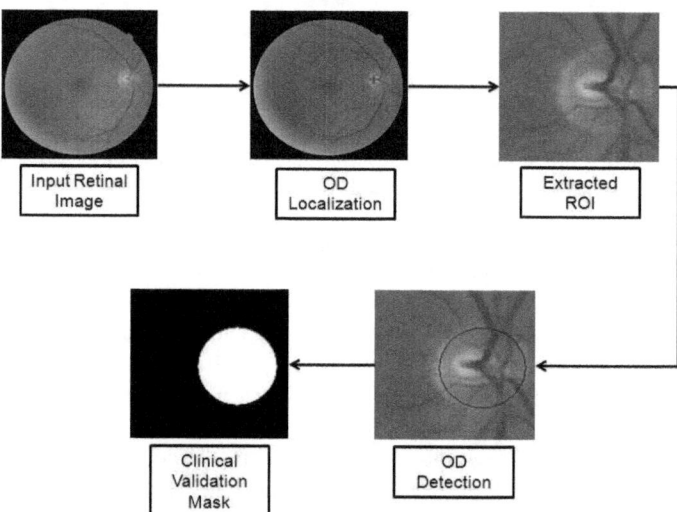

Fig. 3. Retinal images with lesions and distorted shape of optic disk

2.1   Optic Disk Localization

Localization of the optic disc is the first step in our technique. The purpose of localization is to have low processing rate and less computational cost in further steps. Steps for OD detection are as follows:

1. Image is preprocessed by averaging mask of size 31x31, given in eq. 1, in order to remove the background artifacts which can cause false localization.

$$R = \frac{1}{961} \sum_{i=1}^{961} Z_i \qquad (1)$$

   where z's are values of image gray levels and R is the smoothed image.
2. Detect maximum gray values in an image histogram because the gray values of optic disc are brighter than the background values.

Figure 4 shows the images after applying above steps.

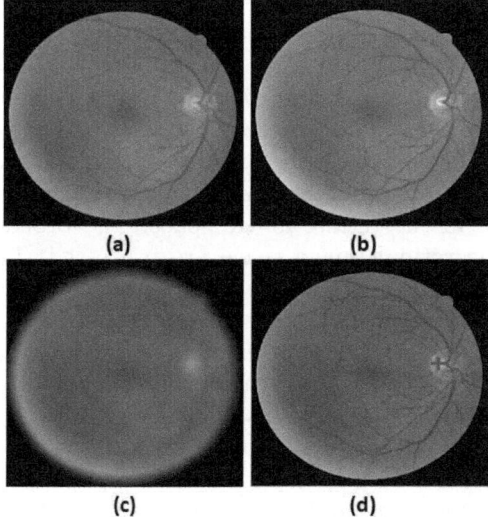

Fig. 4. Optic disk localization: (a) Original retinal image from DRIVE database; (b) Green channel image; (c) Averaged image; (d) Localized optic disk

2.2   Optic Disk Detection

After localizing the optic disc we have to define the region of interest (ROI) for increasing the performance of optic disc detection. After smoothing the size of ROI was set to 130 x 130. After extraction of ROI, we have to detect the optic disc boundary using Hough Transform [15]. The Hough Transform is used to identify the locations and orientations of retinal image features. This transform consists of parameterized description of a feature at any given location in the original image space. It can be used for representing objects that can be parameterized mathematically as in our case, a circle, can be parameterized by equation 2.

$$(x - a)^2 + (y - b)^2 = r^2 \tag{2}$$

Where (a,b) is the coordinate of center of the circle that passes through (x,y) and r is its radius. From this equation, it can be seen that three parameters are used to formalize a circle which means that Hough space will be 3D space for this case. The steps for detecting optic disc boundary by Hough Transform are below:

Edge Detection. In order to calculate the Hough Transform, the edge of the OD's circular shape is needed. Canny Edge detection operator is applied to the image as a first step in this process. This removes most of the noise due to its fine texture leaving only the required edges of the OD. Experimentally, we found that the Canny Operator with the parameters $\sigma = 1$ and window size 5 x 5 gives best results.

First Approximation of Optic Disc. For the rough calculation of OD in this step, the accumulator parameter array is filled where each array is composed of cells for (x,y) coordinates of the center of circle. The edge image is scanned and all the points in this space are mapped to Hough space using equation (2). A value in particular point in Hough space is accumulated if there is a corresponding point in the retinal image space. The process is repeated until all the points in the retinal image space are processed. The resulting Hough transform image was scaled so all the values lie between 0 and 1. Then it was threshold to leave only those points with high probability of being the centers which are then labeled with different numbers. Afterwards the different regions were matched by different circles and the output image is computed by drawing circle with these points and adding this to the input image.

Detecting Best Circle. In this step, the set of approximated circles from step 2 will be compared. The best circle of this set would be that one that fits most of the OD edges. In this step, numbers of pixels which are in the vicinity of detected circle's edge are counted. A mask of a ring shaped is put on the binary edge image on the same location of each of the detected circle. Number of edge pixels under this mask will be counted and compared for all the detected circles. The best circle shows the location of the detected optic disc.

The results were also clinically validated. All images in our test set are sent to ophthalmologist to identify the OD manually. All the OD's which are automatically detected by our system are then compared with clinician's hand-drawn ground truth. Figure 5(d) and 5(e) shows an example of our detected optic disk and manually segmented mask by ophthalmologist respectively.

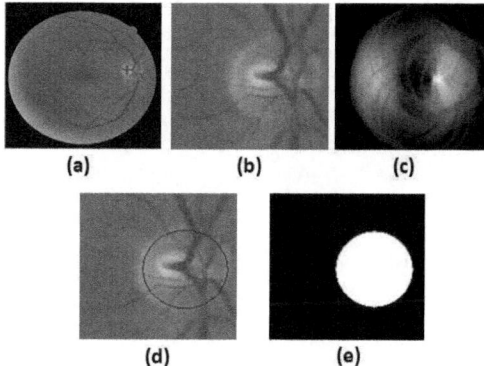

Fig. 5. Optic disk detection: (a) Localized optic disk; (b) Extracted ROI; (c) Hough transform circles; (d) Detected optic disk; (e) Manually segmented mask for clinical validation

3 Experimental Results

We have extensively tested our optic disk localization and detection technique on standard diabetic retinopathy databases. We have used four publicly available datasets, DRIVE, STARE, diaretdb0 and diaretdb1. The DRIVE [12] database consists of 40 RGB color images of the retina. The images are of size 768584 pixels, eight bits per color channel. The STARE [13] database consists of 20 RGB color images of the retina. The images are of size 605700 pixels, 24 bits per pixel (standard RGB). Diaretdb0 [14] database contains 130 retinal images while diaretdb1 [14] database contains 89 retinal images. These databases contain retinal images with a resolution of 1500 x 1152 pixels and of different qualities in terms of noise and illumination. The decision for successful localization or failed localization is based on human eye observation. For the verification of optic disk detection results, optic disks are manually labeled by the ophthalmologists for each image. The manually segmented optic disks by human observer are used as ground truth. The true positive fraction is the fraction of number of true positive (pixels that actually belong to optic disk) and total number of optic disk pixels in the retinal image. False positive fraction is calculated by dividing false positives (pixels that don't belong to optic disk) by total number of non optic disk pixels in the retinal image. Table 1 summarizes the results of optic disk localization for all four databases. It shows the accuracy (fraction of successful localized OD) of proposed algorithm for each database. Table 2 summarizes the results of optic disk detection for all databases. It shows the results in term of Az, average accuracy and their standard deviation for different datasets. Az indicates the area under the receiver operation characteristics curve and accuracy is the fraction of pixels correctly detected.

Figure 6 illustrates the optic disk localization results for retinal images taken from different databases. Red cross sign shows the optic disk location in each

Table 1. Optic disk localization results

Database	Total images	Successful localization	Failed localization	Accuracy (%)
DRIVE	40	40	0	100
STARE	81	76	5	93.8
Diarectdb0	130	126	4	96.9
Diarectdb1	89	87	2	97.7
Overall	340	329	11	96.7

Table 2. Optic disk detection results

Database	Az	Average Accuracy
DRIVE	0.971	0.955
STARE	0.943	0.921
Diarectdb0	0.957	0.932
Diarectdb1	0.961	0.937
Overall	0.958	0.936

Table 3. Comparison Results for OD localization

Method	Total images	Successful localization	Failed localization	Accuracy (%)
Hoover et al.	81	72	9	88.8
Proposed Method	81	76	5	93.8

image. These results support the validity of our technique and show that our technique gives good results for localization even for those images where it is difficult to locate optic disk. Figure 7 compares the results of proposed OD localization technique against Hoover et al. [6] localization method. It shows the successful localization results for our method.

Our method achieved an accuracy of 93.8% against 88.8% achieved by Hoover et al.[6] and is summarized in table 3.

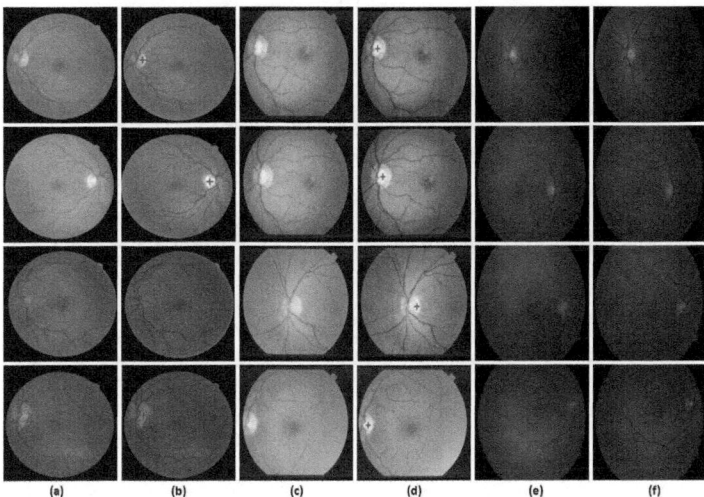

Fig. 6. Experimental results: (a) Retinal images from DRIVE database; (b) Optic disk localization results; (c) Retinal images from STARE database; (d) Optic disk localization results; (e) Retinal images from Diarectdb0 and Diarectdb1 database; (f) Optic disk localization results

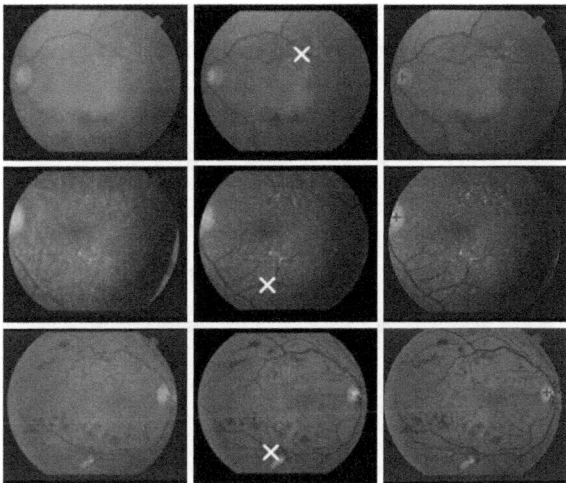

Fig. 7. Comparison results: Column 1: Original retinal images from STARE database, Column 2: Wrongly localized optic disks using Hoover et al [6] method, Column 3: Correctly localized optic disks using proposed method

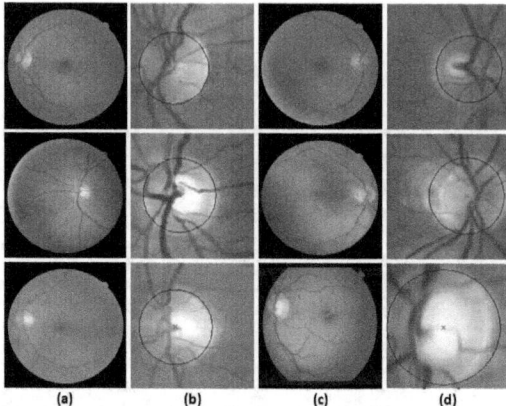

Fig. 8. Experimental results for optic disk detection: Column (a) and (c) Retinal images from different databases; Column (b) and (d) Optic disk detection results

Figure 8 shows the optic disk detection results based on Hough transform. Red cross shows the centroid of optic disk where as blue boundary encloses the optic disk.

4   Conclusion

Our proposed approach for automated optic disk localization and detection is effective in handling retinal images under various conditions with reasonable accuracy and reliability for medical diagnosis. The problem with retinal images is that the visibility and detection of optic disk are usually not easy especially in presence of some lesions. In this paper, retinal images are preprocessed and ROI is extracted prior to optic disk detection. Optic disk is localized using average masking and histogram and it is detected using Hough transform. We have tested our technique on publicly available DRIVE, STARE, diaretdb0 and diaretdb1 databases of manually labeled images. The experimental results demonstrated that our method performs well in locating and detecting optic disk.

References

1. Susman, E.J., Tsiaras, W.J., Soper, K.A.: Diagnosis of diabetic eye disease. JAMA 247(23), 3231–3234 (1982)
2. Effective Health Care - Complications of diabetes: Screening for retinopathy and Management of foot ulcers., vol. 5(4). Royal Society of Medicine Press (1999) ISSN 0965-0288
3. Osareh, A., Mirmehdi, M., Thomas, B., Markham, R.: Automated Identification of Diabetic Retinal Exudates in Digital Colour Images. British Journal of Ophthalmology 87(10), 1220–1223 (2003)

4. Narasimha-Iyer, H., Can, A., Roysam, B., Stewart, C.V., Tanenbaum, H.L., Majerovics, A., Singh, H.: Robust Detection and Classification of Longitudinal Changes in Color Retinal Fundus Images for Monitoring Diabetic Retinopathy. IEEE transactions on biomedical engineering 53(6), 1084–1098 (2006)

5. Sinthanayothin, C., Boyce, J.A., Cook, H.L., Williamson, T.H.: Automated localisation of the optic disc, fovea, and retinal blood vessels from digital colour fundus images. British Journal of Ophthalmology 83, 902–910 (1999)

6. Hoover, A., Goldbaum, M.: Locating the optic nerve in a retinal image using the fuzzy convergence of the blood vessels. IEEE Trans. on Medical Imaging 22(8), 951–958 (2003)

7. Mendels, F., Heneghan, C., Thiran, J.P.: Identification of the optic disc boundary in retinal images using active contours. In: Proc. IMVIP conference, pp. 103–115 (1999)

8. Osareh, A., Mirmehd, M., Thomas, B., Markham, R.: Comparison of colour spaces for optic disc localisation in retinal images. In: 16th International Conference on Pattern Recognition, vol. 1, pp. 743–746 (2002)

9. Li, H., Chutatape, O.: Automated feature extraction in color retinal images by a model based approach. IEEE Trans. on Biomedical Engineering 51(2), 246–254 (2004)

10. Li, H., Chutatape, O.: Boundary detection of optic disk by a modified ASM method. Pattern Recognition 36(9), 2093–2104 (2003)

11. Lalonde, M., Beaulieu, M., Gagnon, L.: Fast and robust optic disc detection using pyramidal decomposition and Hausdorff-based template matching. IEEE Trans. Med. Imag. 20(11), 1193–2001 (2001)

12. Niemeijer, van Ginneken, B.: (2002), http://www.isi.uu.nl/Reseach/Databases/DRIVE/results.php

13. Hoover, STARE database, http://www.ces.clemson.edu/ahoover/stare

14. Machine Vision and Pattern Recognition Research Group, Standard diabetic retinopathy database, http://www.it.lut.fi/project/imageret/

15. Sekhar, S., Al-Nuaimy, W., Nandi, A.K.: Automated localisation of retinal optic disk using Hough transform. In: 5th IEEE International Symposium on Biomedical Imaging: From Nano to Macro, pp. 1577–1580 (2008)

Using Retinex Image Enhancement to Improve the Artery/Vein Classification in Retinal Images

S.G. Vázquez[1], N. Barreira[1], M.G. Penedo[1], M. Saez[2], and A. Pose-Reino[3]

[1] Varpa Group, Department of Computer Science University of A Coruña, Spain
{sgonzalezv,nbarreira,mgpenedo}@udc.es
[2] GRECS Group, Department of Economics, University of Girona, Spain
marc.saez@udg.edu
[3] Service of Internal Medicine, Hospital de Conxo, Santiago de Compostela, Spain
antoniopose@telefonica.net

Abstract. A precise characterization of the retinal vessels into veins and arteries is necessary to develop automatic tools for diagnosis support. As medical experts, most of the existing methods use the vessel lightness or color for the classification, since veins are darker than arteries. However, retinal images often suffer from inhomogeneity problems in lightness and contrast, mainly due to the image capturing process and the curved retina surface. This fact and the similarity between both types of vessels make difficult an accurate classification, even for medical experts. In this paper, we propose an automatic approach for the retinal vessel classification that combines an image enhancement procedure based on the retinex theory and a clustering process performed in several overlapped areas within the retinal image. Experimental results prove the accuracy of our approach in terms of miss-classified and unclassified vessels.

1 Introduction

Several pathologies, such as hypertension, arteriosclerosis, or diabetic retinopathy, change the retinal vessel tree. Moreover, some of them affect differently veins and arteries. For example, the retinal venular widening was associated with diabetes [1], whereas the arteriolar narrowing was considered an early sign of hypertension retinopathy. Therefore, the automatic distinction of retinal vessels into veins and arteries is necessary in order to develop automatic tools for diagnosis.

A previous stage in the classification process is the location of the vessels within the image. There are several methods in the literature that segment the retinal vascular tree accurately [2,3]. However, the vessel classification task is still an open issue due to three main problems. On one hand, the differences between veins and arteries decrease with the vessel size. On the other hand, the intra-image uneven lightness due to biological characteristics, such as pigmentation, makes impossible to set two different global color ranges for both types of vessels. Finally, the inter-image lightness and contrast variability also hinders the correct vessel classification within the image, since all the vessels located at dark areas can be classified as veins and the vessels located at bright areas, as arteries.

A. Campilho and M. Kamel (Eds.): ICIAR 2010, Part II, LNCS 6112, pp. 50–59, 2010.

In the literature, the techniques for the retinal vessel classification can be divided into tracking-based and color-based methods. The former are mainly semiautomatic, i.e. an expert labels the main vessels and after that the expert classification is propagated through the vessel tree by means of tracking techniques [4,5,6]. The color-based methods are automatic and are based on a clustering algorithm that divides the vessels into two categories using the color of the vessel pixels as the classification features since veins are darker than arteries.

Among the automatic methods, Simó and de Ves proposed a Bayesian classifier which segments the image in fovea, background, arteries, and veins [7]. Li et al. [8] developed a supervised classifier that uses the vessel central reflex as the classification feature and represents the vessels by means of two Gaussian functions. Grisan and Ruggeri [9] divided the fundus image into four quadrants and used the pixel color to classify the vessels in each quadrant by means of a fuzzy c-means. Moreover, they normalized the color image in order to prevent the effect of the variability in the lightness [10]. Finally, Jelinek et al. [11] tested several classifiers and features chosen from different color models. However, the available results show high error rates and a high number of unclassified vessels.

The aim of this work is the development of an automatic methodology to classify the retinal vessels that takes into account the influence of the uneven lightness. To this end, we apply a well known color image enhancement, the retinex [12,13], previous to the vessel classification. This technique has been successfully applied to other kinds of medical images and experimental results indicated that it achieves a better contrast enhancement than other algorithms such as homomorphic filtering or histogram equalization. Our classification strategy is an improvement of the Grisan and Ruggeri's method [9], that groups the closest vessels and applies a clustering algorithm to each group independently in order to reduce the effect of the uneven lightness.

The remainder of this paper is concerned with the details of the proposed technique and the obtained results. In the next section, the retinex method is explained. Section 3 presents the details of the retinal vessel classification methodology. Section 4 shows the experimental results of this work and compares the performance of our methodology with the classifications of several experts. Finally, in Section 5, we present the conclusions and future work.

2   Image Enhancement

The color constancy is the ability of the human visual system to rule out the illumination effects from the image. However, the retinal images, as other images captured by cameras, are influenced by the light source positions, shadows, and the object surfaces. Thus, the inhomogeneity in lightness and contrast is common in retinal images mainly due to the image acquisition process under irregular illumination and the spherical surface of the eye. In this sense, the retinex theory tries to recreate a model of the human visual system to get color constancy.

In the image formation model, a color component of the image is the product of the illumination and the reflectance. The basis of the retinex methods is that the illumination varies slowly, so its frequency spectrum is assumed to be distributed at low frequencies. This way, the illumination is approximated and the output image is obtained by subtracting this estimation from the original image.

We propose to apply the retinex techniques [12,13] to achieve color constancy and enhance the retinal image before the vessel classification. We have developed two approaches based on the Single-Scale Retinex (SSR) [12] and the Multi-Scale Retinex (MSR) [13], both proposed by Jobson et al.

In the Single-Scale Retinex (SSR), the illumination is estimated by means of a Gaussian form and, then, it is subtracted from the original image to obtain a description invariant to illumination. This is given by:

$$R_i(x, y) = \log I_i(x, y) - \log[F(x, y) * I_i(x, y)] \tag{1}$$

where $R_i(x, y)$ is the retinex output, $I_i(x, y)$ is the original image in the i-th spectral band, '$*$' denotes the convolution operation and $F(x, y)$ is a surround function

$$F(x, y) = K e^{\frac{-x^2 + y^2}{\sigma^2}} \tag{2}$$

where σ is the scale that controls the extend Gaussian surround, and $K = \frac{1}{(\sum_x \sum_y F(x,y))}$

The Multi-Scale Retinex (MSR) [13] is simply a weighted sum of several different SSR outputs as follows:

$$R_{MSR_i} = \sum_{n=1}^{N} w_n R_{n_i} \tag{3}$$

where R_{MSR_i} is the MSR output in the i-th color component, N is the number of scales, R_{n_i} is the SSR output in the i-th color component on the n-th scale and w_n is the weight of the output of the n-th scale.

The Multi-Scale Retinex came up to get simultaneously the dynamic range compression of the retinex in small scale and the tonal rendition of the big-scale retinex. Experimentally, it was achieved that equal weighting of the scales $w_n = 1/3$, with $N = 3$ was enough for most of the applications.

Figure 1 shows the image results after applying retinex with different scales. We can observe the smaller the scale, the smaller the illumination variation, but more color information loss.

3   Vessel Characterization Method

In this section, we describe the proposed methodology for the classification of the retinal vessels into veins and arteries.

First, we extract the blood vessels in the interest areas. After that, we apply a specific process to classify the vessels using a clustering algorithm and the feature vectors selected in an experimental phase. First, we segment the blood

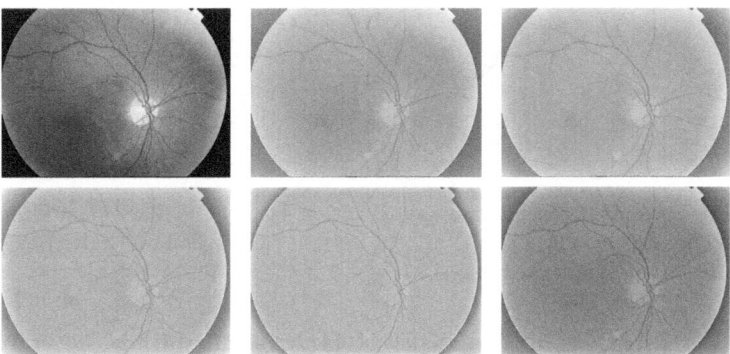

Fig. 1. Retinex enhancement on a retinal image. From left to right and from top to bottom: original image, SSR $\sigma_1 = 576$, SSR $\sigma_2 = 288$, SSR $\sigma_4 = 144$, SSR $\sigma_8 = 72$ and MSR with scales $\sigma_1 = 576$, $\sigma_2 = 288$ and $\sigma_8 = 72$.

vessels within the retinal image by means of a crease extraction algorithm [14]. If we consider the retinal image as a landscape, blood vessels can be seen as ridges or valleys, that is, regions that form an extreme and tubular level on their neighborhood. Once the creases are extracted from the retinal image, we define two concentric circumferences around the optic disc and a specific deformable model is adjusted to the vessel boundaries between both circumferences. The crease points are used as the seed of the deformable model. As a result, we obtain a set of vessel contour points, that forms a parallelogram, and a set of crease points, that represents the vessel center line (see Fig. 2 left). Figure 3 shows the results of the vessel segmentation process in three different analysis radii.

The feature vectors to perform the classification in the final stage, are based on the pixel intensities within the parallelograms computed in the previous stage, this is, the segmented vessels. We have analyzed several options to obtain the most distinctive vessel feature vectors.

Since the feature vectors depend on the pixel intensities, we have analyzed several color spaces (RGB, HSL, gray level) in order to identify the most suitable feature vectors [8,9,11]. In retinal images, the blue component (B), the saturation (S), and lightness (L) are not appropriate for the classification due to their little contrast. Therefore, the feature vectors are made up from the remaining color components.

The feature vectors are based on the concept of *profile*. A *profile* is a 1 pixel thick segment perpendicular to a vessel as Fig. 2 (right) shows. The number of profiles traced in each vessel, t, depends on the size of each parallelogram. Thus, we define the profile set, PR, for all detected vessels as follows

$$PR = \{PR_{ij}, \, i = 1 \ldots n, \, j = 1 \ldots t\} \tag{4}$$

$$PR_{ij} = \{p_k, \, k = 1 \ldots m\} \tag{5}$$

$$p_k = (x_k, \, y_k) \tag{6}$$

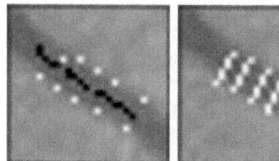

Fig. 2. Left: Detected vessel in the segmentation stage. The black points represent the crease set whereas the white points are the vessel contour points. Right: Several profiles extracted from the segmented vessel.

where n is the number of detected vessels, t the number of the profiles in the i-th vessel, m is the number of points in the profile PR_{ij}, and (x_k, y_k) are the image coordinates of the profile point p_k.

We have defined five feature vectors based on the R component, the G component, the union of the R and G components, the hue (H) component, and the gray level. In this case, there is a feature vector per profile point p_k. Also, we have defined another feature vector as the mean of the H component and the variance of the R component in the profile PR_{ij} [9]. Moreover, we have considered the mean and the median of each profile PR_{ij} in every color component in order to minimize the effect of outliers.

Finally, we apply a clustering process to classify the vessels into arteries and veins. The uneven intra-image contrast and lightness makes difficult the use of a supervised algorithm. We have chosen the k-means algorithm [15] due to its simplicity, computational efficiency and absence of parameter tuning. The centroids of each class are initialized to the minimum and the maximum of the k-means input set, respectively, since there are only two classes whose cluster centers should be as far as possible.

The k-means computes the cluster centers for both artery and vein classes. Then, the feature vectors are classified using the Euclidean distance. The empirical probability of a vessel v_i to be vein ($P^v(v_i)$) or artery ($P^a(v_i)$) is computed as follows:

$$P^a(v_i) = \frac{n_a}{n_a + n_v} \quad , \quad P^v(v_i) = \frac{n_v}{n_a + n_v} \tag{7}$$

where n_a is the number of feature vectors that were classified as artery and n_v is the number of feature vectors that were classified as vein.

We have developed three approaches to apply the clustering algorithm. In the first approach, we apply the k-means algorithm to all the feature vectors found in the retinal image. In the second approach, we divide the image into four quadrants centered at the optic disc. Then, we apply the k-means in each quadrant separately. This way, we try to minimize the effect of the uneven intra-image lightness. Additionally, in the third approach, we rotate the four quadrants (see Fig. 3) and we apply the clustering algorithm to the feature vectors found in each rotated quadrant. The rotation angle is set to a value that allows overlapped areas. Therefore, a vessel can be classified several times. These classification results are combined to make the final decision so that the influence of outliers is reduced.

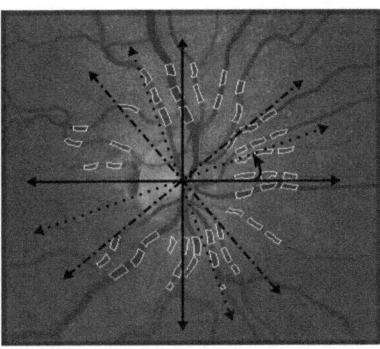

Fig. 3. In the third clustering approach, we divide the image into four quadrants centered at the optic disc and we apply the k-means algorithm to the feature vectors found in each quadrant separately. Then, we rotate the quadrants and we apply the clustering process again. These two steps are repeated in order to cover the whole image. After that, the classification results for each vessel are combined to provide the final decision.

In the third approach, the final vessel probabilities P^v and P^a are the mean of the vessel probabilities P_q^v and P_q^a in all the quadrants q where the vessel was found. The vessel is classified in the class with the highest probability. If the probability values P^v and P^a are the same, we do not classify the vessel.

4   Experimental Results

First, we have analyzed experimentally the three classification approaches with the feature vectors proposed in the previous section without applying the retinex method. The train set that we have used for this purpose consists of 20 images with a resolution of 768×576 pixels and centered at the optic disc. These images have been acquired from a Cannon CR6-45NM non-mydriatic retinal camera and labeled manually by an expert. The lowest training errors (12.32% and 10.88% with analysis radius of 2.5 and 2 times the optic disc radius, respectively) have been obtained using the third approach and the median of the G component, $\tilde{x}(G(PR_{ij}))$, as feature vector. Moreover, we have concluded that a rotation angle of 20° is suited because it balances the number of unclassified vessels and the number of necessary rotations (classifications).

After the training phase, we test the third approach in the public VICAVR data set [16] using the median of the G component of the profiles, $\tilde{x}(G(PR_{ij}))$, or the G component value of each profile point, $G(p_k)$, as feature vectors. This data set is made up of 58 images labeled manually by three different experts. Table 1 shows the performance of our methodology classifying the vessels by means of the aforementioned feature vectors. This results are obtained with respect to the ground truth of each expert individually and with respect to the agreement among groups of experts. Thus, the disagreement between the automatic system and the three expert agreement is 12.63% for $\tilde{x}(G(PR_{ij}))$, similar to the error rate of the training phase.

Table 1. Vessel classification performance with respect to the gold standard obtained from each single expert or the agreement between experts. We have used the third approach without retinex enhancement and the median of the G component $(\tilde{x}(G(PR_{ij})))$ or the value of the G component in each profile point $(G(p_k))$ as features to classify the vessels detected in the VICAVR data set. The percentage of miss-classifications and unclassified vessels by the system is calculated with respect to the second row, that is the number of vessels labeled by a single expert or the number of agreement vessels between experts.

Experts		1	2	3	1&2	1&3	2&3	1&2&3
Vessels Labeled		2472	2782	2950	2239	2247	2555	2116
$\tilde{x}(G(PR_{ij}))$	Error(%)	13.28	16.64	17.99	12.92	12.71	15.61	12.63
	Unclass.(%)	1.29	1.29	1.25	1.16	1.25	1.21	1.18
$G(p_k)$	Error(%)	14.63	18.15	19.58	14.30	14.04	17.15	13.95
	Unclass.(%)	0.73	0.75	0.81	0.67	0.76	0.74	0.71

With the aim of improving the results avoiding the effect of the non lightness-color constancy, we have applied the SSR and MSR image enhancement techniques to the VICAVR image database. We have tested SSR and MSR with different scales in function of the image size, $min(\text{height}, \text{width})/d$, where $d = 1, 2, 4, 8, 16$. Since the test images have a resolution of 768×576 pixels, the scales were $\sigma_1 = 576$, $\sigma_2 = 288$, $\sigma_4 = 144$, $\sigma_8 = 72$, and $\sigma_{16} = 36$ pixels. Figure 4 show some of the best classification results obtained using SSR and MSR with different scales before applying each labeling approach. These figures show the *accuracy rate*, this is, the percentage of the vessel segments which have been correctly classified penalizing with the vessels unclassified by the system.

$$\text{Accuracy rate} = \frac{n_{\text{corrected-classifications}}}{n_{\text{vessels}}} * 100 \tag{8}$$

where $n_{\text{corrected-classifications}}$ is the number of vessels, veins and arteries, correctly classified and n_{vessels} is the total number of detected vessels that includes the number of vessels which have not been able to classify.

There is not much difference between SSR and MSR results with the scales shown. Nevertheless, with the smallest scales, the image loses tonal rendition and the number of miss-classifications increases. It is worth pointing out that applying retinex, the feature formed by the G component value in each profile point, $G(p_k)$, discriminates better than the $\tilde{x}(G(PR_{ij}))$ in most of cases. This is due to the fact that the retinex is similar to the median function as it deletes the outlier values.

The best error percentages obtained applying the retinex technique were 9.49% and 8.38% for $\tilde{x}(G(PR_{ij}))$ and $G(p_k)$ respectively. Thus, the use of retinex produces a significant improvement in the results. Comparing these results with the results of the table 1 the improvement is about the 3.14% using $\tilde{x}(G(PR_{ij}))$ and 5.57% using the $G(p_k)$ as feature.

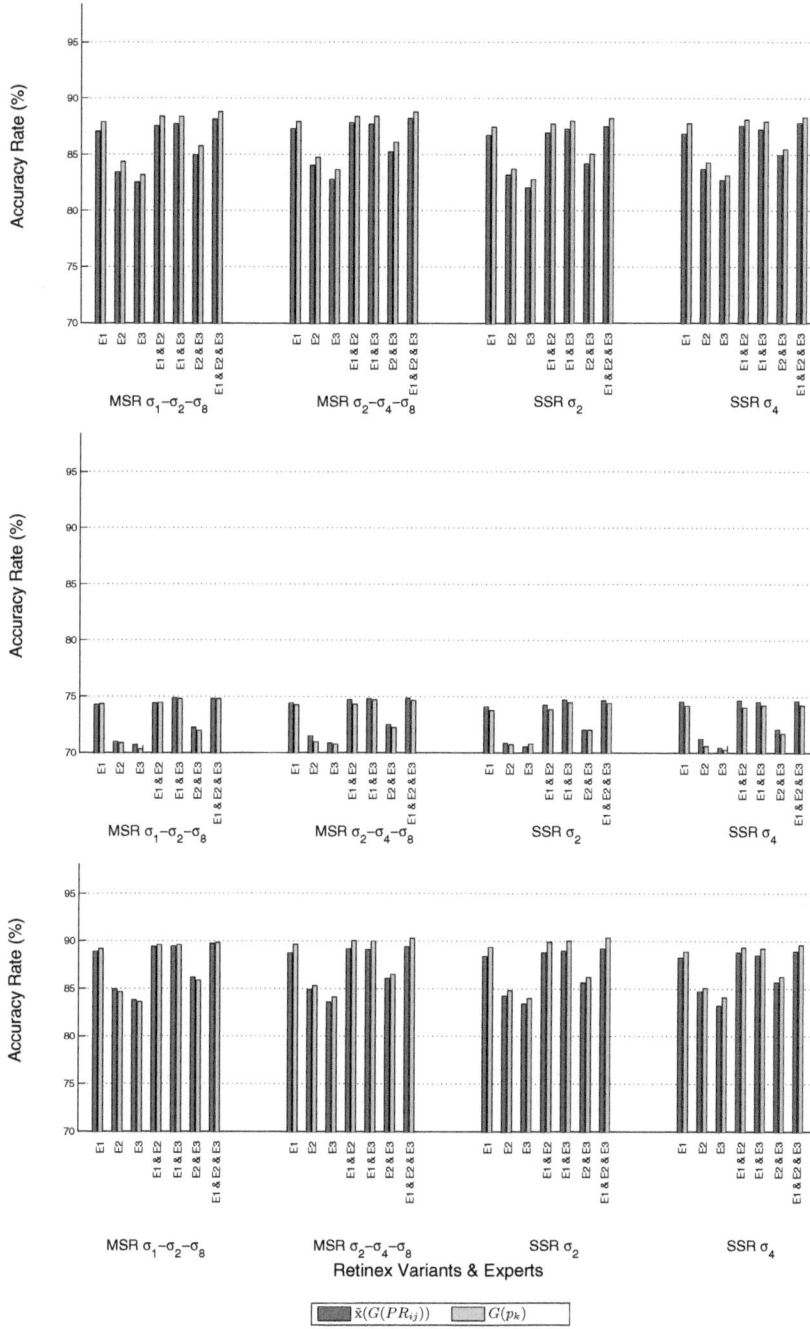

Fig. 4. Classification performance with respect to the gold standard obtained from each single expert or the agreement between experts, Retinex variants and $\bar{\mathrm{x}}(G(PR_{ij}))$ and $G(p_k)$ features using first (top), second (middle) and third (bottom) approaches in the VICAVR data set

5 Conclusions

In this paper, we have proposed a new methodology to classify the retinal vessels into veins and arteries taking into account the effect of the inhomogeneity lightness. This methodology applies the retinex image enhancement method and a clustering strategy based on labeling overlapped groups of closer vessels independently. These partial classifications are combined in order to make the final decision.

We have also tested two other approaches to study the effect of the lightness variability in the labeling and to prove the accuracy of the proposed strategy.

The results show that these approaches are not enough to classify correctly the retinal vessels, even after applying a normalization method as retinex. Also, the results prove that the retinex technique improves the artery/vein classification as the error rate falls down to 9%. However, the color similarity between veins and arteries makes difficult to establish a lightness range for each type of vessel, making necessary the division of the retinal image into regions to classify the vessels. Thus, the proposed clustering strategy achieves the best results since it minimizes the miss-classification error and maintains the number of unclassified vessels low.

Acknowledgments. This paper has been partly funded by Ministerio de Ciencia y Tecnología–Instituto de Salud Carlos III (Spain) through the grant contract PI08/90420 and FEDER funds.

References

1. Nguyen, T.T., Wang, J.J., Wong, T.Y.: Retinal Vascular Changes in Pre-diabetes and Prehypertension: New findings and their research and clinical implications. Diabetes Care 30(10), 2708–2715 (2007)
2. Soares, J.V.B., Leandro, J.J.G., Cesar Jr., R.M., Jelinek, H.F., Cree, M.J.: Retinal Vessel Segmentation Using the 2-D Gabor Wavelet and Supervised Classification. IEEE Transactions on Medical Imaging 25(9), 1214–1222 (2006)
3. Salem, S.A., Salem, N.M., Nandi, A.K.: Segmentation of retinal blood vessels using a novel clustering algorithm (RACAL) with a partial supervision strategy. Medical and Biological Engineering and Computing 45(3), 261–273 (2007)
4. Aguilar, W., Martínez-Pérez, M.E., Frauel, Y., Escolano, F., Lozano, M.A., Espinosa-Romero, A.: Graph-based methods for retinal mosaicing and vascular characterization. In: Escolano, F., Vento, M. (eds.) GbRPR. LNCS, vol. 4538, pp. 25–36. Springer, Heidelberg (2007)
5. Chrástek, R., Wolf, M., Donath, K., Niemann, H., Michelsont, G.: Automated Calculation of Retinal Arteriovenous Ratio for Detection and Monitoring of Cerebrovascular Disease Based on Assessment of Morphological Changes of Retinal Vascular System. In: IAPR Workshop on Machine Vision Applications, Nara, Japan, vol. 11-13, pp. 240–243 (2002)
6. Rothaus, K., Jiang, X., Rhiem, P.: Separation of the retinal vascular graph in arteries and veins based upon structural knowledge. Image and Vision Computing 27(7), 864–875 (2009)

7. Simó, A., de Ves, E.: Segmentation of macular fluorescein angiographies. A statistical approach. Pattern Recognition 34(4), 795–809 (2001)
8. Li, H., Hsu, W., Lee, M.L., Wang, H.: A piecewise Gaussian model for profiling and differentiating retinal vessels. In: ICIP03, vol. 1, pp. 1069–1072 (2003)
9. Grisan, E., Ruggeri, A.: A divide et impera strategy for automatic classification of retinal vessels into arteries and veins. In: Proceedings of the 25th Annual International Conference of the IEEE Engineering in Medicine and Biology Society 2003, vol. 1, p. 890 (2003)
10. Foracchia, M., Grisan, E., Ruggeri, A.: Luminosity and contrast normalization in retinal images. Medical Image Analysis 9, 179–190 (2005)
11. Jelinek, H.F., Lucas, C., Cornforth, D.J., Huang, W., Cree, M.J.: Towards vessel characterization in the vicinity of the optic disc in digital retinal images. In: McCane (ed.) Proceedings of the Image and Vision Computing New Zealand (2005)
12. Jobson, D.J., Rahman, Z., Woodell, G.A.: Properties and performance of a center/surround retinex. IEEE Transactions on Image Processing 6(3) (1997)
13. Jobson, D.J., Rahman, Z., Woodell, G.A.: A Multiscale Retinex for Bridging the Gap Between Color Images and the Human Observation of Scenes. IEEE Transactions on Image Processing 6(7) (1997)
14. Caderno, I.G., Penedo, M.G., Barreira, N., Mariño, C., González, F.: Precise Detection and Measurement of the Retina Vascular Tree. Pattern Recognition and Image Analysis: Advances in Mathematical Theory and Applications (IAPC Nauka/Interperiodica) 15(2), 523–526 (2005)
15. Gan, G., Ma, C., Wu, J.: Data Clustering Theory, Algorithms and Applications. ASA-SIAM Series on Statistics and Applied Probability. SIAM, Philadelphia (2007)
16. VICAVR, VARPA Images for the Computation of the Arterio/Venular Ratio, database (2009), http://www.varpa.org/RetinalResearch.html

Automatic Corneal Nerves Recognition for Earlier Diagnosis and Follow-Up of Diabetic Neuropathy

Ana Ferreira[1], António Miguel Morgado[1,2], and José Silvestre Silva[2,3]

[1] IBILI – Institute of Biomedical Research in Light and Image, Faculty of Medicine,
University of Coimbra, Portugal
[2] Department of Physics, Faculty of Sciences and Technology,
University of Coimbra, Portugal
[3] Instrumentation Center, Faculty of Sciences and Technology,
University of Coimbra, Portugal
`alferreira@ibili.uc.pt, miguel@fis.uc.pt,`
`jsilva@ci.uc.pt`

Abstract. Peripheral diabetic neuropathy is a major cause of chronic disability in diabetic patients. Morphometric parameters of corneal nerves may be the basis of an ideal method for early diagnosis and assessment of diabetic neuropathy. We developed a fully automatic algorithm for corneal nerve segmentation and morphometric parameters extraction. Luminosity equalization was done using local methods. Images structures were enhanced through phase-shift analysis, followed by Hessian matrix computation for structure classification. Nerves were then reconstructed using morphological methods. The algorithm was evaluated using 10 images of corneal nerves, by comparing with manual tracking. The average percent of nerve correctly segmented was 88.5% ± 7.2%. The percent of false nerve segments was 3.9% ± 2.2%. The average difference between automatic and manual nerve lengths was -28.0 ± 30.3 μm. Running times were around 3 minutes. The algorithm produced good results similar to those reported in the literature.

Keywords: Corneal nerves, image segmentation, diabetic neuropathy, confocal microscopy.

1 Introduction

The prevalence of Diabetes Mellitus is dramatically increasing worldwide, having developed to epidemic proportions. Consequently, there will be a substantial increase in the prevalence of chronic complications associated with diabetes [1].

Peripheral diabetic neuropathy is the major cause of chronic disability in diabetic patients. It affects 50% of the patients within 25 years of diagnosis [2]. In long term, undetected and untreated neuropathy can lead to foot infections that not heal, foot ulcers and, in many cases, amputation. Diabetic neuropathy is implicated in 50-75% of non-traumatic amputations [3].

Early diagnosis and accurate assessment of peripheral neuropathy are important to define the higher risk patients, decrease patient morbidity and assess new therapies [4].

A. Campilho and M. Kamel (Eds.): ICIAR 2010, Part II, LNCS 6112, pp. 60–69, 2010.
© Springer-Verlag Berlin Heidelberg 2010

However, early diagnosis of diabetic neuropathy is not easy. In otherwise healthy individuals, symptoms may be dismissed or attributed to incidental factors. The diagnosis often fails or occurs only when patients became symptomatic due to the non-availability of a simple non-invasive method for early diagnosis [5].

Diabetic neuropathy is currently quantified through electrophysiology and sensory tests [6]. However, these tests cannot detect small-fiber damage which is the most important component of nerve damage in diabetes [7]. Skin biopsy, which has been used for early diagnosis of neuropathy, can detect minute nerve damage [8], but is an invasive procedure.

The cornea is one of the most densely innervated tissues in the humans. It is possible to image *in vivo* the sub-basal plexus, using corneal confocal microscopy (CCM), a non-invasive method of examining the living cornea that produces high resolution images of thin corneal layers, enabling the identification of structures without depending on conventional histology techniques [9]. This method has been used to assess the sub-basal plexus of healthy and diabetic corneas.

Malik et al. [10] studied the corneas of diabetic patients with neuropathy and demonstrated that CCM can accurately define the extent of corneal nerve damage and repair, through the measurement of fiber density and branching. Therefore, it can be used as a measure of neuropathy in diabetic patients. Kallinikos et al. [11] showed that CCM allows evaluation of corneal nerve tortuosity and that this parameter relates to the severity of neuropathy.

Our group has done research on CCM. We demonstrated that the number of fibers in the corneal sub-basal nerve plexus of diabetic patients was significantly lower than in healthy humans, even for short diabetes duration [12]. This opens the possibility of using the assessment of corneal innervation by CCM for early diagnosis of peripheral neuropathy. This result was later confirmed by other authors, using different morphometric parameters like the number of nerve branches, the number of beadings and the tortuosity of the fibers [13].

Accurate measurement of corneal nerve morphometry requires accurate representation of the nerve network. The corneal nerves analysis is based on a tedious process of manual tracing of the nerves [10], using confocal microscope built-in software [10, 14-15], commercial programs or software specifically developed for the purpose [11].

Ruggeri et al. proposed automatic methods for the recognition and tracing of the corneal nerve structures [16]. The nerves were recognized by a tracing algorithm based on filtering and pixel classification methods with post-processing to remove false recognitions and link sparse segments into continuous structures. Automatic and manual length estimations on the same image were very well correlated.

The purpose of our work was to develop an algorithm for automatic analysis of CCM images of sub-basal nerve plexus and to extract morphometric parameters from these images for early diagnosis and staging of diabetic peripheral neuropathy.

2 Methods

In a CCM image (see Fig. 1), the corneal nerves appear as bright structures on a dark background. A typical corneal confocal microscope has an effective focal depth greater than 20 μm [17]. Therefore sub-basal nerve plexus images may include non-nerve structures, as basal epithelial cells (see Fig. 1a) or keratocytes (Fig. 1b).

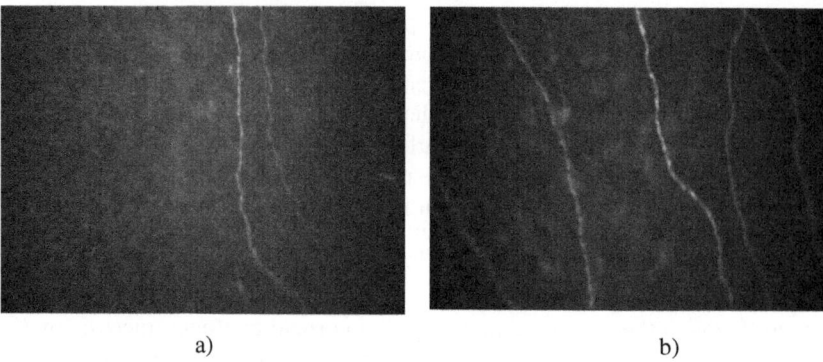

a)   b)

Fig. 1. Corneal confocal microscopy images of a corneal sub-basal nerve plexus region. The structures in the background of image a) are basal epithelial cells. In image b) it is possible to see stromal keratocytes.

The automatic algorithm includes three stages: contrast enhancement and illumination correction, identification of line segments representing the corneal nerves and reconstruction for correcting discontinuities and misclassified pixels.

2.1 Pre-processing

Before applying the segmentation algorithm it was necessary to pre-process the images due to their non-uniform contrast and luminosity. CCM images show brighter areas in the central region, mainly due to corneal curvature, which causes non-uniform light reflection.

Contrast-limited adaptive histogram equalization (CLAHE) [18] was applied to the original images for contrast uniformization using an 8×8 pixels square mask.

Phase symmetry was used to enhance the boundaries of structures in the image (see Fig. 2). The phase symmetry is based on the analysis of local frequency information, overcoming the main shortcomings of symmetry detection algorithms: the need to segment the objects first and not providing any absolute measure of the degree of symmetry at any point in the image [19].

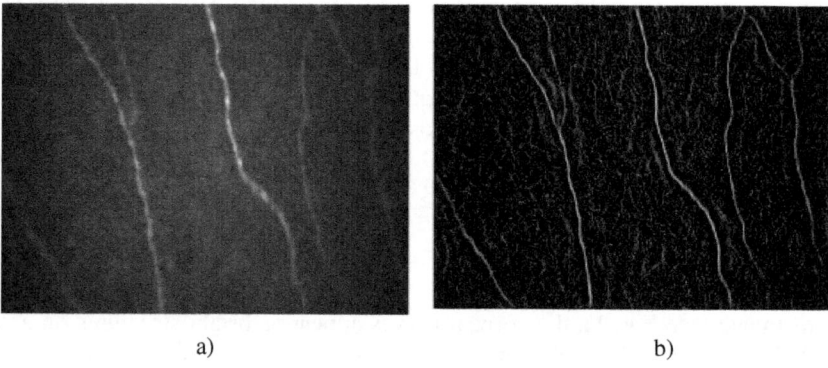

a)   b)

Fig. 2. a) Original CCM image. b) After pre-processing. The phase shift procedure also enhances structures that do not belong to the nerve plexus.

2.2 Nerves Searching

Nerves searching was accomplished using a method based on the Hessian matrix:

$$H = \begin{bmatrix} I_{xx} & I_{xy} \\ I_{yx} & I_{yy} \end{bmatrix} \tag{1}$$

where I_{xy} is the convolution of the input image with the second order derivative of the scaled Gaussian function:

$$I_{xy} = I * \left(\frac{\partial^2}{\partial x \partial y} G \right) \tag{2}$$

with the scaled Gaussian function expressed by:

$$G(x, y; \sigma) = \frac{1}{2\pi\sigma^2} \exp(-\frac{x^2 + y^2}{2\sigma^2}) \tag{3}$$

Sigma (σ) is a length scale factor that controls the scale of the extracted nerves. The information of the second order derivative of a Gaussian function at a scale σ generates a probe kernel that measures the contrast between the regions in the interval $\sigma_{min} \leq \sigma \leq \sigma_{max}$. These values are fixed according to the sizes of the smallest and the largest nerve to be detected in the image.

Let λ_1 and λ_2 be the eigenvalues of the Hessian matrix H used as a vesselness measure for 2D images to describe the curvature at each point in the image. The idea behind eigenvalues analysis is to extract the principal directions in which the local second-order structure of the image can be decomposed. The parameter R_B was proposed to discriminate between blob-like structures and ridge-like structures [20] and is defined as the quotient between λ_1 and λ_2.

Table 1. Structures and their relations to eigenvalues in 2D image, where $\lambda_1 < \lambda_2$

λ_1	λ_2	Orientation
Low	High Negative	Bright tubular structure
Low	High Positive	Dark tubular structure
High Negative	High Negative	Bright blob-like structure
High Positive	High positive	Dark blob-like structure

Table 1 summarizes the relation between eigenvalues and structures in a tissue. Note that we are only interested in tubular structures as they represent nerves. Accordingly, since our main concern was nerves recognition and extraction, we have analyzed the eigenvalues for $\lambda_1 < \lambda_2$, low λ_1 and high negative λ_2.

As Fig. 3 shows, the output of this process is a set of nerve fragments that require further processing to achieve full reconstruction of each nerve.

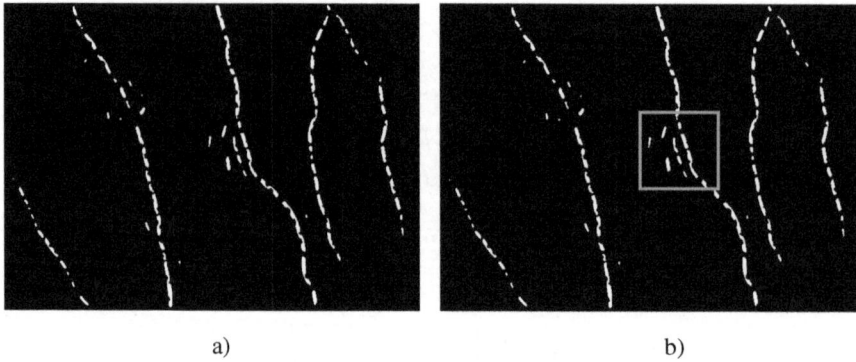

a)   b)

Fig. 3. Result of the nerves searching procedure for the enhanced image in Fig. 2b). Most of the non-nerve fragments were removed. a) Enhanced image, after pre-processing; b) enhanced image, indicating the region to zoom, to show in detail the procedures for post-processing.

2.3 Nerves Reconstruction

Segmented images of corneal nerves exhibit misclassified pixels and visible nerve discontinuities that are not real. Minimum cost path or clustering algorithms, as the ones described by [21-22] may be a suitable basis for automatic reconstruction of discontinued corneal nerves. Although it is usual for these methods to apply region growing from its centroids, this approach failed on corneal nerve images either due to the presence of similar-shaped structures external to the nerve plexus, like stromal keratocytes or basal epithelial cells, or due to nearby structures belonging to a different nerve fiber, as seen in Fig. 4.

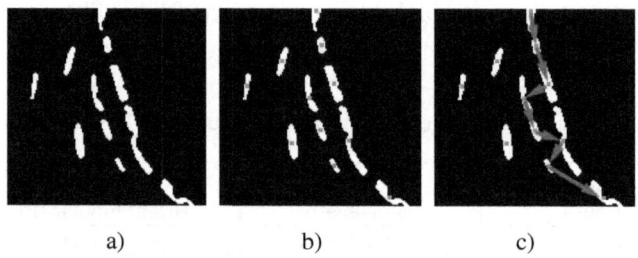

a)   b)   c)

Fig. 4. Example of the post-processing algorithm based on region growing from centroids: a) region of interest (ROI); b) centroids of each segment of the ROI; c) centroids connection using the shortest distance between centroids as a criteria

To overcome this, we used a region growing approach based on its more peripheral points. Our post-processing procedure starts by identifying all segments in the image. For each segment S_0 we compute the morphological skeleton and calculate the number of branches, using morphological operations. When two or more branches are found, erosion is applied to the endpoints of the skeleton until only one branch is obtained. Then, we compute the two extreme points, P_1 and P_2, of each segment as the two opposite points that belong to the major axis of the segment. These points correspond to the endpoints of the morphological skeleton of this segment (Fig. 5a). The next step is to perform a region growing algorithm, using as seed points P_1 and P_2 until this region reaches another segment. This way, we identify the nearest segment neighbor S_1 that suggests a connection between S_0 and S_1. We repeat this procedure, starting in P_2, for obtaining another neighbor segment S_2, suggesting also a connection between S_0 and S_2 (see Fig. 5b and 5c).

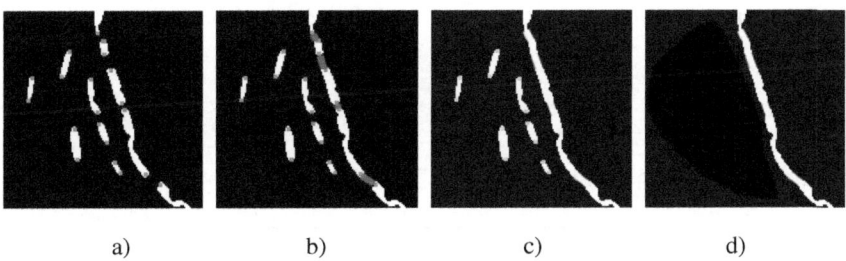

a)   b)   c)   d)

Fig. 5. a) For each segment, the two extreme endpoints are identified; b) starting in the segment at the top of the image, its extreme endpoint would be connected to the nearest endpoint of the another segment; in this new segment, its second extreme endpoint would be connected to the nearest endpoint of another segment and so on until it reached the bottom of the image; c) image with all the segments connected, establishing a continuity between the top and the bottom parts of the image d) final image; the segments that were not connected were discarded.

After processing every segment, connections are discarded if some of the following conditions occur: segments S_i are connected with S_j but S_j does not connect with S_i. This occurs when the nearest segment S_i is S_j but the nearest segment of S_j is S_k; the connections with distance between segments higher than twice the median length of the major axis of the segments; S_0 connects to S_1 and S_2 but $S_1=S_2$.

The connection of adjacent segments, S_0 and S_1, is established using a straight line, which is valid when the distance between these two segments is small, usually less than the length of the major axis of one segment. This straight line is dilated by the same size of the segments diameter, which corresponds to the lengths of the second major axis of the segments (Fig. 5d).

With this algorithm, each nerve is reconstructed using the information of all their segments.

2.4 Morphometric Parameters

Once the nerves are segmented it is possible to compute several morphometric parameters such as nerves length, density, tortuosity and diameter.

2.4.1 Calculation of Length and Density

The lengths (in µm) of the nerve structures were calculated by computing the distance between adjacent pixels along the skeleton of the nerve. The nerve density is calculated by dividing the sum of the nerve lengths by the image area (µm/mm^2).

2.4.2 Calculation of Tortuosity

The Tortuosity Coefficient (TC) intends to convey information on the frequency and magnitude of nerve curvature changes. We followed an approach based on considering each nerve as a mathematical function on the image space and computing the function first and second derivatives [11].

In order to treat each nerve as a mathematical function, we find its endpoints, draw a straight line between them and rotate the image, aligning the straight line with the x-axis. The first and second derivatives, and the TC are calculated using:

$$f'(x_i, y_i) = \frac{y_{i+1} - y_i}{dx} \tag{4}$$

$$f''(x_i, y_i) = \frac{y_{i+1} - 2y_i + y_{i-1}}{dx^2} \tag{5}$$

$$TC = \sqrt{\sum_{i=1}^{N-1} \left((f'(x_i, y_i))^2 + (f''(x_i, y_i))^2 \right)} \tag{6}$$

The step size dx is the distance between the projections on the x-axis of two nerve consecutive pixels and is equal to 1, N is the number of pixels of the nerve skeleton.

2.4.3 Calculation of Nerve Diameter

To compute the nerve diameter, we locate all the segments that belong to a given nerve and compute, for each segment, its two main axes. As most segments have a rectangular shape, we assume that the major axis corresponds to the length of the nerve and the minor one corresponds to the diameter of the nerve.

3 Results and Discussion

The algorithms described in the previous section were applied to 10 images of corneal nerves available online [16]. These images were acquired *in vivo* from non-diabetic patients, using a corneal confocal microscope (ConfoScan4, Nidek Technologies, Padova, Italy), with a 460×350 μm field of view using a 40X objective, and compressed in JPEG monochrome format, with a size of 768×576 pixels.

Best pre-processing results were obtained using an 8×8 pixels mask. For the identification of tubular structures, λ_1 was set to equal or less than 20% of its minimum and R_B was set to be equal or greater than 1. We used $2 \leq \sigma \leq 5$ with steps of 1. The dilation was set to 10 pixels and areas less than 3100 pixels were removed.

Figs. 6 and 7 show examples of the results obtained with the corneal nerves segmentation algorithm. The algorithm still has to be improved particularly in dealing with nerve branching and in rejecting non-nerve structures as can be seen in Fig. 7.

Algorithm performance can be evaluated by comparing the nerve length correctly recognized by the algorithm, with the length of manually traced nerves on the same image. For manual nerve tracing and length measurement, we used the Simple Neurite Tracer plug-in developed for Image J by Mark Longair[1]. The average percent of

---

[1] http://homepages.inf.ed.ac.uk/s9808248/imagej/tracer/

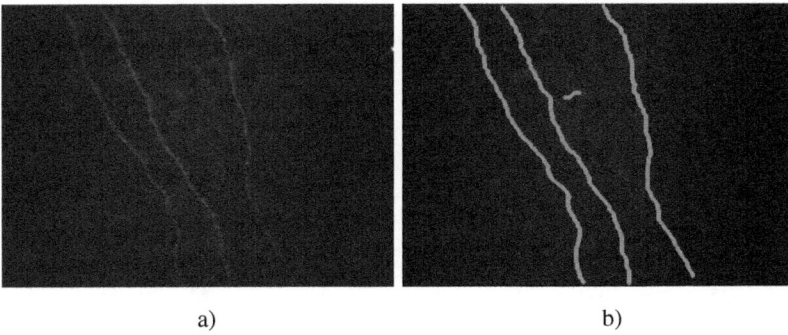

a)   b)

Fig. 6. Representative example of the results obtained for an image with few nerves: a) original image, b) after processing

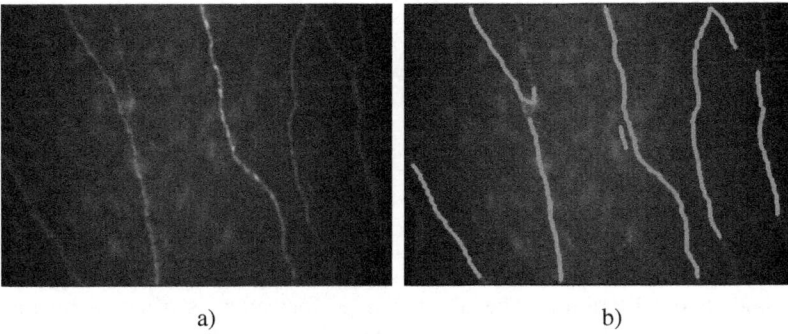

a)   b)

Fig. 7. Representative example of the results obtained for an image with nerve branching: a) original image, b) after processing

nerve correctly segmented by the program was 88.5% ± 7.2% (range: 78.0% - 99.8%). On five images structures were falsely reported as nerves by the algorithm. The percent of false nerve segments on the total segmented length was 3.9% ± 2.2% (range: 2.1% - 6.8%).

Fig. 8 shows a Bland and Altman [23] plot for the comparison between automatic and manual nerve length measurement. The average difference between automatic and manual nerve lengths was -28.0 ± 30.3 μm. This means that, in 95% of the cases, the difference between nerve lengths measured automatically and manually will lie between - 87.4 and 31.3 μm. These limits, as well as the average difference, are shown in the plot. These results are similar to the ones reported in the literature and also show a tendency for underestimation by the automatic method [16].

From the nerves representation obtained through automatic segmentation we have extracted morphometric parameters such as tortuosity, nerves density (μm/mm^2) and segment diameter (μm). The average value of the TC was 21.7 ± 5.6 (range: 16.9 - 32.7). This value agrees with those previously reported, using the same definition of tortuosity, for non-diabetic and mild-neuropathy diabetic individuals. The proposed algorithm for nerve identification was fully automatic, requiring no user intervention.

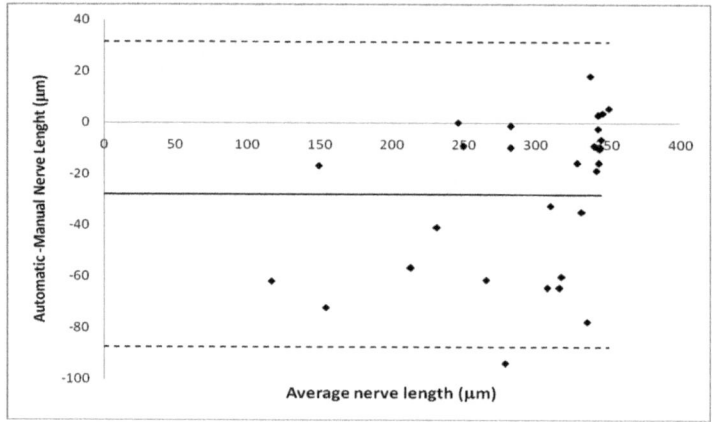

Fig. 8. Comparison between nerve lengths measured by the automatic algorithm and by manual tracking. The solid line represents the average difference between nerve lengths by the two methods. The broken lines are the limits of concordance for a 95% level of confidence.

4   Conclusion

The developed algorithm produced good results, in terms of percentage of nerves detected and nerve length measurement, similar to those reported for a different corneal nerve segmentation method, and yields Tortuosity Coefficients in agreement to those found in the literature. The algorithm performance is affected by the image quality, mainly by the presence of non-nerve structures such as stromal keratocytes or basal epithelial cells. The issues related to non-uniform contrast and luminosity were successfully solved by pre-processing the images with local equalization and phase shift based methods. There is room for improvement particularly when dealing with images containing nerve branches.

In our opinion, the pressing need of having a simple, non-invasive technique, capable of accurately documenting the extent of nerve damage and repair, for early diagnosis of peripheral diabetic neuropathy, can be addressed through the evaluation of corneal nerve morphology from images obtained through CCM. In this work we presented our approach to an automatic algorithm for analysis of corneal sub-basal nerve plexus images. This work is part of a broader project that aims to develop a noninvasive technique for early diagnosis and monitoring of diabetic neuropathy.

References

1. Mokdad, A.H., Ford, E.S., Bowman, B.A., Nelson, D.E., Engelgau, M.M., Vinicor, F., Marks, J.S.: Diabetes trends in the US: 1990-1998. Diabetes Care 23, 1278–1283 (2000)
2. Gooch, C., Podwall, D.: The Diabetic Neuropathies. The Neurologist 10, 311–322 (2004)
3. Vinik, A.I., Park, T.S., Stansberry, K.B., Pittenger, G.L.: Diabetic neuropathies. Diabetologia 43, 957–973 (2000)
4. Park, T.S., Park, J.H., Baek, H.S.: Can diabetic neuropathy be prevented? Diabetes research and clinical practice 66, S53–S56 (2004)

5. Rahman, M., Griffin, S.J., Rathmann, W., Wareham, N.J.: How should peripheral neuropathy be assessed in people with diabetes in primary care? A population-based comparison of four measures. Diabetic Medicine 20, 368–374 (2003)
6. Sullivan, K.A., Feldman, E.L.: New developments in diabetic neuropathy. Current Opinion in Neurology 18, 586–590 (2005)
7. Malik, R.A., Veves, A., Walker, D., Siddique, I., Lye, R.H., Schady, W., Boulton, A.J.M.: Sural nerve fibre pathology in diabetic patients with mild neuropathy: relationship to pain, quantitative sensory testing and peripheral nerve electrophysiology. Acta Neuropathologica 101, 367–374 (2001)
8. Holland, N.R., Stocks, A., Hauer, P., Cornblath, D.R., Griffin, J.W., McArthur, J.C.: Intraepidermal nerve fiber density in patients with painful sensory neuropathy. Neurology 48, 708–711 (1997)
9. Masters, B.R., Bohnke, M.: Three-dimensional confocal microscopy of the living human eye. Annual Review of Biomedical Engineering 4, 69–91 (2002)
10. Malik, R.A., Kallinikos, P., Abbott, C.A., van Schie, C.H.M., Morgan, P., Efron, N., Boulton, A.J.M.: Corneal confocal microscopy: a non-invasive surrogate of nerve fibre damage and repair in diabetic patients. Diabetologia 46, 683–688 (2003)
11. Kallinikos, P., Berhanu, M., O'Donnell, C., et al.: Corneal Nerve Tortuosity in Diabetic Patients with Neuropathy. Invest. Ophthalmol. Vis. Sci. 45, 418–422 (2004)
12. Popper, M., Quadrado, M.J., Morgado, A.M., Murta, J.N., Van Best, J.A., Muller, L.J.: Subbasal nerves and highly reflective cells in corneas of diabetic patients: In vivo evaluation by confocal microscopy. Investigative Ophthalmology & Visual Science 46, 879 (2005)
13. Midena, E., Brugin, E., Ghirlando, A., Sommavilla, M., et al.: Corneal diabetic neuropathy: A confocal microscopy study. Journal of Refractive Surgery 22, S1047–S1052 (2006)
14. Grupcheva, C.N., Wong, T., Riley, A.F., et al.: Assessing the sub-basal nerve plexus of the living healthy human cornea by in vivo confocal microscopy. Clinical and Experimental Ophthalmology 30, 187–190 (2002)
15. Patel, D.V., McGhee, C.N.J.: Mapping of the Normal Human Corneal Sub-Basal Nerve Plexus by In Vivo Laser Scanning Confocal Microscopy. Invest. Ophthalmol. Vis. Sci. 46, 4485–4488 (2005)
16. Scarpa, F., Grisan, E., Ruggeri, A.: Automatic Recognition of Corneal Nerve Structures in Images from Confocal Microscopy. Investigative Ophthalmology & Visual Science 49, 4801–4807 (2008)
17. McLaren, J.W., Nau, C.B., Kitzmann, A.S., et al.: Keratocyte Density: Comparison Of Two Confocal Microscopes. Eye Contact Lens 31, 28–33 (2004)
18. Gonzalez, R.C., Woods, R.E., Eddins, S.L.: Digital Image Processing Using Matlab. Gatesmark Publishing (2009)
19. Kovesi, P.: Symmetry and Asymmetry from Local Phase. In: Tenth Australian Joint Conference on Artificial Intelligence, pp. 185–190 (1997)
20. Frangi, A.F., Niessen, W.J., et al.: Multiscale vessel enhancement filtering. In: Wells, W.M., Colchester, A.C.F., Delp, S.L. (eds.) MICCAI 1998. LNCS, vol. 1496, pp. 130–137. Springer, Heidelberg (1998)
21. Beichel, R., Pock, T., Janko, C., Zotter, R., et al.: Liver Segment Approximation in CT Data for Surgical Resection Planning. In: SPIE Medical Imaging 2004: Image Processing, vol. 5370, pp. 1435–1446 (2004)
22. Salem, N.M., Salem, S.A., Nandi, A.K.: Segmentation of Retinal Blood Vessels Based on Analysis of the Hessian Matrix and Clustering Algorithm. In: 15th European Signal Processing Conference, pp. 428–432 (2007)
23. Bland, J.M., Altman, D.G.: Statistical methods for assessing agreement between two methods of clinical measurement. Lancet 1, 307–310 (1986)

Fusing Shape Information in Lung Segmentation in Chest Radiographs

Amer Dawoud

School of Computing
University of Southern Mississippi, Hattiesburg, MS 39406

Abstract. This paper presents an algorithm for the segmentation of lung fields by fusing shape information priors into intensity-based thresholding in an iterative framework. The main contribution is to maximize information utilization by effectively combining intensity information with shape priors. Experimental results performed on publicly available database demonstrate the effectiveness of the algorithm in comparison with other algorithms.

1 Introduction

Chest radiology remains the most common procedure to detect chest diseases such as lung cancer and tuberculosis. This is due to its advantages: it is the most cost-effective, the most routinely available, and the most dose-effective diagnostic tool. This explains why the detection of subtle or early-stage of these diseases in chest radiographs is one of the outstanding challenges in the field of medical diagnosis. In the past few years, there has been an intensive research in the area of automation of disease detection in chest radiographs, as it assists in earlier detection and achieving better prognosis for the patient.

The first step in these detection systems is the segmentation of lung fields in the chest radiographs in order to restrict the processing area of subsequent detection algorithms. The accuracy by which this step is executed is critical to the overall performance of the system, as inaccurate lung segmentation would increase the false-positive and false-negative errors.

Many techniques have been proposed in the literature for the segmentation of lung fields from posterior–anterior chest radiographs, and they are surveyed in an interesting study [8]. These methods can be classified into four categories: 1) rule-based techniques have been used to detect the outline of ribcage or the diaphragm [1], [5]; 2) pixel-based techniques were proposed to classify each pixel of an image into either lung field or background based on a filter bank of Gaussian derivatives and a K-NN classifier [10], [12]; 3) hybrid techniques were formulated by combining rule-based techniques and pixel-based classifications for lung field segmentation [7]; and 4) deformable model-based techniques, such as active shape model (ASM) and active appearance model (AAM) [3][2][4][17], have been successfully applied in lung field segmentation. In ASM, the statistics of image intensities and gradients along the profiles of contour points are used to drive the contour toward the boundary of the object, and PCA-based shape statistics is used to constrain the contour. Many improvements have been made to improve the performance of the matching strategy. For example, Seghers *et al.* [12] proposed the simultaneous optimization of shape and gray-level appearance models based on non-iterative dynamic programming. Here, the shape of object lungs is described by multiple landmark-specific statistical models that capture local dependencies between adjacent landmarks on the shape. Gleason *et al.* [11] had developed a probabilistic-based deformable model which simultaneously optimizes a single objective function generated by the global and local shape model, as well as the gray-level appearance model. Here, the local shape model is introduced to preserve the local shape information around each user-defined critical landmark points. Other researchers suggest that the local gray-level appearance model should be created by using some optimal features, in order to drive contour points to desired object boundaries. For example, Ginneken *et al.* [6] determined the distinct set of optimal features by means of machine learning. Recently, Seghers *et al.* [14] proposed a generic model-based segmentation algorithm is presented, which can be trained from examples akin to the active shape model (ASM) approach in order to acquire knowledge about the shape to be segmented and about the gray-level appearance of the object in the image. Shi *et al.* [15] proposed a deformable model using both population-based and patient-specific shape statistics. Ginneken *et al.* [9] proposed an automatic method for detection of abnormalities in chest radiographs by using local texture analysis.

This paper is organized as follows. Next section describes the proposed algorithm in detail. Section 3 summarizes experimental results demonstrating the algorithm's effectiveness. Section 4 concludes.

2 Proposed Method

This research proposes an algorithm for the segmentation of lung fields by fusing shape information into intensity-based iterative thresholding. The main contribution is to maximize information utilization by effectively combining intensity information with shape priors. Firstly, a statistical model for the lung shape is extracted from large database; features including size, orientation, major and minor ellipse lengths, eccentricity, and centroid locations for the right and left lung fields are computed from a database of manually segmented lung fields by expert radiologists. This model is then used to optimize the iterative thresholding for the segmentation of lung fields in test images. This is achieved by making sure that shape resulting from this iterative binarization is similar to the statistical model; Mahalanobis distance, which measures the similarity between the shape model statistics and the binarization output, is used. Finally, in the postprocessing stage, the optimized binarization output contour is further adjusted using Active Shape Model techniques. The following is a detailed description of the proposed method.

2.1 Lung Field Statistical Model

To extract a statistical shape model for the lungs, the Japanese Society of Radiological Technology (JSRT) chest radiographs database [15] was used. This is a publicly available database with 247 chest radiographs collected from 13 institutions in Japan and one in the United States. The images were scanned from films to a size of

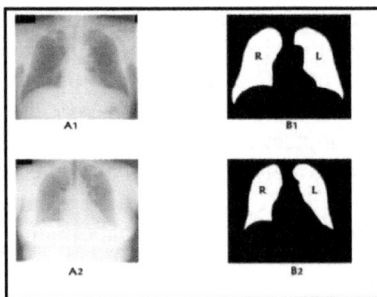

Fig. 1. JSRT database image samples. (A1-A2) Original images. (B1-B2) Manual segmentation of left and right lung fields by expert radiologists.

2048 × 2048 pixels, a spatial resolution of 0.175 mm/pixel and 12 bit gray levels. The JSRT database also includes manual segmentation results of experienced radiologists. Figure 1 shows two JSRT sample images and the manual segmentation of left and right lung fields by expert radiologists.

Using these manual JSRT segmentations, a statistical model of the lungs' shape is derived, which includes the following features:

- Area – Scalar; the actual number of pixels in the left and right lung regions.
- Major Axis Length – Scalar specifying the length (in pixels) of the major axis of the ellipse that has the same normalized second central moments as the left and right lung regions.
- Minor Axis Length – Scalar specifying the length (in pixels) of the minor axis of the ellipse that has the same normalized second central moments as the left and right lung regions.
- Centroid – 1-by-2 vector (x, y coordinates) that specifies the center of mass of the left and right lung regions.
- Eccentricity – Scalar that specifies the eccentricity of the ellipse that has the same second-moments as the left and right lung regions. The eccentricity is the ratio of the distance between the foci of the ellipse and its major axis length. The value is between 0 and 1. (0 and 1 are degenerate cases; an ellipse whose eccentricity is 0 is actually a circle, while an ellipse whose eccentricity is 1 is a line segment.)
- Orientation – Scalar; the angle (in degrees ranging from -90 to 90 degrees) between the x-axis and the major axis of the ellipses that has the same second-moments as the left and right lung regions.
- Solidity – Scalar specifying the proportion of the pixels in the convex hull that are also in the left and right lung regions. Computed as Area/Convex Area.
- Extent – Scalar that specifies the proportion of the pixels in the bounding box that are also in the left and right lung region. (Computed as the Area divided by the area of the Bounding Box, where then Bounding Box is the smallest rectangle containing the left and right lung regions).

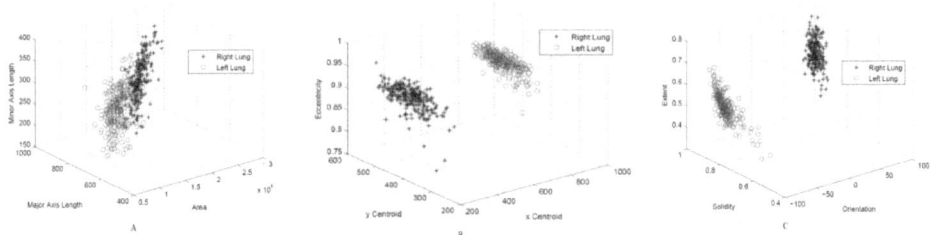

Fig. 2. (A) Area, Major Axis Length, and Minor Axis Length features' samples from JSRT database. (B) x Centroid, y Centroid, and Eccentricity features' samples from JSRT database. (C) Orientation, Solidity, and Extent features' samples from JSRT database.

Figure 2 shows these features' samples from the JSRT database. The mean vectors μ_l and μ_r and the covariance matrices $\sum_l$ and $\sum_r$ are calculated for left and right lungs, respectively. They will be used in the next step of optimizing the iterative binarization.

2.2 Optimizing Iterative Binarization

The process of lung segmentation in a test radiograph image starts with iterative binarization at equally spaced thresholds (TH_i), where i is the iteration number. This process extracts objects that will be classified as lung or non-lung objects. In figures 5 and 6, which demonstrate this process, the red and blue areas are the left and right lung objects (*LLO*) and (*RLO*) and the pink areas are the non-lung objects (*NLO*). The optimization of the iterative binarization involves the following steps:

1. Initially, LLO_0, RLO_0 and NLO_0 are empty.
2. Image is binarized at (TH_i) and using 8-connected component analysis and all objects are identified. The difference between two consecutive thresholds was arbitrary chosen to be 16 gray-levels, which has produced satisfactory results.
3. Objects smaller in size than a predefined limit (*PL*) are ignored. This allows concentrating on larger objects only, which is necessary to reduce run time.
4. Assuming that lungs are located at the in middle of the image, objects touching the boarders or their centroids are very close to the boarders are identified as NLO_i. LLO_i and RLO_i are in the left and right halves of the image.
5. Objects identified as LLO_i, RLO_i and NLO_i are not allowed to merge at iteration $i+1$, but are allowed to grow.
6. In some cases, the trachea, which appears as dark area between the lungs, has to be identified as NLO_i, as shown in Figure 6. This is necessary to prevent the two lungs from touching each other through flooding of the trachea area.
7. Estimate the similarity between the extracted LLO_i and RLO_i and the statistical models of Section 3.1. This is done by calculating the Mahalanobis distance DL_i for the left lung and DR_i for the right lung, according to the following equations:

$$DL_i = (XL_i - \mu_l) \sum_l{}^{-1} (XL_i - \mu_l)^T \tag{1}$$

$$DR_i = (XR_i - \mu_r) \sum_r{}^{-1} (XR_i - \mu_r)^T, \tag{2}$$

where XL_i and XR_i are the feature vectors of LLO_i and RLO_i, respectively.

8. In the initial iterations, there will be a significant dissimilarity between the LLO_i and RLO_i and their corresponding statistical models. Therefore, we expect that XL_i and XR_i to be high. But as the iterative binarization progresses and LLO_i and RLO_i become closer to the lung shapes, XL_i and XR_i should decrease. Figures 7 and 8 shows the resulting DL_i and DR_i for the cases in Figures 3 and 4, respectively.

9. The iterative binarization is repeated, starting from step 2, till a bottom is reached for DL_i and DR_i, independently. In Figures 5(A) and 5(B), the bottoms for DL_i and DR_i were reached at iterations 6 and 4, respectively. Therefore, LLO_6 and RLO_6 are the segmentation outputs for Figure 3, and LLO_4 and RLO_4 are the segmentation outputs for Figure 4.

2.3 Postprocessing

This postprocessing step is necessary because the iterative binarization, described in previous section, produces global solution that needs refinement using local information. In the post-processing stage, the optimized binarization output contour is further adjusted using Active Shape Model technique [3]. The following is a brief description of the method, and for more details, please see the reference [3][6].

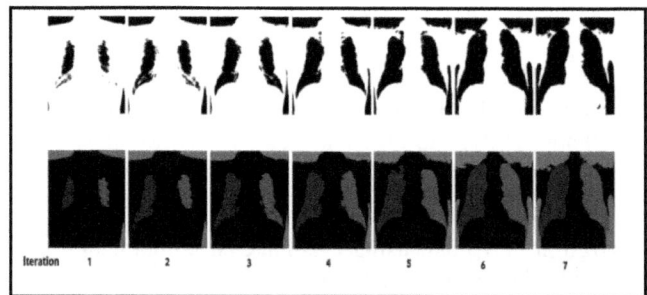

Fig. 3. Top row of images demonstrates iterative binarization of Figure1-A1 image at consecutive thresholds. Bottom row of images shows the resulting LLO (red), RLO (blue) fields, and the NLO (pink).

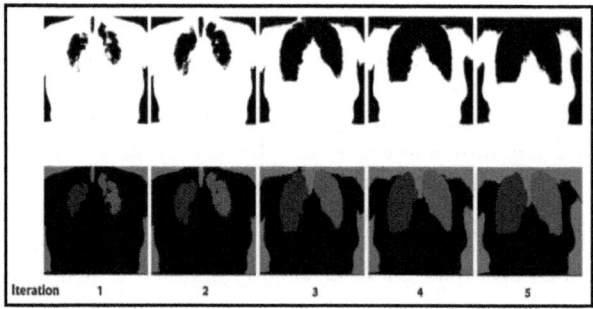

Fig. 4. Top row of images demonstrates iterative binarization of Figure1-A1 image at consecutive thresholds. Bottom row of images shows the resulting LLO (red), RLO (blue) fields, and the NLO (pink). This case demonstrates identifying the trachea area between the lungs as NLO.

An object is described by n points, referred to as landmark points. The landmark points are determined in a set of s training images.

The landmark points (x_1, y_1), ..., (x_n, y_n) are stacked in shape vectors

$$x = (x_1, y_1, \ldots, x_n, y_n)^T \qquad (3)$$

Shapes are fitted in an iterative manner, starting from the mean shape. Each landmark is moved along the direction perpendicular to the contour to n_s positions on either side, evaluating a total of $2n_s+1$ positions. The landmark is put at the position with the lowest Mahalanobis distance. After moving all landmarks, the shape model is fitted to the points, yielding an updated segmentation. This is repeated a fixed number N of times at each resolution, from coarse to fine.

Figure 6 shows that landmark position and directions perpendicular to the contours for the output of iterative binarization of the cases in Figures 3 and 4.

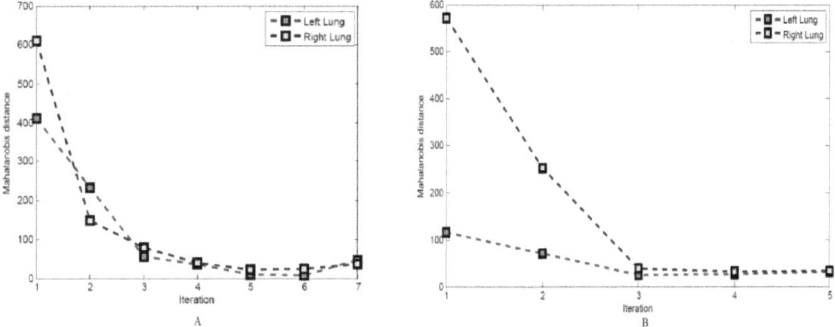

Fig. 5. Mahalanobis distance of segmented *LLO* and *RLO* at the iterations of (A) Figure 3 and (B) Figure 4

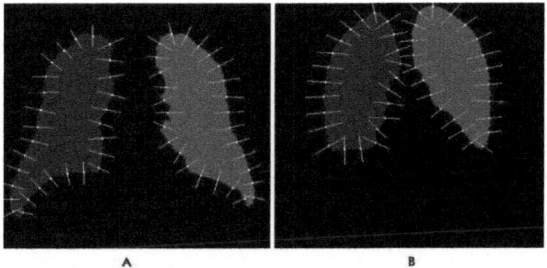

Fig. 6. Post-processing by ASM technique of the output of iterative binarization of A) Figure 5 case, and B) Figure 6 case

The following table summarizes the parameter settings for the ASM post-processing. Note that the number of resolution levels is 1, and this is because it is directly applied to the output of the iterative binarization step.

Table 1. Summary of parameter setting for the post-processing

n number of landmark points	22 and 28
Threshold f_v	0.995
Model parameter range limit	$\pm 2\sqrt{\lambda_i}$
k points in profile on either side of the landmark point	4
Resolution levels	1
n_s Points to evaluate on either side of the landmark point	6
N iteration per level	20

3   Comparative Performance Evaluation

To validate the efficiency of the proposed method, we compared its performance with other well-established methods. For that purpose, two quantitative measures comparing segmentation results of various methods with the manual segmentations (ground truth) are calculated. The first one is the global overlap percentage

$$\Omega = TP / (TP+FP+FN) \tag{4}$$

where TP stands for true positive (area correctly classified as object), FP for false positive (area incorrectly classified as object), and FN for false negative (area incorrectly classified as background). A second validation criterion measures the distance between the two curves. For each point on the automated contour, the closest distance to the manual segmentation contour is computed. The averaged value along the contour yields the 1-D validation measure δ.

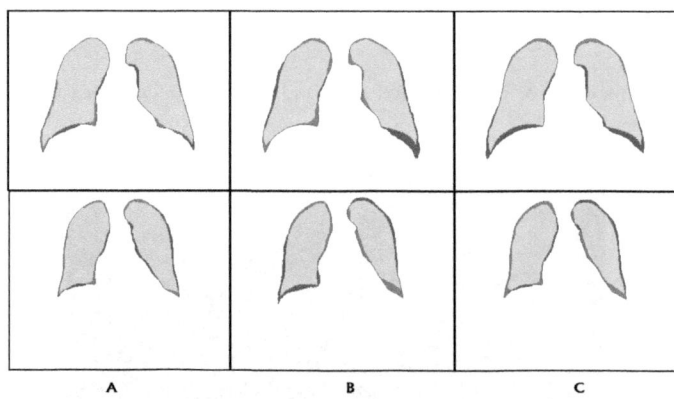

Fig. 7. Lung segmentation results of A) proposed method, B) ASM method, C) AAM method. First row: Case of Figure 5, second row: Case Figure 6. Yellow areas: True positive (TP) area correctly classified as lung. Red areas: False positive (FP) area incorrectly classified as lung. Blue areas: False negative (FN) area incorrectly classified as background.

Figure 7 shows the lung segmentation results of the proposed method, ASM method [3], and AAM method [16] for the cases of Figures 3 and 4. The parameter setting for ASM method are the same as of Table 1, except the number of levels, which is set to 4. The proposed method produced better results by having less FP and FN classification errors.

Table 2 summarizes the overlap percentage and contour distance (in pixels) for the three methods for the 247 images of JSRT database. The proposed method produced better results for both measures. A weakness of the ASM segmentation method is that it uses an iterative optimization process that requires proper initialization in order to converge to the correct solution. During search, landmark points are displaced on lines perpendicular to the current contour. Convergence to the true solution is likely to fail if the distance between the true location of a landmark and its corresponding search line is too large. Therefore, initialization using the iterative binarization method caused ASM to produced better results because it reduced the distance to the true landmark positions.

A Matlab implementation was used on a 2.8-GHz Intel Pentium processor. Our algorithm needed approximately 12 and 14 seconds to segment one image using 22 and 28 landmarks, respectively, compared to 18 and 22 seconds for ASM using 22 and 28 landmarks, and 24 seconds for AAM.

Table 2. Overlap percentage Ω and contour distance δ of the segmentation outputs of the proposed method, ASM method (default initialization using mean shape) [3], and AAM method [16] for the 247 images of JSRT database

	Overlap Percentage Ω	Contour Distance in pixels δ
Ground Truth	1 ± 0.0	0.0 ± 0.0
Proposed method: postprocessing with 22 landmarks Proposed method: postprocessing with 28 landmarks	0.92 ± 0.063 0.94 ± 0.053	3.71 ± 2.35 2.46 ± 2.06
ASM method with 22 landmarks ASM method with 28 landmarks	0.90 ± 0.086 0.92 ± 0.057	4.11 ± 2.21 3.23 ± 2.21
AAM method	0.85 ± 0.095	5.10 ± 4.44

4 Conclusions

We presented an algorithm for the segmentation of lung fields by fusing shape information priors into intensity-based thresholding in an iterative framework. The main contribution is to maximize information utilization by effectively combining intensity information with shape priors. Experimental results demonstrate the effectiveness of the algorithm in comparison with other algorithms.

References

[1] Brown, M.S., Wilson, L.S., Doust, B.D., Gill, R.W., Sun, C.: Knowledge-based method for segmentation and analysis of lung boundaries in chest x-rays images. Comput. Med. Imag. Graph. 22, 463–477 (1998)
[2] Cootes, T.F., Edwards, G.J., Taylor, C.: Active appearance models. IEEE Trans. Pattern Ana. Mach. Intell. 23(6), 681–685 (2001)

[3] Cootes, T.F., Taylor, C.J., Cooper, D.H., Graham, J.: Active shape models — their training and application. Comput. Vis. Image Understand. 61(1), 38–59 (1995)

[4] Cootes, T.F., Taylor, C.J.: Statistical models of appearance for computer vision Wolfson Image Analysis Unit. Univ. Manchester, Manchester (2001)

[5] Duryea, J., Boone, J.M.: A fully automatic algorithm for the segmentation of lung fields in digital chest radiographic images. Med. Phys. 22(2), 183–191 (1995)

[6] Ginneken, B.V., Frangi, A.F., Staal, J.J., Romeny, B.M.T.H., Viergever, M.A.: Active shape model segmentation with optimal features. IEEE Trans. Med. Imag. 21(8), 924–933 (2002)

[7] Ginneken, B.V., Haar Romeny, B.M.: Automatic segmentation of lung fields in chest radiographs. Med. Phys. 27(10), 2445–2455 (2000)

[8] Ginneken, B.V., Haar Romeny, B.M., Viergever, M.A.: Computer-Aided Diagnosis in Chest Radiography: A Survey. IEEE Trans. Med. Imag. 20(12), 1228–1241 (2001)

[9] Ginneken, B.V., Katsuragawa, S., Haar Romeny, B.M., Doi, K., Viergever, M.A.: Automatic Detection of Abnormalities in Chest Radiographs Using Local Texture Analysis. IEEE Trans. Med. Imag. 21(2), 139–149 (2002)

[10] Ginneken, B.V., Stegmann, M.B., Loog, M.: Segmentation of anatomical structures in chest radiographs using supervised methods: A comparative study on a public database. Med. Image Anal. 10(1), 19–40 (2006)

[11] Gleason, S.S., Paulus, M., Johnson, D., Sari-Sarraf, H., Abidi, M.A.: Statistical-based deformable models with simultaneous optimization of object gray-level and shape characteristics. In: Proc. 4th IEEE Southwest Symp. Image Anal. Interp., pp. 93–95 (2000)

[12] McNitt-Gray, M.F., Huang, H.K., Sayre, J.W.: Feature selection in the pattern classification problem of digital chest radiographs segmentation. IEEE Trans. Med. Imag. 14(3), 537–547 (1995)

[13] Seghers, D., Loechx, D., Maes, F., Suetens, P.: Image segmentation using local shape and gray-level appearance models. In: Proc. SPIE, vol. 6144, p. 614401

[14] Seghers, D., Loeckx, D., Maes, F., Vandermeulen, D., Suetens, P.: Minimal Shape and Intensity Cost Path Segmentation. IEEE Trans. Med. Imag. 26(8), 1115–1129 (2007)

[15] Shi, Y., Qi, F., Xue, Z., Chen, L., Ito, K., Matsuo, H., Shen, D.: Segmenting Lung Fields in Serial Chest Radiographs Using Both Population-Based and Patient-Specific Shape Statistics. IEEE Trans. Med. Imag. 27(4), 481–494 (2008)

[16] Shiraishi, J., Katsuragawa, S., Ikezoe, J., Matsumoto, T., Kobayashi, T., Komatsu, K., Matsui, M., Fujita, H., Kodera, Y., Doi, K.: Development of a digital image database for chest radiographs with and without a lung nodule: Receiver operating characteristic analysis of radiologists' detection of pulmonary nodules. Amer. J. Roentgenol. 174, 71–74 (2000)

[17] Stegmann, M.B., Ersboll, B.K., Larsen, R.: FAME—A flexible appearance modeling environment. IEEE Trans. Med. Imag. 22(10), 1319–1331 (2003)

A 3D Tool for
Left Ventricle Segmentation Editing

Samuel Silva, Beatriz Sousa Santos, Joaquim Madeira, and Augusto Silva

Department of Electronics, Telecommunications and Informatics,
Institute of Electronics Engineering and Telematics,
University of Aveiro, Portugal
`sss@ua.pt`

Abstract. Image segmentation has a very important role in many application areas, such as medical imaging. Even robust segmentation methods cannot deal with the wide range of variation observed, for example, in shape and orientation of an anatomical structure. Given the need to accomplish accurate segmentations in order to perform quantitative measurements or compare structures in different time instances, it is important to have tools which allow easy segmentation editing/correction by experts.

In 3D images (e.g., obtained using CT scanners) performing segmentation editing of regions which span several slices might be a tiresome task if it has to be done slice-by-slice with a 2D tool.

This article presents a 3D segmentation editing tool, to be applied to left ventricle segmentations, which enables radiographers to correct segmentations provided by an automatic method.

1 Introduction

Image segmentation is a common step in numerous application areas. Segmentation in medical imaging is quite challenging as completely automatic methods fail to work in every possible situation due to the natural biological variation. Therefore, as important as image segmentation algorithms are the tools which allow user interaction (a survey can be found in Olabarriaga et al. [1]) to guide the method or correct the results. Segmentation editing is one of those features which quite often do not deserve much attention as the focus is, in general, on the automatic steps of the segmentation algorithm [2]. Even though, in practice, editing tools are provided for all segmentation tasks, their suitability is sometimes neglected and, although they allow corrections to be performed, they require a large amount of work and time. One such example is editing 3D segmented regions on a slice-by-slice basis by contour manipulation [3].

Following on work carried out [4], concerning left ventricle (LV) segmentation from cardiac angiography exams obtained using multiple detector-row computerized tomography (MDCT), a software application [5] integrating the segmentation method and providing left ventricle function analysis is being developed using the Medical Imaging Toolkit (MITK) [6]. The automatic segmentations

A. Campilho and M. Kamel (Eds.): ICIAR 2010, Part II, LNCS 6112, pp. 79–88, 2010.

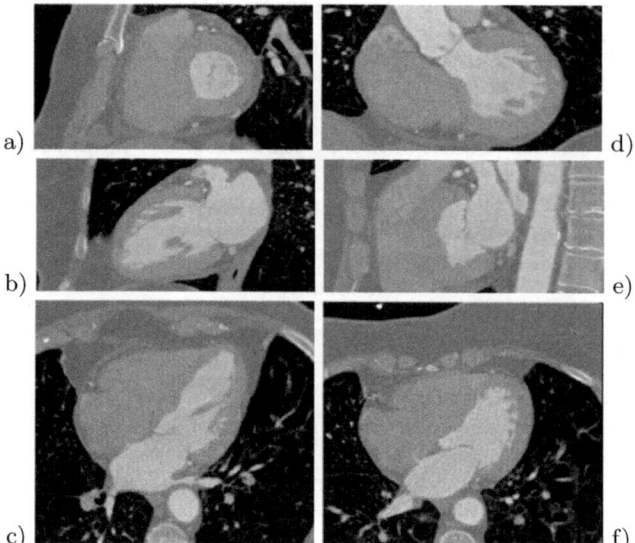

Fig. 1. Common view planes used by the radiographers to analyse LV data: a) LV short-axis, b) 2-chambers (left atrium and ventricle) and c) 4-chambers (both ventricles and both atria) compared with the the usual orthogonal view planes d) coronal, e) sagital and f) axial

provided sometimes need to be corrected. In the presence of poor segmentations, some high level parameters can be used to correct them but the most common problems can be easily solved with some editing (as reported by radiographers on a preliminary evaluation of the segmentation method presented in [4]). Thus, an editing tool must be provided which is easy and intuitive to use and is well suited for the task at hand. This must consider the 3D characteristics of the segmented LV and the kind of LV analysis performed by the radiographers using view planes which differ from the usual orthogonal planes (axial, sagital and coronal) as depicted in figure 1.

The MITK library already provides a segmentation editing tool which presents two limitations: it only supports editing on the usual orthogonal planes (axial, sagital and coronal) and is a 2D tool (i.e., editing only possible slice-by-slice). Since each image volume used for left ventricle segmentation is approx. 512 × 512 × 256 and the LV spans over a considerable number of slices, it would be very tiresome to perform slice-by-slice segmentation editing.

Editing is also important in our case since, for each cardiac exam, the left ventricle must be segmented for 12 phases along the cardiac cycle (i.e., 12 image volumes taken from systole to diastole): in case of a segmentation problem, if the user is allowed to edit a first automatic segmentation of one of those phases, that information can be used to improve the remaining 11 segmentations, therefore reducing the amount of user intervention needed.

We present a 3D tool for left ventricle segmentation editing of first segmentations provided by a segmentation method developed by the authors [4]. This tool is included in a segmentation protocol used to obtain validated left ventricle data for analysis. Even though this tool was developed (and is presented) with a particular application to left ventricle segmentations it can be of use in any situation which deals with similar segmented regions.

The following section presents the editing tool developed. Section 3 presents a simple evaluation performed to assess if the tool does bring some advantages to the editing task when compared with a 2D tool. Finally, some conclusions are presented in section 4.

2   3D Editing Tool

When designing the presented 3D editing tool two main goals were taken into consideration: first, the tool needed to allow easy correction of the typically detected problems (see figure 2 for some examples), and second, the tool should support real-time interaction. This second goal limited the complexity of the operations performed during editing in order to minimize the associated computational cost. Furthermore, the tool should provide editing in any view plane and work by using the mouse as interaction device.

2.1   Voxel Mask Editing

This method works at voxel level by adding/removing voxels from the segmented region. The editing brush has a spherical shape thus providing 3D editing.

Two editing modes are available: ADD voxels to region and REMOVE voxels from region. Editing mode selection is performed automatically according to the voxel value at the center of the brush when the editing operation is started. If it is an active voxel (i.e., a voxel part of the current segmented region), the tool is set to ADD mode (figure 3a) and if it is an inactive voxel (i.e., not part of the current segmented region) the tool is set to REMOVE mode (figure 3b). The tool keeps the current mode until the editing step is finished (i.e., mouse

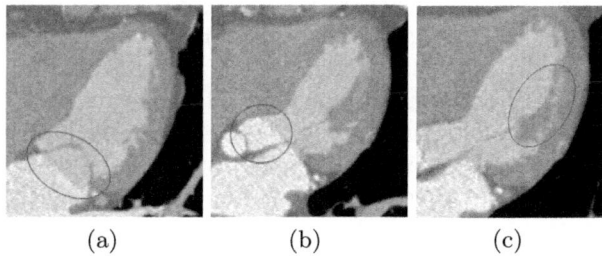

(a)   (b)   (c)

Fig. 2. Common segmentation problems detected in the automatic segmentation method output: a) segmentation beyond the mitral valve; b) no inclusion of the outgoing tract and c) papillary muscle not included in the blood pool

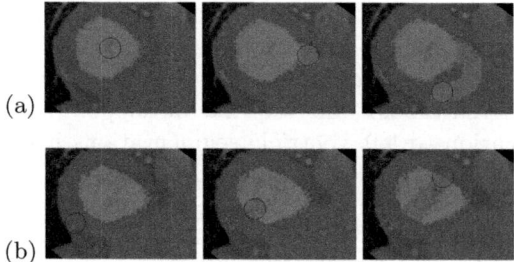

Fig. 3. Two modes available for bit mask editing. In the first mode (a), ADD, the editing operation starts on an active region and voxels are added to the object as the tool is moved. In the second mode (b), REMOVE, the editing operation starts on an inactive region and voxels are removed as the tool is moved.

button is released). This automatic selection mode is based on the observation that when the user wants to add voxels to the border of the segmented volume it is natural to start the editing from inside the object and the contrary, i.e., from outside the object, when the goal is to remove voxels. This also guarantees (although some awkward situations may arise) some continuity of the segmented region. Since we know the left ventricle is a closed region the user must always travel from within it to where he wants to add voxels. This does not, however, solve the problem of isolated regions resulting from voxel removal around them which is dealt, in our case, by a post-processing step.

To speed the voxel mask modification operation a neighborhood iterator is used, provided by the ITK library [7]. It allows the definition of a 3D neighborhood region (a cube centered on the desired voxel) in which it is possible to activate only the neighborhood voxels we are interested in visiting. In our case, voxels are activated to obtain a sphere shaped neighborhood (according to the editing tool radius chosen). Since the neighborhood is defined using offsets towards the central voxel it can be easily relocated without having to re-set the active neighborhood voxels. Therefore, the iterator initialization needs only to be performed once for each tool radius desired.

The tool radius can be chosen from a limited set of values established based on the characteristics of the left ventricle and image resolution. Given that there are no particularly thin regions there is no need for very small radius tools. It should only be guaranteed that a small enough tool exists to be used on the outgoing tract. On the other hand, a large radius tool could be used to edit the segmentation on the mitral valve region but, if image resolution is high (with a very small voxel size), the number of voxels in the neighborhood starts to increase rapidly with the radius, thus influencing both the initialization time of the tool (to set the spherical neighborhood) and its interactivity during editing.

2.2   Surface Editing

Another option for editing is to work with the surface of the segmented region obtained, for example, by using the marching cubes algorithm [8] provided by the

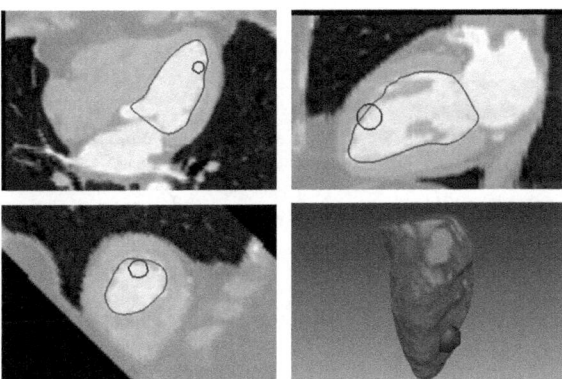

Fig. 4. Segmentation editing by surface deformation. The editing tool is a sphere and can be used in any of the view planes. It appears in different sizes depending where the view plane intersects the sphere. On the bottom right image a 3D view of the surface (in red) and sphere (in blue) are shown.

VTK library [9] thus obtaining a polygonal representation of the surface. This has the advantage that it does not occlude the image with the segmentation mask as much as the voxel mask and does not suffer from isolated regions being created when separated from the main region by removing voxels in-between. Intersecting the surface of the segmented region with different view planes produces contours which can be intuitively adjusted by the user.

Notice that generating contours (instead of a surface) would also be possible by applying an image cutter to the voxel volume with the desired orientation and extracting the contours but, as the user must be able to choose and change to any view plane she/he finds necessary, it would result in a method of much higher complexity.

In this case, the editing tool consists of a sphere, moved by the user, with its center at cursor position, which deforms the polygonal surface when pressed against it. It appears in each visualization plane as a circle (see figure 4) and can be used freely in any of them. Surface deformation is obtained by changing the polygonal surface vertices position in order to keep them always at no less than sphere radius distance from the tool center (i.e., outside the sphere).

When the editing sphere approaches the surface, the closest vertex (if any at less than the sphere *radius* distance) to the sphere center is determined and all its neighbors which are also found inside the sphere are determined. Since computing vertex neighborhoods has a high computational cost, particularly if the sphere radius is large, the first seven neighborhoods (one-ring, two-ring, ...) are pre-computed and stored in memory. This has also the advantage of being a lot faster than the individual neighborhood computations since previously computed neighborhoods for some of the vertices can be used to speed the computation of the remaining neighborhoods.

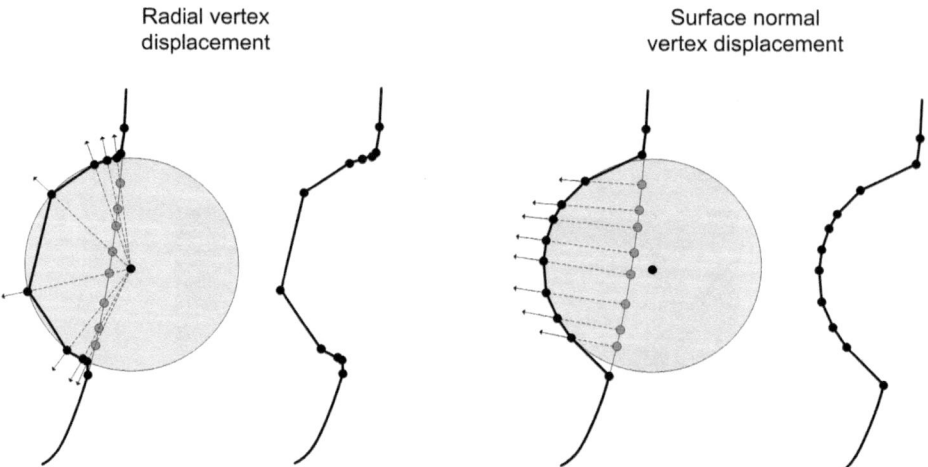

Fig. 5. Two different options for defining vertex displacement direction: on the left, vertices are displaced radially in a direction defined by the sphere center and their initial position resulting in vertex dispersion; on the right, vertices are displaced in the direction given by the original surface normal at their original position which keeps vertex density

After finding the list of vertices that must be displaced it is necessary to determine the direction of their displacement. A simple idea is to displace them radially, i.e., using the normals to the sphere surface (figure 5a). This method has the disadvantage of creating a low density of vertices on the surface region which suffers the largest displacement. It is easy to understand that, very rapidly, the vertex density in that region will be so low that further displacements will be impossible. It is also important to notice that with only a few vertices it is harder to properly deform the surface to accurately contain the desired region. This method will also easily allow surface self intersections as the vertices will move freely in every possible direction.

Thus, it is important to design a vertex displacement method which reasonably preserves vertex density and reduces the chance of surface self-intersections, at least for small correcting operations. This was accomplished by displacing the vertices along the corresponding surface normal, i.e., the vertex is always moved following the direction of the original surface normal at its position. As can be seen in figure 5b, this method results in better vertex density preservation and is less prone to surface self-intersection although sharp edges might be a problem. As the vertices always move in the same direction it is possible to easily revert any of the editing operations by applying the sphere from the opposite side.

Since the vertex neighborhoods are pre-computed the impact of a large sphere radius on interactivity is small (within reasonable radius limits) and, thus, the user is allowed to use a wide range of radius values suited for each situation as illustrated in figure 6.

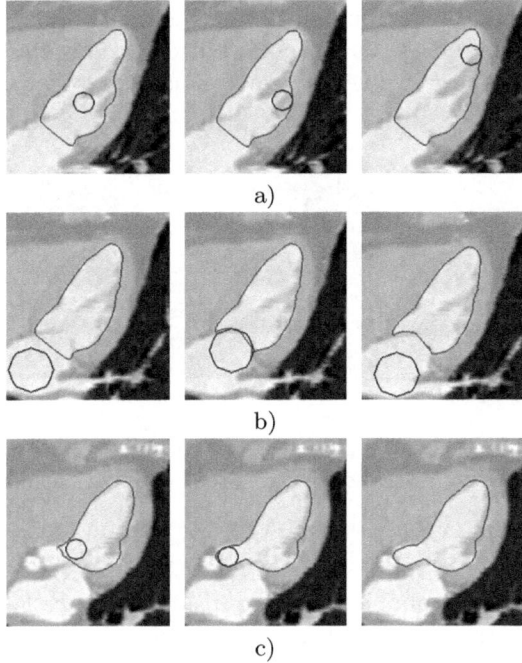

a)

b)

c)

Fig. 6. Different editing situations using surface deformation: a) Correction of the segmentation to include the papillary muscles inside the surface; b) a larger tool radius has been chosen to correct the segmentation on the mitral valve region and c) tool radius has been reduced to an adequate size to correct the outgoing tract

It is important to notice that in our particular situation the left ventricle will never need significative corrections: if a really bad segmentation is performed the user can change some high level parameters to obtain better results before editing. For directly editing really bad segmentations this method would be harder to use, e.g., if the surface had to be corrected to a much larger or less smooth surface, as the number of vertices is fixed.

2.3   Mixed Editing

The methods presented above can also be used interchangeably, if necessary, although changing between them implies some delay (around 4 seconds on a common desktop computer). If initially editing the voxel mask, when changing to surface editing, a new surface must be generated from the current voxel mask and vertex neighborhoods must be computed. When reverting from surface editing to voxel editing the voxel mask contained inside the current surface must be computed.

Computing a new surface from the current voxel volume is accomplished, as previously mentioned, by using the marching cubes algorithm. To obtain the voxel mask contained inside a surface a method provided by VTK is used to

obtain a stencil from the polygonal data which is then applied to a voxel volume
the size of the original image, as a "cookie cutter", to obtain the desired result.

Figure 7 shows some examples of conversion from voxel mask to polygonal
surface and vice-versa.

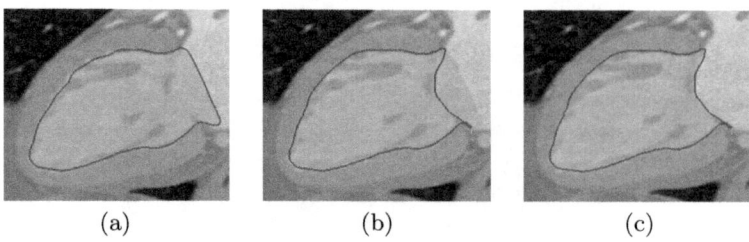

(a)   (b)   (c)

Fig. 7. (a) Surface generated from the voxel mask using the marching cubes algorithm;
(b) the surface is deformed using the presented tool; (c) the voxel mask contained inside
the surface is computed

Computing the voxel volume contained inside a surface is also important to
compute features of the segmented region such as regional and global blood volumes (by counting the voxels and multiplying by voxel volume) for left ventricle
function analysis.

3   Evaluation

To assess the applicability and quantify the advantages of the proposed tool a
simple first evaluation has been conducted concerning the time taken to perform
the same editing task using voxel mask (3DV) and surface editing (3DS) and
the 2D editing tool provided by MITK.

Considering the common editing operations which will be performed by the
radiographers (figure 2), two tasks have been chosen for this evaluation: task 1
consisted in adjusting the segmentation mask to the mitral valves which implied
removing voxels from the initial segmentation; task 2 consisted in adjusting the
segmentation mask to the LV wall adding voxels to the initial segmentation.

Three users participated in this evaluation. In a first stage, they received a
short explanation about the different tools and were allowed to use them (same
amount of time for each tool, per user; user A was allowed to train longer) to get
acquainted with the different features. On a second stage, the users were asked
to perform both tasks with the three available methods.

As can be observed in table 1 the developed tool clearly allows performing the
considered tasks in less time than a typical 2D editing tool available in MITK.
This shows that with a simple 3D tool, focused on the characteristics of the
region and tasks to perform, a considerable gain was attained. For the evaluated
tasks the average time taken to accomplish them was significantly greater for the
2D tool and smaller for both 3D editing modes, which yield similar task times.

Table 1. Time (in seconds) taken to complete an editing task using the presented tool
(3DV - voxel editing and 3DS - surface editing) and a 2D tool

User	Task 1			Task 2		
	2D	**3DV**	**3DS**	**2D**	**3DV**	**3DS**
A	236	35	50	333	42	61
B	594	78	99	554	88	121
C	600	105	107	768	119	90
Avg.	**477**	**73**	**85**	**552**	**83**	**91**

During and after the evaluation users were invited to comment on the different tools. They were unanimous in stating that having to set the editing mode (add or subtract) in the 2D tool was confusing and preferred our tool which sets the mode based on cursor position. Even though times were similar for both 3D tools users preferred voxel editing (3DV). This was probably due to occasional latency in surface editing resulting from a greater computational load.

User A trained longer and managed to accomplish all tasks in less time than the other users. This gives some sign that, with training, performance with our tool can be further improved to even smaller task times. Nevertheless, a reduced number of users was involved in this study and further evaluation is necessary to clarify this aspect.

4   Conclusions

This article presents a simple 3D tool for editing left ventricle segmentations. Its purpose is to allow radiographers to rapidly correct first segmentations provided by an automatic segmentation method. This approach, when compared with the usual 2D slice-by-slice method, has the advantage of being faster and more reliable as it reduces the probability of uncorrected slices being left behind and ensures a higher degree of coherence from slice to slice.

The tool has been developed paying attention to the characteristics of the target data (left ventricle segmentations) and necessary editing operations allowing for simplicity and good suitability, thus presenting a smaller semantic distance towards the tasks envisaged by the user.

Voxel mask editing is very fast, thanks to the neighborhood operator, and allows good interactivity and results. By applying a post-processing filter at the end it is possible to discard isolated voxels missed by the user.

The surface based method works well for small and medium scale modifications, such as those presented in figure 6, but larger modifications or a large number of editing operations around the same region might sometimes result in awkward situations as the one illustrated in figure 8. Apart from possible surface self-intersections this can be just a consequence of vertices below the current view plane starting to appear in it, due to a large displacement or their normal having an orientation that favors such an event. Nevertheless, it has a bad impact on users and should be avoided.

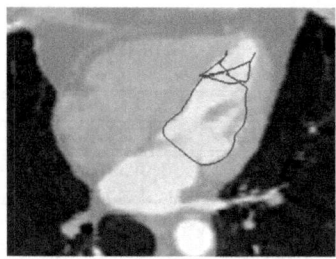

Fig. 8. A worst case example of a problem which might occur after a modification using an editing tool with a very large radius

Smoothing might improve the results for surface editing and some experiments with simple smooth operations (interactivity must be kept) applied to the affected neighborhood (or its border) have already been performed with some promising results, but it is still not clear when and where to apply them and how to detect problematic regions. So, even though the current surface editing feature is already useful its robustness and performance can still be improved.

Acknowledgements

The first author's work is funded by grant SFRH/BD/38073/2007 awarded by the portuguese Science and Technology Foundation (FCT).

References

1. Olabarriaga, S.D., Smeulders, A.W.M.: Interaction in the segmentation of medical images: A survey. Medical Image Analysis 5(2), 127–142 (2001)
2. Kang, Y., Engelke, K., Kalender, W.: Interactive 3D editing tools for image segmentation. Medical Image Analysis 8(1), 35–46 (2004)
3. Jolly, M.P.: Automatic segmentation of the left ventricle in cardiac MR and CT images. International Journal of Computer Vision 70(2), 151–163 (2006)
4. Silva, S., Madeira, J., Silva, A., Sousa Santos, B.: Left ventricle segmentation from heart MDCT. In: Araujo, H., Mendonça, A.M., Pinho, A.J., Torres, M.I. (eds.) IbPRIA 2009. LNCS, vol. 5524, pp. 306–313. Springer, Heidelberg (2009)
5. Silva, S., Madeira, J., Sousa Santos, B., Silva, A.: A software application to support the development and evaluation of visualization methods for left ventricle analysis. In: Proc. 15th Portuguese Conference on Pattern Recognition (2009)
6. MITK: Medical imaging toolkit (March 2010), http://www.mitk.org
7. ITK: Insight segmentation and registration toolkit, http://www.itk.org/ (March 2010)
8. Lorensen, W., Cline, H.: Marching cubes: A high resolution 3D surface construction algorithm. Computer Graphics 21(4), 163–169 (1987)
9. Schroeder, W., Martin, K., Lorensen, B.: The Visualization Toolkit An Object-Oriented Approach to 3D Graphics, 4th edn. Kitware, Inc. (2004)

Myocardial Segmentation Using Constrained Multi-Seeded Region Growing

Mustafa A. Alattar[1], Nael F. Osman[1,2], and Ahmed S. Fahmy[1]

[1] Center of Informatics Science, School of Communication and Information Technology,
Nile University, Cairo, Egypt
[2] Radiology Department, Johns Hopkins University, Baltimore, Maryland USA
{Mustafa.Elattar,ASFahmy}@nileu.edu.eg, Nael@jhu.edu

Abstract. Multi-slice short-axis acquisitions of the left ventricle are fundamental for estimating the volume and mass of the left ventricle in cardiac MRI scans. Manual segmentation of the myocardium in all time frames per each cross-section is a cumbersome task. Therefore, automatic myocardium segmentation methods are essential for cardiac functional analysis. Region growing has been proposed to segment the myocardium. Although the technique is simple and fast, non uniform intensity and low-contrast interfaces of the myocardium are major challenges of the technique that limit its use in myocardial segmentation. In this work, we propose a modified region growing technique that ensures reliable and fast myocardial segmentation of short-axis images. The proposed technique initializes the region growing process using different seed points. Then two types of spatial constraints are used to guarantee fast and accurate segmentation. The technique has been tested and validated quantitatively using a large number of images of different qualities. The results confirm the reliability and accuracy of the proposed technique.

Keywords: Region Growing, Segmentation, Cardiac MRI, Left Ventricle.

1 Introduction

Imaging the heart using standard cine MRI sequences is an important tool to evaluate the cardiac global and regional function. This includes estimating the ejection fraction, left ventricle (LV) mass and volume, wall-thickness and wall-thickening. These parameters are usually estimated from datasets that typically include 3-6 short-axis slices of the heart acquired over the entire cardiac cycle with frame rate equal to 20-35 image/cycle [1]. In these images, the LV appears as a doughnut-shape gray area enclosing a brighter region of the blood and surrounded by a number of regions of different intensities (e.g. lung, liver, RV cavity) as shown in figure (1).

Manual segmentation of the contours in all images through different slices is a cumbersome task. Therefore, methods were proposed to automatically or semi-automatically segment the contours from short-axis images. In literature, a number of LV segmentation techniques have been proposed. This includes region growing, active deformable models and clustering techniques, etc [2, 3]. In this work, we focus on the region growing technique as a powerful, classic and simple technique for myocardial segmentation.

A. Campilho and M. Kamel (Eds.): ICIAR 2010, Part II, LNCS 6112, pp. 89–98, 2010.

1.1 Standard Region Growing Algorithm (RG)

In standard region growing, an initial region composed of one seed point starts to grow iteratively by adding more neighboring pixels that satisfy some predefined criterion. This criterion can be based on intensity, texture, or edge information. One simple yet popular criterion is the intensity similarity among the region pixels. Let R_i be a set of pixels composing the growing region at the ith iteration. Initially, the region is composed of a single seed point. That is,

$$R_0 \equiv \{(x^0, y^0)\} \tag{1}$$

Then at ith iteration, the region is given by this equation,

$$R_i = \begin{cases} R_{i-1} \cup \{(x,y)\} & if \ |\mu_{R_{i-1}} - I(x,y)| < T \\ R_{i-1} & otherwise \end{cases} \tag{2}$$

Where T is a predetermined threshold, $I(x, y)$ is the intensity of the candidate pixel (x, y) and μ_R is the mean intensity of the pixels inside region R defined as,

$$\mu_R = \frac{1}{n} \sum_{\forall (x,y) \in R} I(x,y) \tag{3}$$

Where n is the cardinality of the set R.

1.2 Limitations of Region Growing

Despite its simplicity and speed, a major limitation of the technique occurs at elevated noise levels and/or intensity nonuniformity of the region to be segmented [2]. In myocardial cine MR images, severe intensity variation are frequently encountered due to field inhomogeneity at the myocardium-lung and/or the flow and respiration artifacts. This was one of the reasons that region growing has not been used in left ventricle segmentation [4].

For example, figure (1.a) shows two images, the first is artifact free where the intensity variation inside the myocardium is very limited. On the other hand, figure (1.b) shows large intensity variation inside the myocardium due to respiratory motion. This significantly degrades the performance of RG as will be shown later.

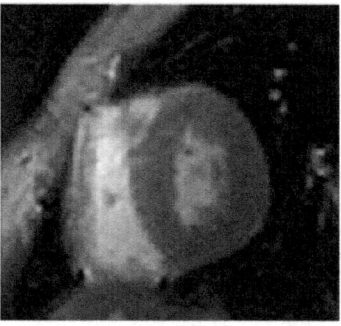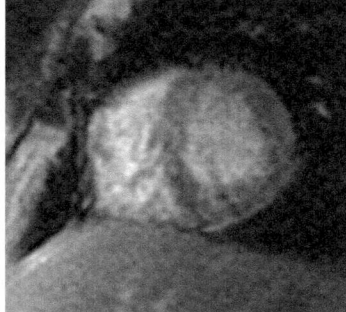

Fig. 1. Short-axis image of the heart with a good quality (Left) Another image with inhomogeneity artifacts (Right)

Another limitation of using region growing in myocardial segmentation is the low-contrast interface between the LV and the liver which causes the growing region to extend beyond the actual borders of the myocardium [5].

In this work, we propose to initialize the region growing with a number of seed points spread inside the myocardium at equi-angle separations. This guarantees correct segmentation even if the myocardial pixel intensities are severely non uniform. In order to reduce the computation time, the growing region of each seed is constrained by an automatically determined surrounding sector of the myocardium.

The sectors of the different seed points are overlapping to guarantee the continuity of the extracted myocardium segment. In addition, using a priori knowledge of the intensity profile along radial lines of the myocardium, control points are set automatically to determine the epicardium near low-contrast interfaces (e.g. with liver).

2 Methods

2.1 Multi-Seeded Region Growing (MSRG)

In MSRG, m seed points are used to initialize m small regions whose union would form the segmented myocardium at the end of the growing operation.

For a given time frame, the seed points are selected automatically from an estimate of the interior myocardial contour. The latter can be estimated as the mean of the epi and endocardium contours of the previous time frame. We choose to select the seed points at equi-angles on the estimated interior contour as shown in figure (2).

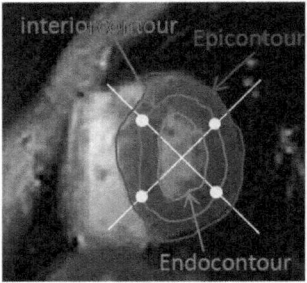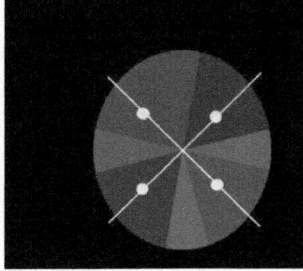

Fig. 2. Short axis image with epi, interior and endocontour are drawn (Left) 4-constraining overlapped masks with their seed points (Right)

To avoid long computation time, constraining masks, CM^j, have been used to limit the circumferential and radial growing of the different regions, R^j. The masks are taken as overlapping sectors covering the entire area of the LV with one seed point lies inside each sector. Each region is thus allowed to grow according to the following equation,

$$R_0^j = \{(x_0^j, y_0^j)\}, \tag{4}$$

$$R_i^j = \begin{cases} R_{i-1}^j \cup \{(x,y)\} & if \; |\mu_{R_{i-1}^j} - I(x,y)| < T \\ & and \; (x,y) \in CM^j \\ R_{i-1}^j & otherwise \end{cases} \qquad (5)$$

Where (x_0^j, y_0^j) is the seed for region R^j and $j=1: m$. After the termination of the iterations, the segmented myocardium is taken as the union of the individual regions.

2.2 Epicardial Control Points

The resulting region from the MSRG technique may include non-myocardial tissues due to low contrast interfaces. This problem occurs mainly at the outer boundary. To avoid this problem, control points on the true interface boundaries are identified by means of feature matching and used to constrain the outer contour to the real epicardium. Thirty four datasets have been analyzed offline to learn the true location of the epicardial points at low contrast interfaces as follows.

First, the outer interface between the myocardium and all other tissues were delineated manually and the intensity profiles along radial lines at these interfaces were plotted. Then, these profiles have been classified into three main classes: myocardium-lung, myocardium-liver and myocardium-blood profiles. The mean profile of each class (shown in figure 3.c) is taken as a template for subsequent profile

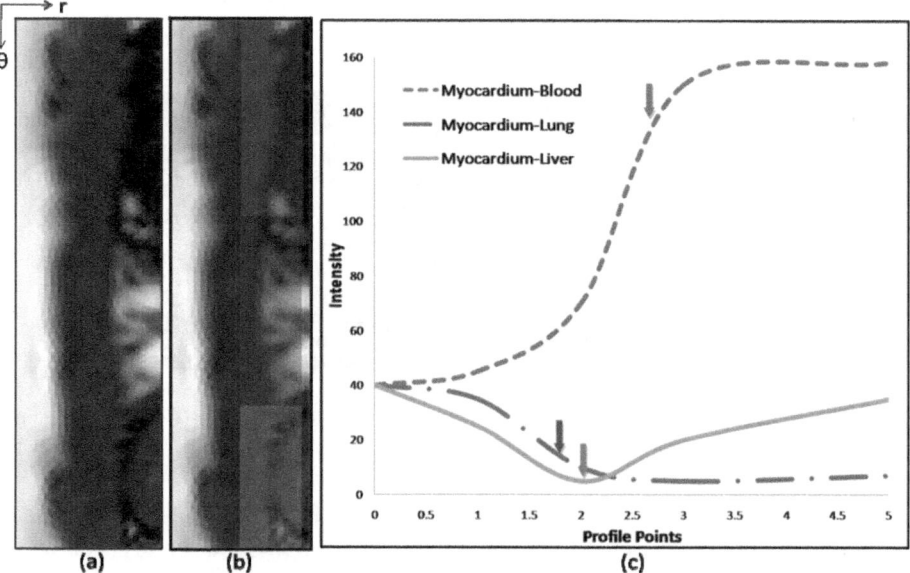

Fig. 3. (a) A myocardium image shown in the polar representation showing different types of interfaces (b) Three regions are highlighted with the same color of their types in the next chart (c) Chart represents the three types of features

matching. The correct interface point along each interface was determined according to a certain criterion depending on the interface type. For the myocardium-liver interface, whose profile appears as inverted Gaussian, the minimum intensity index is set as the true interface point. For the myocardium-lung and myocardial-blood interfaces, the true interface points are determined as the first point on the profile satisfying intensity value smaller than 33% and 66% of the maximum profile intensity, respectively. Those values (33% - 66%) have been concluded from studying the manual segmented contours and founding their positions on the profiles.

After determining interface profile templates and the rules for selecting the true interface points from these templates, the output of the MSRG technique is then processed as follows. The intensity profiles of all points on the outer contour are matched with the three interface templates. If matched with one of the templates, then the type of the interface and thus the location of the true interface points are estimated. This leaves the outer contour of the MSRG segmented region irregular and thus needs some smoothing.

2.3 Refinement of the Contours

After determining the control points, the outer and inner contours of the MSRG segmented region are then refined to smooth sharp bending segments by using a few iterations of standard active contour model [6]. For the outer contour, however, the locations of the control points are enforced unchanged. The complete proposed algorithm is summarized in the flowchart in figure (4).

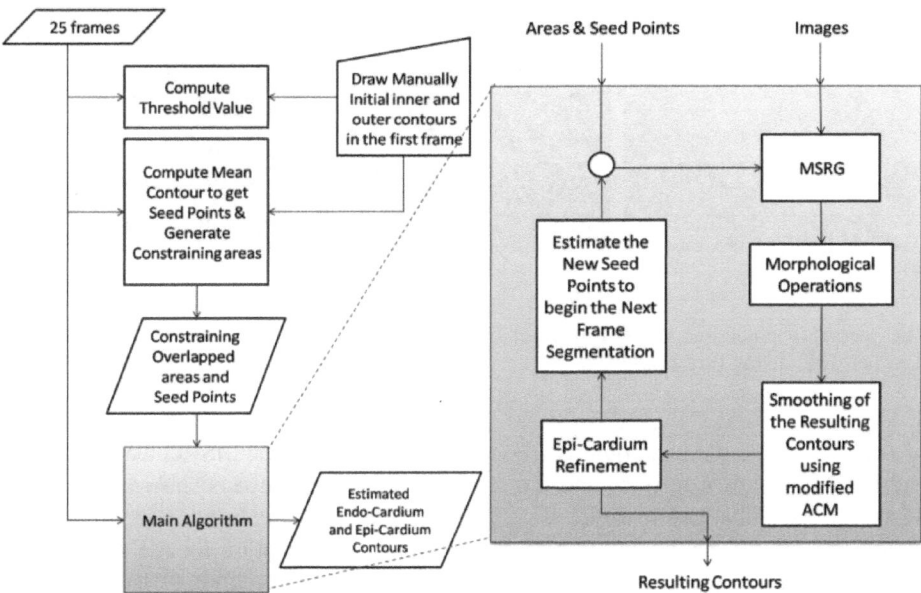

Fig. 4. Flowchart of the whole proposed algorithm (Left) Detailed flowchart of the main algorithm block (Right)

3   Results and Discussion

3.1   Using Simple Region Growing

Figure (5) shows the result of applying simple region growing on a mid-slice short-axis image using threshold value equal to 25. It is obvious that the original image has a good quality and high homogeneity, so that the result is good in the right image but there are some small holes inside the myocardium.

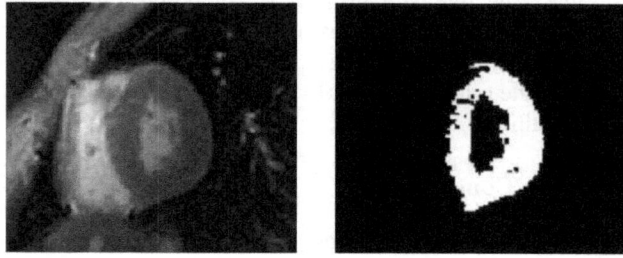

Fig. 5. Short-axis image (Left) Result of applying the simple RG (Right)

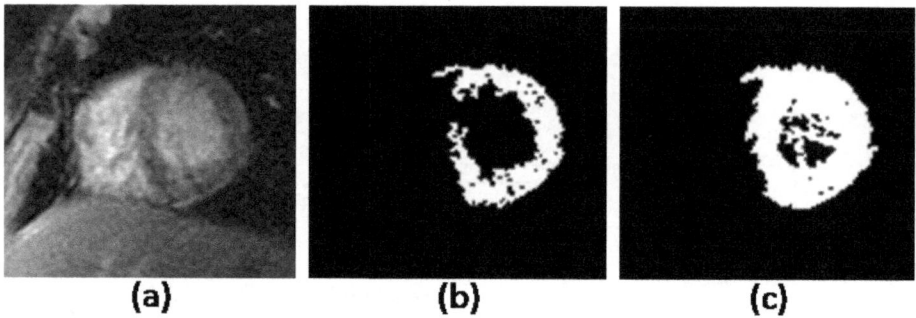

Fig. 6. (a) Original image (b) Result of applying region growing with threshold = 27 (c) Another result using threshold = 52

Another image with bad quality is shown in figure (6.a), the MSRG technique gave bad results as shown in (6.b) and the region discontinuity also is shown. To recover this error the only way to route is to increase the threshold value and the resulted region from this raise is shown in (6.c), there is over estimation for the myocardium tissue and the technique begin to identify the blood in the cavity as tissue. Then the raising in the threshold value will not solve the inhomogeneity problem.

3.2 Using Multi Seeded Region Growing Constrained by Overlapped Sectors

The final result of MSRG is the union of the multiple sectors m resulting as stated before. Then internal and external contours can be segmented using boundary tracing. The result of applying MSRG is shown in figure (7) using no of sectors = 10, overlapping ratio = 0.1 and two different threshold values 21, 29.

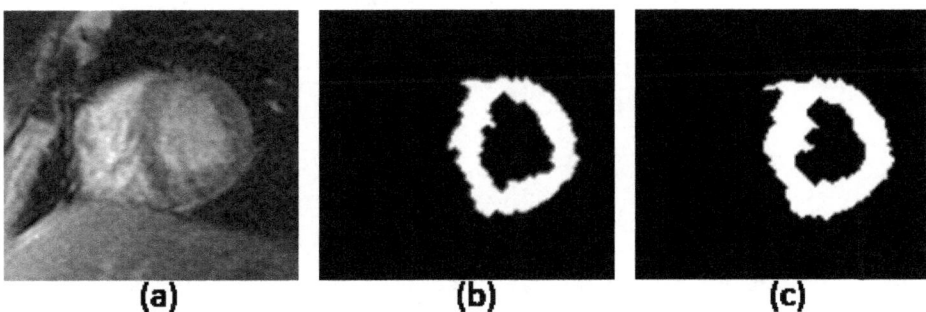

Fig. 7. (a) Original image (b) Result of MSRG with threshold = 21, no. of sectors = 10 and Overlapping Ratio = 0.1 (c) result using threshold = 29

The problem of discontinuity which was caused by image in-homogeneity has been solved but there were some black holes that have been removed using some morphological operation. There are some parameters that affect the performance of MSRG and they are the number of constraining sectors and overlapping ratio between sectors. To study the effect of these parameters on MSRG performance, the MSRG technique was tested with all different and possible values of no. of sectors and overlapping ratio using one hundred selected image [7] representing different qualities, SNRs and homogeneities. True positives and false positives have been computed for each result of RG and MSRG to compare between them.

The segmented inner and outer contours which attached with datasets are available at York University website [7]. It has been found that the mean performance of the MSRG is better than the RG in terms of TP and FP in a specific range of threshold values, as shown in figure (8) and it enables us to pick the best threshold values to perform well. Also the best values deduced from the figure for the overlapping percentage is 10% because it gives partially higher TP and low FP. Another study has been made on the effect of the number of sectors and the overlapping ratio on the MSRG performance.

It has been found that increasing the number of sectors improves the performance and it reaches the steady state at no. of sectors $\geq$ 10 as shown in figure (9). FP will increase if we increase the overlapping percentage and we found that the optimal overlapping ratio from figure (8) and (9) is nearly 10% and more than this percentage it begins to perform worse. Figure (10) shows the final result of applying the proposed algorithm and these results are for 4 series representing four different qualities.

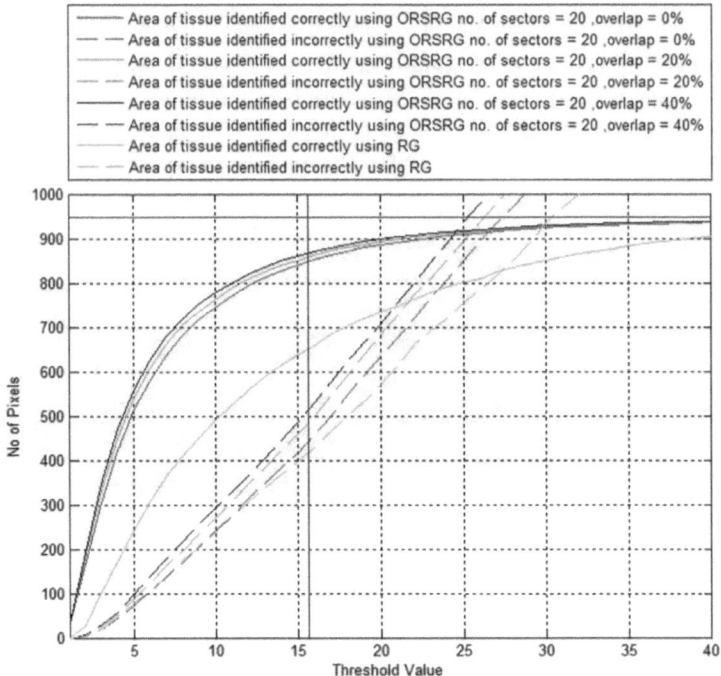

Fig. 8. Performance curves of the RG and MSRG

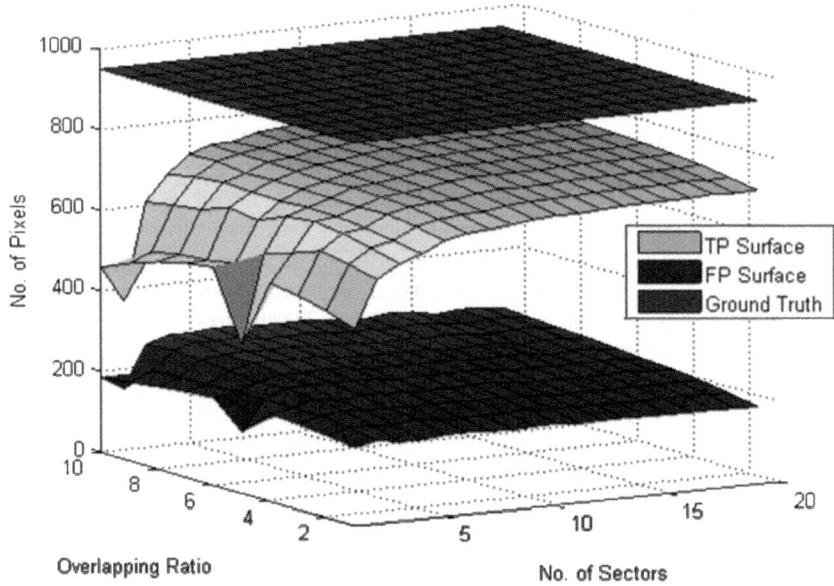

Fig. 9. Three surfaces are representing the ground truth, TP and FP from upper to lower

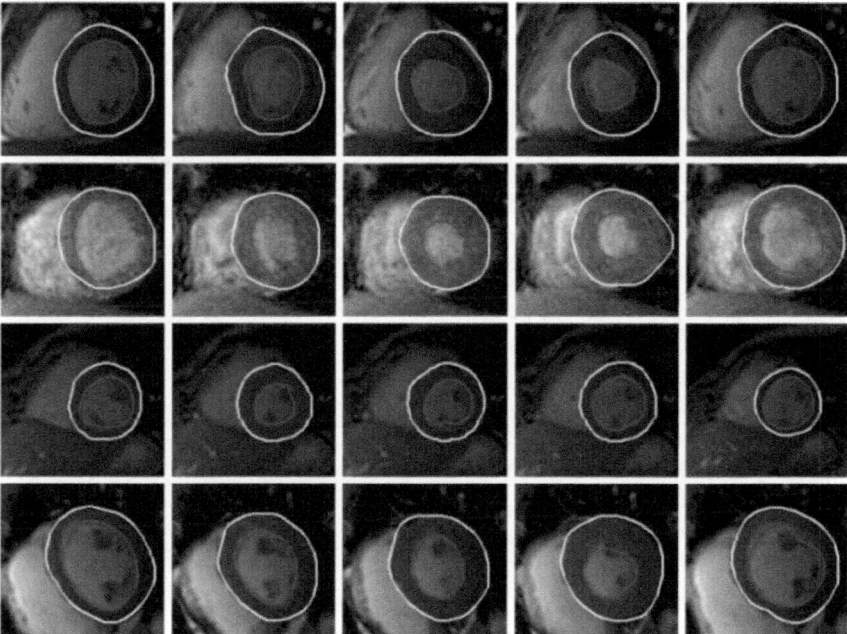

Fig. 10. Final results of the whole algorithm are shown where the white and red contours are representing epicardium and endocardium contour

4 Conclusion

A new method for segmenting the myocardium in CINE SSFP MR Images has been proposed. The MSRG technique takes 112 ms/frame using Matlab code running on 2.26MHz Core2Due processor. The method was tested using 100 images with various qualities. The results show the feasibility of the proposed method for fast and reliable myocardial segmentation.

References

1. Azhari, H., Sideman, S., Weiss, J.L., Shapiro, E.P., Weisfeldt, M.L., Graves, W.L., Rogers, W.J., Beyar, R.: Three dimensional mapping of acute ischemic regions using MRI: wall thickening versus motion analysis. AJP - Heart and Circulatory Physiology (1990)
2. Pham, D.L., Xu, C., Prince, J.L.: A Survey of Current Methods in Medical Image Segmentation. Tech. rep., The John Hopkins University, The John Hopkins University, Baltimore (Janauary 1998)
3. Suri, J.: Computer Vision, Pattern Recognition and Image Processing in Left Ventricle Segmentation: The Last 50 Years. Pattern Analysis and Applications (September 2000)
4. Garcia, A., Vachier, C., Rosset, A., Vallée, J.P.: Multi-criteria seeded region growing for multi-contrast MRI

5. Cerqueira, M.D., Weissman, N.J., Dilsizian, V., Jacobs, A.K., Kaul, S., Laskey, W.K., Pennell, D.J., Rumberger, J.A., Ryan, T., Verani, M.S.: Standardized Myocardial Segmentation and Nomenclature for Tomographic Imaging of the Heart, American Heart Association Writing Group on Myocardial Segmentation and Registration for Cardiac Imaging
6. Kass, M., Witkin, A., Terzopoulos, D.: Snakes - Active Contour Models. International Journal of Computer Vision 1(4), 321–331 (1987)
7. http://www.cse.yorku.ca/~mridataset/

A Level Set Segmentation Method of the Four Heart Cavities in Pediatric Ultrasound Images

Sofia G. Antunes[1], José Silvestre Silva[1,2], and Jaime B. Santos[3]

[1] Department of Physics, Faculty of Sciences and Technology, University of Coimbra, Portugal
[2] Instrumentation Center, Faculty of Sciences and Technology, University of Coimbra, Portugal
[3] Mechanical Engineering Center, Faculty of Sciences and Technology,
University of Coimbra, Portugal
sofigantunes@gmail.com, jsilva@ci.uc.pt,
jaime@deec.uc.pt

Abstract. Echocardiography is the most used medical imaging in pediatric cardiology. It is a fundamental tool to analyze the major heart disease and abnormalities since it is non invasive and simple to use for physicians even when the children are wiggle. Ultrasound images are very noisy, making the segmentation a difficult, not accurate and time consuming task. In this work we propose an automatic segmentation method to extract the four heart cavity boundaries using a new pre-processing algorithm, based on phase symmetry. Experimental results using real echocardiographic images of children show good performance of the proposed method, providing a reliable tool to segment the heart walls that can be helpful for clinical practice.

Keywords: Heart segmentation, echocardiographic images, level set, phase symmetry.

1 Introduction

Medical ultrasonography is an important tool used in newborns due to its non-invasive and fast application. Heart disease can be diagnosed using an echocardiographic examination where pediatricians may visualize the whole heart to analyze its proportions and morphology. Nowadays the best choice is a three dimensional view, which helps the understanding of congenital malformations and defects. An important step in achieving 3D heart reconstruction is the segmentation of the cardiac cavities and the enclosing of their respective volumes. There are several ultrasound segmentation methodologies focusing on techniques developed for medical B-mode ultrasound images [1].

Echocardiographic images have multiplicative noise (mainly speckle noise), artifacts such as shadowing from the lungs, and attenuation which can complicate the analysis task. Thereby the human interactivity is needed to select the correct regions of interest (ROI).

Several pre-processing methods, retaining as much as possible the clinical details, have been proposed to reduce the speckle noise: adaptive filtering [2], Bayesian

A. Campilho and M. Kamel (Eds.): ICIAR 2010, Part II, LNCS 6112, pp. 99–107, 2010.

methods [3], adaptive wavelet thresholding [4] and other approaches such as the wiener filter [5]. It has also been shown that local phase-based methods are a good choice for the pre-processing of ultrasound images due to the absence of speckle or low contrast nature affection [6].

The Active Contours also known as snakes are today one of the most used tools for medical image boundary detection because they tend to smooth the speckle-induced error. Kass *et al.* [7] were the first to propose this method, based on an energy minimization scheme. This classical approach lacks the capability of splitting or merging, which is relevant for the detection of more than one target object in an image. The Geometric models, are a more flexible and convenient solution to overcome the restrictions given by the parametric approach. They were initially proposed by Caselles *et al.* [8] and Malladi *et al.* [9], also known as gradient based algorithms. The level set approach, an implicitly formulation of such models can handle complex shapes and topological changes. It was proposed by Osher *et al.* [10] and has been used intensively since then. In the literature there exist work that use edge based approach using a stopping criterion, region based approach [11-13] which takes into account the statistical information of the image intensity to minimize a global energy and also a combination of both, edge and region based [14-16] that combines edge detection with intensity homogeneity.

The human heart is divided in two ventricles and two auricles. We can find several methods to segment the left ventricle, the most important heart cavity. But, the segmentation of all four chambers is more complex, however an essential element for the 3-D reconstruction. In the present work, we propose a segmentation method to identify simultaneously the four heart chambers. In our method, the echocardiographic images are pre-treated with a phase symmetry algorithm. As far as we know, it is the first time that this pre-processing and the level set segmentation are used in the simultaneous segmentation of the four cardiac cavities, in a successful way. Then, we implement a geometric deformable model with an alternative stopping function that efficiently and shortly identifies the four chambers. Finally we evaluate the performance of the proposed method by comparison with pediatrician's manual segmentation.

The outline of the paper is as follows. In Section 2, we outline the basic level set equations. Concepts and descriptions of our segmentation approach are presented in section 3. Results and discussion of the method applied to individual two dimensional (2D) slices of children heart, and comparison with expert's contours are presented in section 4. Finally, conclusions are presented in Section 5.

2   Mathematical Formulation

In this section a few level set formulations are described which performance was evaluated in order to select the best one to be used with the proposed algorithm.

The main purpose of level set model is to minimize a function, solving the corresponding partial differential equation (PDE) using as numerical method the level set evolution equation. The basis is to evolve a contour (or surface) implicitly by manipulating a higher dimensional function $\phi(x,t)$ where the zero set is used to extract the evolving contour $C = \{x \mid \phi(x) = 0\}$. It is based on geometric measures and the general curve evolution PDE in the level set framework [12] is:

$$\frac{\partial \phi}{\partial t} = | \nabla \phi | F \tag{1}$$

The velocity term F has different evolutions dependent on the authors. In the geometric active contour model [12] F depends on an edge indicator function:

$$F = [div(\frac{\nabla \phi}{|\nabla \phi|}) + v]P \tag{2}$$

where P is the image gradient functional responsible for pushing the model towards image boundaries, attracting in that way the contour. The term v, an outward growing force, provides a faster convergence [14]. F still depends on the curvature function $div(\frac{\nabla \phi}{|\nabla \phi|})$.

Due to the edge based segmentation drawbacks [12], region information of the target objects are used in equation (3) without image gradient related terms. Based on region piecewise constant segmentation [11], it is possible to separate the image background from the image foreground. The author used the regularized Dirac functional (always different from zero) instead $|\nabla \phi|$ in equation (1) to remain close to the minimization problem.

$$F = \lambda_2 (I - \mu_{out})^2 - \lambda_1 (I - \mu_{in})^2 - \alpha + \beta (div(\frac{\nabla \phi}{|\nabla \phi|})) \tag{3}$$

The first two terms in equation (3) measure variations inside/outside the active contour, the area inside the contour is given by the third term; the length of the curve is measured by the fourth term. The last two are regularization terms. In medical images with complex backgrounds, μ_{in} and μ_{out} can fail in performance due to their global nature since they are based on the assumption that image intensities are statistically homogeneous in each region.

The method proposed by Li et al. [13], which F is shown in equation (4) is based on a region-scalable fitting (RSF) energy functional that locally approximate the image intensities on both contour sides. The minimization of the energy is achieved by minimizing the integral over all center points in the image and smoothing the contour by penalizing its length. Note, once more, that $|\nabla \phi|$ in equation (1) is replaced by the smoothed Dirac function.

$$F = [v(div(\frac{\nabla \phi}{|\nabla \phi|})) - (\lambda_1 e_1 - \lambda_2 e_2)] + \mu(\nabla^2 \phi - div(\frac{\nabla \phi}{|\nabla \phi|})) \tag{4}$$

The term $v(div(\nabla\phi/|\nabla\phi|))$ has a smoothing effect and is fundamental to maintain a regularized contour. The term $(\lambda_1 e_1 - \lambda_2 e_2)$, responsible for driving the contour toward object boundary, is called the data fitting term. The last term, called level set regularization term, looks for the regularity of the function.

Zhang et al. [14] proposed for the evolution equation a hybrid method integrating both boundary and region information:

$$F = \alpha(I - \mu) + \beta(div(P\frac{\nabla \phi}{|\nabla \phi|})) \qquad (5)$$

The first term encourages the contour to enclose the regions with gray-levels greater than a specific value. The second term aids the contour to attach the areas with high image gradients.

3   Segmentation Method

The method described in the present work starts with a pre-processing algorithm that detects low-level features in ultrasound images. Then, the signed distance function is iteratively modified applying shrink/expansion operations according to the level set curvature and stopping function to obtain the zero curves. Finally, small regions are removed in post-processing remaining only the four contours of the heart chambers.

3.1   Phase Symmetry

Phase-based symmetry detection (PSD) is an illumination and contrast invariant measure of symmetry in an image, useable as a line or blob detector, to analyze and understand shapes identifying the structure of objects in the frequency space. The components of the signal are analyzed using log Gabor wavelet and the results point out each component that exhibit symmetry or partial symmetries. Six orientations were considered for symmetry searching in 2D images [17].

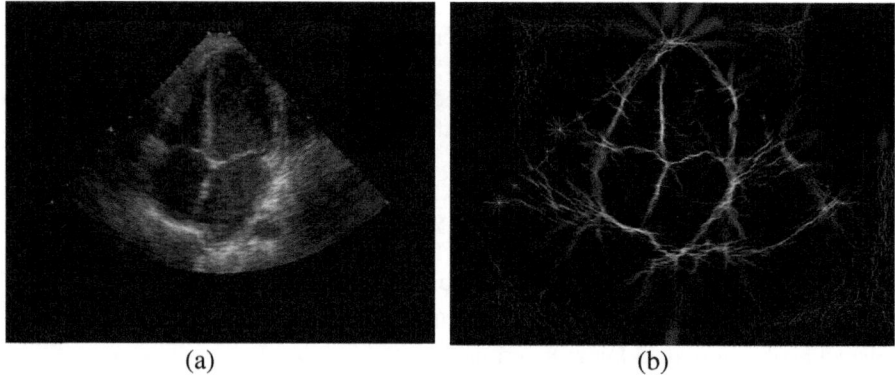

(a)   (b)

Fig. 1. Echocardiographic image: (a) Original image and (b) image after pre-processing with phase symmetry

As we want to delineate the heart walls, (partial) symmetry information shows good results in attenuating ultrasound image noise (see figure 1).

3.2   Level Set Evolution

The different approaches presented in section 2 were firstly implemented. The analysis led to conclude that additional terms for F appearing in eq. 3, 4 and 5 do not produce considerable improvements compared to the results of the edge based formulation. Thus, considering the best results provided by the level set described by eq. 2, combined with its lower complexity when compared with the other three formulations, led us to select it as the methodology to follow in this work, according to the implementation in [18]. The initial level set ϕ_0 is a signed distance function in the Euclidean metric where the central pixel has the largest value and decrease at each deviated element, ending with value zero at the four corners of image. For the stopping function P, we use a logarithmic variation that adjusts itself to the regions to be segmented and where additional terms to impose convergence or fast evolution are not needed; in the following form:

$$P = \log(|\frac{I - \varepsilon}{\gamma}| + 1) \tag{6}$$

where, I is the image, ε the average intensity value of the image and γ the dynamic range of the region that we are searching for.

3.3 Post-processing

Due to the B-mode echocardiographic image characteristics and the level set nature, several boundaries are detected (see figure 2a) most of them are noise that need to be eliminated to put in evidence only the four cardiac cavities. The post-processing is necessary to discard these unwanted small regions. Since the result of the segmentation is a binary image, the background pixels were ignored (with zero value) and all contours identified as follows. For each contour an initial pixel is selected, and then one proceeds through its adjacent pixels until the starting pixel is reached again. This procedure is repeated for each contour and the respective area is calculated. The contours corresponding to the four biggest areas are the four heart chambers. Due to the often existing irregularities in the detected contours, they were smoothed by using morphological operations of dilation and erosion, as a final step (see figure 2b).

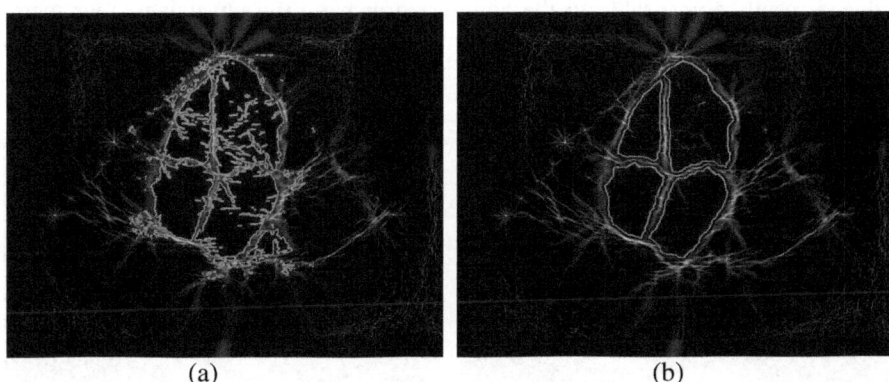

(a)   (b)

Fig. 2. Image of the final step of our method (a) before and (b) after post-processing

3.4 Validation

In order to evaluate the performance of the presented method, the resulting contours are compared with the ones manually drawn by the expert physicians, using images randomly selected. First, the interior of the two contours are white filled giving rise to the areas A1 and A2 illustrated in figures 3a and 3b. Then, the pixels shown in figure 3c are computed using the *xor* operation (D = A1 *xor* A2). The resulting pixels are the ones that belong to one image and do not belong to the other image. The number of "white" pixels in this new image D, is a measure of similarity between the two contours (see figure 7 and section 4 for results explanation).

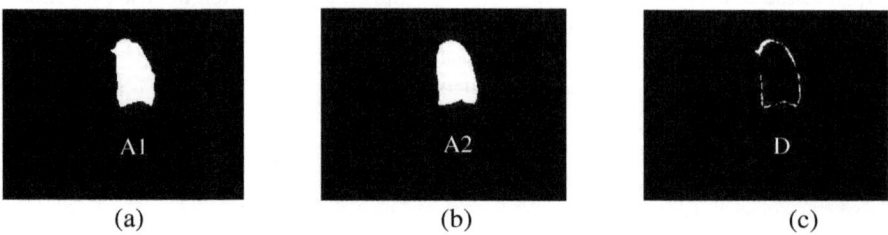

Fig. 3. The left ventricle white filled contour: (a) produced by the proposed segmentation method, (b) drawn by an expert physician and (c) mismatch pixels for the compared images

4   Results and Discussion

Several frames were randomly extracted from echocardiographic videos collected by two different ultrasound equipments using four children (two children for each an equipment). The condition to extract the selected frame is that the boundaries of the four heart cavities are all visible. If the region to be segmented was out of the field of view, the frame would be rejected and another one was randomly selected. For each child, the four best frames were chosen, corresponding to sixteen images of 576×720 pixels.

The proposed method detected simultaneously the four heart chambers, and produced results comparable to the contours drawn by the physicians. Our method needs no user intervention and convergence is achieved after five iterations. Figures 4 and 5 show two of the processed images.

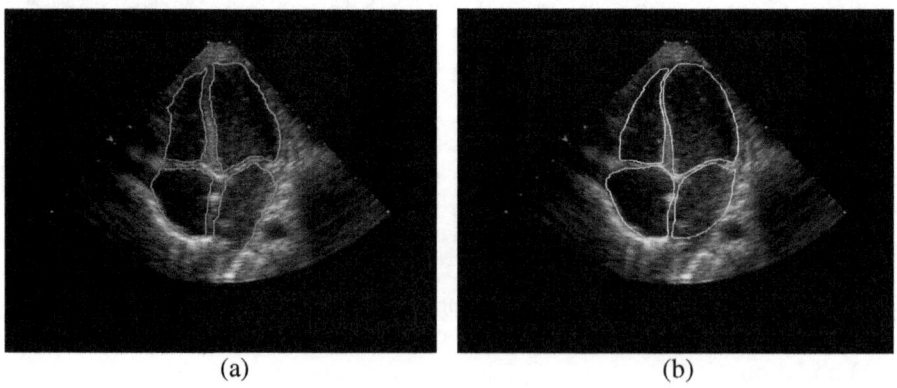

Fig. 4. Echocardiographic image from equipment A: (a) contour produced by the proposed method and (b) contour drawn by the physician

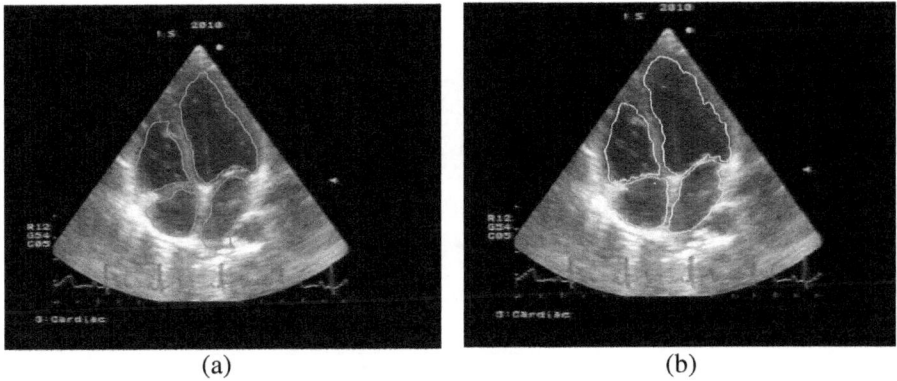

(a)   (b)

Fig. 5. Echocardiographic image from equipment B: (a) contour produced by the proposed method and (b) contour drawn by the physician

Sometimes it was not possible to get a good automatic segmentation. Figure 6 shows two examples illustrating the worst boundary extraction cases, for both automatic and manual procedure. The inadequate segmentation is due to the right ventricle's moderator band that justifies the reason why the level set detects an incorrect boundary (see arrow in figure 6a). Figure 6b illustrates a manually drawn contour, where the physician considers part of the pulmonary vein as being left auricle.

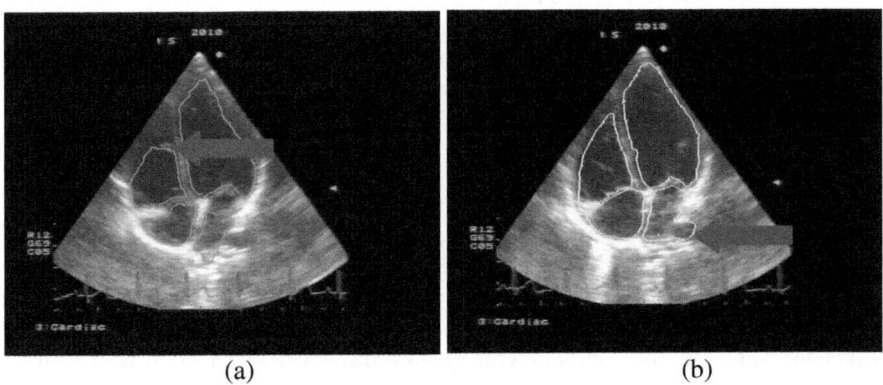

(a)   (b)

Fig. 6. Inadequate image segmentation: (a) algorithm failure due to the presence of artifacts and (b) consideration of the pulmonary vein in left auricle by the physician

The segmentation error was calculated using the validation procedure explained in section 3.4. The deviation from the regions drawn by the experts was calculated for each one of the four regions separately. The results are illustrated in figure 7. The numerical quantity lies between zero and one hundred. The zero value corresponds to the total matching of both regions, i.e., when no pixels are out of the common region. If the percentage is equal to one hundred, both regions do not have any pixels in the common region, meaning that both images are completely different.

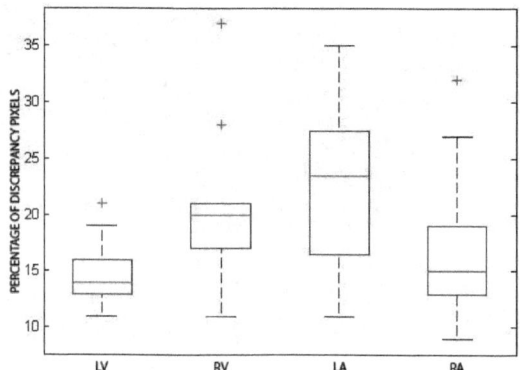

Fig. 7. Box-plots: LV-left ventricle, RV-right ventricle, LA-left auricle and RA-right auricle. Percentage of discrepancy pixels when the reference region and the algorithm produced region are overlapped.

As outlined in figure 7, the smallest error occurs for the left ventricle, where the mean error (median) value of the region is about 14%. The most problematic cavities are the right ventricle and the left auricle. This fact is essentially due to the window view of the heart. It is not always possible to correctly identify the cavity boundaries because of the noise and the attenuation in the heart nearby organs. The left auricle is the more problematic one, since it is hidden behind the pulmonary veins in most of the views. It is important to mention that the values shown in table 1 are based on references that also present errors and depend strongly on the physician's interpretation. In fact, sometimes the physicians need to presume where to place the correct contour.

5   Conclusions

We propose a method that efficiently segments the four heart cavities in the echocardiographic images. A pre-processing step calculates the phase symmetry of the ultrasound image. Then, the level set methodology is used to locate all contours in the image. The final step consists of a post-processing procedure to clean the four cavities and delineate the contours. Our method has shown very good results in noise reduction verified in the ultrasound images. The geometric approach has been chosen for the level set evolution because the model is simple to compute and reveal good performance. The stopping function used, is more robust than traditional Gaussian or exponential functions. The proposed method is capable to segment the four heart chambers, which is an advantage compared to other available methods (see references) that only succeeded in segmenting one heart cavity. The proposed method needs no user intervention and convergence is achieved after few iterations.

The positive results obtained with this work, acts as motivation to improve this method and/or to develop new methods that can simultaneously segment the four heart chambers, and to assist the clinical diagnosis in the identification of various congenital heart diseases: septal defects, valve defects, coarctation of the aorta.

References

1. Noble, J.A., Boukerroui, D.: Ultrasound Image Segmentation: A Survey. IEEE Transactions on Medical Imaging 25, 987–1010 (2006)
2. Nillesen, M.M., Lopata, R.G.P., Gerrits, I.H., Kapusta, L., Huisman, H.J., Thijssen, J.M., Korte, C.L.d.: Segmentation of the Heart Muscle in 3-D Pediatric Echocardiographic Images. Ultrasound in Medicine & Biology 33, 1453–1462 (2007)
3. Nascimento, J., Sanches, J.: Ultrasound imaging LV tracking with adaptive window size and automatic hyper-parameter estimation. In: 15th IEEE International Conference on Image Processing, pp. 553–556 (2008)
4. Sudha, S., Suresh, G., Sukanesh, R.: Speckle Noise Reduction in Ultrasound Images Using Context-based Adaptive Wavelet Thresholding. IETE Journal of Research 55, 135–143 (2009)
5. Nadernejad, E., Karami, M.R., Sharifzadeh, S., Heidari, M.: Despeckle Filtering in Medical Ultrasound Imaging. Contemporary Engineering Sciences 2, 17–36 (2009)
6. Rajpoot, K., Noble, J.A., Grau, V., Rajpoot, N.: Feature Detection from Echocardiography Images Using Local Phase Information. In: Proceedings of 12th Medical Image Understanding and Analysis (2008)
7. Kass, M., Witkin, A., Terzopoulos, D.: Snakes: Active contour models. International Journal of Computer Vision 1, 321–331 (1988)
8. Caselles, V., Kimmel, R., Sapiro, G.: Geodesic Active Contours. International Journal of Computer Vision 22, 61–79 (1997)
9. Malladi, R., Sethian, J.A., Vemuri, B.C.: Shape Modeling with Front Propagation: A Level Set Approach. IEEE Transactions on Pattern Analysis and Machine Intelligence 17, 158–175 (1995)
10. Osher, S., Sethian, J.A.: Fronts Propagating with Curvature Dependent Speed: Algorithms Based on Hamilton-Jacobi Formulations. J. Comput. Phys. 79, 12–49 (1988)
11. Mumford, D., Shah, J.: Optimal Approximations by Piecewise Smooth Functions and Associated Variational Problems. Communications on Pure and Applied Mathematics 42, 577–685 (1989)
12. Chan, T.F., Vese, L.A.: Active Contours Without Edges. IEEE Transactions on Image Processing 10, 266–277 (2001)
13. Li, C., Kao, C.-Y., Gore, J.C., Ding, Z.: Minimization of Region-Scalable Fitting Energy for Image Segmentation. IEEE Transactions on Image Processing 17, 1940–1949 (2008)
14. Zhang, Y., Matuszewski, B.J., Shark, L.-K., Moore, C.J.: Medical Image Segmentation Using New Hybrid Level-Set Method. In: Proceedings of the 2008 Fifth International Conference BioMedical Visualization: Information Visualization in Medical and Biomedical Informatics, pp. 71–76 (2008)
15. Linguraru, M.G., Kabla, A., Marx, G.R., Nido, P.J.d., Howe, R.D.: Real-Time Tracking and Shape Analysis of Atrial Septal Defects in 3D Echocardiography. Academic Radiology 14, 1298–1309 (2007)
16. Mosaliganti, K., Smith, B., Gelas, A., Gouaillard, A., Megason, S.: Level Set Segmentation: Active Contours Without Edge. The Insight Journal 1 (2009)
17. Kovesi, P.: Symmetry and Asymmetry from Local Phase. In: Tenth Australian Joint Conference on Artificial Intelligence, pp. 185–190 (1997)
18. Silva, J.S., Santos, B.S., Silva, A., Madeira, J.: A Level-Set Based Volumetric CT Segmentation Technique: A Case Study with Pulmonary Air Bubbles. In: Campilho, A.C., Kamel, M.S. (eds.) ICIAR 2004. LNCS, vol. 3212, pp. 68–75. Springer, Heidelberg (2004)

Improved Technique to Detect the Infarction in Delayed Enhancement Image Using K-Mean Method

Mohamed K. Metwally[1], Neamat El-Gayar[1], and Nael F. Osman[1,2,*]

[1] Center for Informatics Sciences, Nile University, Egypt
[2] Radiology Department, School of Medicine, Johns Hopkins University, USA
{mohamed.ali,nelgayar,nosman}@nileuniversity.edu.eg

Abstract. Cardiac magnetic resonance (CMR) imaging is an important technique for cardiac diagnosis. Measuring the scar in myocardium is important to cardiologists to assess the viability of the heart. Delayed enhancement (DE) images are acquired after about 10 minutes following injecting the patient with contrast agent so the infracted region appears brighter than its surroundings. A common method to segment the infarction from DE images is based on intensity Thresholding. This technique performed poorly for detecting small infarcts in noisy images. In this work we aim to identify the best threshold value to segment the infarction in case of segmentation using simple Threshold and propose a modified technique to improve the segmentation in noisy images. Our proposed technique is based on enhancing Thresholding using k-means clustering. We test our proposed model using computer simulated and real images with different contrast-to-noise ratio (CNR). We used F-score, which is a combined measure of the precision and sensitivity, to determine the performance of the proposed technique versus simple Thresholding. The results show that the proposed technique outperforms existing methods.

Keywords: Cardiac Magnetic resonance, Delayed Enhancement, k-means clustering technique.

1 Introduction

Myocardial Infarction (MI) is one of the most significant causes of death; therefore, cardiologists are keen in finding ways to identify the infarcted tissue (fibrotic or scarred tissue) in the cardiac muscle in order to define the degree of cardiac viability to plan the proper treatment. Delayed Enhancement (DE) Magnetic Resonance (MR) images are considered the gold standard images to identify infarcted regions. The images are acquired about 10 minutes after injecting the patient with a gadolinium-based contrast agent, such as gadolinium diethyltriaminepentaacetic acid (Gd-DTPA)[2, 5]. The contrast perfuses the myocardium and after the 10 minutes it finishes withdrawal from the normal tissues in a faster rate than the withdrawing from the scarred tissue. Inversion recovery (IR) imaging pulse sequence produce images in which the residual contrast agent that remains inside the dead tissues (the infarction) appears brighter than its surroundings; hence, they are called DE images.

---

* Corresponding author.

A. Campilho and M. Kamel (Eds.): ICIAR 2010, Part II, LNCS 6112, pp. 108–119, 2010.

There are various methods of image segmentation, such as thresholding, region growing, artificial neural network, deformable models as active contour, and clustering as k-means and ISODATA algorithms [1] that can be used. However, it is important to objectively determine the extent of the hyperenhanced region based on the intensity of the images in order to segment the scarred tissue, which is our main target in this work. In literature of segmenting the infarction from DE images, the most common technique was using simple Thresholding. The threshold value was determined based on the mean and the standard deviation of the intensity of healthy tissues or the blood pool [2, 3]. Another method used is the watershed approach [4], however it is reported to be time consuming.

In this work, we aim to 1) Identify, experimentally, the best threshold value to segment the infarction; and 2) enhance the results of threshold segmentation using k-means where we apply the clustering technique on the resulted pixels out from simple Thresholding step taking in consideration the intensity and spatial information of those pixels. Because of the difficulty of determining the true status of the tissue other than pathology labs, we use numerical simulations to produce realistic DE images while providing a ground truth reference. We also investigate the choice of the thresholding level of the current methods as it seems to be *ad hoc* and anecdotal.

The paper is organized as follows: Section 2 describes the DE simulator used and reviews the simple Thresholding method to segment infracted regions. Section 3 reviews the K-means clustering algorithm and introduces the enhanced thresholding technique that is based on K-means. In section 4, we outline the experiments conducted; performance measures used and present the results together with a quantitative analysis. The paper is summarized and concluded in Section 5.

2 Segmentation of Infarction Using Simple Thresholding

2.1 DE Image Simulators

A DE image simulator was built using MATLAB (The MathWorks, Inc) to investigate the performance of different segmentation techniques for different contrast-to-noise ratio (CNR)s. The simulator simulates short axis image that shows the tissue of left ventricle (LV), blood inside LV, and the infarction (gray zone and core). Fig. 1 shows a schematic drawing to simulated image.

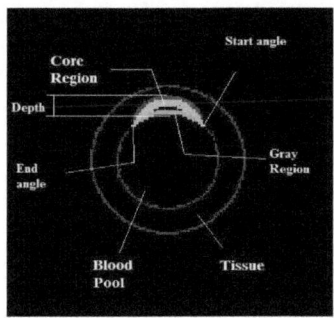

Fig. 1. Schematic drawing to simulated image shows tissue, blood, and infarcted regions

We set the intensity values for normal tissue, blood pool, and the infarction region and the dimensions of the left ventricle (LV) and the infarction. There are 3 parameters, which specify the dimension of infarcted region: the start angle, the end angle, and the depth of which is the maximum radial distance between the endocardium, which is the inner wall or circle of LV, and a point on the boarder of infarcted region. Fig.2 shows simulated images containing infarction regions that have different depths or different start angles.

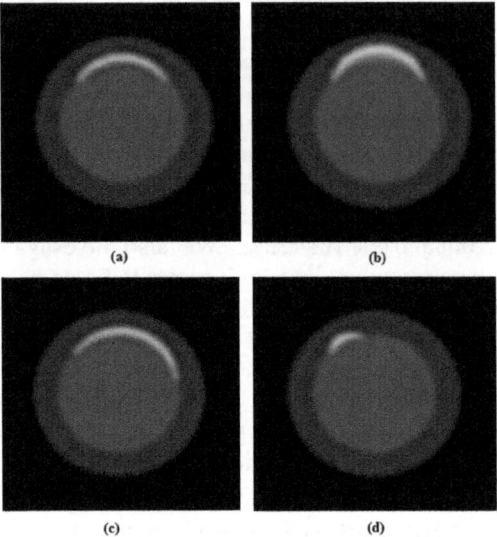

Fig. 2. Simulated image (a) with infarction depth = 0.3 cm, (b) with infarction depth = 0.6 cm, (c) with start angle = 15°, and (d) with start angle = 100°

Our simulated patterns to the infarction are consistent with the infarction patterns in [6]. We apply white Gaussian noise with different noise variance and take the absolute of the image to avoid existence of negative values because the real MR images are magnitude images and as a result to the absolute operation, the background, such as air, has Rayleigh distribution for the background noise model in cardiac MR images [7]. After applying the Gaussian noise, we have applied second order Butterworth low pass filter [8] to achieve partial volume effect, which appears as smoothing to the image.

2.2 Contrast-to-Noise Ratio (CNR)

In image processing, CNR is a ratio between the contrast between intensities of two objects and the noise in the image [14]. It's calculated by

$$CNR = \frac{|I_A - I_B|}{\sigma_0} ,$$ (1)

where I_A and I_B are the average intensities of the objects A and B respectively. In our case the object (A) is the infarction and the object (B) is the normal tissue. σ_0 is the standard deviation of the noise. This standard deviation is calculated by selecting a region in background and calculating the standard deviation of the intensities in this region. The division by the standard deviation of the noise is used instead of the mean intensity of the noise because it's assumed theoretically that the noise has zero mean value.

2.3 Segmentation Using Simple Threshold Technique

The most common way to select the threshold intensity is done, after segmenting the LV manually, by selecting a region of interest in the normal myocardial tissue, and the threshold intensity (τ) is then defined as follows:

$$\tau = \mu (\text{NT}) + m \times \sigma(\text{NT}) , \qquad (2)$$

where NT is the set of intensities of normal tissue that the user selects, μ is the mean intensity and σ is the standard deviation of those intensities. Usually m is chosen to be 2 [3]. In this study we aim to investigate the best multiple (m) of standard deviation which achieves best performance. We used F-score as our measure of performance (refer to section 4.1 for more elaboration on this measure). We experimented with different variations of $m \in \{1, 1.5, 2, 2.5, 3, 3.5, 4\}$.

3 Enhanced Thresholding Technique Using K-Mean

3.1 K-Means Clustering

The K-means method is a simple and popular method that aims to identify groups (clusters) of data points in a multidimensional space [5]. The cluster consists of data points whose inter-point distances are small with respect to the distances to points outside of the cluster. The data point considered here in this work is 2-dimesional representing two features: the intensity of the pixel, and the distance between the pixel and the spatial central pixel as it will be illustrated in section 3.2. We fixed the number of clusters to be 2 clusters; one of them represents the infarction cluster, which has pixels that are close to each other and have higher intensity values than the second cluster, which represents noisy pixels.

3.2 Proposed Thresholding Technique Using K-Means

Applying the Thresholding technique as described in Eq. (2) can result in identifying some scattered noisy pixels as infarction. We assume that the pixels in infarction region are very close in addition to their high intensities with respect to surrounding pixels. Our proposed "enhanced threshold technique" attempts to use k-means to cluster pixels resulting from simple Thresholding technique. The features used for clustering are the distance between the pixel and the spatial central pixel and the intensity. We can define the set of locations of pixels that result from the Thresholding step (PLS) as follow

$$PLS = \{(x,y)| \ x, y \in R, IM(x,y) > \tau\} \ , \tag{3}$$

and the set that contains their corresponding intensities (PIS) is

$$PIS = \{IM(x,y)|IM(x,y) > \tau\} \ , \tag{4}$$

where IM is the DE image. The distance feature for each pixel in PLS is the Euclidian distance between the pixel and the spatial central pixel (SCP) of the pixels in PLS,

$$SCP = \frac{\Sigma_1^N PLS}{N}, \tag{5}$$

where N is the number of elements in the set PLS. Now we calculate the distances set (Dis) as follow,

$$Dis = \{\|PLS - SCP\|^2\} \ . \tag{6}$$

The k-means divides the feature space (intensity and distance), which is described by equations 4,6 respectively into 2 clusters the first cluster represents high intensity pixels with small distance to the spatial central pixel (the infarction group) and the second cluster is normal tissue or noisy pixels.

4 Experiments and Results

Experiments were conducted on computer simulated and real images. In this section we present the results on both data sets using the simple Thresholding method and the enhanced K-means Thresholding technique. We also investigate results under different variations of *m* for specifying the threshold value, which is a common hurdle in similar segmentation technique. As follow we present performance measure (F-score) to evaluate our experiments. Sections 4.2 and 4.3 present the results on simulated and real data respectively.

4.1 F-Score

In our work, precision and sensitivity measurements are equally important to compare among different segmentation methods. Therefore we used the F-score as a measurement that combines the precision and the sensitivity [9, 10]. F-score is the harmonic mean for precision and sensitivity, and it is calculated by the following formula

$$F = 2 * \frac{\text{Precision} \times \text{Sensitivity}}{\text{Precision} + \text{Sensitivity}} \ , \tag{7}$$

The F value always lies between zero and one. A low F value indicates that both of precision and sensitivity have low values or even one of them has extremely low value, while a high value of F indicates a good value of precision and sensitivity. On the other hand, precision is a ratio that indicates the robustness of the segmentation technique to noise. The technique that has higher precision is more robust to noise. The precision is calculated by

$$\text{Precision} = \frac{TP}{TP+FP} \ , \tag{8}$$

where TP is the number of detected pixels that are truly infarcted, and FP is the number of pixels that are not infarcted but are detected as infarction. Sensitivity is a ratio that indicates how well the segmentation technique is able to detect all infarcted pixels. The sensitivity is calculated by

$$\text{Sensitivity} = \frac{\text{TP}}{\text{TP+FN}}, \tag{9}$$

where FN is the number of infarcted pixels that are not detected.

4.2 Simulated Data Results

We simulated DE images with CNR=32.8146 ± 7.58. The actual CNR in DE images using 1.5T scanners is 21.4±9 and 32±13 for images obtained by a high field, 3T MRI scanners [12, 13]. Fig. 3 shows 2 examples of our simulated images: one at CNR=45.26 and the other at CNR=31.46. The simulated images have following parameters: image dimensions are 128x128 pixel, field of view is 10 cm, the radius of the epicardium, which is the outer wall or circle of LV, is 3.5 cm while that of the endocardium is 2.5 cm, the maximum depth of infarction is 0.3 cm, the mean intensities of normal tissue, gray zone, hyper enhanced region and blood pool are 20, 150, 200 and 35, respectively.

We segmented the simulated images using simple Thresholding technique (Eq. 2) at different multiples (m) of standard deviation of normal tissues' intensities. We compared the F-score value for different segmentation results. Fig. 4 shows the relation between the CNR of the simulated images and the F-scores that were resulted from applying simple Thresholding technique at different multiples (m) of standard deviation. From the depicted results in Fig. 4 it can be seen the best combination between the precision and sensitivity is achieved by using threshold intensity at 3 multiples of standard deviation, especially at the lower CNRs, which indicate very noisy images, because this curve (the green curve with circular markers) has the highest values of F-scores. It should be noted that with increased number of (m) of standard deviation, the simple Thresholding technique shows more robustness to noise. However the sensitivity decreases rapidly because the threshold value discards part from the infarction region due to its relative low intensity. Fig. 5 shows the results of segmentation of the 2 images that are in Fig. 4 for different multiples (m) of standard deviation.

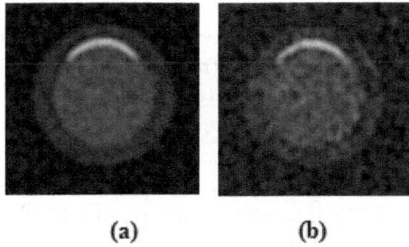

(a)   (b)

Fig. 3. Anatomy DE image (a) at CNR=45.26, (b) at CNR=31.46

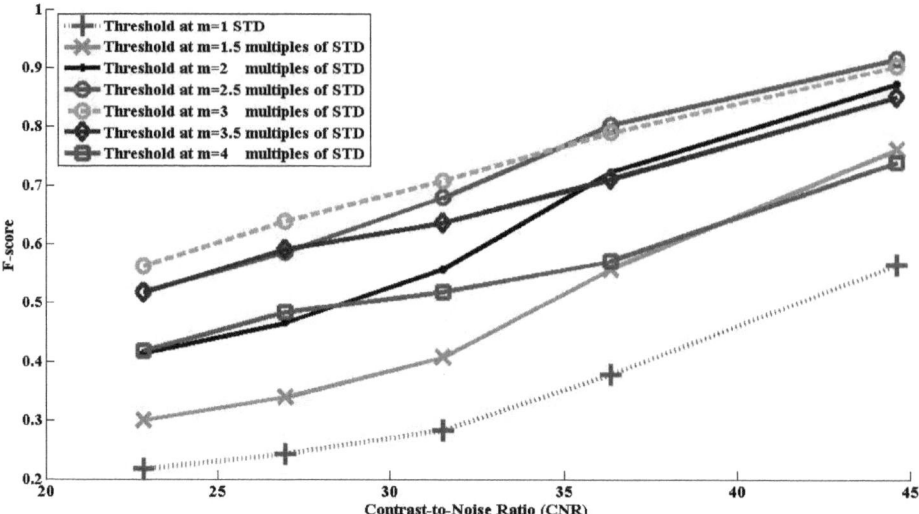

Fig. 4. The F-scores of segmentation using simple Thresholding technique at different multiples (m) of standard deviation at different CNRs. The x-axis is CNR and y-axis is F-scores. (STD: standard deviation).

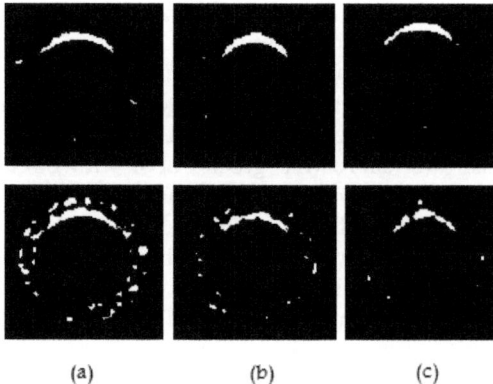

Fig. 5. 1st and 2nd rows show segmentation results corresponding to the 2 shown images that have CNR = 45.26 and 31.46 respectively. The columns from left to right are the results of segmentation using simple Thresholding at (a) 2, (b) 2.5, and (c) 3 multiples of standard deviation of the normal tissues' intensities.

We applied the enhanced thresholding technique using K-means with 2 features, which are the distance and the intensity of the pixels, to eliminate the pixels that are falsely classified as infarction. Fig. 6 shows the relation between the CNR and the F-scores that result from applying k-means after simple Thresholding step. It can be seen that when we applied the minimum value of threshold intensity, we enforced a

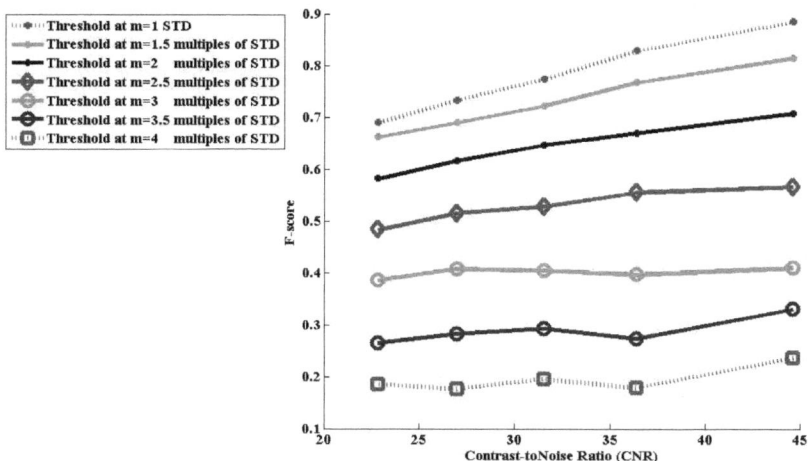

Fig. 6. The F-scores of the proposed technique at different (*m*) multiples of standard deviation followed at different CNR. The x-axis is CNR and y-axis is F-scores. (STD: standard deviation).

large number of pixels to be labeled as infarctions' pixels. However, applying k-means with 2 features on the results filtered those pixels and selected the infarctions' pixels that are close to each other and have high intensity. At this point we can state that the proposed method has high precision while maintaining acceptable sensitivity values. Fig. 7 shows the results of segmentation of the 2 images that are in Fig. 4 with simple Thresholding technique using 1, 2, 2.5, and 3 multiple of standard deviation and followed by k-mean. Fig. 8 depicts the best curves from Fig. 4 and Fig. 6 in one graph and table 1 shows the F-scores that are corresponding to each segmentation method that was mentioned in Fig. 8 over different CNRs.

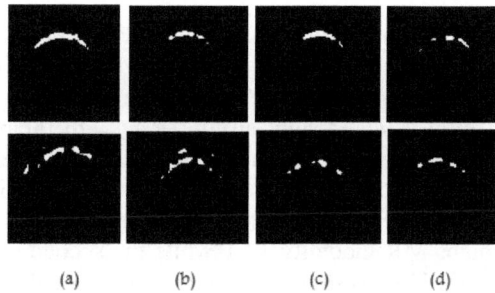

Fig. 7. 1st and 2nd rows show segmentation results corresponding to the 2 shown images that have CNR = 45.26 and 31.46 respectively. The columns are the results of segmentation using simple Thresholding at (a) 1, (b) 2, (c) 2.5, and (d) 3 multiples of standard deviation of the normal tissues' intensities and followed by k-mean.

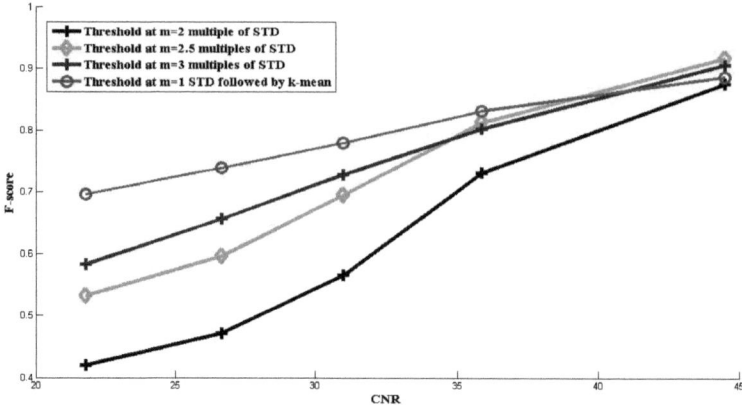

Fig. 8. The best segmentation methods in terms of F-scores (STD: standard deviation)

Table 1. Quantitative measurements to the best methods in terms of F-scores at different CNRs

CNR / Segmentation Method	44.6	36.3	31.5	26.9	22.8
Simple Thresholding (*m=2*)	88.4 %	71.8 %	56.8 %	46.8 %	43.6 %
Simple Thresholding (*m=2.5*)	92.6 %	81 %	68.2 %	59.2 %	57 %
Simple Thresholding (*m=3*)	91.6 %	82.4 %	72.8 %	62.2 %	59 %
Enhanced Thresholding technique using k-means (*m=1*)	88.6 %	83.2 %	77.6 %	73 %	70 %

4.3 Results on Real Data

We applied the best four methods, which are summarized in Table 1 and Fig. 8, to segment the infarcted region in real images. The ground truth was provided by experts who selected the infarction manually. Fig. 9 shows the real images that we used. The short axis images show the LV and right atrium (RA) with infarction in the wall of the LV. The white arrow points to the infarcted region. The real images have the following specifications: field of view 15.6 cm × 19.2 cm, slice thickness = 8 mm, the image dimensions are 192 pixels × 156 pixels for each case. The first and third cases were acquired by Siemens MR modality 1.5T while the second case was acquired by Siemens MR modality 3T. Fig. 10 shows the segmentation results of applying the four segmentation methods on the 3 cases.

The quantitative measurements of the results in terms of sensitivity, precision and the corresponding F-scores are summarized in table 2. Results of table 2 emphasize again that applying k-means enhances the precision of the resulting images in contrast to simple Thresholding significantly. However, the technique may affect the sensitivity.

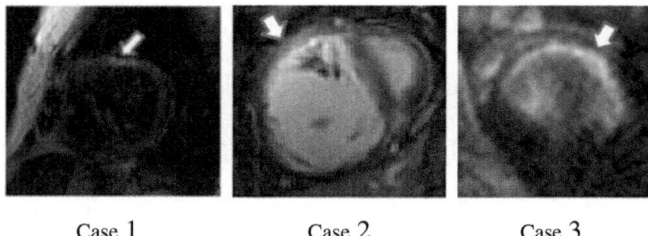

Case 1   Case 2   Case 3

Fig. 9. Real images show the infarction in LV

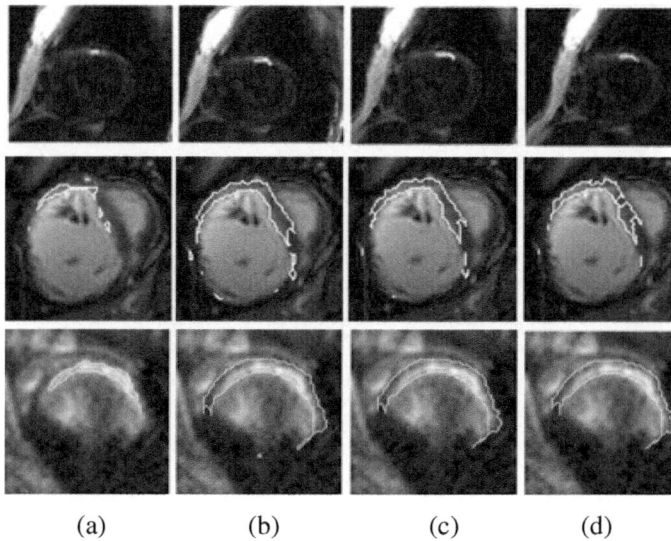

(a)   (b)   (c)   (d)

Fig. 10. Results of segmentation using (a) threshold intensity at 1 standard deviation followed by k-mean, simple Thresholding at (b) 2, (c) 2.5, and (d) 3 multiples of standard deviation

Table 2. The quantitative measurements of the applied methods in terms of sensitivity, precision and F-core

		Threshold at 1 STD followed by k-mean	Threshold at 2 multiples of STD	Threshold at 2.5 multiples of STD	Threshold at 3 multiples of STD
Case 1	Sensitivity	75 %	87.5 %	87.5 %	87.5 %
	Precision	83.3 %	42.16 %	47.29 %	57.37 %
	F-score	79 %	56.9 %	61.4 %	69.3 %
Case 2	Sensitivity	89.23 %	100 %	100 %	100 %
	Precision	72.5 %	21.67 %	24.71 %	28.14 %
	F-score	80 %	35.6 %	39.6 %	43.9 %
Case 3	Sensitivity	73.25 %	100 %	100 %	100 %
	Precision	100 %	56.58 %	59.07 %	60.32 %
	F-score	84.56 %	72.27 %	74.27 %	75.27 %

5 Summary and Conclusions

In this work we aimed to identify the best threshold value to detect the infarction in case of segmentation using simple Thresholding. We proposed a modified technique to improve the segmentation of noisy images. Our proposed technique was based on enhancing Thresholding using K-means clustering. We tested our proposed model using computer-simulated and real images with various contrast-to-noise ratios (CNR). We used F-score - a combined measure of precision and sensitivity - to determine the performance of the proposed technique versus simple Thresholding. The results showed that the proposed technique outperforms existing methods with respect to the precision measure. It was shown that, in simple Thresholding, $\mu+3\sigma$ (the mean intensity of normal tissue plus 3 multiples of standard deviation of the normal tissues' intensities) is a better choice of the threshold to detect the infarction in DE images. This conclusion was found to be consistent with that in [2]. On the other hand we showed that the proposed technique - which applied a $\mu+\sigma$ threshold (mean plus one standard deviation of normal tissue's intensities) followed by K-means on the feature space that encompasses the pixel intensities and their distances from their spacial central pixel results in better performance in terms of F-score when compared to the results of the common simple Thresholding technique. The precision showed higher contribution to the F-score than to the sensitivity due to the strong effect of spatial information which means that the technique is more robust to noise than the current simple Thresholding technique. For future work, we intend to investigate more clustering techniques for segmentation of infarction in DE images. We aim at using cluster validity measures to validate and enhance clustering results. Also, it is interesting to investigate, for any studied techniques, what the minimum detectable size of infarctions is.

References

1. Pham, D.L., Xu, C., Prince, J.L.: Current Methods in Medical Image Segmentation. Annual Reviews 2, 315–338 (2000)
2. Hsu, L.-Y., Natanzon, A., Kellman, P., Hirsch, G.A., Aletras, A.H., Arai, A.E.: Quantitative Myocardial Infarction on Delayed Enhancement MRI. Part I: Animal Validation of an Automated Feature Analysis and Combined Thresholding Infarct Sizing Algorithm. J. Magn. Reson. Imaging 23(3), 298–308 (2006)
3. Kolipaka, A., Chatzimavroudis, G.P., White, R.D., ODonnell, T.P., Randolph, M.: Segmentation of non-viable myocardium in delayed enhancement magnetic resonance images. The International Journal of Cardiovascular Imaging (formerly Cardiac Imaging) 21, 303–311 (2005)
4. Doublier, C., Couprie, M., Garot, J., Hamam, Y.: Computer assisted segmentation, quantification and visualization of an infarcted myocardium from MRI images. Biomedsim. 3, 151–156 (2003)
5. Duda, R.O., Hart, P.E., Stork, D.G.: Pattern Classification, pp. 526–528. Wiley, New York (1973)
6. Hood, M.: Overview of delayed myocardial enhancement in cardiac MRI (February 12, 2009), http://www.ismrm.org/smrt/09/mock_clinical_poster_Mo.pdf

7. Sijbers, J., Den, D.A.J., Van, D.D., Raman, E.: Estimation of signal and noise from Rician distributed data. In: Proc. Int. Conf. Signal Proc. and Comm., Spain, pp. 140–142 (1998)

8. Gonzalez, R.C., Wood, R.E.: Digital Image Processing, pp. 147–175. Prentice-Hall, Englewood Cliffs (2002)

9. Beitzel, S.M.: On Understanding And Classifying Web Queries, Chicago, Illinois (May 2006)

10. Zumel, N.: Statistics to English Translation, Part 1: Accuracy Measures (November 3, 2009),
    http://win-vector.com/dfiles/
    StatisticsToEnglishPart1_Accuracy.pdf

11. Marina, S., Nathalie, J., Stan, S.: Beyond Accuracy, F-score and ROC: a Family of Discriminant Measures for Performance Evaluation. In: Australian Conference on Artificial Intelligence (2006)

12. Sievers, B., Rehwald, W.G., Albert, T.S., Patel, M.R., Parker, M.A., Kim, R.J., Judd, R.M.: Respiratory Motion and Cardiac Arrhythmia Effects on Diagnostic Accuracy of Myocardial Delayed-enhanced MR Imaging in Canines. Radiology 247, 106–114 (2008)

13. Bernhard, K., Michael, F., Tobias, H., Uwe, H., Albertus, S., Claus, C., Stephan, M.: Assessment of myocardial viability using delayed enhancement magnetic resonance imaging at 3.0 tesla. Investigative Radiology 41, 661–667 (2006)

14. Thompson, M.: November 21 (2003),
    http://www.phys.cwru.edu/courses/p431/
    notes-2003/node123.html

Detection of Arterial Lumen in Sonographic Images Based on Active Contours and Diffusion Filters

Amr R. Abdel-Dayem

Department of Mathematics and Computer Science, Laurentian University,
Sudbury, Ontario, Canada
aabdeldayem@cs.laurentian.ca

Abstract. This paper presents a scheme for extracting carotid artery contours from ultrasound images using a modified active contour model. The scheme uses a single seed point as an input. A complex diffusion filter is used to provide a robust estimation of the image's edge map. This edge map is used to define the external energy function for the proposed active contour. The scheme produces accurate results compared to the gold standard images. Moreover, the proposed snake model was compared to two snake models found in literature. While the first model uses Canny edge detector, the second employs the Sobel operator to calculate the image's edge map. Experimental results over a set of 40 images show that the proposed model outperforms the other two models. Finally, sensitivity analysis over the entire set of test images revealed that the scheme is insensitive to the seed point location, as long as it is located inside the artery area.

Keywords: Segmentation, parametric active contours, complex diffusion, carotid artery lumen, ultrasound image.

1 Introduction

Vascular plaque, a consequence of atherosclerosis, results in an accumulation of lipids, cholesterol, smooth muscle cells, calcifications and other tissues within the arterial wall. It reduces the blood flow within the artery and may completely block it. As plaque layers build up, it can become either stable or unstable. Unstable plaque layers in a carotid artery can be a life-threatening condition. If a plaque ruptures, small solid components (emboli) from the plaque may drift with the blood stream into the brain. This may cause a stroke. Early detection of unstable plaque plays an important role in preventing serious strokes.

Currently, carotid angiography is the standard diagnostic technique to detect carotid artery stenosis and the plaque morphology on artery walls. This technique involves injecting patients with an X-ray dye. Then, the carotid artery is examined using X-ray imaging. However, carotid angiography is an invasive technique. It is uncomfortable for patients and has some risk factors, including allergic reaction to the injected dye, renal failure, the exposure to ionic radiation, as well as arterial puncture site complications, e.g., pseudoaneurysm and arteriovenous fistula formation.

A. Campilho and M. Kamel (Eds.): ICIAR 2010, Part II, LNCS 6112, pp. 120–130, 2010.
© Springer-Verlag Berlin Heidelberg 2010

Ultrasound imaging provides an attractive tool for carotid artery examination. The main drawback of ultrasound imaging is the poor quality of the produced images. It takes considerable effort from clinicians to assess plaque build-up accurately. Furthermore, manual extraction of carotid artery contours generates a result that is not reproducible. Hence, a computer aided diagnostic (*CAD*) technique for segmenting carotid artery contours is highly needed.

Abolmaesumi *et al.* [1] proposed a scheme for tracking the centre and the walls of the carotid artery in real-time using an improved star algorithm with temporal and spatial Kalman filters. The scheme depends on the estimation of the Kalman filter's weight factors, which are estimated from the probability distribution function of the boundary points. In practice, this distribution is usually unknown.

Hamou *et al.* [2] proposed a segmentation scheme for carotid artery ultrasound images based on Canny edge detector [3]. This scheme has shortcomings dealing with noisy images, leading to contour bleeding in such cases.

Da-Chuan *et al.* [4] introduced a dual dynamic programming method to detect arterial wall in ultrasound images. Some progress has been achieved in reducing the sensitivity to speckle noise. However, the computational complexity of the proposed method is questionable. Moreover, it requires the user to manually select the region of interest for further processing.

Abdel-Dayem *et al.* proposed many schemes for segmenting carotid artery ultrasound images, including the watershed based segmentation [5][6], fuzzy region growing based segmentation [7][8], fuzzy c-means based segmentation [9], graph-based segmentation [10], and complex diffusion based segmentation [11].These schemes provide satisfactory performance (overlap with the clinician-segmented images) in most cases.

All methods, described so far, may fail to produce accurate contours in some challenging cases (images with shadowing effects, high noise levels, partially occluded or incomplete contours). This performance pitfall hinders the applicability of the proposed schemes in real clinical trials. Active contours (snakes) are good candidates, if properly tuned, to overcome some of these shortcomings particularly, the incomplete contour problem.

Active contours [12,13,14,15] are widely used in various computer vision applications to locate object boundaries. They are divided into two main categories, according to their representation and implementation: the parametric and the level set active contours. Both categories suffer the following shortcomings:

a) the difficulties encountered in progressing into boundary concavities
b) the sensitivity to contour initialization
c) poor convergence to object boundaries when dealing with noisy images.

The first is usually insignificant in the medical arena, as biological objects usually exhibit smooth structure, with no deep cavities. Whereas, estimating *accurate* initial contours is a tedious and time consuming task for clinicians, who prefer systems with *minimal* inputs. Finally, the presence of high level of speckle noise in ultrasound images severely degrades the performance of active contour models on extracting carotid artery boundaries, as noise-corrupted pixels may influence the contour progression.

Mao *et al.* [16] proposed a scheme for extracting carotid artery walls from ultrasound images using a deformable model. The model's external force is defined in terms of the gradient image, which is highly influenced by the noise level within the original image. As a result, this scheme is susceptible to poor convergence to artery boundaries.

Da-chuan *et al.* [17] proposed a modified snake model for automatic detection of intimal and adventitial layers of the common carotid artery wall in ultrasound images using a snake model. The proposed model modified the Cohen's snake [18] by adding spatial criteria to obtain the contour with a global minimum cost function. However, this scheme has the same pitfall as [16], where the gradient image is used to calculate the model's external energy. Moreover, the computational time for the proposed model was significantly high.

Based on their initial contribution in [2], Hamou *et al.* [19] used Canny edge detector [3] to provide more robust estimation of the image's edge map. Then, a parametric active contour model is used to extract the artery boundaries. Due to the higher accuracy of Canny edge detector, compared to simply using the gradient image as in [16] and [17], some improvements have been achieved. Similar to [2], Canny edge detector generates a lot of false edges which influence the progression of the active contour. Moreover, the scheme requires the user to provide an accurate initial contour, which limits the use of the proposed scheme in clinical trials.

This paper proposes a snake-based scheme to extract carotid artery contours from ultrasound images. The proposed scheme overcomes most of the shortcomings of the previous work in [16,17, and 19]. First, a single seed point is needed to initialize the snake. This saves considerable clinician time and effort, making the system more attractive for real clinical applications. Second, a robust edge map estimator is used to define the snake's external force. This edge map estimator is based on our previous contribution [11]. As a result, the influence of noisy pixels and false edges on the snake's progression is diminished.

The rest of this paper is organized as follows. Section 2 describes the proposed scheme in details. Section 3 and Section 4 present the experimental setup and the obtained results, respectively. Finally, Section 5 offers the conclusions of this paper.

2   The Proposed Solution

This paper presents a *two-stage* scheme to extract carotid artery boundaries using ultrasound images. The former stage employs a complex diffusion filter to produce a robust estimation of the artery boundary. Then, the latter stage utilizes a snake model, where the snake's external energy function is measured by the Euclidean distance map of the extracted boundary. In the following subsections, a detailed description of each stage is introduced.

2.1   Robust Estimation of Artery Boundary

The diffusion equation (Equation 1) is widely used to model various physical phenomena such as wave propagation, gas dynamics, heat transfer, and mass transfer.

$$\frac{\partial}{\partial t} u(x,t) = \frac{\partial}{\partial x} (D \times \frac{\partial}{\partial x} u(x,t)) \tag{1}$$

where, u is the concentration of matter (e.g. mass or heat,..etc.), and D is the diffusivity (describes how fast or slow an object diffuses).

Diffusion processes are widely used in various image processing and computer vision applications, such as image de-noising, smoothing, segmentation, optical flow and stereo vision [20,21,22]. In image processing context, pixel intensity can be considered as the concentration of mass (or heat) which diffuses following the diffusion equation (Equation 1). This leads to various types of diffusion filters based on the diffusivity D. The evolution of the original image $u_o(x,y)$ with respect to the time parameter t into a steady state solution $u_{ss}(x,y)$ is equivalent to the filtering process.

Various research studies [23,24,25,26,27] introduced different classes of diffusion filters, based on the selection of the diffusivity function D. Among these studies is the pioneer model of nonlinear diffusion, introduced by Perona and Malik [23]. In this model, the diffusivity function is reduced at those locations having a larger likelihood to be edges. This likelihood can be measured by ∇u. Perona and Malik [23] demonstrated that edge detection based on their nonlinear diffusion model outperforms Canny edge detector [3].

Gilboa et al. [28] proposed a complex shock filter based on a complex diffusion equation. This equation can be viewed as a generalization of the traditional real-valued one. They followed the Perona-Malik model [23] with the adoption of the complex time τ:

$$\tau = e^{j\theta} \times t \tag{2}$$

where $\theta \in (-\frac{\pi}{2}, \frac{\pi}{2})$, $j = \sqrt{-1}$, and t is the time parameter.

Furthermore, Gilboa et al. [28] mathematically proved that the imaginary part of the complex diffusion coefficient approximates a smoothed second derivative of the image as the parameter θ (Equation 2) approaches zero. Hence, it can be used as an efficient edge detector in noisy images. They demonstrated that edge detection by their proposed complex diffusion is superior to the real-valued diffusion, proposed by Perona and Malik [23]. The detailed proof is outside the scope if this paper. Interested readers may consult Gilboa et al. [28] for more details.

In this stage, a complex shock filter [28] is used to provide a robust estimation of the image's edge map. Then, the user is asked to select a seed point within the artery area to focus on the region of interest (ROI) and to neglect all other edges. The contour enclosing the seed point will be used as a robust estimation of the artery contour. Please, note that the same seed point will be used during the next stage to initialize a snake. Finally, the distance transform is used to compute the Euclidean distance map relative to the extracted edge map. This map represents the Euclidean distance between each pixel and the nearest contour pixel. This map is used to define the external energy for the proposed snake model.

It is worth mentioning, that employing a complex diffusion filter to extract carotid artery contours was explained in more details and was experimentally validated in our previous contribution [11].

2.2 Snake Segmentation

2.2.1 Initialization
As discussed in the introduction section, introducing a system with minimal clinician interaction is a major design decision in this scheme. As a result, only a single seed point is required from the user to initialize the snake. The scheme uses the same seed point, specified by the user during the first stage. From this seed point, a circle with radius r is automatically generated and used as an initial snake. Then, this initial snake evolves to minimize the snake's energy function. The value of r was set to 30 pixels, which is believed to represent a close approximation of the size of a typical carotid artery. The scheme's sensitivity to the seed point selection was experimentally studied, and the results are reported in Section 4.1.

2.2.2 Snake Model
A snake is a parametric curve $X(s) = [x(s), y(s)]$, where the parameter $s \in [0,1]$. It moves through the spatial domain of an image to minimize a specified energy function. This energy function is defined in terms of both internal and external energy functions:

$$E = \int_{s=0}^{s=1} (E_{int}(s) + E_{ext}(s))\, ds \qquad (3)$$

where E_{int} and E_{ext} are the internal and external energy functions, respectively.

The objective of the internal energy function E_{int} is to force the snake to have a smooth shape. As a result, the internal energy is defined as a function of both the snake's tension and rigidity:

$$E_{int} = \alpha |X'(s)|^2 + \beta |X''(s)|^2 \qquad (4)$$

where α and β are weighting parameters that control the snake's tension and rigidity, respectively. X' and X'' are the first and second derivatives of $X(s)$ with respect to the parameter s.

The external energy function E_{ext} should be defined in terms of the image features under consideration. In our model, E_{ext} is defined as:

$$E_{ext} = \gamma [dist(X(s))]^2 \qquad (5)$$

where γ is a weighting parameter, and $dist(X(s))$ is the Euclidean distance at point $X(s)$, which is computed during the first stage of the proposed scheme (Section 2.1).

We used a snake model with the internal and external energy functions defined in Equation 4 and Equation 5, respectively. The snake evolves from its initial configuration (Section 2.2.1) to minimize the total energy, by balancing internal and external energy functions on each vertex. The weighting parameters α, β and γ have great influence on the snake's convergence to the desired boundary. The values of these parameters were adjusted through a training process, which is explained in section 3.2.

3 Experimental Setup

Our proposed scheme was tested using a set of 40 B-mode ultrasound images. These images were obtained using ultrasound acquisition system (Ultramark 9 HDI US machine and L10-5 linear array transducer) and were digitized with a video frame grabber. These images were carefully inspected by an experienced clinician and artery contours were manually highlighted to represent gold standard images. These gold standard images are used to validate the results produced by our proposed scheme.

3.1 Objective Analysis Metric

To compare the output of the proposed scheme to the gold standard images, we define the overlap ratio as:

$$Overlap\,ratio = \frac{TP}{FN + TP + FP}, \qquad (6)$$

Fig. 1 shows the definition of the *true positive* (*TP*), *false positive* (*FP*), *true negative* (*TN*) and *false negative* (*FN*) terms.

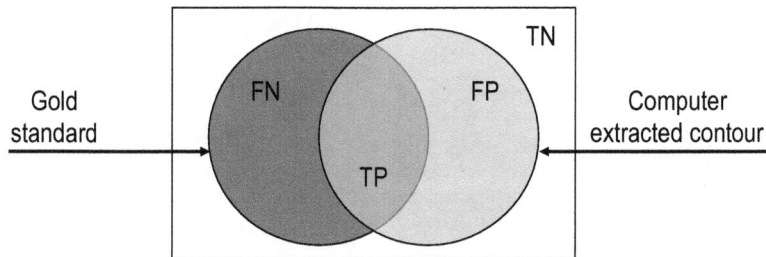

Fig. 1. The definition of the *true positive* (TP), *false positive* (FP), *true negative* (TN) and *false negative* (FN) terms, used to calculate the *overlap ratio*

3.2 Parameter Tuning

As explained in Section 2.2.2, the parameters α, β and γ of Equation 4 and Equation 5 have great influence on the snake's convergence to the desired boundary. To adjust theses parameters, a training set of 10 B-mode ultrasound images were collected and were manually segmented. This training set is different from the 40 images, used to evaluate the performance of the proposed scheme. This is to ensure independence between training and testing data. Then, the unit interval [0,1] is discretized into sub-intervals of length 0.1. The scheme was tested for all possible combinations of the parameters α, β and γ in the discrete range. The average percentage overlap ratio (Equation 6) with the gold standard images was calculated for every combination. Finally, the parameters α, β and γ are set to the values that maximize the average overlap ratio over the entire training set. The training process revealed that, setting the parameters α, β and γ to 1, 0.9, and 0.2, respectively, produces the maximum average overlap ratio over the entire training set.

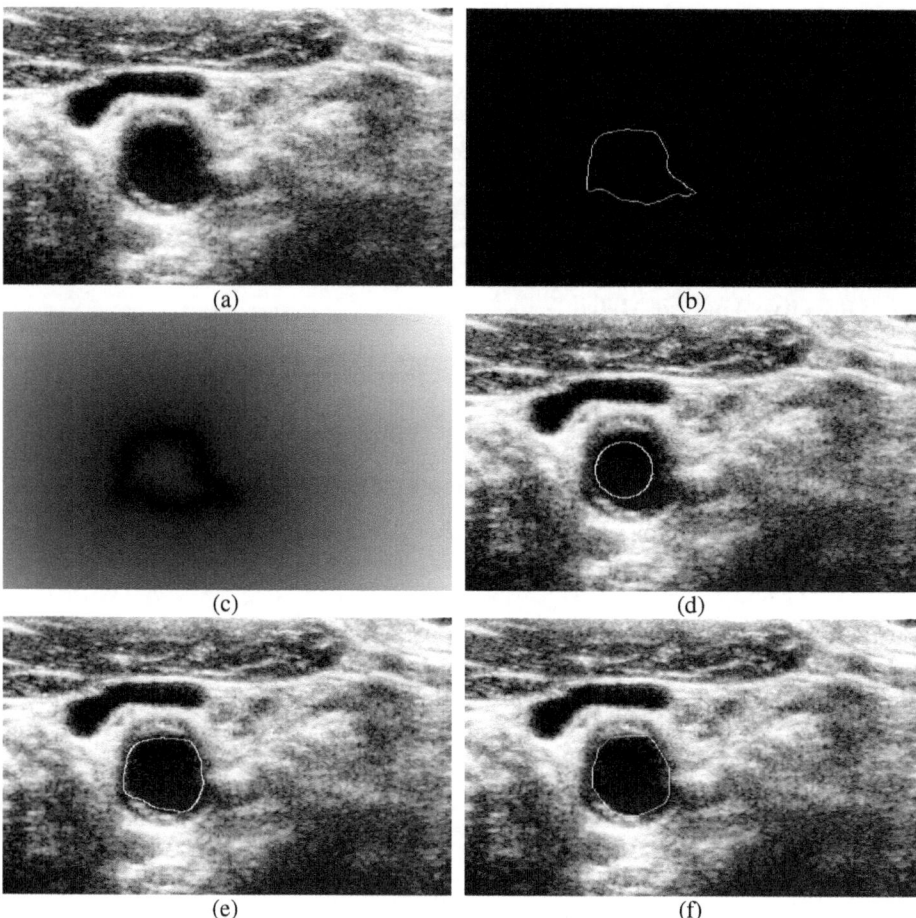

Fig. 2. Experimental results: (a) Original ultrasound image; (b) The edge map, produced by the first stage of the proposed scheme "Section 2.1 - Robust Estimation of Artery Boundary"; (c) The Euclidean distance map relative to the extracted edge map shown in (b); (d) The initial snake; (e) The final out of the proposed scheme; (f) The clinician segmented image (gold standard image)

4   Results

We used the image shown in Fig. 2(a) to demonstrate the output produced by one of our experiments. This image is a typical carotid artery ultrasound image.

Fig. 2(b) shows the edge map, produced by the first stage of the proposed scheme (Section 2.1 - Robust Estimation of Artery Boundary). Whereas, Fig. 2(c) shows the Euclidean distance map relative to the extracted edge map. This map represents the Euclidean distance between each pixel and the nearest contour pixel.

Fig. 2(d) shows the initial snake, defined as a circle, where a single seed point is required to represent the centre of the circle. Fig. 2(e) shows the final output of the proposed scheme after the snake's convergence. Comparing Fig. 2(e) to the gold standard image (Fig. 2(f)) shows that the proposed scheme accurately highlights the artery lumen.

The performance of the proposed system over the entire set of 40 images was objectively compared to the gold standard images, using the *overlap ratio* (Equation 6) as a performance metric. On average, our proposed scheme produces an *overlap ratio* of 0.766.

Two further experiments were conducted to evaluate the improvement achieved by the modified external energy function, employed in the proposed scheme. The two experiments used the entire 40 test images (same data set), the same seed point and snake initialization. However, the first experiment used Canny edge detector [3] to calculate the image's edge map. Whereas, Sobel operator was employed in the second experiment. The two experiments produces overlap ratios of 0.654 and 0.578, respectively; see Table 1 and Fig. 3 for detailed comparison results. This comparison shows that our proposed snake model surpasses the traditional snake models, under the same testing conditions.

Table 1. The performance measure of our experiments over the entire set of images

	Proposed Scheme	Snake with Canny	Snake with Sobel
Average overlap ratio	0.766	0.654	0.578
Standard deviation	0.113	0.179	0.217
95% confidence interval	[0.735, 0.797]	[0.605, 0.704]	[0.517, 0.638]

4.1 Sensitivity Analysis

Since, the seed point represents the only input from the user, it is crucial to analyze the proposed scheme sensitivity to the seed point selection. For this analysis, we used the entire 40 test images. For each image, four seed points were randomly selected *inside the artery*. The artery region was segmented for each selected seed point. Then, these segmented binary images were added up to produce a grayscale image that demonstrates the overlapping areas between segmented regions generated by the four seed points. Finally, the *percentage overlap* between segmented areas (the number of pixels having a value of 4 over all non-zero pixels) was calculated.

The statistical analysis over the entire 40 test images revealed that, on average, the proposed scheme achieved a *percentage overlap* equal to 94.9%. Hence, we can conclude that the proposed scheme is insensitive to the selected seed point, as long as it is located inside the artery area. Note that, selecting a seed point within the artery area is a trivial process even for ordinary user. Hence, the proposed scheme provides accurate results which are independent of the clinician's level of expertise.

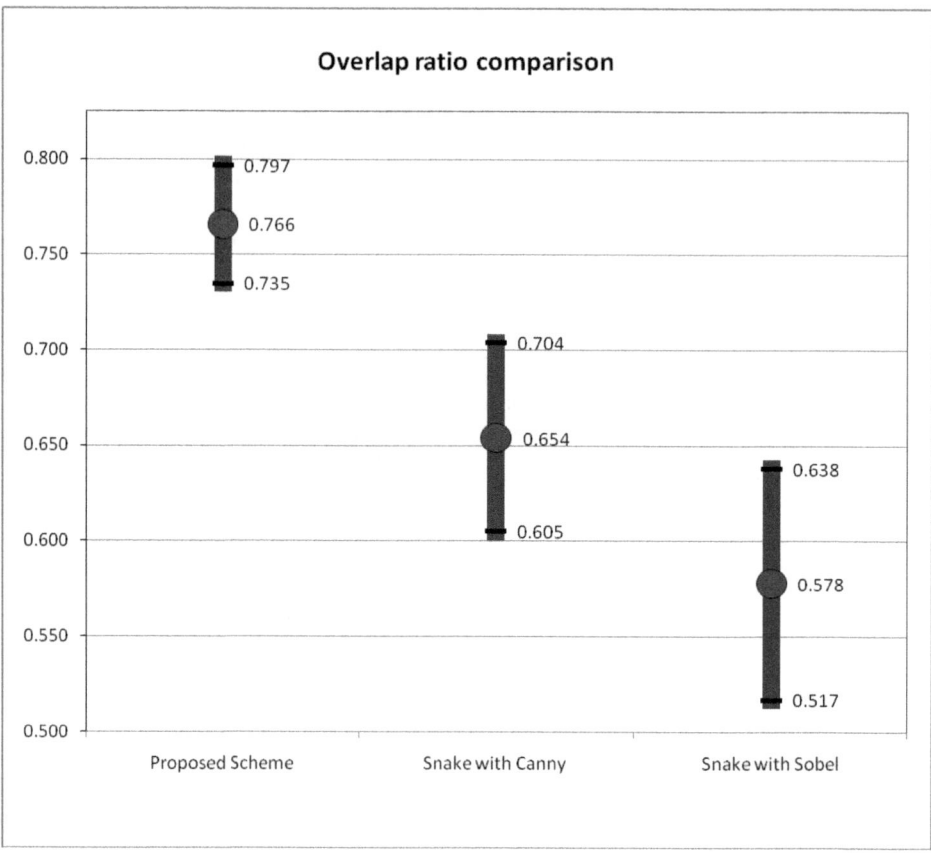

Fig. 3. The 95% confidence interval of the overlap produced by the proposed scheme, snake model with edge map extracted by Canny edge detector, and snake model with edge map extracted by Sobel operator

5 Conclusion

In this paper, a modified snake model is introduced to extract carotid artery contours from ultrasound images. The snake's energy functions are designed to force the snake to converge to a robust edge map, which is produced by employing complex diffusion-based filtering scheme. Experimental results demonstrate the efficiency of the proposed scheme in producing accurate artery contours. Furthermore, our modified snake was experimentally compared to two different snake models, found in literature. The two models force the snake to converge to the image's edge map which is produced by either Sobel or Canny edge detector. Comparative studies, using identical testing conditions, show that incorporating the complex diffusion filtering into our modified snake model outperforms the other two snake models. Finally, sensitivity analysis over the entire set of test images revealed that the scheme is insensitive to the seed point location, as long as it is located inside the artery area.

References

1. Abolmaesumi, P., Sirouspour, M., Salcudean, S.: Real-time extraction of carotid artery contours from ultrasound images. In: Proc. of the 13th IEEE Symposium on Computer-Based Med. Systems, June 2000, pp. 81–186 (2000)
2. Hamou, A., El-Sakka, M.: A novel segmentation technique for carotid ultrasound images. In: Proc. of the IEEE Int. Conf. on Acoustics, Speech and Signal Processing, May 2004, vol. 3, pp. 521–424 (2004)
3. Canny, J.: Computational Approach To Edge Detection. IEEE Trans. on Pattern Anal. and Machine Intell. 8(6), 679–698 (1986)
4. Da-chuan, C., Xiaoyi, J.: Detections of Arterial Wall in Sonographic Artery Images Using Dual Dynamic Programming. IEEE Trans. Information Tech. in Biomedicine 12(6), 792–799 (2008)
5. Abdel-Dayem, A., El-Sakka, M., Fenster, A.: Watershed segmentation for carotid artery ultrasound images. In: Proc. of the IEEE Int. Conf. on Computer Systems and Applications, January 2005, pp. 131–138 (2005)
6. Abdel-Dayem, A., El-Sakka, M.: Carotid Artery Contour Extraction from Ultrasound Images Using Multi-Resolution-Analysis and Watershed Segmentation Scheme. ICGST Int. J. on Graphics, Vision and Image Processing 5(9), 1–10 (2005)
7. Abdel-Dayem, A., El-Sakka, M.: Carotid Artery Ultrasound Image Segmentation Using Fuzzy Region Growing. In: Kamel, M.S., Campilho, A.C. (eds.) ICIAR 2005. LNCS, vol. 3656, pp. 869–878. Springer, Heidelberg (2005)
8. Abdel-Dayem, A., El-Sakka, M.: Multi-Resolution Segmentation Using Fuzzy Region Growing for Carotid Artery Ultrasound Images. In: Proc. of the IEEE Int. Computer Engineering Conf., December 2006, 8 pages (2006)
9. Abdel-Dayem, A., El-Sakka, M.: Fuzzy c-means clustering for segmenting Carotid Artery Ultrasound Images. In: Kamel, M.S., Campilho, A. (eds.) ICIAR 2007. LNCS, vol. 4633, pp. 933–948. Springer, Heidelberg (2007)
10. Abdel-Dayem, A., El-Sakka, M.: Segmentation of Carotid Artery Ultrasound Images Using Graph Cuts. Int. J. for Computational Vision and Biomechanics (in press)
11. Abdel-Dayem, A., El-Sakka, M.: Diffusion-based Detection of Carotid Artery Lumen from Ultrasound Images. In: Kamel, M., Campilho, A. (eds.) ICIAR 2009. LNCS, vol. 5627, pp. 782–791. Springer, Heidelberg (2009)
12. Kass, M., Witkin, A., Terzopoulos, D.: Snakes: Active Contour models. International Journal of Computer Vision 1, 321–331 (1988)
13. Caselles, V., Kimmel, R., Sapiro, G.: On geodesic active contours. International Journal of Computer Vision 22(1), 61–79 (1997)
14. Malladi, R., Sethian, J.A., Vemuri, B.C.: Shape modeling with front propagation: A level set approach. IEEE Trans. Pattern Anal. and Machine Intell. 17, 158–175 (1995)
15. Chan, T., Vese, L.: Active Contours Without Edges. IEEE Trans. Image Processing 10(2), 266–277 (2001)
16. Mao, F., Gill, J., Downey, D., Fenster, A.: Segmentation of carotid artery in ultrasound images. In: Proc. of the 22nd IEEE Annual Int. Conf. on Engineering in Medicine and Biology Society, July 2000, vol. 3, pp. 1734–1737 (2000)
17. Da-chuan, C., Schmidt-Trucksass, A., Kuo-Sheng, C., Sandrock, M., Qin, P., Burkhardt, H.: Automatic detection of the intimal and the adventitial layers of the common carotid artery wall in ultrasound B-mode images using snakes. In: Proc. of the Int. Conf. on Image Analysis and Processing, September 1999, pp. 452–457 (1999)

18. Cohen, L.: On active contour models and balloons. Computer Vision, Graphics, and Image Processing: Image Understanding 53(2), 211–218 (1991)
19. Hamou, A., Osman, S., El-Sakka, M.: Carotid Ultrasound Segmentation Using DP Active Contours. In: Kamel, M.S., Campilho, A. (eds.) ICIAR 2007. LNCS, vol. 4633, pp. 961–971. Springer, Heidelberg (2007)
20. Weickert, J.: Anisotropic Diffusion in Image Processing, ch. 1. ECMI Series, pp. 1–53. Teubner-Verlag, Stuttgart (1998)
21. Romeny, B.: Geometry-Driven Diffusion in Computer Vision. Computational Imaging and Vision 1, 39–71 (1994)
22. Saint-Marc, P., Chen, J., Medioni, G.: Adaptive smoothing: a general tool for early vision. IEEE Trans. Pattern Anal.&Machine Intell. 13(6), 514–529 (1991)
23. Perona, P., Malik, J.: Scale-space and edge detection using anisotropic diffusion. IEEE Trans. Pattern Anal. & Machine Intell. 12(7), 629–639 (1990)
24. Osher, S., Rudin, L.: Feature-oriented image enhancement using shock filters. SIAM J. on Numerical Analysis 27(4), 919–940 (1990)
25. Catté, F., Lions, P., Morel, J., Coll, T.: Image selective smoothing and edge detection by nonlinear diffusion. SIAM J. on Numerical Analysis 29(1), 182–193 (1992)
26. Alvarez, L., Mazorra, L.: Signal and image restoration using shock filters and anisotropic diffusion. SIAM J. on Numerical Analysis 31(2), 590–605 (1994)
27. Black, M., Sapiro, G., Marimont, D., Heeger, D.: Robust anisotropic diffusion. IEEE Trans. Image Processing 7(3), 421–432 (1998)
28. Gilboa, G., Sochen, N., Zeevi, Y.: Image enhancement and Denoising by Complex Diffusion Processes. IEEE Trans. Pattern Anal.&Machine Intell. 26(8), 1020–1036 (2004)
29. Gonzalez, G., Woods, E.: Digital image processing, 3rd edn. Prentice Hall, Englewood Cliffs (2008)
30. Dargherty, E., Lotufo, R.: Hands–on morphological image processing. The society of Photo-Optical Instrumentation Engineers (2003)

Classification of Endoscopic Images Using Delaunay Triangulation-Based Edge Features

M. Häfner[1], A. Gangl[2], M. Liedlgruber[3], A. Uhl[3], A. Vécsei[4], and F. Wrba[5]

[1] Department for Internal Medicine, St. Elisabeth Hospital, Vienna
[2] Department of Gastroenterology and Hepatology, Medical University of Vienna, Austria
[3] Department of Computer Sciences, Salzburg University, Austria
[4] St. Anna Children's Hospital, Vienna, Austria
[5] Department of Clinical Pathology, Medical University of Vienna, Austria
{mliedl,uhl}@cosy.sbg.ac.at

Abstract. In this work we present a method for an automated classification of endoscopic images according to the pit pattern classification scheme. Images taken during colonoscopy are transformed using an extended and rotation invariant version of the Local Binary Patterns operator (LBP). The result of the transforms is then used to extract polygons from the images. Based on these polygons we compute the regularity of the polygon positions by using the Delaunay triangulation and constructing histograms from the edge lengths of the Delaunay triangles. Using these histograms, the classification is carried out by employing the k-nearest-neighbors (k-NN) classifier in conjunction with the histogram intersection distance metric.

While, compared to previously published results, the performance of the proposed approach is lower, the results achieved are yet promising and show that a pit pattern classification is feasible by using the proposed system.

1 Introduction

Today, the third most common malignant disease in western countries is colon cancer. Therefore a regular colon examination is recommended, especially for people at an age of 50 years and older. Currently the gold standard for colon examination is colonoscopy, which is performed by using a colonoscope. Modern colonoscopes are able to take pictures from inside the colon which allows to obtain images for a computer-assisted analysis with the goal of detecting tumorous lesions. To get highly detailed images a magnifying endoscope is used [1]. Such an endoscope represents a significant advance in colonoscopy as it provides images which are up to 150-fold magnified, thus uncovering the fine surface structure of the mucosa as well as small lesions.

In Sect. 2 we review the classification of pit patterns of the colonic mucosa. Section 3 describes the feature extraction process, including image transformation using a LBP extension, polygon extraction, Delaunay-based feature computation, histogram creation, and the classification. Experimental results and configuration details of the classification system proposed are given in Sect. 4. Section 5 concludes the paper.

A. Campilho and M. Kamel (Eds.): ICIAR 2010, Part II, LNCS 6112, pp. 131–140, 2010.

2 Pit Pattern Classification

Polyps of the colon are a frequent finding and are usually divided into metaplastic, adenomatous, and malignant. As resection of all polyps is time-consuming, it is imperative that those polyps which warrant endoscopic resection can be distinguished: polypectomy of metaplastic lesions is unnecessary and removal of invasive cancer may be hazardous. For these reasons, assessing the malignant potential of lesions at the time of colonoscopy is important.

The most commonly used classification system to distinguish between non-neoplastic and neoplastic lesions in the colon is the pit pattern classification, originally reported by Kudo et al. [2]. This system allows to differentiate between normal mucosa, hyperplastic lesions (non-neoplastic), adenomas (a pre-malignant condition), and malignant cancer based on the visual pattern of the mucosal surface. Thus this classification scheme is a convenient tool to decide which lesions need not, which should, and which most likely can not be removed endoscopically. The mucosal pattern as seen after dye staining and by using magnification endoscopy shows a high agreement with the histopathologic diagnosis. Due to the visual nature of this classification it is also a convenient choice for an automated image classification.

As illustrated in Fig. 1(a)-(f) in this classification scheme exist five main types according to the mucosal surface of the colon. Type III is divided into types III-S and III-L, designating the size of the pit structure. It has been suggested that type I and II pattern are characteristic of non-neoplastic lesions (benign and non-tumorous), type III and IV are found on adenomatous polyps, and type V are strongly suggestive of invasive carcinoma, thus highly indicative for cancer.

Furthermore lesions of type I and II can be grouped into non-neoplastic lesions and types III to V can be grouped into neoplastic lesions. This allows a grouping of lesions into two classes, which is more relevant in clinical practice as indicated in a study by Kato et al. [3].

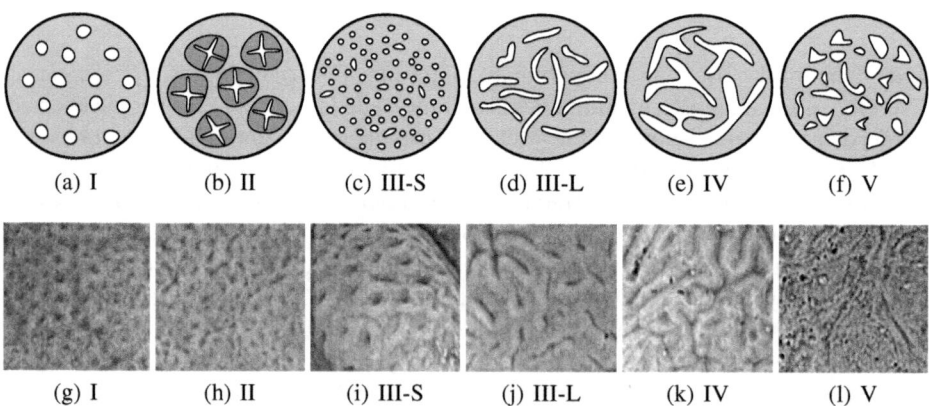

(a) I	(b) II	(c) III-S	(d) III-L	(e) IV	(f) V
(g) I	(h) II	(i) III-S	(j) III-L	(k) IV	(l) V

Fig. 1. Pit pattern classification according to Kudo et al. (a)-(f) Schematically and (g)-(l) example images for the respective classes taken from the available image database

Using a magnifying colonoscope together with indigo carmine dye spraying, the mucosal crypt pattern on the surface of colonic lesions can be observed [4]. Several studies found a good correlation between the mucosal pit pattern and the histological findings, where especially techniques using magnifying colonoscopes led to excellent results [3].

From Fig. 1 we notice that pit pattern types I to IV can be characterized fairly well, whereas type V is a composition of unstructured pits. At a first glance this classification scheme seems to be straightforward and easy to be applied. But it needs some experience and exercising to achieve fairly good results [5].

As evident from Fig. 1(g)-(l), pit pattern types I and II are regular to some extent and the pits are distributed more tightly. Types III to V in contrast are more irregular in terms of the pit distribution, showing a lower pit density or even a complete absence of pits. These observations are the basis for the method presented in the following sections.

3   Proposed Approach

In the past we have already shown that an automated classification of endoscopic images based on the pit pattern scheme is feasible. But in our previous work we mainly focused on general purpose features describing texture properties (e.g. [6,7,8]), dealing with the two-classes case as well as with the six-classes case. By contrast, the method proposed in this work aims at distinguishing between non-neoplastic and neoplastic images only. It is furthermore based on high level features obtained by measuring the density of pits visible within the images. This is inspired by the fact that the pit distributions in non-neoplastic images are more dense than in case of the neoplastic ones, as already pointed out above. An overview of the feature extraction process is shown in Fig. 2.

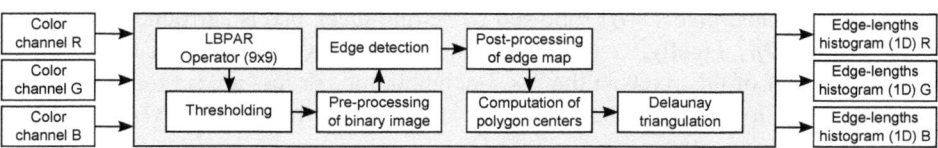

Fig. 2. This figure illustrates the different parts of the feature extraction process. The parts within the gray box are carried out for each color channel under consideration separately.

3.1   Local Binary Patterns

Prior to any further processing all color channels of the input images are transformed separately using a modified Local Binary Patterns operator (LBP) based on block averaging which we already used successfully to classify endoscopic images [8]. In contrast to the standard operator, which is described in more detail in [9], we compute the average over neighboring blocks and compare the average of the center block against the averages of the neighboring blocks to obtain the LBP number. By adjusting the block size used it is possible to find a trade-off between noise-suppression (higher block widths) and detail preservation (smaller block widths). Throughout this work we used a rather high block width of 9 pixels to suppress noise which otherwise would have had a negative influence on the subsequent edge detection.

The motivation behind extracting edges from LBP transformed channels is that the pit structures we try to locate can be identified more easily since pits usually are surrounded by brighter areas. In terms of LBP searching for pits thus corresponds to locating LBP numbers above some certain threshold. Another advantage of the LBP operator is that it is known to be invariant against global illumination changes in images.

Furthermore, we achieve rotation invariance by circularly rotating each LBP number obtained until the minimum is reached [10]. This way we are able to cope with changes in the direction of illumination across different images. In the remaining part of this work this combination of averaged LBP blocks and rotation invariance is abbreviated with LBPAR.

3.2   Polygon Extraction

In order to extract polygons from a LBP channel we first apply a global thresholding. The choice for the threshold used throughout this work is motivated by the appearance of an ideal pit ($t = 127$). It is chosen such that at least 7 of the 8 neighbor block averages must be higher than the center block average for a pixel to be assumed to be part of a pit.

Prior to edge detection we pre-process the binary image by using a set of six different morphological operators (O_C, O_B, O_I, O_H, O_M, and O_R). First, we apply a closing (O_C) - using a disk of radius 1 as structuring element - to remove small "holes". The small radius has been chosen to not disturb the shape of the pits too much but to only fill small holes and cancel out small notches eventually present along the borders of pit areas. Then we bridge (O_B) unconnected pixels by setting pixels to white which lie between two unconnected, white neighbors (using a 3×3-neighborhood). Furthermore we remove isolated pixels (O_I) followed by setting black pixels surrounded by white ones to white (O_H). Finally, we cancel out pixels which have less than five white neighbors if only half of the pixels in the 3×3-neighborhood or less are set to white (O_M). This step helps to minimize the number of small spurs which might eventually have endured the previous steps.

To extract edges from the resulting binary image we use the Canny edge detector without multi-resolution feature synthesis [11] which may produce polygons having a boundary with gaps. Thus we post-process the edges by using the morphological operators from above, except for the closing (in the same order). To obtain the final edge map we remove all interior pixels of the white areas (O_R).

This processing of the binary map and edges ensures that we end up with closed polygons only, which are smooth and free of unwanted artifacts. We get only closed polygons since when applying the filling of closed areas (O_H) polygons previously not closed are not affected (not filled) as can be seen in Fig. 3(e). By subsequently applying O_M these polygons are removed, which can be noticed from Fig. 3(f).

Apart from that we see from Fig. 3(a) that some images contain ridges which can be considered to be artifacts. By applying the post-processing steps to the edges ridges touching the image border are removed, thus reducing the number of these artifacts.

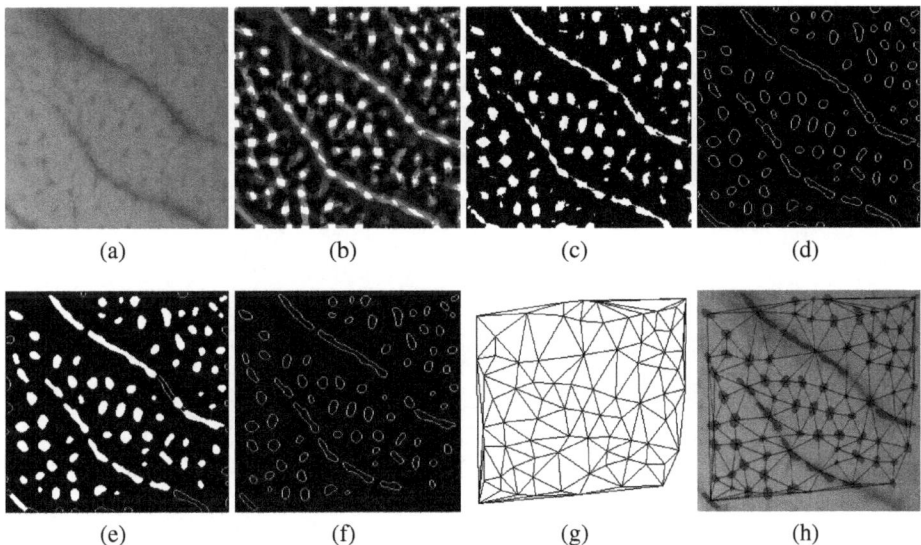

Fig. 3. Some steps from the process of obtaining polygons from a color channel of a pit pattern type I image (a) the red color channel of the input image, (b) the according LBPAR transformed image, (c) the result of thresholding, (d) result of the Canny edge detector, (e) the pits (white and filled) and the parts which get discarded by the edge map post-processing (not filled), (f) the final edge map, (g) the respective Delaunay triangulation, and (h) the triangulation with the pits overlayed to the original color channel

After tracing the edges of the connected components we determine the polygon center for each polygon as the mean position of all edge pixels belonging to the polygon. Some of these steps are illustrated in Fig. 3(a)-(f).

3.3   Delaunay Triangulation

To measure the density of pits within an image we aim at constructing a mesh from the previously extracted polygon centers. Then we deduce the density of the pits from the edge lengths within the mesh. For this purpose we employ the Delaunay triangulation based on the Quickhull algorithm [12].

This algorithm basically transforms the 2D points to 3D (lifted to a paraboloid), computes the convex hull in 3D, and projects the lower part of the hull back to 2D to obtain the triangulation. This way we get a set of non-overlapping triangles with the minimum of the inner angles maximized. An example triangulation for a pit type I image is shown in Fig. 3(g).

Figure 4 shows sample images from our image database along with the respective Delaunay triangulations and the detected pits. From this figure we notice that non-neoplastic images exhibit a higher density with respect to the arrangement of the detected pits. But we also notice that in case of the non-neoplastic images ridges have a negative influence on this density in some parts of the images.

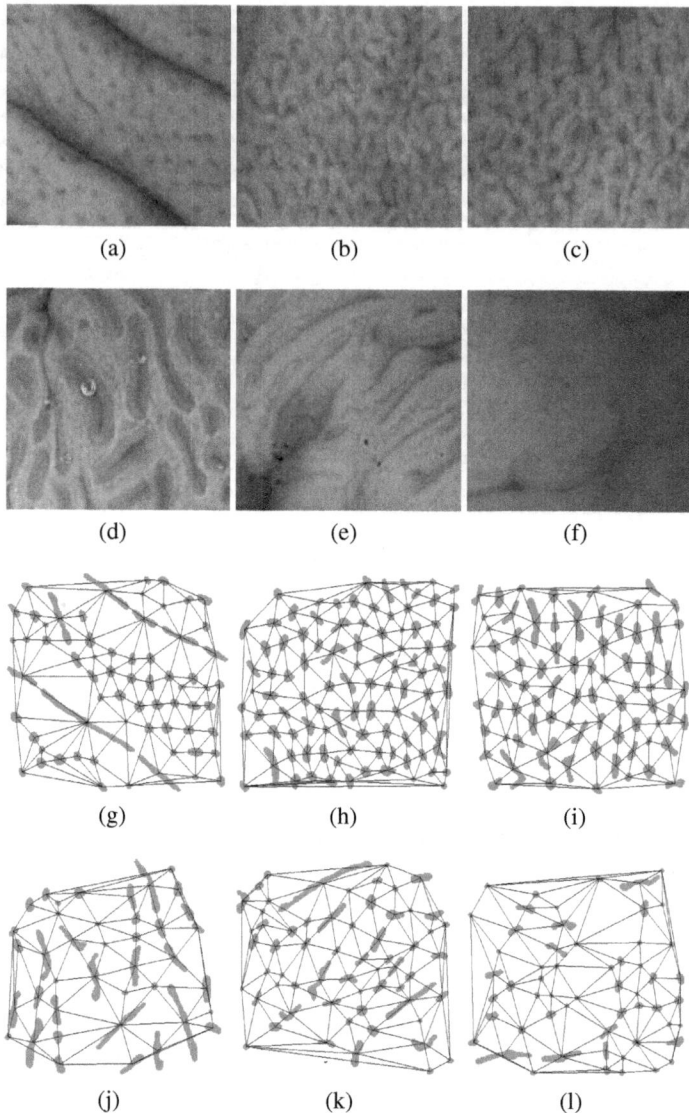

Fig. 4. Results of the Delaunay triangulation along with the detected pits. (a)-(c) example images from the non-neoplastic class (red channel), (d)-(f) neoplastic images, and (g)-(l) the according Delaunay triangulations along with the detected pits.

3.4   Histogram Creation and Classification

Based on the triangulations we create 1-dimensional histograms from the edge lengths of all triangles for each color channel of an image separately. To concentrate on triangles not located on the border of the triangulation we iterate over all triangles and use each edge of each triangle to update the histogram. This way edges shared by two triangles

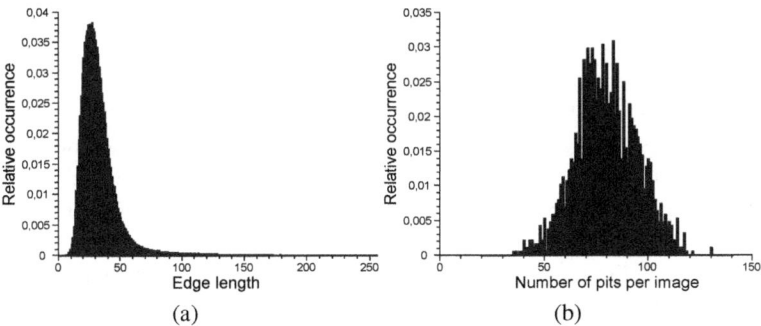

Fig. 5. (a) Relative occurrences of the different edge lengths in our image database and (b) the relative occurrences of the different number of detected pits across all images

contribute to the histogram twice, while edges located on the border of the triangulation result are used only once.

If only a few pits are detected within an image the respective edge lengths will be rather high. For a high number of detected pits (higher density) the distances between them will get smaller, hence lowering the respective Delaunay edge lengths too.

Since the number of edges between images most likely will vary we normalize each histogram such that the histogram bins sum up to 1. This makes the histograms comparable during the classification process. Moreover, since all our images have a dimension of 256×256 pixels the upper limit for an edge length is $\sqrt{256^2 + 256^2} \approx 362$ (corresponding diagonal). But it is very unlikely that pits are only detected in the image corners. This implies that it is also unlikely that the maximum possible edge length occurs. Apart from that, the more pits we detect the more likely it is that the distances between neighboring pits get smaller.

The images used throughout our experiments show a maximum edge length of approximately 249, but most edge lengths lie between 10 and 100, as can be observed from Fig. 5(a). We also detect a rather high number of pits in each of our images (between 35 and 130), as can be seen from Fig. 5(b).

Based on these observations we consider the range for the edge lengths between 1 and 256 as a reasonable choice and therefore use this range throughout our experiments.

For the classification of unknown images we employ the k-NN classifier along with the histogram intersection distance metric, defined as

$$d(H_i, H_j) = \sum_{k=1}^{B} \min\left(H_{i,k}, H_{j,k}\right), \tag{1}$$

where H_i and H_j are two normalized histograms, B denotes the number of bins used in our histograms, and $H_{i,k}$ and $H_{j,k}$ represent the value of the k-th bin of histogram H_i and H_j, respectively. We also carried out experiments using the Euclidean distance metric and the Bhattacharyya distance metric but the histogram intersection always yielded a slightly better classification performance.

To combine multiple color channels we compute the distances for each color channel separately and multiply them to obtain the final distance D. This can be formulated as

$$D(I_a, I_b) = \prod_{i=1}^{C} d\left(H_i^{(a)}, H_i^{(b)}\right), \qquad (2)$$

where I_a and I_b denote two images, C is the number of color channels considered for combination, and $H_i^{(a)}$ and $H_i^{(b)}$ represent the histograms for the i-th color channel considered of image I_a and I_b, respectively. There are also other possibilities for a combination, for example summing up the distances instead of multiplying them by replacing the product in (2) by a sum. But since the product is more tolerant against outliers - one similar color channel in terms of histogram distance leads to a very small total distance between two images already - we favor the product instead of a sum.

4 Experiments

4.1 Settings

The image database used throughout our experiments consists of 627 images acquired between the years 2005 and 2008 at the Department of Gastroenterology and Hepatology (Medical University of Vienna) using a zoom-colonoscope (Olympus Evis Exera CF-Q160ZI/L) with a magnification factor set to 150.

Lesions found during colonoscopy have been examined after application of dye-spraying with indigocarmine as routinely performed in colonoscopy. Biopsies or mucosal resection have been performed in order to get a histopathological diagnosis. Biopsies have been taken from type I, II, and type V lesions, as those lesions need not to be removed or cannot be removed endoscopically. Type III and IV lesions have been removed endoscopically. Out of all acquired images, histopathological classification resulted in 178 non-neoplastic and 449 neoplastic cases which is used as ground truth for our experiments.

Using leave-one-out cross-validation, 626 out of 627 images are used as training set. The remaining image is then classified. This process is repeated for each image.

To find the optimal values for B (number of histogram bins used) and k for the k-NN classifier we carry out a naive search testing all possible combinations for $k = 1, \ldots, 25$ and $B = 16, \ldots, 256$ (for different color channel combinations).

4.2 Results

From the results shown in Table 1 we see that the proposed method achieves very promising results – in particular when combining two or more color channels. The best result has been obtained by combining the red and the blue channel, resulting in an overall classification accuracy of 93,3%. But also combining all color channels available yielded a high result of 93%.

From the results we also see that in case of the single channel results the green channel yielded the worst results. Also in case of combined channels the results always drop as soon as the green channel is taken into consideration.

Despite the high overall classification results we also notice that there is an imbalance between the two classes. While the results for the neoplastic images are always above

Table 1. Overall classification rates obtained by different color channel combinations along with the respective choices for k and B (compared to the results published in [8])

	Non-neoplastic	Neoplastic	Total	k	B
R	61,2	98,0	87,6	5	249
G	57,9	94,9	84,4	11	243
B	74,2	91,3	86,4	8	179
R+B	83,2	97,3	93,3	15	202
R+G	68,0	98,2	89,6	7	249
G+B	78,7	92,9	88,8	10	220
R+G+B	77,5	99,1	93,0	11	53
[8]	98,3	99,5	99,2	–	–

90% the results for the first class vary between approximately 58% and 83% only. This effect is especially apparent in case of the single channel results. When considering the ground truth we notice that the number of neoplastic images is about 1.6 times higher compared to the other class, which is one reason for this behavior.

Compared to the results we published in [8] we see that especially in case of the non-neoplastic images the results of the proposed approach are still very low. This is most possibly due to ridges, which – although not characteristic for non-neoplastic images – sometimes appear in these images too (see Fig. 3). As we also notice from Fig. 3(g) these ridges have a noticeable influence on the triangulation result.

Additional problems arise from image artifacts and noise which are quite frequently misinterpreted as being pits. As a consequence neoplastic images get more similar to non-neoplastic ones in terms of the Delaunay edge length histograms which makes misclassification of such images more likely. Although this problem exists, especially in case of neoplastic images, this is not evident from Table 1 due to the imbalance between the two classes.

5   Conclusion and Future Research

In this work we presented a method for an automated pit pattern classification system which - in contrast to all our previously published methods - is strongly linked to the visual appearance of the pits on the colonic mucosa. Although, compared to previously obtained results, this method still delivers lower recognition rates, the results we currently achieve are very promising already - especially when combining different color channels for the classification.

We also identified ridges as a potential problem being very likely one cause for a lowered classification performance. In future work we will therefore focus on minimizing the effect of ridges to a maximum possible extent. Besides that we will also have to investigate other features in order make the system work in the six-classes case as well. This case has been neglected completely in this work due to the nature of the features used, since these rely on differences in the density of pit distributions across different image classes. In the six-classes case this is unfortunately not sufficient since these differences are not that distinct between all of the six classes. We will also have to make the pit detection more robust to improve the discrimination between the image classes.

Another interesting possibility will be to use the method proposed as part of an ensemble classifier, since this method works completely different compared to our previous approaches.

Acknowledgements

This work is partially funded by the Austrian Science Fund (FWF) under Project No. L366-N15 and by the Austrian National Bank "Jubiläumsfonds" Project No. 12514.

References

1. Bruno, M.J.: Magnification endoscopy, high resolution endoscopy, and chromoscopy; towards a better optical diagnosis. Gut 52(4), 7–11 (2003)
2. Kudo, S., Hirota, S., Nakajima, T., Hosobe, S., Kusaka, H., Kobayashi, T., Himori, M., Yagyuu, A.: Colorectal tumours and pit pattern. Journal of Clinical Pathology 47, 880–885 (1994)
3. Kato, S., Fu, K., Sano, Y., Fujii, T., Saito, Y., Matsuda, T., Koba, I., Yoshida, S., Fujimori, T.: Magnifying colonoscopy as a non-biopsy technique for differential diagnosis of non-neoplastic and neoplastic lesions. World Journal of Gastroenterology: WJG 12(9), 1416–1420 (2006)
4. Kudo, S., Tamura, S., Nakajima, T., Yamano, H., Kusaka, H., Watanabe, H.: Diagnosis of colorectal tumorous lesions by magnifying endoscopy. Gastrointestinal Endoscopy 44(1), 8–14 (1996)
5. Hurlstone, D.: High-resolution magnification chromoendoscopy: Common problems encountered in "pit pattern" interpretation and correct classification of flat colorectal lesions. American Journal of Gastroenterology 97, 1069–1070 (2002)
6. Häfner, M., Gangl, A., Liedlgruber, M., Uhl, A., Vécsei, A., Wrba, F.: Pit pattern classification using multichannel features and multiclassification. In: Exarchos, T.P., Papadopoulos, A., Fotiadis, D.I. (eds.) Handbook of Research on Advanced Techniques in Diagnostic Imaging and Biomedical Applications, pp. 335–350. IGI Global, Hershey (2009)
7. Häfner, M., Gangl, A., Liedlgruber, M., Uhl, A., Vecsei, A., Wrba, F.: Combining Gaussian Markov random fields with the discrete wavelet transform for endoscopic image classification. In: Proceedings of the 17th International Conference on Digital Signal Processing (DSP'09), Santorini, Greece (2009)
8. Häfner, M., Gangl, A., Liedlgruber, M., Uhl, A., Vécsei, A., Wrba, F.: Pit pattern classification using extended local binary patterns. In: Proceedings of the 9th International Conference on Information Technology and Applications in Biomedicine (ITAB'09), Larnaca, Cyprus (November 2009)
9. Ojala, T., Pietikäinen, M., Harwood, D.: A comparative study of texture measures with classification based on feature distributions. Pattern Recognition 29(1), 51–59 (1996)
10. Mäenpää, T., Pietikäinen, M.: Texture analysis with local binary patterns. In: Handbook of Pattern Recognition and Computer Vision, 3rd edn., pp. 197–216. World Scientific, Singapore (2005)
11. Canny, J.: A computational approach to edge detection. IEEE Transactions on Pattern Recognition and Machine Intelligence 8(6), 679–698 (1986)
12. Barber, C., Dobkin, D., Huhdanpaa, H.: The Quickhull algorithm for convex hulls. ACM Transactions on Mathematical Software 2(4), 469–483 (1996)

A Framework for Cerebral CT Perfusion Imaging Methods Comparison

Miguel Moreira[1], Paulo Dias[1,2], Miguel Cordeiro[3,5],
Gustavo Santos[4], and José Maria Fernandes[1,2]

[1] Institute of Electronics and Telematics Engineering of Aveiro
[2] Dept. of Electronics, Telecommunications and Informatics
University of Aveiro, Campus Universitário de Santiago, 3810-193, Aveiro, Portugal
{miguelmoreira,paulo.dias,jfernan}@ua.pt
[3] Faculty of Medicine, University of Coimbra, Coimbra, Portugal
[4] Dept. of Neurology and [5] Neuroradiology
Hospitais Universitários de Coimbra, Coimbra, Portugal

Abstract. Stroke is among the most frequent cause of death around the world and the decision to treat and final outcome is highly dependent on the quality of diagnosis. Recently, cerebral perfusion tomography have been used with promising results in the stroke evaluation mainly because this technique gives further information about the hemodynamic changes within the stroke area. However many different parameters are actually used to analyze the CT perfusion results, trying to integrate the temporal information it contains. Some of these parameters are Blood Volume, Blood Flow or Transit Time for example. This paper reviews the most relevant methods used to calculate perfusion related parameters and describes our framework that defines a reproducible processing pipeline that supports visual and quantified comparison between them.

Keywords: Stroke, computed tomography, perfusion CT, brain imaging, blood volume, blood flow, transit time.

1 Introduction

Stroke [1] is one of the major causes of death around the world. A stroke happens when there's a sudden vessel occlusion – usually with a blood clot – which results in inefficient blood supply and leads to poor oxygenation of brain cells. As a result, cellular activity is perturbed and can lead to cellular death if early recanalization does not occur.

However it is possible to distinguish two different areas in stroke. One is the infarct penumbra where cells are affected by the lack of oxygen but still intact and possible to recover with a fast reperfusion, the other is called infarct core where there's a cell death and no recovery is possible. Within a limited timeframe (around 3 hours from the stroke) it is possible to recover the brain tissue in the penumbra with an injection of a tissue plasminogen activator to destroy blood clots – thrombolysis – and avoid total tissue loss [2-5]. Discriminating the penumbra from the unrecoverable area is, for that reason, the main clinical issue in the acute stroke management [6-9].

A. Campilho and M. Kamel (Eds.): ICIAR 2010, Part II, LNCS 6112, pp. 141–150, 2010.

Several imaging tools are currently used to support such decision namely Computed Tomography (CT) and Magnetic Resonance Image (MRI) [10]. In the current paper the focus is on cerebral perfusion CT (PCT – Perfusion Computed Tomography).

Cerebral perfusion uses a contrast material that is injected in brain vessels that enables tracing the blood flows in cerebral vessels along time [11]. Both PCT and MRI perfusion modalities measure the concentration of contrast material along time in the tissues generating a Time-Concentration Curve for each voxel (Fig. 1) from time of injection to time of contrast material leave the system resulting in tridimensional perfusion maps for overall brain perfusion.

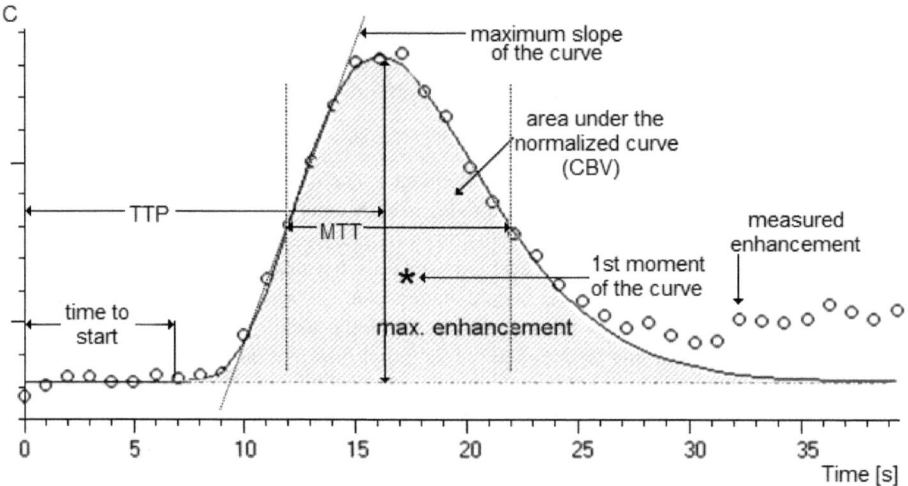

Fig. 1. Time concentration curve – The curve is obtained by measuring the concentration (C) of the contrast material in a given brain position (voxel) along time. [6, 12-13]

From the Time-Concentration curve several parameters can be extracted to characterize the hemodynamic blood flow. The usual parameters are Cerebral Blood Flow (CBF), Cerebral Blood Volume (CBV), Mean Transit Time (MTT) and Time To Peak (TTP) [14]. CBV is the percentage of blood per unit volume of tissue. In infarct penumbra CBV is usually normal or high due to auto-regulation mechanism but low in the infarct core. CBF represents the time a certain amount of blood takes to pass the cerebral blood vessels and arrive to the veins, in infarct penumbra and infarct core this value is low because of the artery obstruction. MTT is the time between the inflow and outflow blood flow in the brain. TTP is the time elapsed between the injection of the contrast material and the appearance of a maximum concentration in the cerebral blood vessels [6-8, 13, 15-17].

Regardless of the clinical relevance of these measures [6, 15], their actual use lacks an independent and reproducible validation namely because the measurement process and algorithms often depends on closed commercial applications use or in human expertise making it difficult to compare objectively the different methods [18-20].

The objective of this work is to present the most relevant methods used to calculate perfusion related parameters and describe our framework that defines a reproducible and traceable processing pipeline that can support visual and quantified comparison between them.

2 Methods

In this section we present the methods described in the literature for calculating Cerebral Blood Volume (CBV), Cerebral Blood Flow (CBF), Mean Transit Time (MTT) and Time To Peak (TTP).

2.1 Cerebral Blood Volume (CBV)

Cerebral Blood Volume is the percentage of blood per unit volume of brain tissue, according to Axel [21] the CBV can be determined using the following equation:

$$CBV = \frac{\int_0^\infty C_t(t)dt}{\int_0^\infty C_a(t)dt} = \frac{\int_0^\infty C_t(t)dt}{\int_0^\infty C_v(t)dt} = \frac{\sum_{t_0}^\infty C_t(t)}{\sum_{t_0}^\infty C_a(t)} = \frac{\sum_{t_0}^\infty C_t(t)}{\sum_{t_0}^\infty C_v(t)}. \tag{1}$$

The integral of the concentration indicates the fractional vascular volume, this volume represents the ratio between the area under the concentration curve of the contrast material $C_t(t)$ through the brain tissue and the area under the curve of the artery $C_a(t)$ or vein $C_v(t)$, if Blood Brain Barrier (BBB) is still intact the results are the same. The integral can be replaced by a sum as we are in discrete time [21-22].

In another method proposed by Klotz and König [22], CBV is determined using maximum values concentration tissue and vein, according to equation (2):

$$CBV = \frac{\max C_t(t)}{\max C_v(t)}. \tag{2}$$

2.2 Cerebral Blood Flow (CBF)

Cerebral Blood Flow represents the time that a certain amount of blood takes to flow through the brain vessels and arrive to the veins.

CBF can be obtained using Fick's method, calculating the derivation of concentration curve. This method is simple but relies on the assumptions of single blood inflow and outflow [23]. In this method, the CBF is given by equation (3) where parameter t_{max} represents the instant of maximum slope (maximum derivation in upslope segment of the curve, see Fig. 1) of the curve.

If the maximum tissue slope is reached before venous outflow starts $C_v(t_{max}) = 0$ and equation (3) turns into equation (4).

$$CBF = \frac{dC_t(t_{max})/dt}{C_a(t_{max}) - C_v(t_{max})} \tag{3}$$

$$CBF = \frac{dC_t(t_{max})/dt}{C_a(t_{max})} \; . \tag{4}$$

This method does not require correction for recirculation of contrast material and the results derived from a short period of time reducing possible patients movements, however, it is more susceptible to noise and require a pre-processing step to reduce noise in input data [22].

Another method widely used is based on deconvolution. The main advantage is the possibility to reduce administration rates of contrast material since delay and dispersion of the contrast material is corrected using the residue function. The Singular Value Decomposition (SVD) is generally used in most commercial applications as it is less sensitive to variations in vascular anatomy because of the assumption of the single point of input and output blood [23-24].

The variation of contrast concentration tissues can be described in function of arterial input function (AIF), the residue function and CBF (5).

$$C_t(t) = C_a(t) \otimes R(t) \cdot CBF \; . \tag{5}$$

The residue function R(t) represents the fraction of the contrast material that remains in the tissue at time t. The final CBF in each voxel is the maximum value of R(t) [25]. In our implementation, we compute R(t) using SVD. Since the concentration analysis performed for small time intervals Δt, we can consider the residue function and arterial flow as constant and can use the following approach:

$$\Delta t \begin{bmatrix} C_a(t_0) & 0 & \cdots & 0 \\ C_a(t_1) & C_a(t_0) & \cdots & 0 \\ \cdots & \cdots & \ddots & 0 \\ C_a(t_{N-1}) & C_a(t_{N-2}) & \cdots & C_a(t_0) \end{bmatrix} \cdot \begin{bmatrix} R(t_0) \\ R(t_1) \\ \vdots \\ R(t_{N-1}) \end{bmatrix} = \begin{bmatrix} C_t(t_0) \\ C_t(t_1) \\ \vdots \\ C_t(t_{N-1}) \end{bmatrix} \tag{6}$$

For simplification we can assume:

$$A \cdot b = c \; . \tag{7}$$

Where b are the values of the residue function and c is the concentration in tissue.

Equation (7) is solved using SVD. The method uses three matrices: V, W and U^T, where W is a diagonal matrix, V and U^T are orthogonal matrices, U^T denotes a transpose matrix.

$$A \cdot b = c \Leftrightarrow b = A^{-1} \cdot c \; . \tag{8}$$

$$A = U \cdot S \cdot V^T \; . \tag{9}$$

$$A^{-1} = V \cdot 1/S \cdot U^T = V \cdot W \cdot U^T \tag{10}$$

$$b = V \cdot W \cdot U^T \cdot c \; . \tag{11}$$

Ostegaard et al. [26] assumes $C_a(t)$ and $R(t)$ varies linearly with time, and the elements a_{ij} matrix A are:

$$a_{ij} = \begin{cases} \Delta t \big(C_a(t_{i-j-1}) + 4C_a(t_{i-j}) + C_a(t_{i-j+1}) \big)/6 & 0 \le j \le i \\ 0 & otherwise \end{cases}. \tag{12}$$

To minimize the oscillation of $R(t)$, a cutoff threshold level of 20% of the maximum value of the diagonal matrix W is used [26-28].

2.3 Mean Transit Time (MTT)

The Mean Transit Time is the average time necessary for the blood to flow through the brain.

Using the first moment of the curve (equivalent to the center of gravity of the shape defined by the time concentration curve, see Fig. 1 [12]), MTT can be calculated using the equation (13):

$$\bar{t} = \frac{\int_0^\infty t \cdot C(t) dt}{\int_0^\infty C(t) dt} = \frac{\sum_0^\infty t \cdot C(t)}{\sum_0^\infty C(t)}. \tag{13}$$

Another method proposed by Axel [21] is the area of the curve divided by its height according to the equation (14), where height is the difference between Cmax and Cmin for each voxel.

$$\bar{t} = \frac{\int_0^\infty C(t) dt}{height} \Rightarrow \frac{\sum_0^\infty (C(t) - C_{\min}(t))}{C_{\max}(t) - C_{\min}(t)}. \tag{14}$$

In this equation we must translate the curve to zero because in PCT the values of the curve do not start in zero, the concentration in baseline is zero however we have a different contrast value.

Based on central volume principle MTT can be defined by the ratio of CBV and CBF [25].

$$CBF = CBV / MTT \Leftrightarrow MTT = CBV / CBF. \tag{15}$$

Finally, according to Phillips the MTT can also be defined as the width of curve at half of the maximum value [13]. To estimate the width of the curve the average perfusion value between the upward and downward curve slopes is used as reference. This value is used to determine the points in both curve slopes that will be used to calculate the time difference that is the actual estimation of the MTT (see Fig. 1).

2.4 Time To Peak (TTP)

Time to Peak is the time of the contrast material to reach a concentration peak in the cerebral blood vessels after the injection. Phillips [13] ignore the delay from the injection of contrast material until it reaches the cerebral tissue but others authors [6, 12] correct the concentration curve to remove the injection delay.

In our implementation, we determine the instant where the concentration in the artery reaches its peak, after that the curve is back-tracked from the maximum to the arrival time (arrival time is the instant of time corresponding to the arrival of contrast material to cerebral tissue). This implementation corrects the injection delay. TTP is the time between the arrival of contrast material to the tissue and the moment where the concentration is maximum (see Fig. 1).

3   Methods Comparison and Results

Our main objective was to define an automatic processing pipeline that calculates and displays the perfusion related parameters with reduce human input – selection of artery and vein references. This ensures that, regardless of the method considered, the final visualization results will not be user dependent and quantified measure can be mapped directly to original data. The pipeline starts with the selection of both artery and veins two reference voxels that will be used as static reference along the following stages.

3.1   Pre-processing

To minimize noise and smooth the vein and artery contrast concentration curve an initial filtering of the image data using Simple Moving Average (SMA) is applied (Fig. 2). The objective of this step is to minimize the presence of noise in the image data namely due to equipment, patient movements or to the effect of discretization when sampling the data [17, 27].

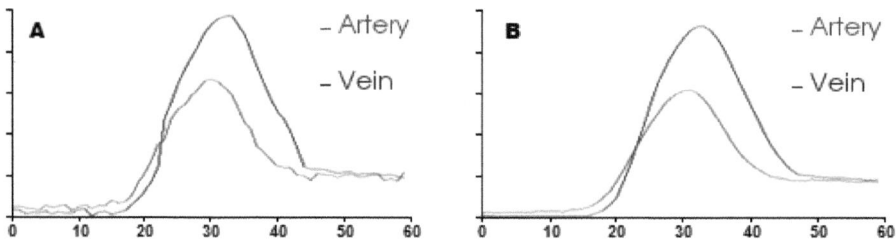

Fig. 2. Time concentration curve for artery and vein, without filter in A and filtered data using SMA in B

SMA filter was chosen among others (Simple Moving Average, Moving Median, smoothing) because it was a good trade-off between simplicity and results. Preliminary tests with a 7 seconds window centered in the current time (t-3, t+3) shows that SMA can remove noise and smooth with low processing time. The perfusion parameters are calculated for each voxel after removing non brain tissues (e.g. bone). Non brain tissues are removed using the minimum and maximum values of Time Concentration Curves as references: all values above the maximum value and below 80% of the minimum value are considered as non brain tissue.

3.2 Post-processing

To ensure a good contrast on the images without any manual adjustment by the users, we applied a window level correction, the image is truncated with the maximum value corresponding to the maximum value obtained in artery voxel (in the current method not in the concentration curve) in the artery and the minimum is a global minimum in the image. By using an automatically generated reference we maintain a clear map between transformed and original perfusion values.

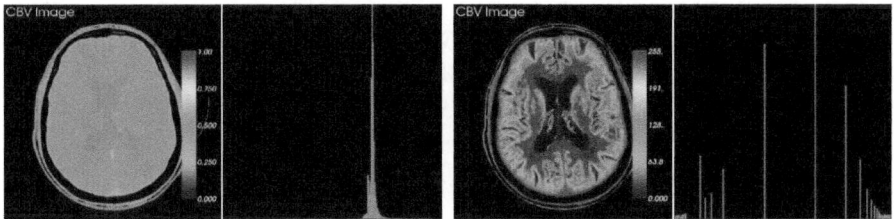

Fig. 3. Comparison between CBV before and after postprocessing: initial CBV image (left) and after hitogram equalization (right). The histogram distribution is also presented for both images.

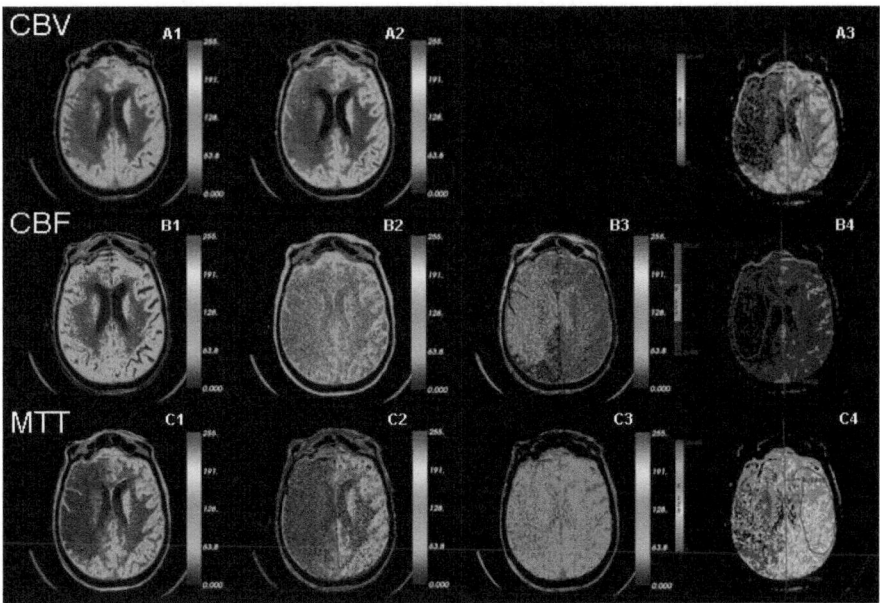

Fig. 4. The perfusion parameters estimation methods for CBV: (A1) using Axel method (equation 1), (A2) using Klotz and König method; for CBF: (B1) using SVD method; (B2) using Fick's method; (B3) using central volume principle; and for MTT: (C1) using Axel method; (C2) using Phillips method; (C3) using central volume principle. Images A3, B4 and C4 are the original image given by medical equipment for each of the parameters.

To enhance visually the different brain tissues (e.g. gray vs. white matter) and structures (e.g. brain tissues and ventricles) - expected to present different values regardless of the perfusion parameter in consideration - we also applied a histogram equalization method. The result of this step is emphasized in Fig. 3. The OpenCV library [29] was used to calculate histogram information as well as SVD (see 2.2) for visualization purposes, VTK was used [30].

3.3 Results

We have tested this pipeline in several perfusion CT exams and the results are presented for the same dataset for the CBV methods, CBF and MTT in Fig. 4 in the next page. All the images presented are the final visualization results with no user intervention except the selection of the voxel references (artery and vein).

4 Conclusion

This paper describes a first step in deploying a framework to perform quantified comparisons between different existing methods used in the literature to estimate perfusion related parameters methods. Our main objective was to define an automatic processing pipeline that calculates and displays the perfusion related parameters with reduce human input – selection of artery and vein references - while keeping a map between the several calculated parameters and the original perfusion data. By using as unique references along all the processed the artery and veins perfusion values to rescale the different methods parameters, a comparison is possible either quantified or visual. Given the clinical relevance of visualization of the results, our method also provides a standardized post processing stage, where regardless of clinicians own preferences (e.g. window levels, lookup tables, scales), we provide a reproducible baseline representation that enables the visual comparison of different methods – critical to have a clinical comparison using standard inter-rater agreements evaluation (e.g. [18]).

This work can have an impact in clinical practice with special emphasis in the acute stroke management by contributing to define which methods are more clinically relevant and, in consequence, quantify relevant stroke related features like the penumbra or unrecoverable brain tissues.

At this stage, some details were overlooked in the present work that are part of planned future work. On the technical side, a more thorough analysis on the filter selection is planned. Currently we use the SMA filtering but other methods exist that may exhibit better results. The delay between injection and arrival of contrast material may also result in inaccurate CBF and MTT, we plan to study its impact using SVD method with delay-corrected SVD (dSVD) namely trying corrections by shifting the concentration curve in time [28]. We also need to quantify the effect of post processing enhancements on the clinical decision process to avoid inducing erroneous clinical interpretations.

The present work will support such comparisons both quantified (by using quantified comparison between different methods features) and clinical to assess the clinical value of the results within a diagnosis context vs. the actual solution based on proprietary systems.

References

1. Lovblad, K.O., Baird, A.E.: Computed tomography in acute ischemic stroke. Neuroradiology (2009)
2. Ebinger, M., et al.: Imaging the penumbra - strategies to detect tissue at risk after ischemic stroke. J. Clin. Neurosci. 16(2), 178–187 (2009)
3. Paciaroni, M., Caso, V., Agnelli, G.: The concept of ischemic penumbra in acute stroke and therapeutic opportunities. Eur. Neurol. 61(6), 321–330 (2009)
4. Ledezma, C.J., Fiebach, J.B., Wintermark, M.: Modern imaging of the infarct core and the ischemic penumbra in acute stroke patients: CT versus MRI. Expert Rev. Cardiovasc. Ther. 7(4), 395–403 (2009)
5. Fisher, M., Bastan, B.: Treating acute ischemic stroke. Curr. Opin. Drug Discov. Devel. 11(5), 626–632 (2008)
6. Tomandl, B.F., et al.: Comprehensive imaging of ischemic stroke with multisection CT. Radiographics 23(3), 565–592 (2003)
7. Radaideh, M., et al.: Correlating the Basic Chronological Pathophysiologic Neuronal Changes in Response to Ischemia with Multisequence MRI Imaging. Neuro Graphics 2 (2003)
8. Latchaw, R.E.: Cerebral perfusion imaging in acute stroke. J. Vasc. Interv. Radiol. 15(1 Pt 2), S29–S46 (2004)
9. Tsurupa, G., Medved, L.: Identification and characterization of Novel tPA- and plasminogen-binding sites within fibrin(ogen) alpha C-domains. Biochemistry 40(3), 801–808 (2001)
10. Liebeskind, D.S.: Imaging the future of stroke: I. Ischemia. Ann. Neurol. 66(5), 574–590 (2009)
11. Wiesmann, M., et al.: Dose reduction in dynamic perfusion CT of the brain: effects of the scan frequency on measurements of cerebral blood flow, cerebral blood volume, and mean transit time. European Radiology 18(12), 2967–2974 (2008)
12. Kudo, K.: Procedure Guidelines for CT/MR Perfusion Imaging. Joint Committee for the Procedure Guidelines for CT/MR Perfusion Imaging (2006)
13. Phillips, M.D.: Brain Perfusion Imaging. Cerebrovascular Diseases and Stroke 1 (2001)
14. Wintermark, M., et al.: Cerebral perfusion CT: technique and clinical applications. J. Neuroradiol. 35(5), 253–260 (2008)
15. Wiesmann, M.: CT Perfusion of the Brain. VISIONS (2006)
16. Sanelli, P.C., Shetty, S.K., Lev, M.H.: Cerebral CT Perfusion: Clinical Pearls and Technical Pitfalls. Neuro Graphics 4 (2005)
17. Konig, M.: Brain perfusion CT in acute stroke: current status. Eur. J. Radiol. 45(suppl.1), S11–S22 (2003)
18. Soares, B.P., et al.: Automated versus manual post-processing of perfusion-CT data in patients with acute cerebral ischemia: influence on interobserver variability. Neuroradiology 51(7), 445–451 (2009)
19. Serafin, Z., et al.: Reproducibility of dynamic computed tomography brain perfusion measurements in patients with significant carotid artery stenosis. Acta Radiol. 50(2), 226–232 (2009)
20. Sanelli, P.C., et al.: Effect of training and experience on qualitative and quantitative CT perfusion data. AJNR Am. J. Neuroradiol. 28(3), 428–432 (2007)
21. Axel, L.: Cerebral Blood Flow Determination by Rapid-Sequence computed Tomography. Neuroradiology (1980)

22. Klotz, E., Konig, M.: Perfusion measurements of the brain: using dynamic CT for the quantitative assessment of cerebral ischemia in acute stroke. Eur. J. Radiol. 30(3), 170–184 (1999)
23. Shetty, S.K., Lev, M.H.: CT perfusion in acute stroke. Neuroimaging Clin. N. Am. 15(3), 481–501 (2005)
24. Eastwood, J.D., et al.: CT perfusion scanning with deconvolution analysis: pilot study in patients with acute middle cerebral artery stroke. Radiology 222(1), 227–236 (2002)
25. van der Schaaf, I., et al.: Influence of partial volume on venous output and arterial input function. AJNR Am. J. Neuroradiol. 27(1), 46–50 (2006)
26. Ostergaard, L., et al.: High resolution measurement of cerebral blood flow using intravascular tracer bolus passages.1. Mathematical approach and statistical analysis. Magnetic Resonance in Medicine 36(5), 715–725 (1996)
27. Wu, O., et al.: Tracer arrival timing-insensitive technique for estimating flow in MR perfusion-weighted imaging using singular value decomposition with a block-circulant deconvolution matrix. Magn. Reson. Med. 50(1), 164–174 (2003)
28. Kudo, K., et al.: Difference in tracer delay-induced effect among deconvolution algorithms in CT perfusion analysis: quantitative evaluation with digital phantoms. Radiology 251(1), 241–249 (2009)
29. Bradski, G., Kaehler, A.: Learning OpenCV: Computer Vision with the OpenCV Library, p. 576. O'Reilly Media, Sebastopol (2008)
30. Schroeder, W., Martin, K., Lorensen, B.: The visualization toolkit: an object-oriented approach to 3-D graphics, p. 528 (2006)
31. Aaslid, R., et al.: Cerebral autoregulation dynamics in humans. Stroke 20(1), 45–52 (1989)

Application of the Laplacian Pyramid Decomposition to the Enhancement of Digital Dental Radiographic Images for the Automatic Person Identification

Dariusz Frejlichowski and Robert Wanat

West Pomeranian University of Technology, Szczecin,
Faculty of Computer Science and Information Technology,
Zolnierska 49, 71-210, Szczecin, Poland
{dfrejlichowski,rwanat}@wi.zut.edu.pl

Abstract. The paper provides some experimental results on medical images enhancement, namely digital dental radiographic images of entire dentition — pantomograms. This problem is a first step in the process of automatic identification of persons basing on the mentioned kind of images. The most crucial task is the emphasizing of some characteristics, e.g. shapes of the teeth and dental fillings. These features are widely used as an input for the methods of automatic dental identification. In the paper the Laplacian pyramid-based image enhancement approach is utilized. This method has been successfully used for other radiographic images — mammograms and computed tomograms (CTs). Exemplary methods of uniform and non-uniform Laplacian pyramid enhancement are presented along with their influence on a typical image.

1 Introduction

Digital radiography has become increasingly popular in the last two decades over its analog counterpart. The usage of computers and electronic detectors in lieu of film speeds up the process of developing photographs, removes the necessity of using possibly harmful chemicals and allows for further image processing. The latter reason is especially significant, since X-ray examination is considered intrusive and allowed only in certain time intervals. Therefore, the ability to increase the legibility of a low-quality image provides an extended margin of error for radiography technicians and in the end, helps the physician (in the described case — a dentist) in making a correct diagnosis. Every commercial dental software program allows, to some degree, for image enhancement. Some popular programs have been described in [1] by Lehmann et al., along with the list of the methods implemented by them. According to the authors all the programs offer contrast and brightness adjustments, scarcely including image filtering and comparison options.

There have been previous attempts at improving the quality of radiograms in general. Most of them focus on decomposing the source image into layers containing a subset of the information derived from the original image. Afterwards,

A. Campilho and M. Kamel (Eds.): ICIAR 2010, Part II, LNCS 6112, pp. 151–160, 2010.

the layers can be processed independently allowing for an improvement of different types of signals — both high- and low-frequency. There are two major approaches to the decomposition of original images. The first one is based on the wavelet transform and the second — on the Laplacian pyramid.

In wavelet-based decomposition, continuous wavelets are used as the basis functions. The restriction of continuity must be upheld in order to prevent discontinuities in the resulting image, thus rendering the popular Haar transform inutile. Wavelet-based approach has been tested in [2] and [3]. Its main drawback, as discussed in [3], is the appearance of ringing artifacts during the reconstruction of the image.

In Laplacian pyramid decomposition, the images are firstly low-pass filtered using a Gaussian filter and downsampled and the achieved result is interpolated to the original size and subtracted from the original image at the end. The result becomes the next layer of the pyramid and the subsampled intermediate image becomes the original image in the next step of the algorithm. This process is reiterated until the image size reaches one pixel. This method is more robust than the wavelet-based one and it will be described more precisely in the following sections. We focus in the paper on the Laplacian-based decomposition and its usefulness in enhancement of dental radiographs.

The described process of enhancement has to be performed in order to improve the quality of the pantomograms before they can be further used in the process of automatic human identification, as described by Jain in [4]. Even though the image enhancement is frequently mentioned as the first step of the human identification process, specific methods used are scarcely mentioned. Zhou and Abdel-Mottaleb ([5]) proposed a rather simple method of using top-hat filtered and bottom-hat filtered versions of the original image in the process of enhancement ([5]):

$$X_E = X_O + X_T - X_B, \tag{1}$$

where:
X_O — the original image,
X_E — the resultant enhanced image,
X_T — top-hat filtered version of the image X_O,
X_B — bottom-hat filtered version of the image X_O.

This approach is sufficient for bitewing images used in the study presented in [5], but proves inefficient when only pantomograms are used for identification.

Pantomograms require relatively low amounts of radiation, taking into consideration the surface that is presented on the radiogram. Therefore, they are considered to be of lower quality than the other two popular types of dental radiographic images: bitewing and periapical. Some additional image enhancement, such as edge sharpening and contrast improvement, is highly recommended and could prove beneficial at the later stages of the process, i.e. image segmentation and feature extraction.

Dental fillings and teeth shapes are two major sources of information used by identification methods based on dental radiograms and any improvement in their legibility will be sought for the most, but amelioration in other aspects, such as trabecular structure visibility, will also be described. An exemplary image being an object of interest in the paper is provided in Fig. 1.

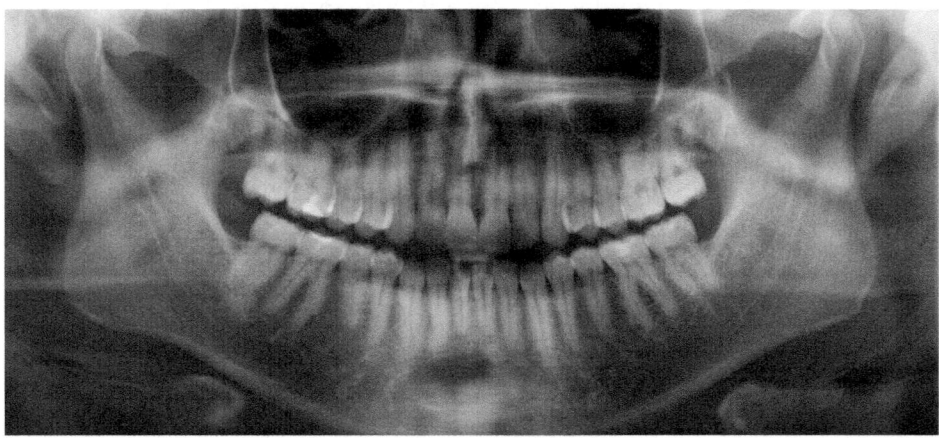

Fig. 1. Sample digital dental radiographic image, used courtesy of Pomeranian Medical University

2   Image Quality Enhancement

The concept of Laplacian decomposition of an image was firstly introduced by P. J. Burt and E. H. Adelson in [6]. Subsequent layers of Laplacian pyramid are calculated by subtracting consequent layers of a Gaussian pyramid. The process can be described using the equation ([6]):

$$
\begin{aligned}
X_k &=\downarrow (\bar{X}_{k-1}),\\
L_k &= X_{k-1} - \uparrow (X_k),
\end{aligned}
\tag{2}
$$

where:
$\downarrow (X)$ and $\uparrow (X)$ represent the process of downsampling and upsampling the image by a factor of 2,
$\bar{X}_k$ — low-pass filtered image X_k (with X_0 denoting the original image)
L_k — the successive layer of the Laplacian pyramid.

Gaussian filter is popularly used as a low-pass filter, but Stahl et al. ([7]) point out that small binomial filter kernels can also be used.

The above decomposition method was later used as a basis for multiscale image enhancement in [4], [7] and [8]. The methods used there were uniform, i.e. applying the same transformation to all the layers, with a small change in the method presented in [8], where the gain parameter could vary depending on the layer. Those methods will be examined later in this paper, after presenting the non-uniform methods, where the layers are processed independently.

Before we describe existing methods of enhancing the quality of a decomposed image, it must be noted that there is no simple quality measure of an image. Whether an image is considered of high or low quality can only be measured by its ability to satisfy some specific needs. Because these needs can be different, as dental radiographs could be used by a dentist as well as a forensic specialist identifying a body, we assumed that a measure of quality is a subjective prediction of how the enhancement could affect further stages of digital image processing, e.g. image segmentation. This measure does not necessarily overlap with the definition of quality agreed upon by experienced physicians, for whom some important minutiae might have been lost in the process of enhancement.

As it was noted, the images that form the Laplacian pyramid contain progressively lower frequencies of the image data. In result the first layer of the pyramid can be instinctively identified with the trabecular structures of the mandible and maxilla. Some of the smaller layers of the pyramid contain unobstructed contours of the teeth and surrounding bones and on the lowest level there is only the mean of the image brightness. Stahl et al. ([8]) noted that because of the downsampling performed after the frequency range is reduced, even though every layer theoretically represents the spatial frequencies of up to half the Nyquist frequency of the previous one, the spatial frequencies contained in the actual layer are on a par with the frequencies of the previous layers. An example of the normalized 4th layer of the sample image is presented in Fig. 2.

Fig. 2. The 4th layer of the Laplacian pyramid decomposition achieved for a pantomogram. The image is eight times smaller than the original one and contains lower frequency signals — edges.

The simplest modification of the processed image can be achieved by changing the value of the only pixel on the last layer of the pyramid, thus changing the bias of the original image or relative brightness. Operating on the layers with low frequency data — usually the 2-3 layers before the last one — allows for easy enhancement of large portions of the original image. This can be especially useful if the image is not evenly developed due to a non-uniform distribution of radiation when the radiograph was taken — a simple averaging filter or even substitution of all coefficients on the layer with the mean value of that layer solves this problem.

Additional enhancement of an image can be achieved by using the unsharp filter on a layer containing high frequency signal, ideally the second or third layer. The first layer contains too much fine detail, including noise, therefore the usage of the unsharp filter on it would give a similar result as using it on the original image. When using it on further layers the sharp edges of teeth and bone structures are enhanced without strong amplification of the noise.

The most significant drawback of the non-uniform methods is that they hardly work automatically. Images of the same type should have the same layer distribution, but it has to be determined beforehand, what could be time-consuming. Uniform layer manipulation methods are free of this problem.

The use of multiscale images in contrast enhancement was introduced in [9] and further developed in [7]. It is also commercially used in Agfa ADC system. Vuylsteke et al. proposed a contrast equalization function given by the formula ([7]):

$$f(x) = a(\frac{x}{|x|})|x|^p, \tag{3}$$

where ([7]): "x are normalized to the range $[-1, 1]$ and the factor a is needed for rescaling the resulting image to the original dynamic range". This operator resembles a standard exponential operator, working for both positive and negative pixel values.

This function is further developed in [8], where Stahl et al. proposed another version of equation 1, with slight modifications ([8]):

$$\begin{cases} r(x) = G \cdot x \cdot (1 - \frac{|x|}{M})^p + x, & if \; |x| \leq M \\ r(x) = x, & elsewhere \end{cases}, \tag{4}$$

where M is the upper limit for linear enhancement and G is a constant gain. This contrast equalization function was also used in [3]. Its main drawback is a significant amplification of image noise. Fortunately it can be easily solved — Stahl et al. proposed an additional method of noise suppression in their model, given by the formula ([8]):

$$S_n(x, y) = b(x, y) \cdot S_f(x, y) + (1 - b(x, y)) \cdot S_o(x, y), \tag{5}$$

where S_n is the final value of the pixel, $b(x, y)$ is the attenuation factor such that it is ([8]): "smaller than 1 in the noise sensitive region and equal to 1 elsewhere", S_o is the original pixel value in the layer being contrast-equalized and S_f is the pixel value after initial equalization.

3   Experimental Results

The various approaches to Laplacian pyramid manipulation, presented in the previous sections, were tested using several pantomograms. The effect of every method will be presented on the exemplary image and some additional images will be shown to demonstrate the effect of the selected best method on other examples.

The first presented example (see Fig. 3) is a result of using the averaging filtration on low-frequencies layers.

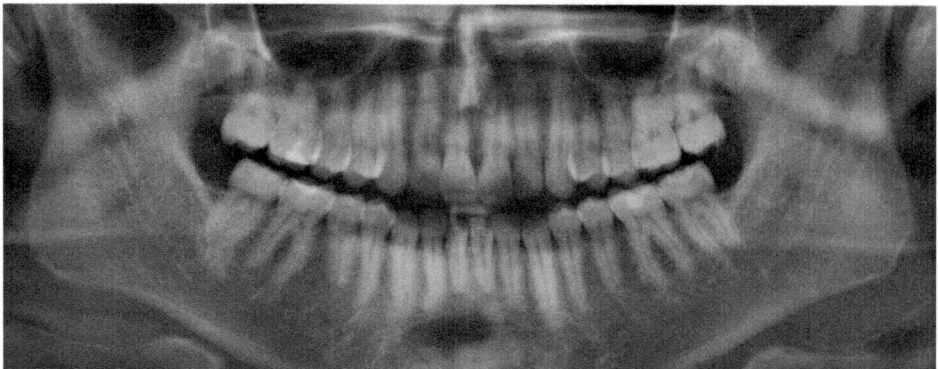

Fig. 3. The result of averaging the low-frequencies layers — smoother background and uniformly colored bones

The appearing discoloration in the general areas of cheeks and lower mandible has been removed and the background behind teeth has been smoothened. That could be helpful during the image segmentation, where smooth background and a uniform underlying bone color would improve the detection of edges belonging to crowns and roots.

The following example (see Fig. 4) shows the effect of using the unsharp filter on the second layer of the Laplacian pyramid decomposition.

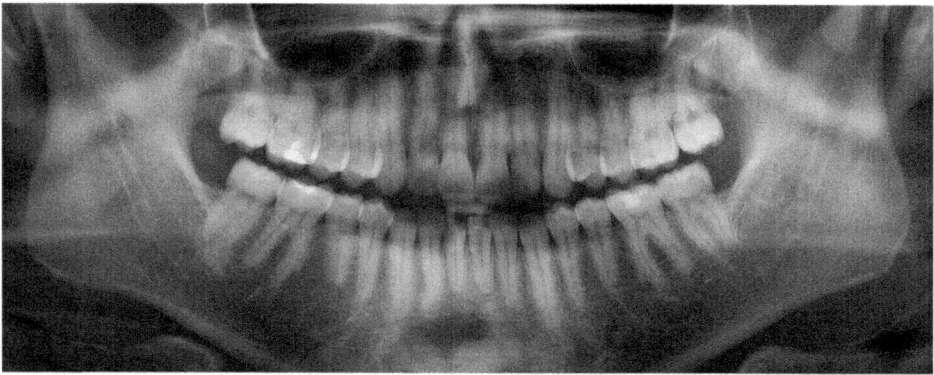

Fig. 4. The result of the application of unsharp filter — the edges of teeth fillings and roots are improved

The all-important improvements can be seen in the areas that lacked sharpness in the original image — the surroundings of the roots, the fillings and the tips of the molars. Trabecular structures also look sharper and can be easily extracted from the image. The noise amplification is not as severe as it is the case of an unsharp filtering for the whole image. The contrast between the teeth and mandible\maxilla is rather low.

The effect of the contrast equalization function (for $p = 0.75$) on layers 3 through 11, achieved using eq. 3 for the sample image can be seen in Fig. 5.

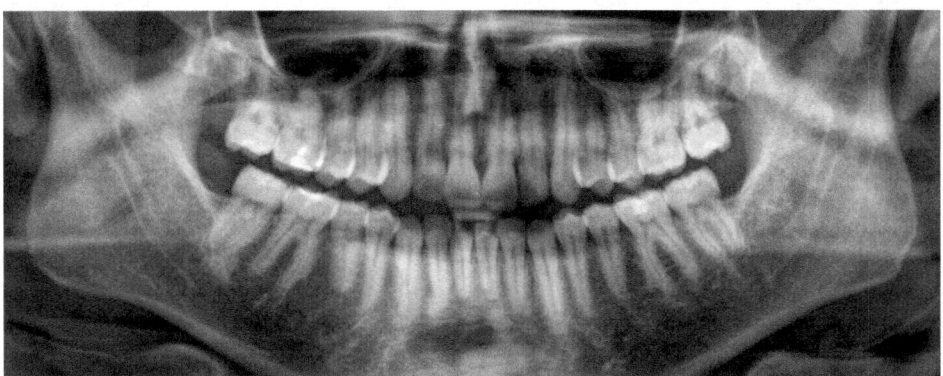

Fig. 5. The result of the contrast equalization of a pantomogram — the edges of fillings, dental pulp and roots are more pronounced

The contrast of the most important areas has been significantly increased. Dark bone areas have become darker and bright teeth and the fillings have become brighter. This also causes the dental pulp to become more distinct, what can be valuable in diagnosing lesions in this part of the tooth. The silhouettes of the roots also have sharper edges, thus simplifying the separation of teeth from bone. Trabecular structure of the bones is also sharper than on the original image.

An example of the second contrast enhancing method, as described in equation [4], with the parameters $M = 0.15$ and $p = 1.5$, is presented in Fig. 6.

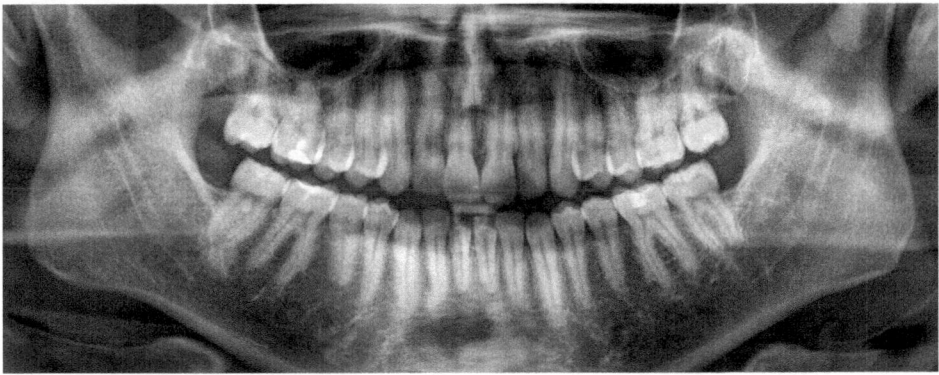

Fig. 6. Exemplary result of the contrast boost. It allows for an easier distinction between fillings and teeth and between teeth and surrounding bones.

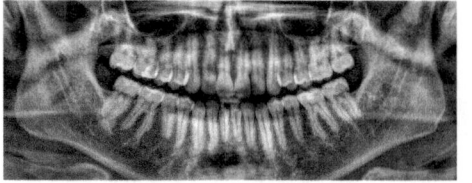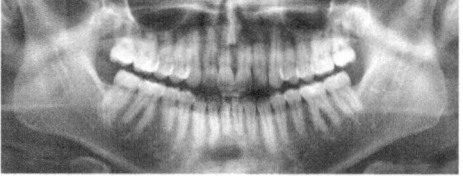

Fig. 7. The result of a combination of three enhancement methods (left) and the same enhancement applied to the image in spatial domain (right)

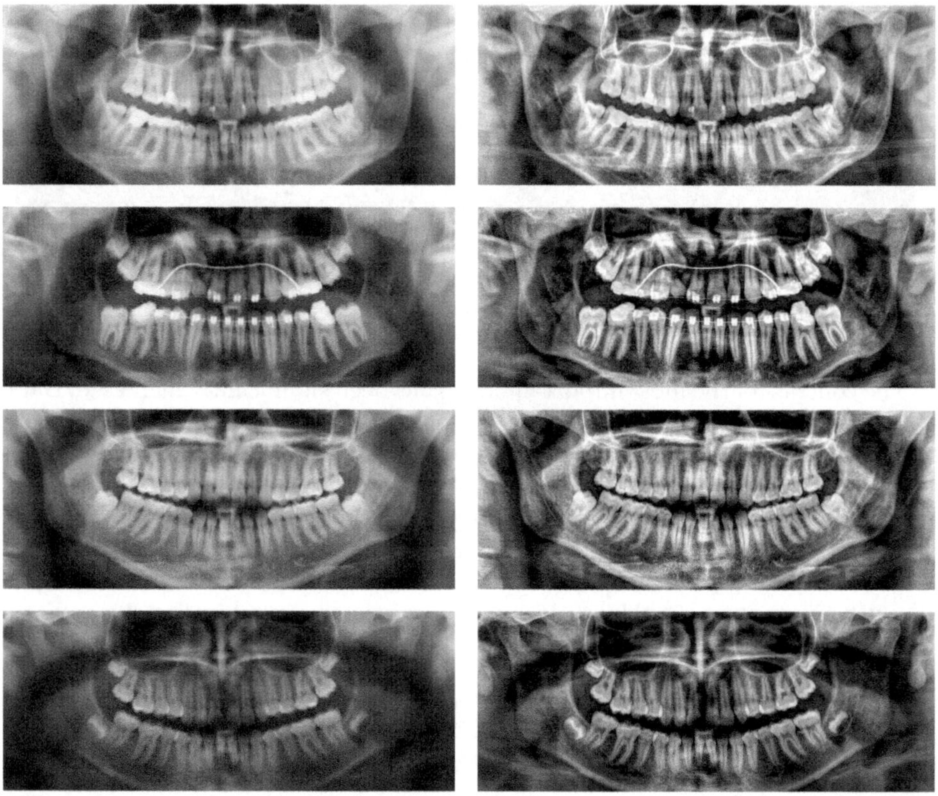

Fig. 8. The comparison between the original (left) and enhanced using the explored approach digital dental radiographic images (right). The enhanced images provide better basis for further image segmentation.

Considering further processing of the resultant image, this achieved image has the highest quality so far, with greatly improved contrast, sharper edges and the easier discernible difference between bone and teeth. Trabecular structures are also sharper than as a result of any method applied so far. The most important is the high contrast between three distinct groups of objects useful in people's identification: the teeth, surrounding bones and teeth fillings.

The last performed by us experiment was based on a combination of three operations on a digital dental radiographic image. We have used the following sequence: the averaging of the two layers next to last, unsharp filter on the second layer, and the second presented in the paper method of contrast enhancement, achieved using eq. 4. The result is provided in Fig. 7. As we can see some very interesting elements are emphasized, e.g. roots of the teeth. The same image was enhanced in spatial domain, using the above methods and is also presented in Fig. 7. The averaging could not be implemented in the spatial domain as it would remove all the details from the image, so a more complex operator would have to be included in order to remove the effects of the unequal exposure of the picture. The use of the unsharp filter used in spatial domain also increased the noise.

At the end of this section, we present some additional results of the explained approach, achieved using the combination of the three methods. The results, compared to the original images, can be seen in Fig. 8.

The images after the enhancement have generally better contrast, easily discernible teeth from surrounding bones and sharper edges. In all of the presented cases, the image enhancement improved the possibilities of successful teeth and fillings segmentation at the cost of amplified noise.

4   Conclusions

The approach presented in this paper covers only one group of existing radiogram enhancement methods. We did not compare methods that do not employ the Laplacian pyramid decomposition, like the mentioned wavelet-based image decomposition ([2,3]) or a method based on manipulation on local standard deviations ([10]). Moreover, the evaluation was purely subjective. However, the obtained enhanced images gave promising results seeing that the regions of an original image, where the contrast was improved the most, were the regions that are the most important in the process of human identification, whether done by a specialist or automatically. Further improvement could be easily achieved at the cost of the automation, which is crucial when the size of an average dental radiograms database is taken into account.

As it was stated in the paper, the quality of an image can only be measured by its ability to satisfy specific needs, thus making the influence of selected methods on the accuracy of a sample persons identification system based on pantomograms the only reliable measure of image quality change. Therefore our future work will be concentrated on experiments exploring this influence on a larger database of pantomograms in order to validate the initial results presented in the paper.

References

1. Lehmann, T.M., Troeltsch, E., Spitzer, K.: Image processing and enhancement provided by commercial dental software programs. Dentomaxillofacial Radiology 31, 264–272 (2002)
2. Lu, J., Healy Jr., D.M.: Contrast enhancement of medical images using multiscale edge representation. In: Proc. of SPIE: Wavelet applications, Orlando, FL (1994)

3. Dippel, S., Stahl, M., Wiemker, R., Blaffert, T.: Multiscale Contrast Ehnahnce-ment for Radiographies: Laplacian Pyramid Versus Fast Wavelet Transform. IEEE Trans. on Medical Imaging 21(4), 343–353 (2002)
4. Jain, A.K., Chen, H.: Automatic Forensic Dental Identification. In: Jain, A.K., Flynn, P., Ross, A.A. (eds.) Handbook of Biometrics, pp. 231–251 (2008)
5. Zhou, J., Abdel-Mottaleb, M.: A content-based system for human identification based on bitewing dental X-ray images. Pattern Recognition 38(11), 2132–2142 (2005)
6. Burt, P.J., Adelson, E.H.: The Laplacian pyramid as a compact image code. IEEE Trans. on Communications 31(4), 532–540 (1983)
7. Vuylsteke, P., Schoeters, E.: Image Processing in Computed Radiography. In: Proc. of International Symposium on Computerized Tomography for Industrial Applications and Image Processing in Radiology, Berlin, Germany (1999)
8. Stahl, M., Aach, T., Buzug, T.M., Dippel, S., Neitzel, U.: Noise-resistant weak-structure enhancement for digital radiography. In: SPIE 1999, vol. 3661, pp. 1406–1417 (1999)
9. Vuylsteke, P., Schoeters, E.P.: Multiscale image contrast amplification (MUSICA). In: SPIE 1994, vol. 2167, pp. 551–560 (1994)
10. Chang, D.-C., Wu, W.-R.: Image contrast enhancement based on a histogram trans-formation of local standard deviation. IEEE Trans. on Medical Imaging 17(4), 518–531 (1998)

Automatic Recognition of Five Types of White Blood Cells in Peripheral Blood

Seyed Hamid Rezatofighi[1], Kosar Khaksari[1], and Hamid Soltanian-Zadeh[1,2]

[1] Control and Intelligent Processing Center of Excellence,
Department of Electrical and Computer Engineering, Faculty of Engineering,
University of Tehran, Tehran 14395-515, Iran
`h.tofighi@ece.ut.ac.ir, hszadeh@ut.ac.ir`
[2] Image Analysis Lab., Department of Radiology, Henry Ford Hospital, Detroit, MI, USA
`hamids@rad.hfh.edu`

Abstract. An automatic system which is capable of recognizing white blood cells can assist hematologists in the diagnosis of many diseases. In this paper, we propose a new system based on image processing techniques in order to recognize five types of white blood cells in the peripheral blood. To segment nucleus and cytoplasm, a Gram-Schmidt orthogonalization method and a snake algorithm are applied, respectively. Moreover, three kinds of features are extracted from the segmented areas and two groups of textural features extracted by Local Binary Pattern (LBP) and co-occurrence matrix are evaluated. Best features are selected using a Sequential Forward Selection (SFS) algorithm and performances of two classifiers, ANN and SVM, are compared. In this application, the best result is obtained using LBP as the textural feature and SVM as the classifier. In sum, the results demonstrate that the methods are accurate and fast enough to execute in hematological laboratories.

Keywords: White blood cell, peripheral blood, segmentation, textural feature, feature selection, classification.

1 Introduction

Recognition and inspection of white blood cells in peripheral blood can assist hematologists in diagnosing many diseases such as AIDS, Leukemia, and blood cancer. Thus, this process is assumed as one of the most salient steps in hematological procedure. This analysis can be accomplished by automatic and manual approaches. Automatic methods usually examine white blood cells just quantitatively but not qualitatively, because they do not benefit from image processing techniques. Applying automatic systems juxtapose to image processing techniques may provide some qualitative evaluation and thus enhanced judgments. Furthermore, some of these tasks such as manually scrutinizing blood cells by experts are tedious and susceptible to error. Therefore, an automatic system based on image processing techniques can help the hematologists and expedite the trend.

A. Campilho and M. Kamel (Eds.): ICIAR 2010, Part II, LNCS 6112, pp. 161–172, 2010.
© Springer-Verlag Berlin Heidelberg 2010

Albeit not extensive, some methods are proposed in the literature for this purpose. Since segmentation is the most challenging step in white blood cells recognition procedure, improvement of nucleus and cytoplasm segmentation is the most widespread effort in many researches. For example, in [1], [2], [3] and [4], the authors suggested several methods to segment nuclei of white blood cells via techniques that can be categorized into color-based methods. These methods are simple but are not capable of segmenting the white blood cells nucleus accurately. In addition, cytoplasm is colorless in most cases. Thus, its boundary is not detectable and cannot be segmented by these methods. Methods based on imaging techniques generate superior results. For example, the method proposed in [5] obtained more acceptable results using multi-spectral imaging techniques. In this method, intensity of each pixel in different spectra is used to construct the feature vectors and a support vector machine (SVM) is used for classification and segmentation. In spite of efficacy of this method for segmenting white blood cells components, this system's implementation is costly and thus cannot be used widely at all laboratories. Cytoplasm and nucleus segmentation via mathematical and contour models is the third method and also the most important one. In this field, some methods such as region growing [6], watershed [7], parametric active contour deformable models [8], and also combination of the watershed technique and a parametric deformable model [9] are introduced in the literature. These methods are more complex and require more processing time in comparison with the first group of methods. However, their advantage is subtle more accurate segmentation. Since morphological and textural features are the features which are elicited from white blood cells by a hematologist, many papers such as [1], [10], [11] use feature extraction methods on the basis of these features. For classification, Bayes classifier [12], different types of artificial neural networks (ANNs) such as feed-forward back-propagation [13] and [14], local linear map [15], and fuzzy cellular neural network [16] are often used in the literature.

In this paper, our purpose is to design a new system based on image possessing methods to classify five major groups of white blood cells in peripheral blood. Therefore, at first, segmentation of white blood cells nuclei is carried out via Gram-Schmidt method. Then, distinguishing basophils from the other samples are performed using features extracted from nucleus areas. As cytoplasm edge is unobservable, a snake algorithm is used after some preprocessing procedures in order to segment the cytoplasm. The features elicited from the nucleus and cytoplasm areas in both steps are categorized into color, morphological, and textural features. Two groups of textural features attained by the Local Binary Pattern (LBP) and the co-occurrence matrix are evaluated. The feature selection step is adjoined to this process for ameliorating the classifier performance and expediting the program trend. Finally, the performance of two different classifiers, SVM and ANN, when using different sets of features is compared. The main difference between this research and other researches is that we propose an accurate and high-speed system for recognition of white blood cells which processes all segmentation, feature extraction and classification steps automatically.

The rest of the paper is organized as follows. In Section 2, we propose a pertinent system for recognition of five types of white blood cells. The experimental results are presented and discussed in Section 3. Finally, Section 4 is appropriated to presentation of the conclusions.

2   System Architecture

Designing an automatic system to recognize five types of white blood cells in a hematological image of the peripheral blood is the main purpose of this work. It is necessary to design a block diagram based on this type of dataset. Fig. 1 illustrates the block diagram of our proposed system.

As shown in this figure, the method has three major phases whose details are explained in the next sections.

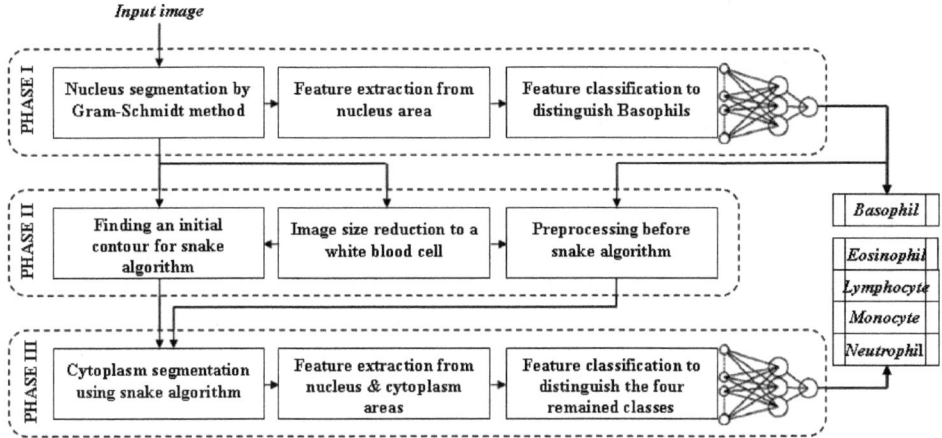

Fig. 1. The block diagram of the proposed system

2.1   Phase I

Since in most of the samples, the boundary between nucleus and cytoplasm of basophils cannot be distinguished visually; these cells should not involve in segmentation of cytoplasm's step. Therefore, they should be recognized from the other samples in this Phase.

Segmentation of nucleus by Gram-Schmidt method: In this method, pixel intensities of the RGB components of the color image of the dataset are considered as 3D vectors. Then, v_1 as a desired vector is obtained by averaging the 3D vectors of the nucleus area in some samples. The v_2 and v_3 as undesired vectors are defined from the areas that are similar to the nucleus but are not the nucleus area. The training samples are selected randomly from one of the samples of each class which are not used in evaluation step. Using the Gram-Schmidt orthogonalization method proposed in [17] and v_1, v_2 and v_3, a weighting vector w is attained whose inner product with the pixel vectors results a composite image with higher intensity in the nucleus area compared to other areas (Fig. 2(a)). Next, by choosing an appropriate threshold based on the histogram information, we segment the image. The final result of the nucleus segmentation is shown in Fig. 2(b).

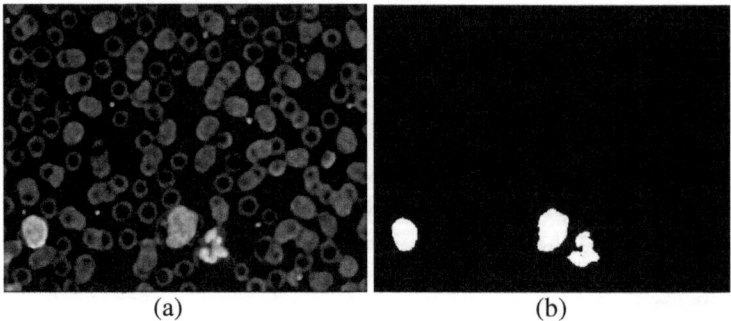

(a)   (b)

Fig. 2. (a) Resulting image with higher intensity in nucleus area relative to the other areas after applying the Gram-Schmidt orthogonalization method in a hematological image. (b) Final result of nucleus segmentation after thresholding.

Feature extraction from nucleus area and classification to distinguish basophils: As previously stated, basophils should be recognized and separated from the other types of white blood cells in this step. Therefore, some morphological features such as nucleus area and perimeter, number of the separated parts of nucleus, mean and variance of the nucleus boundaries and roundness criterion of nucleus are extracted from the segmented area. Color features are the other features extracted as a normalized vector of averaged nucleus color. To extract textural features, the co-occurrence matrix and the local binary pattern are applied and the results are compared.

Co-occurrence Matrix: The co-occurrence matrix is constructed on the basis of gray levels with the distances d and angles φ. In fact, this matrix describes the second order probabilistic features. Fourteen features are extracted from the co-occurrence matrix that explain contrast, homogeneity, entropy and others that properly represent the image textural properties. To make the features rotation invariant, 4 matrices are usually computed at 4 angles and the average of these matrices specifies the 14 features [18].

Local Binary Patterns (LBP): Local Binary Pattern (LBP) is another feature for texture processing. Because LBP analyzes textures in different radii, it can be supposed as a multi-resolution textural feature. Two features are usually extracted for each radius. The first one is LBP^{riu2} which represents the structure of texture and the other one is *VAR* which depicts changes in the gray levels [19].

2.2 Phase II

In this Phase, the main purpose is to prepare the image for the snake algorithm in order to segment the cytoplasm. To this end, image size reduction, preprocessing before the snake algorithm, and finding an initial contour for the snake algorithm are applied. These are explained below.

Image size reduction: Using the nucleus area segmented in the previous phase, we find the center of each nucleus and fit an appropriate window around to get sub images with a complete white blood cell. This trend makes the segmentation process easier. In our research, a 141*141 window is used.

Finding an initial contour for snake algorithm: To find an initial contour for the snake algorithm, the morphological dilation operation is first applied to the segmented nucleus region. The structuring element for the dilation operation is a square with an adaptive size based on the nucleus size. In this step, the boundary of this region is used as the initial contour. Fig. 3 shows this procedure.

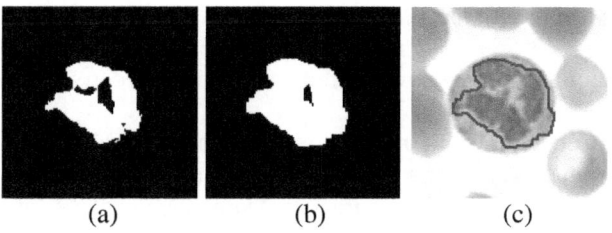

(a)   (b)   (c)

Fig. 3. (a) A segmented nucleus, (b) The image after dilation operation, (c) Initial contour

The adaptive parameter, size of the structuring element, is because of the differences in the white blood cells sizes. To have a congruous initial contour next to the cytoplasm edge, the contour of a small nuclei should be processed with the smaller dilation operator than a large nuclei.

Preprocessing before snake algorithm: Due to high accumulation of the red blood cells, they may touch the cytoplasm of the white blood cells. Thus, the boundary between the cytoplasm and the red blood cells may not be distinguished when the color image is changed into gray-scale. To solve this problem, the image is enhanced by color histogram equalization. Next, the enhanced image is transferred into the Hue-Saturation-Intensity (HSI) space. The final image is attained by extracting the saturation plate from the HSI image. Based on this idea, we have a gray-scale image that has good discrimination between the boundaries of the cytoplasm and the red blood cells. Then, the image is smoothed by a Gaussian kernel to eliminate the cytoplasm cavities and inhomogeneities and ameliorates the image for the snake algorithm (Fig. 4).

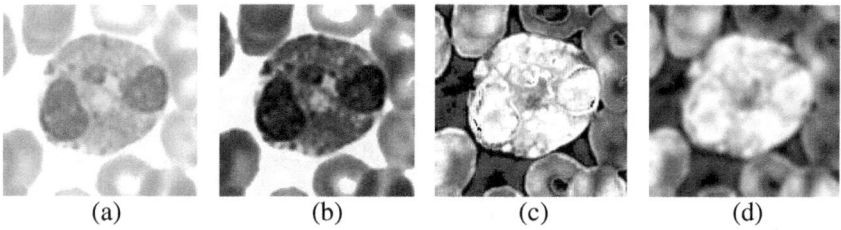

(a)   (b)   (c)   (d)

Fig. 4. (a) A sample image. (b) The image after Histogram Equalization. (c) The image after extracting the saturated plate. (d) The smoothed image by a Gaussian kernel.

2.3 Phase III

The main aim of this phase is to recognize the four remaining classes of the white blood cells. For this purpose, after preprocessing and initial contour detection, the snake algorithm is applied to segment the cytoplasm. After segmentation, textural and morphological features are extracted from both of the nucleus and the cytoplasm and the four cell types are classified.

Cytoplasm segmentation using snake algorithm: The snake algorithm used in this paper is the algorithm proposed in [20].It starts from the obtained initial contour and its parameters are set as $\alpha = 2, \beta = 5, \gamma = 0.7, \tilde{n} = 0.4$. The snake algorithm ends when no snake points move to new positions for four consecutive iterations.

Feature extraction from both nucleus and cytoplasm areas, and classifying the four remaining classes: in this step, features are extracted from the cytoplasm area in combination with the features extracted from the nucleus area in phase I. Retrospectively, these features are categorized into three groups of morphological, textural, and color features. The morphological features are cytoplasm area and whole cell body perimeter, mean and variance of the cytoplasm boundaries, roundness of the whole cell and the ratio between the cytoplasm and nucleus areas. Textural features are also extracted from the cytoplasm area by a co-occurrence matrix and the local binary pattern and their results compared. At the end, a normalized vector of the average cytoplasm color is extracted as color features.

3 Experimental Results

The proposed method was evaluated by 251 blood smear slide images acquired by a light microscope from stained peripheral blood using the Digital Camera-Sony-Model No. SSC-DC50AP with magnification of 100. The images contain 720*576 pixels and were classified by a hematologist into the normal leukocytes: basophil, eosinophil, lymphocyte, monocyte, and neutrophil. Also, the areas related to the nuclei were manually distinguished by an expert.

3.1 Segmentation Results

In order to quantitatively evaluate the results of the nucleus and cytoplasm segmentations, the following similarity measure is defined.

$$T_s = 100 \times \frac{A_{program} \cap A_{expert}}{\max(A_{program}, A_{expert})} \tag{1}$$

where $A_{program}$ is the segmented area by the algorithm and A_{expert} is the segmented area by an expert. When these two areas are the same, T_s is 100. In Table 1, the resulting measures for each kind of the white blood cells and their overall segmentation are presented.

Table 1. Similarity measures (*Ts*) for the segmentation of different types of white blood cells

	Basophil	Eosinophil	Lymphocyte	Monocyte	Neutrophil	Overall
Nucleus	94.7%	90.81%	88.86%	96.7%	94.05%	**93.02%**
Cytoplasm	-	95.55%	93.05%	81.23%	97.25%	**91.79%**
Average	94.7%	93.22%	90.01%	91.23%	96.23%	93.09%

According to Table 1, it can be inferred that the accuracy result of nucleus segmentation for the lymphocyte class is lower than the other classes. The main reason is that the color of the cytoplasm is analogous to the color of the nucleus in many of the lymphocytes samples, especially young ones. Therefore, the segmentation error for this type of the white blood cells is larger than the others. Also, since in most of the cases, the vitreous cytoplasm of monocytes is colorless, even the deformable model with a congruous preprocessing is unable to find the cytoplasm boundaries precisely. Therefore, the accuracy result of the cytoplasm segmentation is worse than those of the other classes.

3.2 Classification Results

In this paper, classification is performed in two sections, discriminating the basophils from the other types of white blood cells in phase I and recognizing the remaining classes in phase III.

To appraise the performance of the classifiers and the result of our proposed algorithms in recognizing the white blood cells, an accuracy criterion is used.

Experimental result for basophil classification: To compare the performance of the textural features, two groups of features, extracted from the nucleus area, are created. These two groups include similar morphological and color features but they are different in textural features. To classify these features, at first, feature dimension is reduced by a SFS algorithm [21] and the results for ANN [22] and SVM [23] are compared by means of an overall accuracy criterion. In this research, a Multi-Layer Perceptron (MLP) [22] ANN is used. Fig. 5(a) and (b) illustrate the results related to the local binary pattern and the co-occurrence matrix, respectively.

Some points are construed from these figures. The first point is that reduction in dimension of features aggravates the classification as expected. The second one is that the overall accuracy does not have considerable escalation after 15 features for both ANN and SVM. This occurs because after selecting 15 features, the differentiation between the features of each class is at a maximum and increasing the features dimension perplexes the classifier. The third one is that the ANN classifier has more fluctuation in the overall accuracy in comparison with the SVM. These changes are because the MLP is not trained well as a result of trapping in a local minimum. The other point is that the ANN and SVM classifiers have similar performances in most of the feature dimensions. But due to the stability of the SVM in training, it may be preferred. The last point is that the groups of the features attained from the co-occurrence matrices have generally better performance in comparison with the groups of features obtained from LBP. However, the time required for calculating the first group of features is significantly higher than the second one.

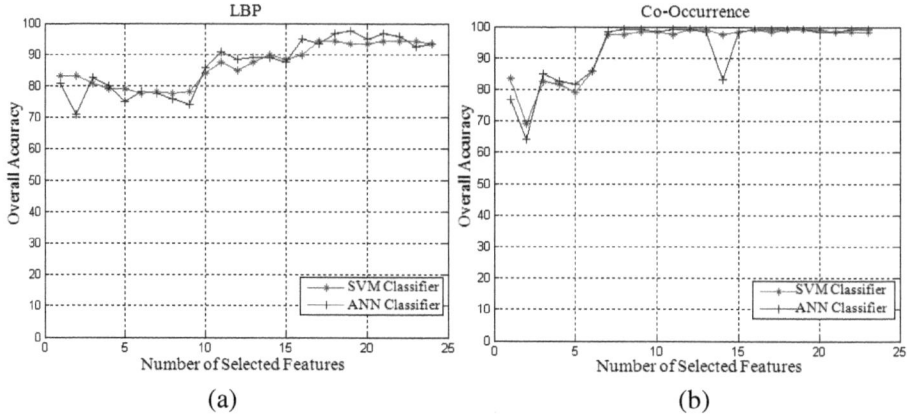

Fig. 5. The overall accuracy results for the ANN and SVM using: (a) LBP (b) Co-occurrence matrix as their textural features, to distinguish basophils

According to the above conclusion, selecting 15 features and utilizing SVM is the best way to have optimal performance in classification results and speed. To compare these two groups of features, Tables 2 and 3 show confusion matrices, the accuracy, and the overall accuracy for these two groups when 15 features are selected and SVM is used as the classifier, respectively.

Table 2. Confusion matrix, Accuracy, and Overall Accuracy for 15 LBP features and SVM classifier

	Recognized Basophil	Recognized Non-Basophil	Accuracy
Basophil	50	0	100%
Non-Basophil	23	150	86.71%
Overall Accuracy			89.69%

Table 3. Confusion matrix, Accuracy, and Overall Accuracy for 15 Co-Occurrence features and SVM classifier

	Recognized Basophil	Recognized Non-Basophil	Accuracy
Basophil	49	1	98%
Non-Basophil	5	168	97.11%
Overall Accuracy			98.64%

According to these tables, the results of classification with the features of the co-occurrence matrix are superior to those of the LBP. However, considering a trade-off between accuracy and processing time, LBP may be preferred. The ratio between the times required for feature extraction using the co-occurrence and LBP methods is 20 to one.

Experimental result for classifying the four remaining groups of white blood cells: Similar to the previous section, the performance of the two groups of features whose textural features are extracted by the LBP and the co-occurrence matrix are compared whilst these features are obtained from both of the nucleus and cytoplasm areas. For a second time, the SFS algorithm is applied in order to select the best features in a prespecified dimension and the results are compared for ANN and SVM by the accuracy criterion. Fig. 6(a) delineates the overall accuracy of the SVM and ANN for the first group of features and Fig. 6(b) demarcates the result for another group of features.

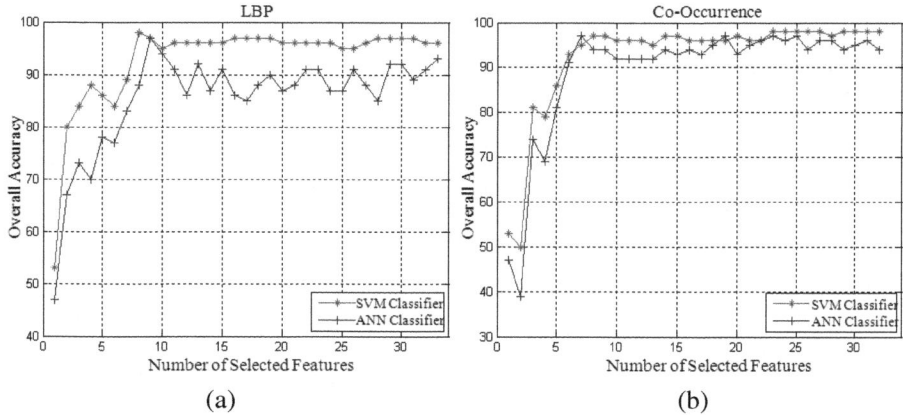

Fig. 6. The overall accuracy results of the ANN and SVM using: (a) LBP; (b) Co-occurrence matrix as their textural features, in order to distinguish the 4 remaining classes

According to the above figures, some points related to the performance of the classifiers are analogous to the conclusion discussed in the previous section. For instance, reduction in dimension of features exacerbates the classifier performance. Furthermore, the oscillation of the overall accuracy of the ANN classifier is considerably more than that of the SVM. Other points inferred from the figures are: after 10 features, curves do not have significant escalation for both of the ANN and SVM, the SVM classifiers have superior performance in this case in comparison with the ANN. In addition, it is obvious that the groups of features obtained from the co-occurrence matrices have generally superior performance again.

In conclusion, considering both of the classification accuracy and processing time, SVM classifiers and feature dimension of 10 can be considered as optimal.

Tables 4 and 5 illustrate the confusion matrix, accuracy, and overall accuracy when 10 features are selected and SVM is used as a classifier with the local binary pattern and the co-occurrence matrix, respectively. According to these tables, in this case, the classification accuracy for the two defined groups of features seems to be equal. However, noting this fact that calculation of the features extracted by the co-occurrence matrix is noticeably more computational, using LBP for the textural features is proposed for this phase again.

Table 4. Confusion matrix, Accuracy, and Overall Accuracy for 10 LBP features and SVM classifier

	Recognized Eosinophil	Recognized Lymphocyte	Recognized Monocyte	Recognized Neutrophil	Accuracy
Eosinophil	19	0	0	0	100%
Lymphocyte	0	27	2	0	93.1%
Monocyte	0	0	23	1	95.83%
Neutrophil	1	0	0	27	96.43%
Overall Accuracy					96%

Table 5. Confusion matrix, Accuracy, and Overall Accuracy for 10 the co-occurrence features and SVM classifier

	Recognized Eosinophil	Recognized Lymphocyte	Recognized Monocyte	Recognized Neutrophil	Accuracy
Eosinophil	18	0	0	1	94.74%
Lymphocyte	0	27	2	0	93.1%
Monocyte	0	0	23	1	95.83%
Neutrophil	0	0	0	28	100%
Overall Accuracy					96%

4 Conclusion

In this paper, we proposed a system in order to recognize five groups of white blood cell in the peripheral blood. The proposed system has a reasonable processing time and is sufficiently accurate. The overall segmentation result of 93% and classification accuracies of 90% and 96% in phases I and III verify the accuracy of the system. Regarding the processing time, the program requires 10 seconds for analyzing a single white blood cell on a Pentium-4 PC, running at 3.2 GHz, with 1 GB of RAM and MATLAB. Hence, differential counting of 100 white blood cells lasts about 16 minutes. As a comparison, an expert requires almost 15 minutes to carry out this process. Thus, this program can be used in the hematological laboratories.

Notwithstanding mentioned advantages, the proposed method may need initial calibration at the start point of the program when new datasets with different characteristics are introduced to the system. This is due to the required alignment of the initial vectors in the Gram-Schmidt method. As a future work, an algorithm can be designed to align the three preceded vectors in the Gram-Schmidt method automatically. In addition, it is cogent to add a new class for the white blood cells that do not belong to those five classes. This is due to the fact that sometimes other cells called Blast appear in the peripheral blood. This cell type is more frequently found in the abnormal blood samples.

References

1. Sabino, D.M.U., Costa, L.D.F., Rizzatti, E.G., Zago, M.A.: A Texture Approach to Leukocyte Recognition. Real-Time Imaging 10(4), 205–216 (2004)
2. Angulo, J., Flandrin, G.: Automated Detection of Working Area of Peripheral Blood Smears Using Mathematical Morphology. Anal. Cell Pathol. 25, 37–49 (2003)
3. Wu, J., Zeng, P., Zhou, Y., Oliver, C.: A Novel Color Image Segmentation Method and Its Application to White Blood Cell Image Analysis. In: 8th International Conference on Signal Processing (ICSP'06), vol. 2, pp. 245–248. IEEE Press, Beijng (2006)
4. Umpon, N.T.: Patch-based White Blood Cell Nucleus Segmentation Using Fuzzy Clustering. ECTI Trans. Electrical Electronic Communications 3(1), 5–10 (2005)
5. Guo, N., Zeng, L., Wu, Q.: A Method Based on Multi-Spectral Imaging Technique for White Blood Cell Segmentation. Comput. Biol. Med. 37(1), 70–76 (2006)
6. Chassery, J.M., Garbay, C.: An Iterative Segmentation Method Based on Contextual Color and Shape Criterion. IEEE Trans. Pattern Anal. Machine Intell. 6(6), 795–800 (1984)
7. Jiang, K., Liao, Q.M., Dai, S.Y.: A Novel White Blood Cell Segmentation Scheme Using Scale-Space Filtering and Watershed Clustering. In: 2nd International Conference on Machine Learning and Cybernetics, vol. 5, pp. 2820–2825 (2003)
8. Zamani, F., Safabakhsh, R.: An Unsupervised GVF Snake Approach for White Blood Cell Segmentation Based on Nucleus. In: 8th International Conference on Signal Processing (ICSP'06), vol. 2. IEEE Press, Beijng (2006)
9. Park, J., Keller, J.M.: Snakes on the Watershed. IEEE Trans. Pattern Anal. Mach. Intell. 23(10), 1201–1205 (2001)
10. Umpon, N.T., Dhompongsa, S.: Morphological Granulometric Features of Nucleus in Automatic Bone Marrow White Blood Cell Classification. IEEE Trans. Inf. Technol. Biomed. 11(3), 353–359 (2007)
11. Thiran, J.P., Macq, B.: Morphological Feature Extraction for the Classification of Digital Images of Cancerous Tissues. IEEE Trans. Bio. Med. Eng. 43(10), 1011–1020 (1996)
12. Hengen, H., Spoor, S.L., Pandit, M.C.: Analysis of Blood and Bone Marrow Smears Using Digital Image Processing Techniques. In: Sonka, M., Fitzpatrick, J.M. (eds.) Proceedings of the SPIE, Medical Imaging 2002: Image Processing, vol. 4684, pp. 624–635 (2002)
13. Umpon, N.T., Gader, P.D.: System-Level Training of Neural Networks for Counting White Blood Cells. IEEE Trans. Syst., Man, Cybern. 32(1), 48–53 (2002)
14. Long, X., Cleveland, W.L., Yao, Y.L.: A New Preprocessing Approach for Cell Recognition. IEEE Trans. Inf. Technol. Biomed. 9(3), 407–412 (2005)
15. Nattkemper, T.W., Ritter, H.J., Schubert, W.: A Neural Classifier Enabling High-Throughput Topological Analysis of Lymphocytes in Tissue Sections. IEEE Trans. Inf. Technol. Biomed. 5(2), 138–149 (2001)
16. Shitong, W., Min, W.: A New Detection Algorithm (NDA) Based on Fuzzy Cellular Neural Networks for White Blood Cell Detection. IEEE Trans. Inf. Technol. Biomed. 10(1), 5–10 (2006)
17. Rezatofighi, S.H., Soltanian-Zadeh, H., Sharifian, R., Zoroofi, R.A.: A New Approach to White Blood Cell Nucleus Segmentation Based on Gram-Schmidt Orthogonalization. In: 2nd International Conference on Digital Image Processing (ICDIP'09), pp. 107–111. IEEE Press, Thailand (2009)
18. Haralick, R., Shanmugam, K., Dinstein, I.: Textural features for image classification. IEEE Trans. Syst. Man Cyber. 3(6), 610–621 (1973)

19. Ojala, T., PietikaÈ inen, M., MaÈenpa, T.: Multiresolution Gray-Scale and Rotation Invariant Texture Classification with Local Binary Patterns. IEEE Trans. Pattern Anal. Machine Intell. 24(7), 971–987 (2002)
20. Kass, M., Witkin, A., Terzopoulos, D.: Snakes: Active contour models. Int. J. of Comput. Vision 1(4), 321–331 (1988)
21. Pudil, P., Novovičová, J., Kittler, J.: Floating Search Methods in Feature Selection. Pattern Recogn. Lett. 15(11), 1119–1125 (1994)
22. Basheer, I.A., Hajmeer, M.: Artificial Neural Networks: Fundamentals, Computing, Design, and Application. J. Microbiol. Meth. 43(1), 3–31 (2000)
23. Burges, C.J.C.: A tutorial on Support Vector Machines for Pattern Recognition. Knowledge Discovery and Data Mining 2(2), 121–167 (1998)

An Application for Semi-automatic HPV Typing of PCR-RFLP Images

Christos Maramis, Evangelia Minga, and Anastasios Delopoulos

Dept. of Electrical and Computer Engineering,
Aristotle University of Thessaloniki, 54124 Thessaloniki, Greece
{chmaramis,evang}@mug.ee.auth.gr, adelo@eng.auth.gr

Abstract. The human papillomavirus, coming in over 100 flavors/types, is the causal factor of cervical cancer. The identification of the types that have infected the cervix of a patient is a very laborious yet critical task for molecular biologists that is still performed manually. HPV-Typer is a novel research software application that assists biologists by analyzing digitized images of electrophorized gel matrices that contain cervical samples processed by the PCR-RFLP technique in order to semi-automatically identify the existing types of the virus. HPVTyper has been designed to be functional under minimum user input conditions and yet to allow the user to intervene in any step of the typing procedure.

Keywords: HPV typing, gel electrophoresis, PCR-RFLP, biomedical image processing, software application.

1 Introduction

The human papillomavirus (HPV) is a double stranded DNA virus that is responsible for many forms of genital dysplasia and neoplasia [1,2] and is considered to be the causal factor for cervical cancer [3,4]. There have been identified more than 100 types of HPV having similar but slightly altered genotypes; more than 40 of these infect the anogenital tract [5]. However, not all of them are associated with the development of malignancies of the cervix [6]; there are HPV types associated with a high risk of malignant progression (high-risk types), types with a low risk of malignant progression (low-risk types) and types whose associated risk has not been determined yet (undetermined-risk types).

Given the above facts, it becomes evident that the discovery of the identity of the HPV type(s) that have infected a patient is crucial for determining the patient's risk of developing cervical lesions and cancer. This identification process is called HPV typing and remains even nowadays an inherently manual procedure. In this paper we introduce a software application that is intended to help molecular biologists in the task of HPV typing.

HPVTyper is a novel research application that has been developed within the Information Processing Laboratory of the Electrical and Computer Engineering

A. Campilho and M. Kamel (Eds.): ICIAR 2010, Part II, LNCS 6112, pp. 173–184, 2010.

Department of the Aristotle University of Thessaloniki. The application has been designed with the collaboration of and is currently under evaluation by the Molecular Biology Laboratory of the Papageorgiou Hospital of Thessaloniki. HPVTyper attempts to semi-automatically identify the types of HPV that have infected a patient by analyzing the image resulting from the gel electrophoresis of material that has been processed by the PCR-RFLP method (see Sect. 2.1 for an explanation of the molecular biology terms). In this effort, many of the steps are performed automatically while others require input from the biologist. However, the user can intervene at every step in order to adjust the miscomputed parameters of the problem.

The paper is structured as follows: In Sect. 2 we present the molecular biology techniques that comprise the current in vitro HPV typing protocol and also cite the related software applications. In Sect. 3 we describe HPVTyper and its components. Finally, in Sect. 4 we discuss the results of the preliminary use and also possible future improvements of HPVTyper.

2   Background

2.1   HPV Typing

In this section we describe step by step the in vitro protocol that is followed by molecular biologists in order to perform HPV typing on human samples.

First of all, a cervical tissue sample is being collected and is amplified with the use of the polymerase chain reaction (PCR) technique [8] by employing an appropriate set of primers. The reaction increases the concentration of any existing viral DNA molecules up to six orders of magnitude. Afterwards, the amplified material is being digested by a carefully selected restriction enzyme, which cuts the genetic material of HPV at positions of specific DNA base sequence; this is the restriction fragment length polymorphism (RFLP) technique [9], which results, due to genotype differences among HPV types, in a – known in advance – set of fragments of different lengths in base pairs (bp) for each virus type.

The next step in the protocol is gel electrophoresis. Solutions containing the genetic material from different samples are marked with a fluorescent dye and loaded into separate wells at the front end of a gel matrix. Then, in the presence of an electric field, the DNA fragments of various sizes are forced to move with different mobilities in a direction parallel to the field: the fragments of large size remain close to the well, while the more agile smaller fragments cover a much larger distance. This way, a number of *lanes*, starting from each well, are formed that contain blobs of DNA fragments of the same size shaped as *bands* perpendicular to the electric field. One or more wells are reserved to include material of known length (usually fragments constantly increasing by 20, 50, or 100 bp). These wells serve as *ladders* that help the biologist estimate the unknown lengths associated with the bands of the other lanes.

After the electrophoresis, a digitized image of the electrophorized gel is acquired by an appropriate digital camera in order to obtain a permanent record of the resulting gel matrix. Figure 1 depicts such an image and also emphasizes the concepts of *lane*, *band* and *ladder*. The electrophoresis image is analyzed by the biologist in order to answer to the following questions for each lane/sample: Is the sample infected by HPV? If so, by which types of the virus?

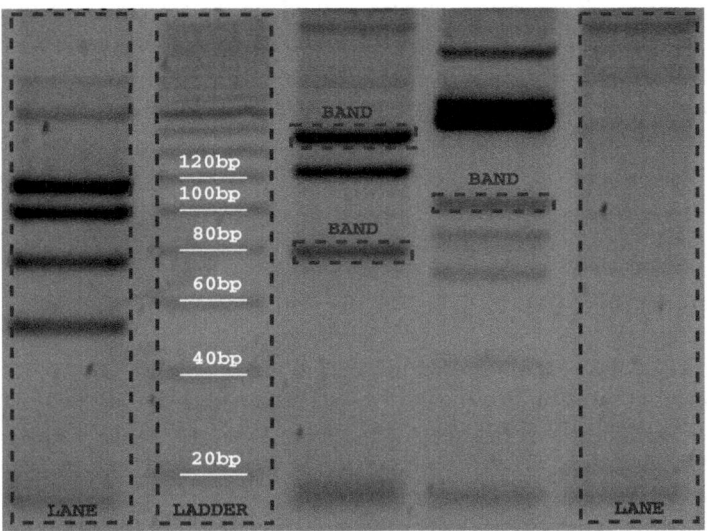

Fig. 1. Typical image of a gel matrix after electrophoresis. Samples of *lanes*, *bands* and *ladder* are enclosed in rectangles.

The first step towards answering the above questions is locating the bands of viral DNA that exist in each lane. Then, the fragment length which corresponds to each band is calculated by comparing its location with the locations of the bands of known length from the ladder(s). This is accomplished through an appropriate interpolation procedure (see Sect. 3.4). The result of this step for each lane is a set of estimated fragment lengths for the viral DNA existing in the sample. At this point, the biologist determines the combination of HPV types which is the most probable to have produced the estimated fragment lengths on each lane, having in mind for each type the set of fragment lengths that result from its digestion by the employed restriction enzyme. This is a tedious and often error-prone procedure.

2.2   Related Work

Gel electrophoresis has been at the forefront of molecular biology for many decades and it remains the most popular technique for separation of macromolecules. Thus, it comes as no surprise that there are plenty of software applications that deal with the processing and analysis of electrophoretic images:

TotalLab Quant [10], GelCompar II [11], Gel-Pro Analyzer [12] – just to name a few. However, all these applications are generic and cannot be employed directly for the typing of HPV. The most a biologist can get out of them is the estimation of the fragment lengths corresponding to the bands of a lane (see the previous section). Still, the actual typing procedure, i.e., the discovery of the combination of types that best explains the estimated lengths has to be performed manually.

On the other hand, there are application-specific programs that analyze electrophoretic images. For example, SAFA [13] which deals with DNA footprinting, GASepo [14] with Epo doping control, GelBandFitter [15] which defines the boundaries of closely spaced bands, etc. However, to the best of our knowledge, there is no software application dealing with HPV typing and this makes HPV-Typer both innovative and useful.

3 System Description

3.1 System Overview

HPVTyper is a standalone software application implemented in C++; for the development of its graphical interface we have employed the cross-platform wxWidgets library [16,17].

HPVTyper handles digital images of electrophorized gel matrices and is parametrized according to the restriction enzyme(s) used by the RFLP technique. The parametrization is accomplished through a configuration file which contains for each HPV type the list of the lengths (in bp) of the DNA fragments that result from the application of RFPL with the employed restriction enzyme(s). Genetic material that have been digested with different restriction enzymes can be analyzed as long as the information of the resulting fragment lengths per virus type for each utilized restriction enzyme is contained in the above configuration file.

Our application consists of three modules. The *Image Processing and Segmentation* module performs all the required image processing operations on the input image and also locates the boundaries of the existing lanes. The *Fragment Mobility Calibration* module deals with the ladder(s) included in the image and employs optimization techniques to map band positions (in pixels) to fragment lengths (in bp). This is achieved by optimally estimating the parameters that determine the mobility of the DNA molecules on the gel from the observed positions of the bands of the ladder(s). The *Band Selection and Type Identification* module performs the actual HPV typing procedure. For each lane, it helps the user select the existing bands of viral DNA and, based on the selected bands, it calculates for each type the probability[1] of the fact that this type is present in the sample loaded on the lane. The system architecture of HPVTyper is given in Fig. 2.

---

[1] It will be explained later that this is not exactly the probability but a compatibility degree.

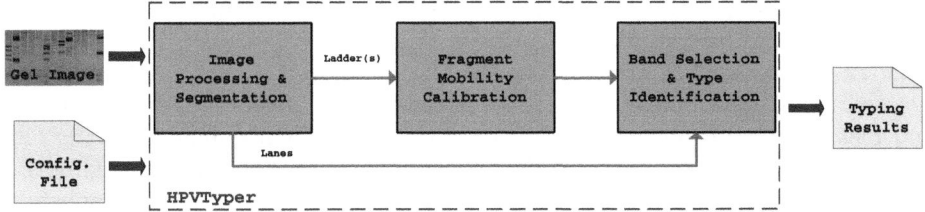

Fig. 2. HPVTyper's system architecture

The application is organized in four tabs that have a serial relation. This means that, in order to perform any action on some tab, the user has to visit the previous tabs and interact with the application so that HPVTyper can set all the prerequisite parameters. Each one of the above modules corresponds to a different tab of the application.

3.2   Image Processing and Segmentation

After loading the digitized image of an electrophoresis experiment into the application, the user can isolate the useful part of it, cropping the blank margins. This part, that contains only the lanes, looks like the one depicted in Fig. 1. Next, the application attempts to correct three types of defects that are apparent on the remaining part of the image. First, the lanes might not be exactly vertical. HPVTyper allows the user to rotate the image by small angles until the lanes are aligned to the vertical axis. Second, there exist dark stains of undetermined shape all over the area of the image due to unavoidable gel impurity. The application tries to eliminate them by applying a 3×3 median filter on the image. Finally, it is often the case that the bands appear dark on lighter background. However, it is visually better for the bands to appear light on darker background. Thus, the image can be subjected to color inversion by a simple click in order to stick to the above color convention.

After the above preprocessing actions, the application is ready to segment the image into lanes, i.e., to attempt to automatically locate the boundaries of the lanes. The only input the user has to provide is the number of the lanes. As the image is now properly oriented, the boundaries are simply vertical lines. The main idea is that, since the lane areas are covered with viral genetic material, each lane area generally appears lighter than the empty gel areas between the lanes. Therefore, we expect high intensity transitions between lanes and background when moving horizontally. This effect is magnified if we consider the entire length of a lane. Thus, the application calculates the discrete intensity derivative in the horizontal direction and sums its value across the vertical direction. The resulting one-dimensional curve has local extrema at the boundaries of the lanes with positive sign at transitions from lane to background area,[2] and

---

[2] When moving from the left to the right of the image.

with negative sign at the inverse transitions. The extraction of the local extrema is performed by the watershed algorithm [18].

However, in the case of noisy images, the discovered extrema do not always correspond to lane boundaries. To overcome this problem, we employ a second idea: The lanes must by design have similar – if not equal – widths. This means that the distance between the left or right boundaries of two neighboring lanes should be almost constant for all lanes. This "equadistance" property is combined with the located extrema positions so as to conclude to the actual boundaries of the lanes.

When the lane boundaries have been determined, the application displays the results by drawing blue dotted lines on the image at the positions of the left boundaries, and red dotted lines at the positions of the right boundaries. The user is allowed to modify the boundary positions by dragging the dotted lines accordingly. A snapshot of the application's tab which is used to perform the actions described in this section is given in Fig. 3.

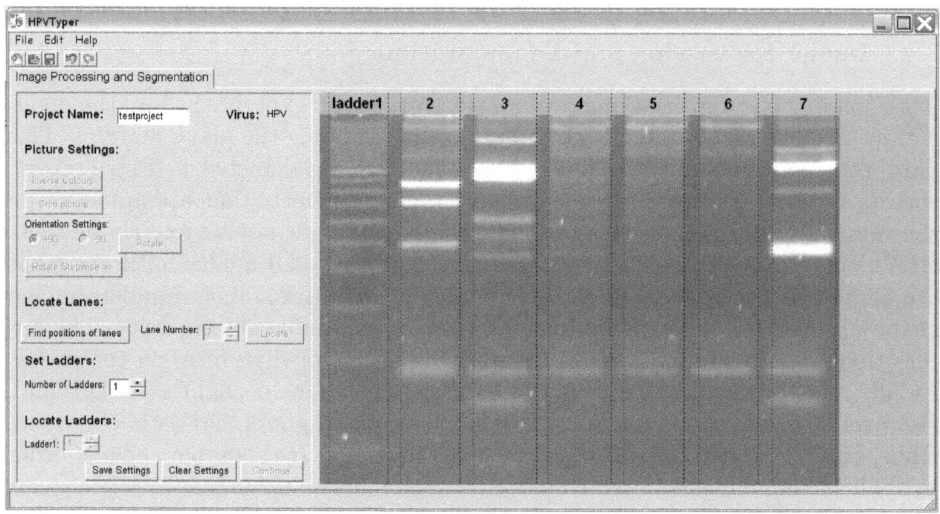

Fig. 3. A snapshot of the Image Processing and Segmentation tab

3.3   Lane Identification

In this tab, the user designates the ladder(s) that exist in the image, thus discriminating them from the other lanes and provides IDs for the lanes that correspond to patient samples. For the subsequent analysis, a ladder is assigned automatically to each lane in order to serve as a ruler of fragment lengths. By default, if there exist more than one ladders, the ladder assigned to a lane is the one closest to it. However, the user can manually alter this assignment.

3.4   Fragment Mobility Calibration

The goal of this module is to perform the mapping of pixel positions on the image to lengths of DNA fragments. This is accomplished by processing the ladder(s) that exist in the image. First, the positions of a number of ladder's bands corresponding to known fragment lengths are located. Then, the extracted pairs of positions on the image and fragment lengths are fitted into a predefined model of DNA mobility on the gel. This analysis is performed individually for each ladder and this also applies to the description that follows.

Before the application takes action, the user has to specify the step of the ladder, i.e., the constant length in bp by which the material loaded in the ladder increases. After that, the average intensity profile along the width of the lane is extracted, and the background intensity is subtracted from it. Next, the application attempts to locate a predefined number of bands on the ladder starting from the band corresponding to the smallest fragment length. These bands are basically local maxima of the extracted one-dimensional profile satisfying the following condition: Between two local maxima corresponding to successive bands the curve must fall below a near-zero intensity threshold. The number of the maxima sought depends on ladder's step. For instance, for a 20 bp ladder the lowermost 10 bands are sought.

The ladder part is detached from the gel image and displayed in horizontal orientation with the estimated positions of the bands indicated as superimposed red lines. The average intensity profile of the ladder with the located maxima is drawn just below the ladder image and serves as a visual aid for the user in case he would like to move some of the band position indicators.

According to [19,20], the theory which best describes the mobility of DNA fragments on gel under electrophoresis is the one claiming that the distance covered by a fragment on the gel is inversely proportional to the logarithm of its length. Hence, if l_i is the length of the DNA fragments forming the i-th of the N bands of a ladder and d_i is the distance they have covered from the start (i.e. the well) of the lane in pixels, then the above statement can be expressed as:

$$d_i = \theta_1 - \theta_2 \log(l_i) \quad \text{for } i = 1, 2, \ldots, N \ . \tag{1}$$

This can be treated as a linear least-squares optimization problem with respect to the unknown parameters θ_1 and θ_2. The extracted set of band positions and their corresponding fragment lengths are used in (1) to estimate θ_1 and θ_2 and when this is accomplished a ruler of fragment lengths is drawn just below the ladder's profile curve. The ruler also depicts the fragment length values that correspond to the bands of the ladder as they are calculated from the estimated mobility parameters.

Since the above estimation procedure is determined by the automatically located band positions on the ladder, it is evidently error-prone. To overcome this problem, the user may alter the band positions, thus invoking a new parameter optimization round as many times as needed. A snapshot of the application's tab in which the actions just described are performed is given in Fig. 4.

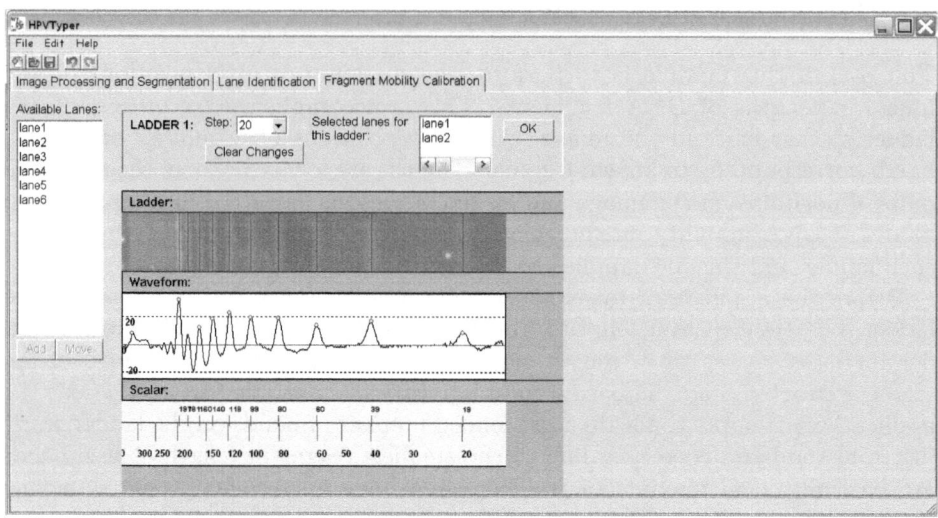

Fig. 4. A snapshot of the Fragment Mobility Calibration tab. This is the case where only one ladder exists in the image.

3.5   Band Selection and Type Identification

The last tab of the application is linked with the *Band Selection and Type Identification* module. Here, the user locates, with the guidance of the application, the bands that exist in each lane and, based on this information, the HPV types that may be present in the sample are identified. This analysis is performed for each lane separately and this also applies to the description that follows. The tab displays from top to bottom:

1. The image of the ladder that is assigned to the lane in horizontal orientation.
2. The image of the selected lane, also in horizontal orientation.
3. The background-free average intensity profile of the lane, which is extracted as explained in the previous subsection.
4. The ruler that has resulted from the estimated mobility parameters of the ladder.

At this point the user has to manually select all the bands that exist in the lane under investigation by clicking on the lane's image. The user selection is marked with a thick red line on the image and with a dotted red line on the profile curve. Moreover, the corresponding fragment length is displayed on the ruler. Although band selection is a manual procedure, HPVTyper assists the user in this task in many ways. First of all, the displayed profile of the lane can be proved very helpful during band selection, especially when the bands are thick or vague. The same holds for the ruler. Moreover, the application can indicate with the click of a button the expected positions of the bands that correspond to all the possible fragment lengths for all the types of the virus. These virtual bands are displayed as dotted blue lines and can guide the user while selecting the band positions.

After the band selection, the HPV typing algorithm takes over. The algorithm aims to answer to the following question for each type of the virus: Is the existence of this type in the sample compatible with the image of the lane? In other words, could the genetic material of this type have caused *some* of the observed bands on the lane? Thus, a compatibility degree is calculated for each type with the algorithm that is described in the following paragraphs. The degree ranges from 0 to 1, with 0 meaning that the type is completely incompatible with the gel image and 1 meaning that the type is fully compatible with it.

Obviously the fragment lengths that are interpolated for the selected bands are not completely accurate. There are plenty of reasons for that: impurities of the gel, imperfections of the capturing device, misplaced selection of the bands by the user, etc. In order to overcome this problem, the algorithm assigns to each selected band, instead of the corresponding estimated fragment length, a range of lengths centered around it. The width of the range is determined by its center. For instance, for fragment lengths lower than 80 bp, the range spans 2 bp on each side of the center, while for fragment lengths greater than 80 bp, it spans 7 bp on each side. This length-dependent assignment of the range's width makes perfect sense if we consider the motion mechanism of the macromolecules on the gel.

Next, for each HPV type, the application counts how many of its expected fragments lengths after digestion belong in the ranges of the observed bands. Only these lengths that can be interpolated by the discovered ladder bands are considered.[3] For example, since for the case of a 20 bp ladder, the mobility parameters estimation algorithm considers the 10 smaller fragment lengths, i.e., from 20 to 200 bp, all the fragment lengths (both observed and expected) that do not belong in this range are ignored. The compatibility degree of a type is the percentage of the type's expected fragments within the considered length range that belong in the ranges of the observed lengths.

In case no type is found to be compatible with degree higher than 0.7, the application suggests that the sample loaded on the lane has no HPV infection. Otherwise, the type(s) that overpass the above threshold appear on the right pane of the tab in order of decreasing compatibility. HPVTyper displays up to 5 HPV type employing a color-coding scheme for their names. Low-risk types are typed in blue font, high-risk types in red font and undetermined-risk types in green font. A report containing both the intermediate and the final (typing) results can be stored in a human-readable format and reloaded by the application at a future time. A snapshot of the application's last tab is given in Fig. 5.

---

[3] Forbidding length extrapolation is the common practice among molecular biologists.

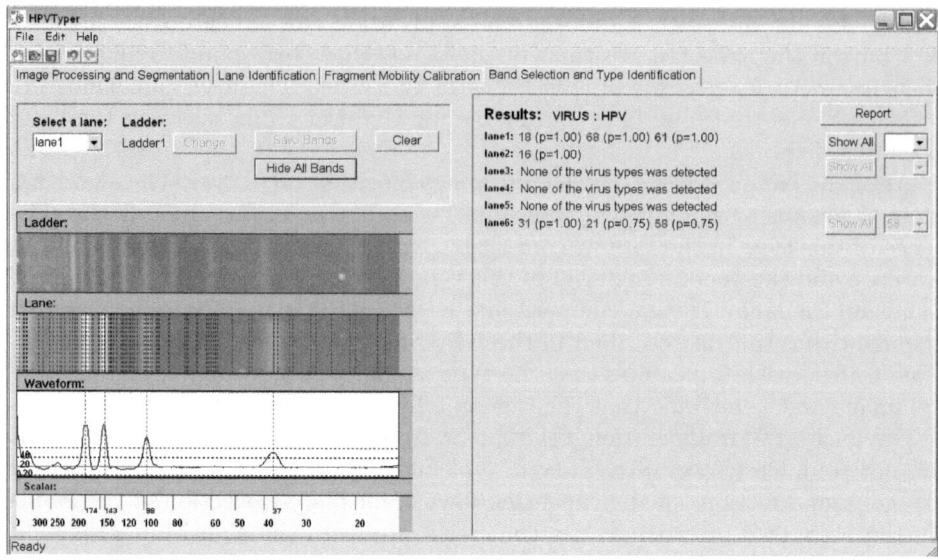

Fig. 5. A snapshot of the Band Selection and Type Identification tab

4 Discussion

HPVtyper was subjected to some early tests in the typing of images of gels that had been produced according to the materials and methods described in [7]. More specifically, cervical tissue samples from 20 individuals were collected, including 4 healthy subjects, 14 single type infections and 2 double type infections. The L1 region of the viral DNA existing in the samples was amplified using MY09/11 (pair of primers) and then was digested by HpyCH4V (restriction enzyme). The material was loaded to non-denaturating polyacrylamide gel for the electrophoresis and each gel matrix included one 20 bp ladder. Only the 41 types and subtypes of HPV given in [7] were considered. Each lane that resulted from the electrophoresis was manually typed by an expert molecular biologist and these were our ground-truth results for the comparison with HPVTyper's outcomes.

The results were very satisfactory. All the types that had been discovered by the expert were also identified by HPVTyper with very high compatibility degrees (ranging from 0.85 to 1). Moreover, all the lanes for which the expert had found no type, were also characterized healthy by the application (i.e. no type had compatibility degree higher than 0.7). At this point, we should mention that there were cases where HPVTyper pointed out as partially compatible types that had not been mentioned by the expert.

We consider HPVTyper as an application that can help molecular biologists in HPV typing as it is now, but also as a basis for a much more powerful application in the future. It is our intention to add new features that will automate some steps of the typing procedure and make more accurate some other steps. First of

all, more efficient strategies for removing the noise from the image and subtracting the background intensity have to be employed during the profile extraction
procedure. Moreover, if more than one ladders exist in the same image, the information extracted from both of them should be combined for the estimation of
the mobility parameters. This can improve the accuracy of length assignment to
bands especially for lanes that lie far from the ladders. Furthermore, the process
of locating the bands that belong to the lanes should be automated by fitting the
extracted profile to a superposition of properly shaped parametric functions (e.g.
superposition of Gaussian or Lorentzian functions). Finally, we have to employ
typing algorithms that are sophisticated enough to actually combine the types
of the virus in order to explain the observed bands on a lane and not just deal
with each type separately. Such algorithms could possibly be based on the use
of more than one restriction enzymes [21].

References

1. Baseman, J.G., Koutsky, L.A.: The epidemiology of human papillomavirus infections. J. Clin. Virol. 32, 16–24 (2005)
2. Wang, S.S., Hildesheim, A.: Viral and host factors in human papillomavirus persi-
   stence and progression. J. Natl. Cancer Inst. Monogr. 31, 35–40 (2003)
3. Bosch, F.X., Lorincz, A., Muñoz, N., Meijer, C.J., Shah, K.V.: The causal relation
   between human papillomavirus and cervical cancer. J. Clin. Pathol. 55(4), 244–265
   (2002)
4. Walboomers, J.M., Jacobs, M.V., Manos, M.M., Bosch, F.X., Kummer, J.A., et al.:
   Human papillomavirus is a necessary cause of invasive cervical cancer worldwide.
   J. Pathol. 189(1), 12–19 (1999)
5. de Villiers, E.M., Fauquet, C., Broker, T.R., Bernard, H.U., zur Hausen, H.: Classification of papillomaviruses. Virology 324(1), 17–27 (2004)
6. Muñoz, N., Bosch, F.X., de Sanjosé, S., Herrero, R., et al.: Epidemiologic classification of human papillomavirus types associated with cervical cancer. N. Engl. J.
   Med. 348, 518–527 (2003)
7. Santiago, E., Camacho, L., Junquera, M.L., Vázquez, F.: Full HPV typing by a
   single restriction enzyme. J. Clin. Virol. 37(1), 38–46 (2006)
8. Mullis, K., Faloona, F., Scharf, S., Saiki, R., Horn, G., Erlich, H.: Specific enzymatic
   amplification of DNA in vitro: the polymerase chain reaction. Cold Spring Harb.
   Symp. Quant. Biol. 51(1), 263–273 (1986)
9. Saiki, R.K., Scharf, S., Faloona, F., Mullis, K.B., Horn, G.T., Erlich, H.A., Arn-
   heim, N.: Enzymatic amplification of beta-globin genomic sequences and restriction
   site analysis for diagnosis of sickle cell anemia. Science 230(4732), 1350–1354 (1985)
10. TotalLab Quant, http://www.totallab.com/products/totallabquant
11. GelCompar II, http://www.applied-maths.com/gelcompar/gelcompar.htm
12. Gel-Pro Analyzer, http://www.mediacy.com/index.aspx?page=GelPro
13. Das, R., Laederach, A., Pearlman, S.M., Herschlag, D., Altman, R.B.: SAFA: Semi-
    automated footprinting analysis software for high-throughput quantification of nu-
    cleic acid footprinting experiments. RNA 11(3), 344–354 (2005)
14. Bajla, I., Holländer, I., Minichmayr, M., Gmeiner, G., Reichel, C.: GASepo—a soft-
    ware solution for quantitative analysis of digital images in Epo doping control.
    Comput. Methods Programs Biomed. 80(3), 246–270 (2005)

15. Mitov, M.I., Greaser, M.L., Campbell, K.S.: GelBandFitter–A computer program for analysis of closely spaced electrophoretic and immunoblotted bands. Electrophoresis 30(5), 848–851 (2009)
16. The wxWidgets Cross-Platform GUI Library, http://www.wxwidgets.org/
17. Smart, J., Hock, K., Csomor, S.: Cross-Platform GUI Programming with wxWidgets. Prentice Hall PTR, Upper Saddle River (2005)
18. Vincent, L., Soille, P.: Watersheds in digital spaces: An efficient algorithm based on immersion simulations. IEEE Trans. Pattern Anal. Mach. Intell. 13(6), 583–598 (1991)
19. Southern, E.M.: Measurement of DNA length by gel electrophoresis. Anal. Biochem. 100(2), 319–323 (1979)
20. Schaffer, H.E., Sederoff, R.R.: Improved estimation of DNA fragment lengths from agarose gels. Anal. Biochem. 115(1), 113–122 (1981)
21. Nobre, R.J., Almeida, L.P., Martins, T.C.: Complete genotyping of mucosal human papillomavirus using a restriction fragment length polymorphism analysis and an original typing algorithm. J. Clin. Virol. 42(1), 13–21 (2008)

Automatic Information Extraction from Gel Electrophoresis Images Using GEIAS

C.M.R. Caridade[1,2], A.R.S. Marçal[2,4], T. Mendonça[2],
A.M. Pessoa[3], and S. Pereira[3]

[1] Instituto Superior de Engenharia de Coimbra, R. Pedro Nunes, Qt.Nora,
3030-199 Coimbra, Portugal
[2] Faculdade de Ciências, Univ. Porto, Dep. Matemática, R. Campo Alegre, 687,
4169-007 Porto, Portugal
[3] Faculdade de Ciências, Univ. Porto, Dep. Biologia, R. Campo Alegre,
46, 4169-007 Porto, Portugal
[4] CICGE & CMUP, Faculdade de Ciências, Univ. Porto

Abstract. This paper presents a method (*GEIAS*) for the automatic
processing of digital images obtained from Gel Electrophoresis. The performance of *GEIAS* was tested using 12 images, obtained from 4 gels
with 3 different exposures with a total of 1082 bands, comparing the
results provided by *GEIAS* and 3 other software tools. The *GEIAS* is
able to fully automatically detect DNA lanes while the other 3 software
tools tested can only do this in a semi-automatic or manual way. For
the correct location of DNA bands, *GEIAS* required a manual correction of the location in 10.0% of the bands, and the other software tools
13.0%, 15.0% and 25.4%. The average error in the estimation of molecular weight was tested using a total of 5443 bands in 12 image using 672
reference/observed lane pairs. The average error was found to be 9.2%
for *GEIAS* and 11.2%, 14.4% and 13.1% for the other software tools
tested.

1   Introduction

In molecular biology laboratories, fluorescent dyes are used for the detection
and sizing of DNA and RNA in agarose gels. The most common dye to visualize
DNA or RNA bands in agarose gel electrophoresis is ethidium bromide, usually
abbreviated as EtBr [1]. It fluoresces under UV light when intercalated into
DNA. Typically DNA bands containing more than ~10ng DNA become visible
in an EtBr-treated gel viewed under UV light and the fluorescent images can
be recorded as photographs or digital images. Under a constant field strength, a
linear duplex DNA molecule migrates through the gel matrix at a rate inversely
proportional to the $\log_{10}$ of their molecular weight (or molecular size expressed
in number of base pairs) and proportional to the applied voltage [2] [3]. However
with higher voltages (5-10 V/cm) the migration of large DNA molecules increase
at a faster rate than small DNA molecules [2]. Electrophoresis in agarose gels
provides a rapid and convenient way to measure the quantity of DNA. Because

A. Campilho and M. Kamel (Eds.): ICIAR 2010, Part II, LNCS 6112, pp. 185–194, 2010.

the amount of fluorescence is proportional to the total mass of DNA, the amount of nucleic acid can be estimated from the intensity of fluorescence emitted by ethidium bromide. The quantity of DNA in the sample can be estimated by comparing the fluorescent yield of the sample with that of a series of standards [3]. DNA molecules are sized by their relative movement through a gel compared to a molecular weight standard, so mobility measurements are critical to size determinations. To compute the size of unknown DNA fragments separated on gels, a standard curve must be created using fragments of known size from the standard molecular weight markers that are run in parallel with the unknown samples during gel electrophoresis. The colors on the Gel Electrophoresis Image (GEI) vary with the dye/stain used, but generally the GEI can be converted to an intensity (or greyscale) image without any loss of information. An example of a greyscale GEI is presented in figure 1 (left). A GEI might contain one or more gels, each with a number of lanes. In this example the image has a single gel with 8 lanes. Each lane has various bands, corresponding to the presence of DNA molecules with a given molecular weight. The intensity of a band depends on the mass (amount, quantity) of DNA present.

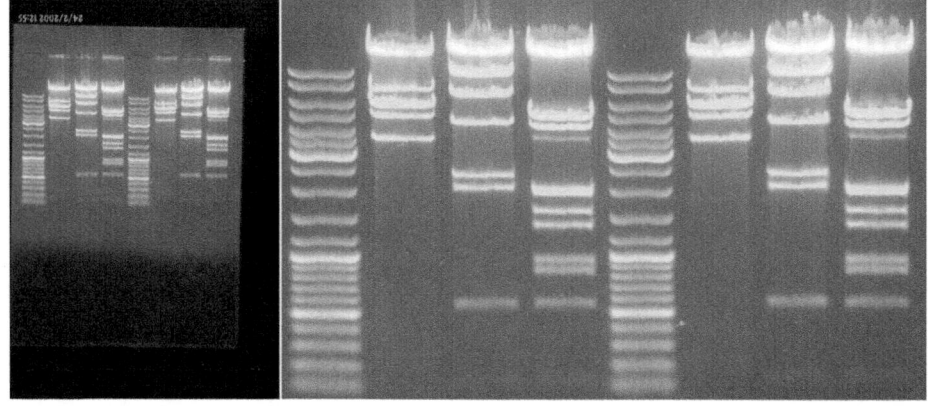

Fig. 1. Example of a Gel Electrophoresis Image (*G1b*). Original image in greyscale (left) and sub-image with the interest area extracted automatically (right).

The calculation of the molecular weight and mass for an observed substance is done using a reference in one of the lanes. The reference is a standard substance, with the molecular profile of the various bands known.

There are a number of software tools available for GEI analysis, such as ImageQuant TL (GE Healthcare, UK), Kodak 1D software (KodaK, USA) and Quantity One (Bio-Rad, USA). However, in all these software packages there are several steps that require a considerable interaction from the operator, including the identification of the exact location of a gel in the image, and very often also the location of individual lanes.

The purpose of this work is to present *GEIAS* - Gel Electrophoresis Image Analysis Software and to compare it with 3 software tools: *Quantity One* version 4.6.3 from Bio-Rad [6], *ImageQuant TL* from GE Healthcare [7] and *Kodak 1D* from Kodak [8]. The current version of GEIAS is an improvement from initial versions presented in [4] and [5], with the inclusion of an automatic correction procedure, which improves the molecular weight profile calculation precision. The performance evaluation of GEIAS and the other software packages include tests on the automatic identification of interested area, image rotation correction, location of lanes and bands, as well as the calculation of the molecular weight profile for each lane observed, given a reference lane.

2   Method

2.1   Pre-processing

Initially, the original GEI image is converted to greyscale (by averaging the RGB components), and from greyscale to binary using the Otsu global threshold method [9]. To eliminate noise, morphological operators [10] are used with a circular structure element of 5 pixels radius (more details in [5]). Cumulative line function are used to identify and extract the gels present in an image and cumulative column function are used to detected the interested area that correspond to non-void pixels in the binary image (more details in [4]). As an example, the result for the image *G1b* after the pre-processing stage is presented in Figure 1 (right), which includes only the area of interest from the original image.

2.2   Automatic Rotation

Once the interest area on the GEI is established, it might be necessary to rotate the image so that the lanes became vertical. Initially, rotated versions of the original image are created, by applying rotation angles from $-10°$ to $10°$ with an increment of $0.5°$ (40 different images). For each rotated image, the pre-processing described in 2.1 is also applied. The number of pixels ON in each column of the binary images created are used to produce a histogram function f. Figure 2 (middle row) shows the function f for the original test image *G1b* and for a rotated version of this image (by $5°$). The basic assumption to estimate the correction for rotation is that when the best alignment of lanes is achieved (lanes near by vertical) the separation between lanes is maximum, which results in the highest number of null values in f. The total number of null values found in f is thus used as a measurement of the lane separation. This value (h) is obtained for each rotated image. The maximum value of h, as a function of the rotation angle, indicates the rotation correction needed for the image. Figure 2 shows the plots of h as a function of θ, for the original image (*G1b*) and for the rotated version of the image (by $5°$).

For a more accurate estimation of the rotation, the process is repeated using increments of $0.1°$ on a range of $1°$ centered around the previous estimate (maximum of h). For the examples presented in Figure 2 the local estimation was

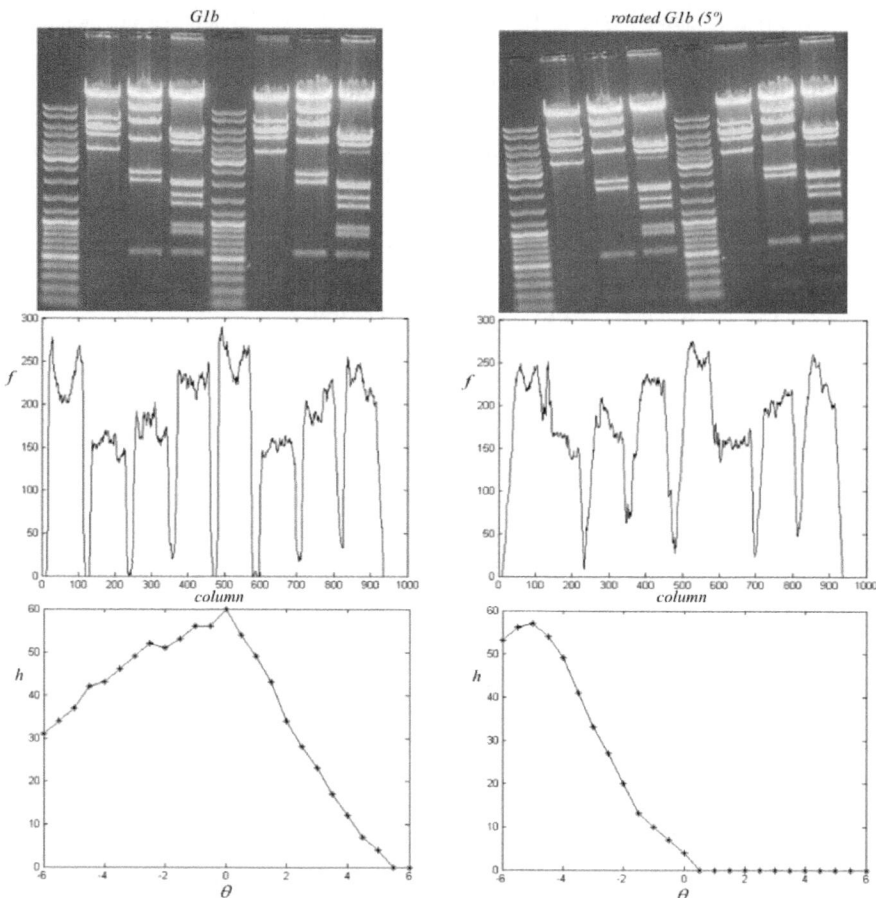

Fig. 2. Original *G1b* image (top left) and rotated version (top right), with the corresponding plots for functions f (middle row) and h (bottom row)

made for θ between $-0.5°$ and $0.5°$ for the left image, and between $4.5°$ and $5.5°$ for the right image. In these examples, the rotation correction for the original image was found to be $0.0°$, and for the rotated image $5.0°$. Once the rotation angle is found, the image is corrected by applying that rotation.

2.3   Lane Detection

The number of lanes is detected using cumulative column function (f) from the interested area. The assumption is that all lanes have equal width (W_n). The image (interested area only) is divided in n equal parts (with n between 3 and 25), providing an estimate for the lane's width. If the estimate is correct, the lane centres i (Equation 1) should have high values of f and in the edges between i and $i+1$ (Equation 2) there should be low values of f ($i = 1, ..., n$).

$$X^i = (i - \frac{1}{2})W_n \tag{1}$$

$$X^{'i} = iW_n \tag{2}$$

Three function are thus constructed: the cumulative average for edges $F_e(n)$ (Equation 3); the cumulative average for centres $F_c(n)$ (Equation 4); and the Lane Detection Index (LDI), represented by ϕ (Equation 5). The final estimation of the number of lanes (n') is obtained by searching for the LDI maximum. A sub-image is obtained for each lane by dividing the original image into n' parts, where n' represent the maximum value of ϕ (number of lanes).

$$F_e(n) = \frac{\sum_{i=1}^{n} \sum_{j=4}^{W_n-3} f(W_n \times (i-1) + j)}{(W_n - 6) \times n} \tag{3}$$

$$F_c(n) = \frac{\sum_{i=1}^{n-1} f(W_n \times i - 1) + f(W_n \times i) + f(W_n \times i + 1)}{3 \times (n-1)} \tag{4}$$

$$\phi(n) = (F_c(n) - F_e(n)) \times \sqrt{n} \tag{5}$$

2.4   Band Extration

A band is an area of a roughly rectangular shape, with a high density of pixels ON in the binary image. Its location corresponds to a local maximum of the histogram function (t) obtained for the number of pixels ON per line. This function is calculated using only the central 2/3 of the lane's width. The local maxima below a certain threshold are ignored, as they represent false bands (more details in [5]). Figure 3 shows a lane sub-image (bottom) and the corresponding plot of t as a function of the image line (top).

2.5   Reference Calibration

In order to compare the performance of the proposed method with existing software tools, 5 standard molecular weight DNA markers were used: MassRuler^{TM} DNA Ladder Mix (A)[1], GeneRuller^{TM} DNA Ladder Mix, ready-to-use (B)[2], Lambda DNA/EcoRI Marker (C)[6], Lambda DNA/ HindIII Marker (D)[6], and Lambda DNA/EcoRI+HindIII Marker (E)[3] (Fermentas, Lithuania). The characteristic of each reference substance is unique. Figure 4 shows the standard signature of the 5 DNA markers used.

The calibration is performed in two steps. The first step matches the strong bands with a linear function. In the second step, the strong bands are used to

---

[1] MassRuler^{TM} DNA Ladders, LabAid^{TM}, Fermentas, 2006
http://www.fermentas.com/pdf/labaids/labaid_massruler2006.pdf
[2] GeneRuller^{TM} DNA Ladders, LabAid^{TM}, Fermentas, 2006
http://www.fermentas.com/pdf/labaids/labaid_generuler2006.pdf
[3] Conventional Lambda DNA Markers, LabAid^{TM}, Fermentas, 2005
http://www.fermentas.com/pdf/labaids/labaid_lambdamarkers2005.pdf

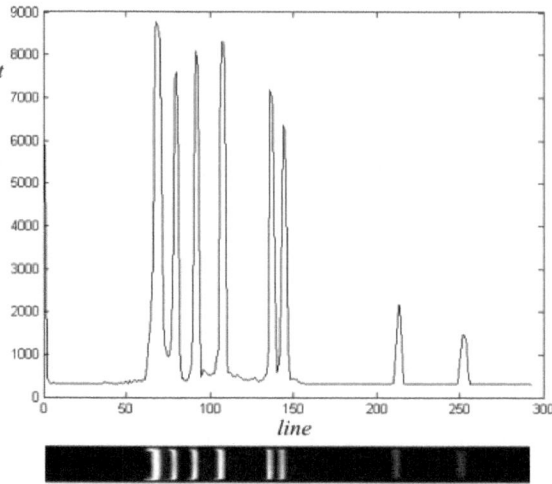

Fig. 3. Example of an histogram function t (top) for a reference lane (bottom)

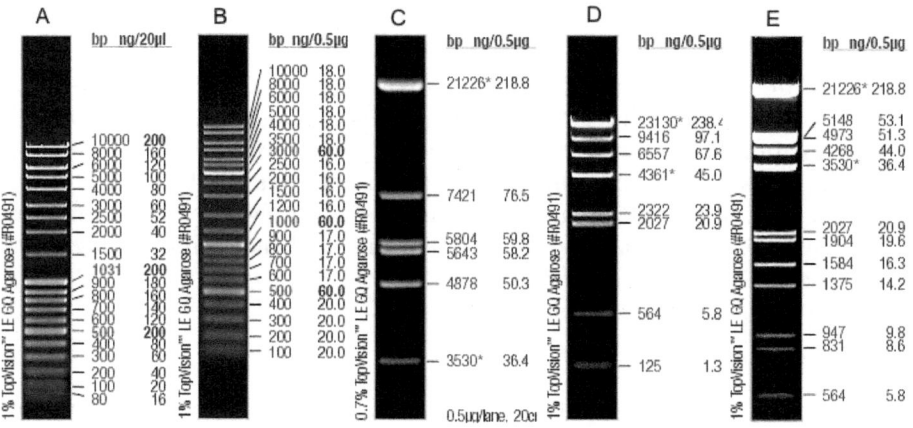

Fig. 4. Reference substances used for the evaluation experiment

predict the location of the weaker ones. If the distance between the predicted and observed bands is less than 10 image lines, the match is accepted, otherwise it is ignored (more details in [5]).

2.6   Lane Analysis

The estimation of the molecular weight of an observed band is obtained by linear interpolation between the closest values along the vertical axis in the reference lane, using Equations (6) and (7)

$$bp = exp[M * (ln(Xband) - ln(XOref_-)) + ln(Xref_-)] \qquad (6)$$

$$M = \frac{ln(Xref_+) - ln(Xref_-)}{ln(XOref_+) - ln(XOref_-)} \tag{7}$$

where $Xband$ is the x-axis position of a band, $XOref_+$ and $XOref_-$ are the x-axis position of the closer observed reference bands and $Xref_+$ and $Xref_-$ are the x-axis positions of the reference bands matching with $XOref_+$ and $XOref_-$ (more details in [5]).

3 Results

3.1 Test Images

DNA electrophoresis experiments were prepared to evaluate the performance of *GEIAS*, comparing it with 3 other software tools, according to standard molecular biology procedures [3]. The 25 mL gels used contained 1% (w/v) agarose (Molecular Biology Grade - Bioron GmbH, Germany) and 0,2 μg/mL Ethidium Bromide (BioRad GmbH, Germany), dissolved in 1X TAE (40 mM Tris, 20 mM Sodium Acetate, 2 mM EDTA (pH=8,0) - BioRad GmbH, Germany). Each gel was loaded with four of the DNA markers twice (1 μg each); the Lambda DNA Markers were mixed with 0,20 volumes of 6X Orange DNA Loading Dye (Fermentas, Lithuania). The electrophoresis were performed for 50 min, using 80V (5 V/cm) and 1X TAE as the running buffer [3] and conducted using the LifeTechnologies Horizon 58 apparatus (GibcoBRL, UK). The GEI were acquired using the Kodak EDAS 290 imaging system and Kodak 1D software v.3.5.4 (Kodak, USA). Four different gels were prepared (*G1*, *G2*, *G3* and *G4*), each with different exposures (*a*, *b*, and *c*), resulting in a total of 12 images (all with 8 lanes). The test images *G1*, *G2* and *G3* have (from left to right) the references substances *B*, *C*, *D*, *E*, *B*, *C*, *D*, *E* and *G4* have reference substances *A*, *B*, *C*, *D*, *A*, *B*, *C*, *D*. Using each lane as reference at a time, a total of 96 ($4 \times 3 \times 8$) different test cases were made available, with the remaining 7 lanes on each image used as observations, resulting in a total of 672 (96×7) lane observations (more details in [5]). The evaluation was made in three issues: (i) the detection of the number and location of lanes in the GEI, (ii) the detection of bands, (iii) the quantitative estimation of the molecular weight for each band. A typical GEI is presented in Figure 1, in greyscale (the original is in 24 bits color).

3.2 Automatic Detection of Lanes

The first step is the identification of the region of interest in the original image, which is basically the removal of the margins around the gel. The automatic detection of the region of interest was successful in all 12 images tested in the *GEIAS*, as well as the automatic rotation. In the other software tools tested, both the identification of the area of interest and the rotation correction have to be performed manually by the user.

The results for the detection of the number of lanes and their location in the image are presented in Table 1. The semi-automatic detection is used when the application did not detected the correct number os lanes. In that case, the user must provide the number of lanes so that the software can detected their correct position in the image. In those cases where the software fails to properly locate the lane's position, there has to be a manual identification of each lane. All 12 images were detected in the semi-automatic made by *ImageQuant* and 11 by *Quantity One*. For *Kodak*, 3 images were detected automatically and the remaining 9 had to be processed manually. The proposed system (*GEIAS*) managed to process all 12 images without any operator intervetion.

Table 1. Evaluation of lane detection (12 images tested)

Detection \ Appl.	Quantity One	ImageQuant	Kodak	GEIAS
Fully Automatic[1]	0	0	3	12
Semi Automatic[2]	11	12	0	0
Manual	1	0	9	0
(1) user indicates number lanes				
(2) user marks 1 or more lanes				

3.3 Automatic Detection of Bands

The results for the detection of band locations in lanes are presented in Table 2. Positional error represents a band detected in the wrong position, false positive (false +) refers to bands that do not exist and false negative (false −) to bands that are not detected. Kodak was the only software with bands detected with a positional error (8 cases). All 4 software tools produced more false + (90 to 239) than false − (8 to 28). Overall, the percentage of bands that required manual correction (bottom row in Table 2) was found to be between 10.0% (*GEIAS*) to 25.4% (*Kodak*).

Table 2. Evaluation of band detection (total of 1082 bands in 12 images)

Error \ Appl.	Quantity One	ImageQuant	Kodak	GEIAS
Positional error	0	0	8	0
False +	124	154	239	90
False −	17	8	28	18
Bands requiring manual correction(%)	13.0%	15.0%	25.4%	10.0%

3.4 Molecular Weight Estimation

The molecular weight (*bp*) estimation was processed separately for each band in the 12 GEI, using each lane as reference at a time ($12 \times 8 = 96$ cases) and the remaining 7 lanes as observations. A total of 672 (96×7) testes where therefore considered (more details in [5]). The relative error in the estimation of *bp* (in %)

for *GEIAS* and the 3 other software tools tested are presented in Table 3. The *bp* error values range from 6.8% (*GEIAS*) to 17.8% (*ImageQuant*). The error is lowest for image *G1b* (6.8% to 7.7%) (Fig. 1) and worst for for *G4b* (10.9% to 17.8%). *GEIAS* was found to have a lower relative error in the estimation of *bp* less than the other software tools for all images tested except for *G1a* and *G3b*. For image *G1a* the best software was Kodak (7.5%) and for image *G3b* *Quantity One* (9.5%). The average error for all 12 images is presented in in the bottom row of Table 3. According to this experimental, the ranking of the 4 software tools tested was: 1st - *GEIAS* (9.2%), 2nd - *Quantity One* (11.2%), 3rd - *Kodak* (13.1%), 4th - *ImageQuan* (14.4%).

Table 3. Relative error in *bp* for all the G1, G2, G3 and G4 with 3 levels of exposure (a, b and c). 1082 band tested in 12 images.

I \ Appl.	Quantity One	ImageQuant	Kodak	GEIAS
G1a	7.6%	12.3%	7.5%	8.4%
G1b	7.5%	7.6%	7.7%	6.8%
G1c	7.3%	12.4%	15.8%	7.2%
G2a	14.4%	16.4%	14.4%	10.2%
G2b	14.4%	19.1%	15.1%	10.0%
G2c	15.7%	16.5%	17.7%	9.2%
G3a	9.5%	12.0%	10.2%	11.5%
G3b	11.4%	13.7%	12.5%	8.1%
G3c	8.7%	12.3%	17.1%	8.5%
G4a	12.0%	16.0%	12.3%	9.0%
G4b	13.0%	17.8%	13.7%	10.9%
G4c	13.0%	17.2%	13.2%	10.4%
Average	11.2%	14.4%	13.1%	9.2%

4 Conclusions

The proposed method (*GEIAS*) for the automatic processing of DNA Gel Electrophoresis Images (GEI) was found to be very efficient in the automatic detection of the number of gels, the region of interest, as well as the number and location of lanes. The method is fully automatic in the gel and lane detection process. For the others software tools tested, these tasks had to be done manually. The automatic correction of the image rotation is only performed by *GEIAS*. The other 3 software tools do not perform any correction for image rotation. The band detection is automatic in all softwares, although manual corrections are required for a number of bands. For *GEIAS* 10.0% of the bands required correction, while for the other software tools the values were higher: 13.0% (*Quantity One*), 15.0% (*ImageQuant*) and 25.4% (*Kodak*). The quantitative calculation of the molecular weight (*bp*) for an observed DNA band depends on the experimental conditions, including the reference substance used, but also on the quality of the resulting GEI (exposure, amount of drag and noise). The average error in the estimation of bp was lowest for *GEIAS* in 10 out of 12 images tested. The overall average in the estimation of bp was 9.2% for *GEIAS*, 11.2% (*Quantity One*), 13.1% (*Kodak*) and 14.4% (*ImageQuant*).

The experiment presented in this paper showed the potential of *GEIAS* as an alternative to existing GEI processing software. Plans for future work include the computation of the mass present on each band as well as the development of an easy to use interface.

References

1. Sharp, P.A., Sugden, B., Sambrook, J.: Detection of two restriction endonuclease activities in haemophilus parainfluenzae using analytical agarose-ethidium bromide electrophoresis. Biochemistry 12, 3055–3063 (1973)
2. Helling, R.B., Goodman, H.M., Boyer, H.W.: Analysis of endonuclease R–EcoRI fragments of DNA from lambdoid bacteriophages and other viruses by agarose-gel electrophoresis. J. Virol. 14, 1235–1244 (1974)
3. Sambrook, R.: Molecular cloning: A laboratory manual, 3rd edn. Cold Spring Harbor Laboratory Press (2001)
4. Caridade, C.M.R., Marçal, A.R.S., Mendonça, T.: Automatic extraction and classification of DNA profiles in digital images. In: Tavares, J., Jorge, N. (eds.) Computational Vision and Medical Image Processing, pp. 111–116. Taylor & Francis, Abington (2008)
5. Caridade, C.M.R., Marçal, A.R.S., Mendonça, T., Pessoa, A.M., Pereira, S.: An automatic Method to identify and extract information of DNA bands in Gel Electrophoresis Images. In: Proceedings of the 31st Annual International Conference of the IEEE EMBS - EMBC 2009, pp. 1024–1027 (2009)
6. Quantity One version 4.6.3 from Bio-Rad Laboratories, Inc., http://quantity-one.software.informer.com (visited August 2009)
7. ImageQuant TL, GE Healthcare, http://www.imsupport.com/ (visited August 2009)
8. KODAK 1D Image Analysis software, KodaK, http://www.carestreamhealth.com/molecular-imaging-software.html (visited August 2009)
9. Otsu, N.: A threshold selection method from gray-level histograms. IEEE Trans. on Systems Man Cybernetics 9(1), 62–69 (2002)
10. Gonzalez, R.C., Woods, R.E.: Digital Image Processing, 3rd edn. Prentice Hall, Upper Saddle River (2008)

Elastography of Biological Tissue: Direct Inversion Methods That Allow for Local Shear Modulus Variations

C. Antonio Sánchez[1], Corina S. Drapaca[2],
Sivabal Sivaloganathan[1,3], and Edward R. Vrscay[1]

[1] Department of Applied Mathematics, Faculty of Mathematics,
University of Waterloo, Waterloo, ON, Canada
[2] Department of Engineering Science and Mechanics, Penn State Center for Neural
Engineering, Pennsylvania State University, University Park, PA, USA
[3] Centre for Mathematical Medicine, Fields Institute, Toronto, ON, Canada
c.antonio.sanchez@gmail.com, csd12@psu.edu, ssivalog@uwaterloo.ca,
ervrscay@uwaterloo.ca

Abstract. In recent years, imaging techniques have been adapted to indirectly measure stiffness of biological tissues, with the hope of using this information to aid in detecting and classifying pathological regions. Several methods have been developed to convert a sequence of strain images into a single elasticity image, but most are based on assumptions that limit the local variability of stiffness in the estimate. In this paper, two direct inversion methods are introduced. The novelty of these methods is that they concurrently solve a system of differential equations for the stiffness, allowing for strong local variations. Some ideas regarding uniqueness of solutions, an issue that is ignored in existing works, are also presented. Preliminary numerical results show that by keeping the differential terms in the tissue model, the new inversion methods can more accurately determine the tissue's stiffness distribution.

1 Introduction

It has been well known for millenia that the presence of abnormally stiff regions in a biological tissue is a strong indicator of damage or disease. Palpation has been the primary diagnostic procedure used since the days of Hippocrates [1], and is still one of the most common and efficient methods for detecting pathology in some soft tissues (such as tumours or cysts in the breast). This method, however, is limited to tissues near the surface of the body. In order to non-invasively examine tissues not accessible by touch, imaging techniques must be employed. Conventional imaging procedures are not able to discern pathological tissue from the surrounding healthy tissue if the two have similar molecular densities. This has led to the development of elastography techniques, which combine imaging with tissue mechanics in order to estimate stiffness values.

Magnetic Resonance Elastography (MRE) is a recently developed medical imaging technique used to non-invasively 'measure' mechanical properties of tissue [2]. The result is an image that quantitatively describes stiffness, and can be

A. Campilho and M. Kamel (Eds.): ICIAR 2010, Part II, LNCS 6112, pp. 195–206, 2010.

used by clinicians to diagnose or classify pathology. The MRE method consists of the following steps: (i) apply a known stress to the tissue, (ii) measure the physical response using MR phase-differencing [2], and (iii) invert a mathematical model that describes the mechanical behaviour of the tissue to find the desired stiffness parameters. Inversion can be performed using the measured data directly in the model [3,4,5], or indirectly with the use of an iterative technique [6,7]. In this paper, direct methods are considered.

The standard model is a system of differential equations that describes tissue motion in response to an applied stress. Inverting such a system can be difficult due to a lack of proper boundary conditions for stiffness, which cannot be measured due to the non-invasive nature of the experiment. To avoid this issue, simplifying assumptions are typically imposed when using direct methods, the most common of which is that the mechanical parameters can be locally approximated by constant functions [3,4,5]. In this case, the system of differential equations is converted into an algebraic system in terms of the unknown mechanical parameters, which can be easily solved using well-known matrix methods. The assumption, known as *local homogeneity*, limits the local variability of the parameters in the estimate. Furthermore, wherever the true parameters do vary significantly, the estimate is invalid. From a clinician's point-of-view, regions where the true stiffness changes rapidly are perhaps the most important, for they indicate the boundaries of any pathology.

In this paper, two novel direct methods are proposed to solve the system of differential equations, rather than the algebraic approximation. In the first method, the differential system is set as an equality constraint, and optimization methods are used to steer any free parameters. The second technique uses Green's functions to solve the system of differential equations. By solving the system of differential equations, a more accurate description of the stiffness parameters can be obtained, allowing for strong local variations. The result is an elastogram of more practical utility.

The structure of the paper is as follows. In Section 2, the tissue model is presented, and a few simplifying assumptions are used to reduce the number of free parameters. In Section 3, the issue of uniqueness of solutions is discussed. The novel methods are described in Sections 4 and 5, and preliminary results are presented for both simulated and measured data. The paper ends with some concluding remarks.

2   The Tissue Model

In this work, the tissue is modelled as a linearly viscoelastic isotropic continuum undergoing harmonic motion. To induce harmonic motion, a sinusoidal shear stress is applied to the boundary of the tissue near the region of interest, and the system is allowed to reach a quasi-steady state. The harmonic assumption is required for the MR phase-difference data acquisition technique [2]. An example of a measured displacement field is shown in Fig. 1. A linear constitutive law is adopted since the displacements have very low amplitudes (on the order of micrometres). With these assumptions, the equations of motion, written in the frequency domain, are the following:

$$-\omega^2 \rho \boldsymbol{U} = \nabla \cdot \left(\mathcal{M} \left[\nabla \boldsymbol{U} + \nabla^{\mathrm{T}} \boldsymbol{U} \right] \right) + \nabla \left(\Lambda \nabla \cdot \boldsymbol{U} \right) \ , \tag{1}$$

where ω is the frequency of excitation, ρ is the density of the tissue, $\boldsymbol{U}$ is the three-dimensional harmonic displacement field, and Λ and $\mathcal{M}$ are complex, frequency-dependent versions of Lamé's first and second parameters. The vector differential operators act in such a way that the dimensions are consistent, and $\nabla^{\mathrm{T}} \boldsymbol{U} = \left[\nabla \boldsymbol{U} \right]^{\mathrm{T}}$. Equation (1) is a system of three differential equations with three unknown functions: ρ, Λ and $\mathcal{M}$. Without specifying some form of boundary or regularizing conditions, these three functions cannot be uniquely determined. The second Lamé parameter, $\mathcal{M}$ (also referred to as the complex shear modulus), is the parameter of interest since it has been experimentally shown to vary strongly with pathology [2,4,5].

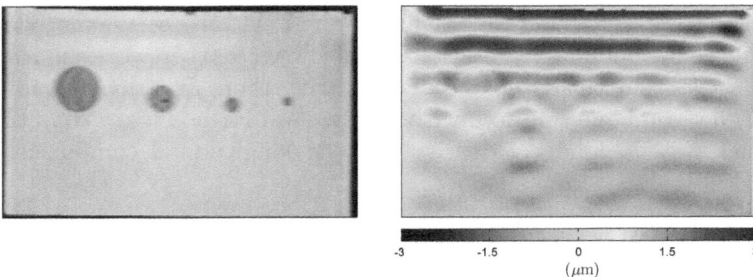

Fig. 1. Harmonic excitation of a gel phantom containing four stiff cylindrical inclusions: MR magnitude image (left) and the displacement pattern estimated from a sequence of MR phase-difference images (right). The experiment was performed in Dr. Richard Ehman's MRE lab at the Mayo Clinic [8].

In order to limit the number of free parameters in the system, two of the unknown functions are eliminated. Soft tissues are mainly composed of water, so the density is usually assumed to be constant with a value of $1 \, \mathrm{g/cm^3}$. The large presence of water also causes the tissue to be nearly incompressible; thus, the longitudinal component, $\nabla \left(\Lambda \nabla \cdot \boldsymbol{U} \right)$, is expected to be negligible. With these two further assumptions, the system is reduced to

$$-\omega^2 \boldsymbol{U} = \mathcal{M} \nabla \cdot \left(\nabla \boldsymbol{U} + \nabla^{\mathrm{T}} \boldsymbol{U} \right) + \nabla \mathcal{M} \cdot \left(\nabla \boldsymbol{U} + \nabla^{\mathrm{T}} \boldsymbol{U} \right) \ , \tag{2}$$

where the only unknown function is the desired shear modulus, $\mathcal{M}$, and the equations have been expanded using a vector calculus identity. If the local homogeneity assumption is imposed, then $\nabla \mathcal{M}$ is forced to be zero everywhere. The result is an algebraic expression for the shear modulus that can be solved using least-squares. This approach is known as Algebraic Inversion of the Differential Equation (AIDE) [5,9]. In this paper, however, the gradient of the shear modulus is not ignored. Two novel techniques are proposed to invert the system of differential equations in (2). If the combination of the three equations in the system does not provide enough information to determine a unique solution, then regularizing assumptions are invoked to isolate a particular one.

3   Uniqueness

If the mathematical model accurately describes the tissue of interest, then there is a theoretical 'true' distribution of the complex shear modulus. The goal of elastography is to determine this modulus using the measured displacement data. If there is not enough information to determine a unique solution, then further assumptions must be made. However, the determined solution will only be accurate if the assumptions are valid. As soon as assumptions like local homogeneity are imposed, there is no guarantee that a good approximation to the true solution will be found.

Uniqueness has not been thoroughly explored in existing MRE literature. It has been strongly suggested that including the $\nabla \mathcal{M}$ term in the inversion procedure requires the boundary conditions to be known [5,9]. However, mathematical analysis shows that this is not necessarily true; there exist local conditions for which (2) has a unique solution, without the need for boundary conditions. The differential system, ignoring density and the longitudinal component, can be expressed in matrix form as follows:

$$\left[\nabla \cdot \left[\nabla U + \nabla^{\mathrm{T}} U \right] \ \left[\nabla U + \nabla^{\mathrm{T}} U \right] \right] \begin{bmatrix} I \\ \nabla^{\mathrm{T}} \end{bmatrix} \mathcal{M} = -\omega^2 U \ , \tag{3}$$

where I is the identity operator. Note that this is a system of three equations involving four unknown terms: $\mathcal{M}$, $\frac{\partial \mathcal{M}}{\partial x}$, $\frac{\partial \mathcal{M}}{\partial y}$, and $\frac{\partial \mathcal{M}}{\partial z}$. Also, note that the coefficients of the differential operator depend on the displacement field, U. A consequence of this is that if a second, linearly independent displacement field is obtained (perhaps by repeating the experiment while changing the location of the shear stress), enough equations are generated to locally estimate all unknown terms. Without repeating the experiment, there are additional conditions that still guarantee uniqueness, at least locally. For example, for all points satisfying

$$\mathrm{rank} \left\{ \left[\nabla U + \nabla^{\mathrm{T}} U \right] \right\} < \mathrm{rank} \left\{ \left[\nabla \cdot \left[\nabla U + \nabla^{\mathrm{T}} U \right] \ \left[\nabla U + \nabla^{\mathrm{T}} U \right] \right] \right\} \ , \tag{4}$$

$\mathcal{M}$ is uniquely determined. This condition allows for $\nabla \mathcal{M}$ to be removed using local Gaussian elimination techniques, leaving an expression involving only $\mathcal{M}$. If $\left[\nabla U + \nabla^{\mathrm{T}} U \right]$ has full rank, then another condition that guarantees local uniqueness is

$$\nabla \times \left[\left[\nabla U + \nabla^{\mathrm{T}} U \right]^{-1} \nabla \cdot \left(\nabla U + \nabla^{\mathrm{T}} U \right) \right] \neq 0 \ . \tag{5}$$

With Condition (5), it can be shown that the only consistent solution to the homogeneous problem is the zero solution, which guarantees uniqueness of the inhomogeneous problem [10]. Thus, for any local regions where either of (4) or (5) hold, the system in Equation (2) can be locally inverted without the need for boundary conditions. In these instances, the system is known as being *completely integrable* in the Frobenius sense [11].

There are cases, however, where the solution is not unique. For example,

$$U = \begin{bmatrix} e^{ikx} \\ e^{iky} \\ e^{ikz} \end{bmatrix} \quad \Longrightarrow \quad \mathcal{M} = \frac{w^2}{2k^2} + \alpha\, e^{-ik(x+y+z)}, \quad \text{for any } \alpha \in \mathbb{C} \ .$$

Thus, there is a family of solutions with one free parameter, α. A regularizing condition must be enforced in order to isolate a particular solution. This will only match the true solution if the assumptions are valid. Physically, the complex shear modulus should satisfy $\mathrm{Re}\{\mathcal{M}\} > 0$ and $\mathrm{Im}\{\mathcal{M}\} \geq 0$, which forces α to be zero. Thus, based on physically justifiable assumptions, the true solution might still be uniquely determined.

The family of solutions to the system in (2) has very few free parameters to begin with, if any. Assumptions like local homogeneity, while convenient, are not always required. By enforcing them, one is limiting the accuracy of the inversion method. To accurately capture strong variations in the shear modulus, it might be best to attempt to solve the differential system instead of the algebraic approximation, then impose regularizing conditions as a last resort to set any free parameters.

4   Direct Inversion Using Optimization Techniques

In order to solve (2) for the complex shear modulus, the problem is re-posed in a constrained optimization framework:

> *Minimize* $\ f(\mathcal{M}) = \|\nabla \mathcal{M}\|^2 \ ,$
>
> *subject to the equality constraint given by* (2) .

Inequality constraints, such as $\mathrm{Re}\{\mathcal{M}\} > 0$ and $\mathrm{Im}\{\mathcal{M}\} > 0$, can also be imposed. The equality constraint guarantees that the the complex shear modulus satisfies the equations of motion. The quadratic minimization functional, f, then sets the free parameters in order to select a particular solution out of the space of admissible ones. Even though the condition that $\|\nabla \mathcal{M}\|^2$ be minimized is similar to the local homogeneity assumption, $\nabla \mathcal{M} = 0$, this optimization method is fundamentally different from AIDE. Here, one is searching for the particular solution among the set of admissible solutions that is the most locally homogeneous. In the AIDE technique, local homogeneity is enforced before the inversion is performed, and the resulting estimate will not be found in the set of admissible solutions if the true shear modulus varies at all.

To minimize computational complexity, inversion is performed on local block domains, similar to Van Houten's subzone approach [6]. On these domains, derivative operators acting on $\mathcal{M}$ are represented by discrete approximations, such as finite-difference matrices, with the boundary conditions left as free parameters. This can be accomplished by setting to zero all rows in the finite-difference matrices that correspond to boundary points. A simple null-space method is used to steer any free parameters if the discretized version of (2) is not invertible. Otherwise, the system is solved using least-squares.

4.1   Results

Simulated Data

Two simulated displacement fields were generated by solving the full three-dimensional forward problem of (2) for U, given a known distribution of the shear modulus, $\mathcal{M}$. In the first simulated test-case, a spherical inclusion was introduced in the centre of the domain, with a stiffness value of $\mathrm{Re}\{\mathcal{M}\} = 7\,\mathrm{kPa}$. The background stiffness was taken to be $3\,\mathrm{kPa}$, and a linear transition was introduced around the boundary of the inclusion to ensure continuity. In the second simulated test-case, a Gaussian bump was added to the centre of the domain with background stiffness of $3\,\mathrm{kPa}$, to have a peak of $7\,\mathrm{kPa}$. Sinusoidal variations were then introduced everywhere to test the impact of local variations. The volume of the simulated tissue is $10 \times 10 \times 4.8\,\mathrm{cm}$, with grid dimensions $128 \times 128 \times 24$. A sinusoidal shear force was applied to one side of the tissue, with an amplitude of $1\,\mu\mathrm{m}$ at a frequency of $100\,\mathrm{Hz}$. Absorbing boundary conditions were implemented using a perfectly matched layer adapted to the elastic wave equation [10,12]. The first harmonic was extracted using a Fast Fourier Transform, and this was fed into the new inversion algorithm.

The inversion was performed on $20 \times 20 \times 20$ blocks, with a 50% overlap. Solutions in overlapping regions were averaged together to ensure smooth transitions. Savitzky-Golay filters were used to estimate derivatives of the data [13]. Fourth-order centred finite difference matrices were used to represent the derivative operators acting on $\mathcal{M}$, with the boundary conditions left free. For these two simulated test-cases, it was found that the discretized version of (2) was a fully determined system, implying the solution is unique. Results are shown in Fig. 2. The shear modulus estimates from the new method are significantly more accurate than those obtained using the traditional AIDE technique. In AIDE, the shear modulus distribution is deformed for both test-cases, and a ringing behaviour is exhibited at the boundary of the spherical inclusion of the first test-case. By allowing local variations in $\mathcal{M}$, the sharp transition about the spherical inclusion can be accurately determined, and the true shapes of the stiffness maps are preserved.

Experimental Data

Measurements of a three-dimensional displacement field were obtained from an experiment on a gel phantom containing four cylindrical stiff inclusions. The background gel was 1.5% Agar, having a stiffness of approximately $2.9\,\mathrm{kPa}$, and the four cylindrical inclusions were composed of 10% B-gel (bovine), with a

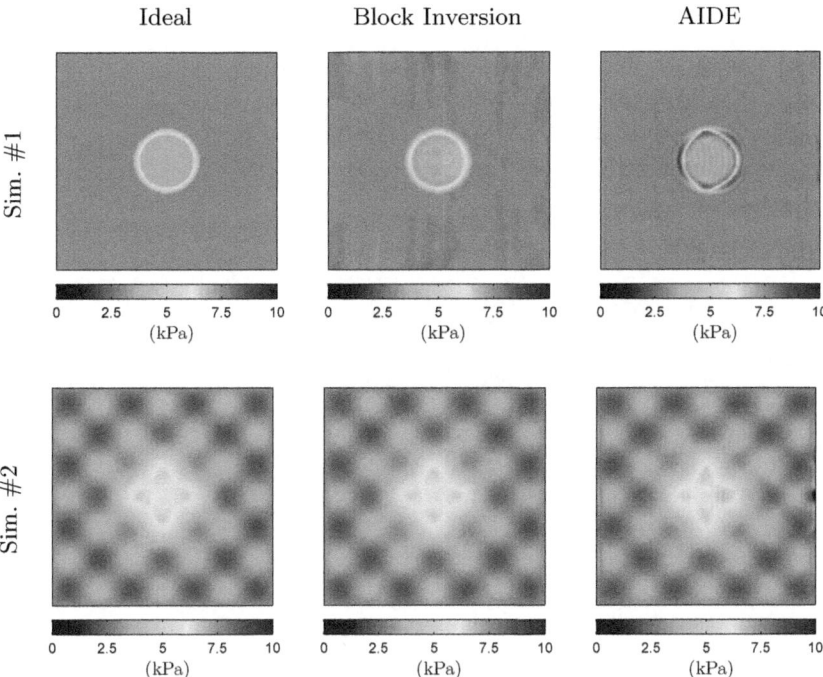

Fig. 2. Centre slice, parallel to the xy-plane, of the real component of the complex shear modulus for two simulated test-cases. Waves are travelling from left to right in the simulated displacement fields.

stiffness of approximately 6.4 kPa. The diameters of the four cylinders were approximately 5, 10, 16 and 25 mm. The field of view was 20×20 cm with a slice thickness of 3 mm. The grid dimensions were $256 \times 256 \times 16$. A sinusoidal force was applied to one side of the domain with an amplitude of $2\,\mu$m at a frequency of 100 Hz. The displacement field was measured using a MR phase-difference technique. This experiment was performed by Dr. Richard Ehman's MRE lab at the Mayo Clinic [8]. The data was preprocessed with a third-order Savitzky-Golay smoothing filter to remove noise, and was trimmed to contain $188 \times 98 \times 14$ points. The first harmonic was extracted using a Fast Fourier Transform.

Inversion was performed on $20 \times 20 \times 14$ blocks, with 50% overlap. Again, Savitzky-Golay filters were used to estimate derivatives of the data, and the derivative operators acting on $\mathcal{M}$ were represented by fourth-order centred finite difference matrices with boundary conditions left as free parameters. Similar to the simulated test-cases, the solution was determined to be unique. The estimates using the new method and AIDE are shown in Fig. 3. Again, one can see that the estimate from the new method is more accurate, particularly in and around the stiff inclusions, than that from AIDE. The AIDE result continues to show a ringing behaviour at sharp transitions of the shear modulus. This is caused by the local homogeneity assumption; in these regions of sharp variation, the assumption breaks down, rendering the shear modulus estimate invalid. The AIDE result, however, is smoother, particularly in the background region. This is because the local homogeneity assumption also regularizes, smoothing the solution. No such regularization is performed in the new block inversion method, making it more sensitive to background noise.

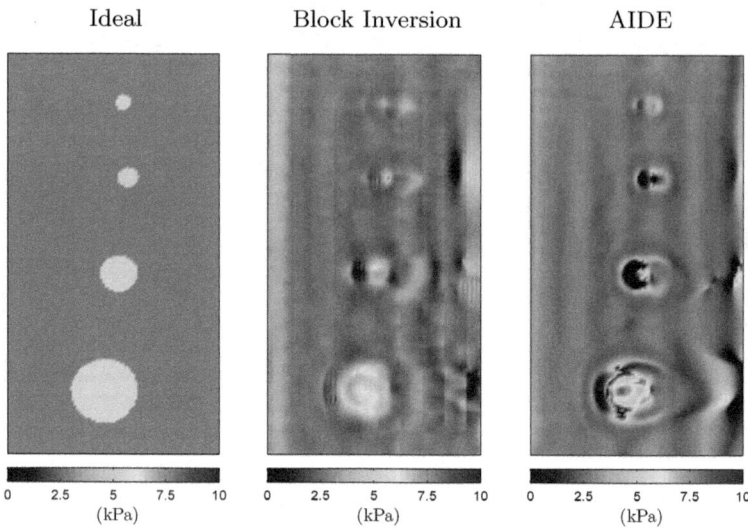

Fig. 3. Centre slice, parallel to the xy-plane, of the real component of the complex shear modulus for a the experimental data. Waves are travelling from left to right in the measured displacement field.

5   Direct Inversion Using Green's Functions

Both the AIDE and block inversion methods require knowledge of second-order derivatives of the displacement data. In practice, the measured data will contain noise, which is amplified by differentiation. This can make inversion methods unstable. The block inversion method presented in the previous section is particularly sensitive to variations in the data, so heavily relies on preprocessing techniques. To reduce this sensitivity, the need to estimate high-order derivatives of the data must be eliminated. This can be accomplished with the use of a Green's function.

Green's functions are used extensively in the study of Partial Differential Equations (PDEs), and are extremely useful in solving inhomogeneous problems subject to known boundary conditions. If a Green's function for a particular differential operator exists, then the PDE can be inverted using a convolution. To demonstrate this concept, consider the following simplified equations:

$$-\omega^2 \boldsymbol{U} = \nabla^2 \left(\mathcal{M} \boldsymbol{U} \right) \ . \tag{6}$$

In order to arrive at this Helmholtz system, it must be assumed that the tissue is locally homogeneous and completely incompressible. However, like the previous method, the complex shear modulus is expressed in a system of differential equations with unknown boundary conditions. The free-space Green's function for the 3D Laplacian operator is given by

$$\mathcal{G}(\boldsymbol{x}) = -\frac{1}{4\pi\|\boldsymbol{x}\|} \ .$$

A convolution with this Green's function will invert a Laplacian provided the function decays rapidly at the extremes. Unfortunately, the convolution is defined on all $\mathbb{R}^3$, but the data is only known in some finite region. Thus, either the Green's function must be modified to account for the finite boundary, or a data window must be introduced. To prevent the need to numerically determine a Green's function based on the data, the latter approach is used. Both sides of (6) are multiplied by a window function, W, that has compact support. A convolution with the Green's function, $\mathcal{G}$, is then performed, resulting in the following system:

$$-\omega^2 \mathcal{G} * (W\boldsymbol{U}) = \mathcal{M} \, W\boldsymbol{U} - \underbrace{\mathcal{G} * \left(2\nabla W \cdot \nabla\left(\mathcal{M}\boldsymbol{U}\right) + \mathcal{M}\boldsymbol{U}\nabla^2 W\right)}_{\mathcal{B}} \ . \tag{7}$$

If W is discontinuous, then derivatives must be considered in the sense of distributions. A typical example of a window is the boxcar function, which has a value of one in some interior region, and zero outside. With a 3D boxcar window, the boundary term, $\mathcal{B}$, depends on the values of $\mathcal{M}$ and its derivatives at the boundaries. Unfortunately, this means boundary conditions on $\mathcal{M}$ are required in order to evaluate the expression given in (7).

As an initial estimate, it is assumed that the complex shear modulus is constant over the boundaries of the data window. With this assumption, $\mathcal{M}$ can be pulled outside of the boundary term, allowing the initial estimate to be expressed as follows:

$$\mathcal{M}^{(1)} = -\frac{\omega^2 \mathcal{G} * (W\boldsymbol{U})}{W\boldsymbol{U} - \mathcal{G} * (2\nabla W \cdot \nabla \boldsymbol{U} + \boldsymbol{U}\nabla^2 W)} \ .$$

This only involves first derivatives of the data, resulting in a more stable system. In this way, three estimates of the shear modulus are obtained, one from each of the three equations of motion. These should be combined using a weighted averaging scheme, like least-squares. The estimate can then be iteratively improved by using the boundary values found in the previous iteration:

$$\mathcal{M}^{(k)} = -\frac{\omega^2 \mathcal{G} * (W\boldsymbol{U})}{W\boldsymbol{U} - \mathcal{G} * \left(2\nabla W \cdot \nabla\left(\mathcal{M}^{(k-1)}\boldsymbol{U}\right) + \mathcal{M}^{(k-1)}\boldsymbol{U}\nabla^2 W\right)} \ ,$$

where k is the iteration number. In this way, it is hoped that the complex shear modulus will converge to a solution that satisfies the system given in (7). If the system of differential equations has any degrees of freedom, then further regularizing assumptions must be included in the iterative procedure to steer any free parameters and guarantee convergence.

Preliminary Results

The Green's function method was applied to the two simulated displacement fields and the one measured displacement field. Derivatives of the data window and Green's function were analytically determined, then discretized. For these preliminary results, only the first estimate, $\mathcal{M}^{(1)}$, is reported. A small window with a $5 \times 5 \times 5$ support was used in order to reduce the impact of the constant boundary assumption. The window was shifted one point at a time in order to simplify the inversion procedure, allowing the estimate to be expressed as a single ratio of two convolutions:

$$\mathcal{M}^{(1)} = -\frac{\omega^2 (W\mathcal{G}) * \boldsymbol{U}}{\nabla^2 (WG) * \boldsymbol{U}} .$$

A slice of the results is shown in Fig. 4.

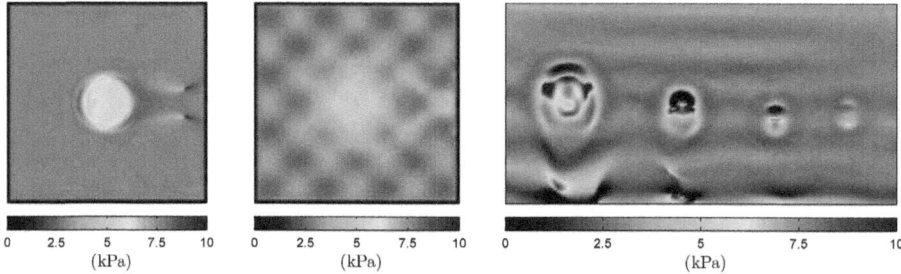

Fig. 4. Centre slice of the real component of the complex shear modulus for the three test-cases. Waves are travelling from left to right in the simulated displacement fields (left, centre), and top to bottom in the measured displacement field (right).

Since the simplified equation of motion is based on the local homogeniety and incompressibility assumptions, the accuracy of the shear modulus estimate is limited. Just as for the AIDE method, the sharp transitions about the inclusions in the first and third test cases cannot be accurately captured; the local homogeneity assumption introduces a ringing artifact in these regions. The incompressibility assumption also affects the estimated stiffness values, particularly within the regions of high stiffness for the two simulated cases. The advantage of this method, however, is that it is incredibly robust. Even with high levels of noise, reasonable estimates of the complex shear modulus can be obtained. The convolution acts as a smoothing operator, averaging noise that falls within the support of the data window.

In order to increase accuracy, the original system of equations, (2), should be used instead of the Helmholtz system. Unfortunately, the Green's function will depend on the displacement field, so will need to be determined numerically. The same iterative procedure can still be used, where an initial estimate of the shear modulus can provide boundary conditions for the next iteration. In this way, the system of differential equations can be solved using a Green's function.

6   Conclusions

Using theory for systems of PDEs, it can be shown that the equations relating the complex shear modulus to the displacement field, (2), can admit unique local solutions, without the need for regularizing assumptions or boundary conditions. When local homogeneity is imposed, the estimate of the shear modulus is rendered invalid in any regions where the true stiffness varies strongly. These regions mark the boundaries of any pathological tissue, so are important from a clinical perspective. By keeping the differential terms, a more accurate estimate of the true stiffness distribution can be obtained.

One approach to solving the system of partial differential equations is to leave the boundary conditions as free parameters, and use constrained optimization techniques to converge to a solution. In this way, the obtained solution is guaranteed to satisfy the original system, even in regions with local stiffness variations. This method has been found to produce more accurate estimates of the complex shear modulus than the traditional AIDE technique. However, the inversion is quite sensitive to noise. To reduce sensitivity, Green's functions can be employed. These functions are applied through a convolution, which has desireable numerical properties; noise is smoothed out over the region of integration, and derivatives of the data can be passed onto the Green's function, which helps stabilize the system. Boundary conditions are determined through an iterative procedure. The result is a very robust method to solve the system of PDEs. The 'best' method to use depends on the quality of the data. If it is relatively noise-free, then the optimization technique is the most straightforward and produces the most accurate estimate. If the data contains a lot of noise, then the more robust Green's function method provides a more reliable estimate of the complex shear modulus. By improving the accuracy of stiffness estimates, the resulting elastograms can be used more reliably in clinical applications, becoming part of a non-invasive tool to aid in the initial diagnosis of tissue pathology, and in tracking the progress of treatment strategies.

Acknowledgements

We gratefully acknowledge the generous support of this research by the Natural Sciences and Engineering Research Council of Canada (NSERC) in the forms of a Discovery Grant (ERV) and a Canada Graduate Scholarship (CAS). We would also like to thank Dr. Richard Ehman and his lab at the Mayo Clinic for the experimental data.

References

1. Bynum, W., Porter, R.: Medicine and the Five Senses. Cambridge University Press, Cambridge (2005)
2. Muthupillai, R., Lomas, D.J., Rossman, P.J., Greenleaf, J.F., Manduca, A., Ehman, R.L.: Magnetic resonance elastography by direct visualization of propagating acoustic strain waves. Science 269, 1854–1857 (1995)
3. Sinkus, R., Lorenzen, J., Schrader, D., Lorenzen, M., Dargatz, M., Holz, D.: High-resolution tensor MR elastography for breast tumour detection. Phys. Med. Biol. 45, 1649–1664 (2000)
4. Manduca, A., Oliphant, T.E., Dresner, M.A., Mahowald, J.L., Kruse, S.A., Amromin, E., Felmlee, J.P., Greenleaf, J.F., Ehman, R.L.: Magnetic resonance elastography: non-invasive mapping of tissue elasticity. Med. Image Anal. 5, 237–254 (2001)
5. Oliphant, T.E., Manduca, A., Ehman, R.L., Greenleaf, J.F.: Complex-valued stiffness reconstruction for magnetic resonance elastography by algebraic inversion of the differential equation. Magn. Reson. Med. 45, 299–310 (2001)
6. Van Houten, E.E., Miga, M.I., Weaver, J.B., Kennedy, F.E., Paulsen, K.D.: Three-dimensional subzone-based reconstruction algorithm for MR elastography. Magn. Reson. Med. 45, 827–837 (2001)
7. Van Houten, E.E., Doyley, M.M., Kennedy, F.E., Paulsen, K.D., Weaver, J.B.: A three-parameter mechanical property reconstruction method for MR-based elastic property imaging. IEEE Trans. Med. Imaging 24, 311–324 (2005)
8. Ehman, R.L.: Mayo Clinic Research: Magnetic Resonance Elastography (January 2007), http://mayoresearch.mayo.edu/ehman_lab/mre.cfm
9. Oliphant, T.E.: Direct Methods for Dynamic Elastography Reconstruction: Optimal Inversion of the Interior Helmholtz Problem. PhD in Biomedical imaging, Mayo Graduate School, Rochester, MN (2001)
10. Sánchez, C.A.: Dynamic Magnetic Resonance Elastography: Improved Direct Methods of Shear Modulus Estimation. MMath in Applied Mathematics, University of Waterloo, Waterloo, ON (2009)
11. Abraham, R., Marsden, J.E.: Foundations of Mechanics, 2nd edn. Addison–Wesley, California (1987)
12. Berenger, J.P.: Perfectly Matched Layer (PML) for Computational Electromagnetics. Morgan & Claypool Publishers, San Francisco (2007)
13. Savitzky, A., Golay, M.: Smoothing and differentiation of data by simplified least squares procedures. Analytical Chemistry 36, 1627–1639 (1964)

Segmentation of Cell Nuclei in Arabidopsis Thaliana Roots

Jonas De Vylder, Filip Rooms, and Wilfried Philips

Department of Telecommunications and Information Processing,
IBBT - Image Processing and Interpretation Group,
Ghent University, St-Pietersnieuwstraat 41, B-9000 Ghent, Belgium
jonas.devylder@telin.ugent.be

Abstract. The work flow of cell biologists depends more and more on the analysis of a large number of images. The manual analysis is a tedious, time consuming and error-prone task. Therefore the aid of automatic image analysis would be beneficial. This however needs segmentation techniques which can handle low contrast noisy images with bleed-through from different fluorescent dyes. In this paper we propose a technique which can cope with these problems by using intensity statistics. The proposed techniques are validated conform the requirements for the ICIAR *Arabidopsis Thaliana Root Cell Segmentation Challenge*, which allows straight forward comparison of different techniques for segmentation of Arabidopsis nuclei.

1 Introduction

The biotechnology industry is a fast growing industry. However, the number of bio-engineered products that reach the market is only a small part of the wide range of ideas tested in research and development labs. Typically, such tests are designed to measure the effect of a given treatment in samples of the specimen of interest. Observable effects include growth, cell division and behavioural changes. The aim of image analysis is to obtain a quantitative description of such variations. Thus, a specialist needs to outline the biological objects in images or video sequences before any measurement can be taken. This is often done manually, which is a tedious, error-prone and time consuming task. Especially, since a large number of images have to be processed to obtain statically significant results. Currently, an increasing number of tests use automatic image acquisition systems, such as high throughput systems. They provide new possibilities for research and development, however, manual process is still the main bottleneck in the sample processing chain.

To overcome this bottleneck, serious efforts have been made toward automatic detection of biological objects such as cells and cell nuclei. For some applications, good contrast images are available, such as DIC or contrast phase micrographs. A wide range of techniques have already been reported in literature [1,2,3,4,5] for the automatic analysis of these kind of images. Other applications might need fluorescence microscopy, where certain biological features, such as DNA or membranes are tagged by a fluorescent dye. It is this dye which is visible in the micrographs. For a more detailed overview

A. Campilho and M. Kamel (Eds.): ICIAR 2010, Part II, LNCS 6112, pp. 207–216, 2010.
© Springer-Verlag Berlin Heidelberg 2010

of fluorescence microscopy, we refer to [6,7]. Automatic analysis of fluorescence microscopy has been investigated in [8,9,10,11]. Although they report good results on the segmentation of cell nuclei, they only consider micrographs with a single channel. So they don't have to cope with the problem of bleed-through, i.e. a fluorescent marker is not only visible in its corresponding channel, but it also affects other channels. In [12] the micrographs analyzed do consist of multiple channels, but the bleed through is limited, which allows them to use only one channel for segmentation, ignoring information coming from other channels. In this paper we propose a technique for the segmentation of cell nuclei in fluorescence images coming from a confocal microscope. The micrographs consists of two channels, one channel corresponding with cell walls and one channel corresponding with the cell nuclei such as can be seen in Fig. 1. In this technique we will use information coming from both channels This paper is part of the *Arabidopsis Thaliana Root Cell Segmentation Challenge*[1].

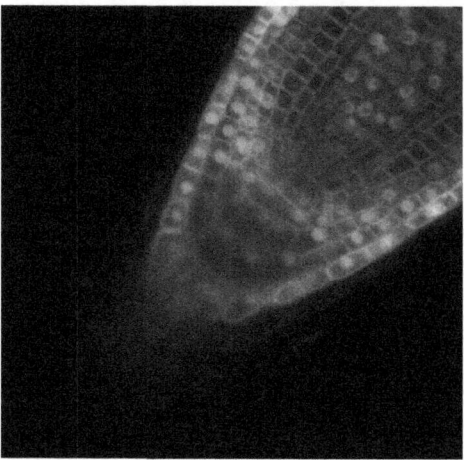

Fig. 1. An example of a fluorescence micrograph of Arabidopsis roots

This paper is organized as follows: In section 2 an overview is given of the proposed segmentation technique. The following four sections explain the main blocks of the segmentation algorithm in more detail. Section 6 discusses the results and section 7 recapitulates and concludes.

2 General Method

In Fig. 2, the main steps of the processing method are summarized. The method consists of four main phases: pre-processing, image fusion, nuclei detection and finally nuclei

---

[1] This is a challenge in order to provide a basis for comparison of different segmentation techniques for this kind of application.
http://paginas.fe.up.pt/~quelhas/Arabidopsis/

segmentation. The pre-processing step takes the green and red channel as input and corrects both images in order to get a more uniform intensity. After this, the root itself is extracted using a simple thresholding technique. Then in a second phase it combines both images to a single image expressing the probability a pixel belongs to a cell nuclei. In the nuclei detection phase, this probability map is split into different regions, where each region contains maximum one nucleus. Then each region is classified as *background* or as *containing a nuclei*. In the final phase each region is segmented. As a last control, each segment is classified as *background* or as *cell nuclei*. The phases are discussed in more detail in the following four sections.

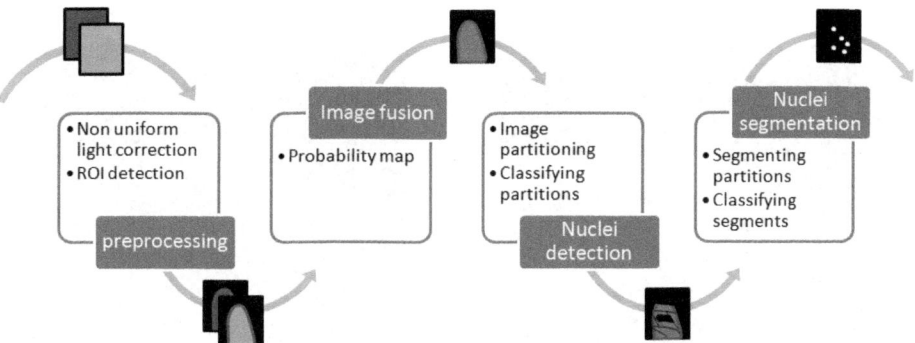

Fig. 2. The main steps of the processing method: In the first phase a region of interest (ROI) is defined for the green and red channel. Both ROI's are corrected for non uniform lightning. Then based on these two ROI's a single probability map is calculated in the second phase. The third phase partitions the ROI in possible regions of nuclei. The final phase detects the actual nuclei in the partitions.

3 Pre-processing

The intensity of a pixel is not only depending on the amount of fluorescent protein in the sample, but also depending on how well the sample is in focus. This can be seen in Fig. 1, where the cell walls and nuclei are much brighter near the border of the root than they are in the centre or the tip of the root. This non uniform light makes it hard to use statistics based on intensity. In order to use this kind of statistics, we correct the images. We use a morphological based method to correct both the red and the green channel independently [13]. The idea is that the background is similar to the image, except for the locally high intensity parts, which correspond to nuclei and cell walls. By using the morphological opening operator on the images, with a structuring element bigger than the foreground object, cell nuclei and cells are replaced by the intensities of the surrounding background. The size of the structuring element can be learned out of a ground truth training set, e.g. calculate the maximum radius of nuclei or a cell. The background is subtracted from the image, resulting in an image with a more uniform intensity. In Fig. 3 the illumination correction is shown for both the green and the red channel.

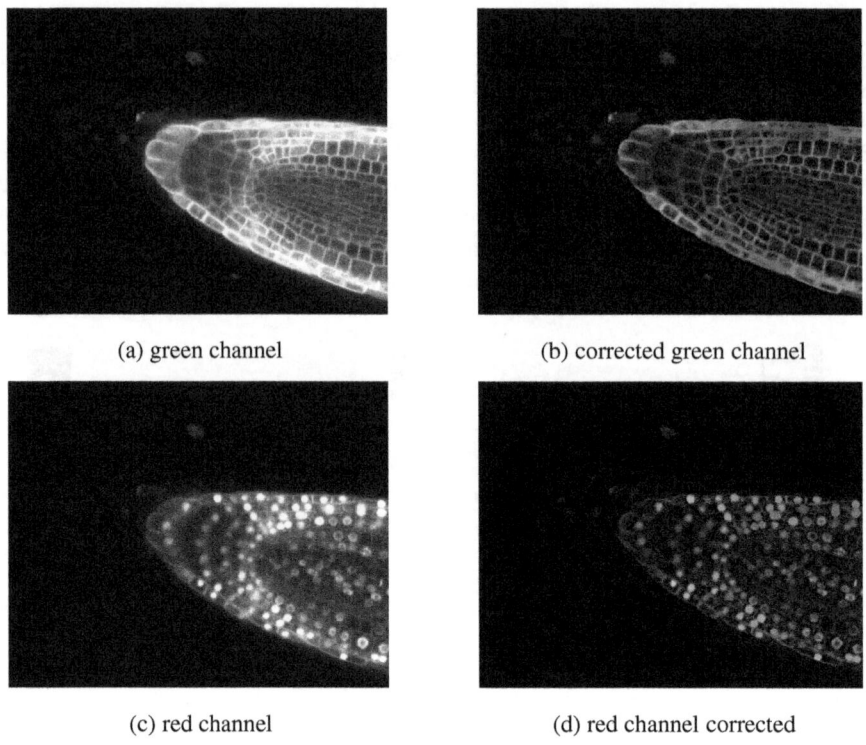

(a) green channel   (b) corrected green channel

(c) red channel   (d) red channel corrected

Fig. 3. Example of correcting non uniform light. On the left are the original green and red micrographs, on the right are the images corrected using morphological operators.

Based on the green image, a region of interest is defined (ROI). We consider each pixel with a green intensity higher than a certain threshold to be part of the ROI. This threshold is chosen manually and is the same for all images. By defining this ROI, only pixels belonging to the root are considered. If it is a cell wall, a nuclei or something else, as long as it belongs to the root, it is considered to be part of the ROI. Note that although the ROI is calculated based only on the green channel, it is the same region for both the green and the red channel.

4 Probability Map

Although only the red channel contains information about the cell nuclei, it is interesting to consider the information in the green channel as well. The first reason is that the green channel influences the red channel, due though bleed-through. This means that there is a certain response in the red channel, not because there's a nucleus in the image, but because there is a cell wall in the image. A second reason is because it provides us with extra knowledge: a nucleus belongs to only one cell, so at places where there is a cell wall, it is unlikely to see a cell nucleus. This property will be used implicitly in our method.

Out of the red and green ROI, we define a probability map. This map will have high values at places where it is likely to find a nucleus and it is near to zero if it is very unlikely that there is a nucleus. Let us define this probability map as follows:

$$P\big(N(x,y) = true \big| R(x,y), G(x,y)\big) \tag{1}$$

where $N(x,y)$ is a function returning $true$ if (x,y) belongs to a cell nucleus, otherwise the function returns $false$. The R and G are the red and green intensities at coordinates (x,y). This formula expresses the probability that (x,y) belongs to a nucleus, given the red and green intensity values. Note that in this formulation, we assume each pixel to be independent, i.e. the probability is only influenced by its own intensity values and not by its surrounding probabilities. Formula 1 can be rewritten using Bayes rule:

$$P\big(N(x,y) = true \big| R(x,y), G(x,y)\big) =$$
$$\frac{P\big(R(x,y), G(x,y) \big| N(x,y) = true\big) P\big(N(x,y)\big)}{P\big(R(x,y), G(x,y)\big)} \tag{2}$$

In the following we discuss the semantics of this equation:

- $P\big(R(x,y), G(x,y) \big| N(x,y) = true\big)$: this is the probability that a given green-red combination appears, if you know that (x,y) belongs to a nucleus. This can be learned from the training data, i.e. consider the green-red probability for all ground truth pixels.
- $P\big(N(x,y)\big)$: this is the probability that a certain pixel belongs to a nucleus. For simplicity reasons, we assume each pixel in our ROI to have the same probability. This assumption results in a constant value for each pixel and so this term can be discarded in Eq. (2).
- $P\big(R(x,y), G(x,y)\big)$: this is the probability that a given green-red combination appears, if no prior knowledge is available about (x,y) belonging to a nucleus or not. This can be learned from the training data set: calculate the ROI in each image, and based on the 2D histograms calculate this probability.

Due to the high amount of shot noise in the micrographs, there is a high amount of shot noise in the probability map. To minimize the influence of this noise the probability map is filtered with a median filter. In Fig. 4 an example is shown of a probability map based on the red and green ROI. As can be seen in Fig. 4.c, there is nearly no influence of bleed-through in the probability map. It is also clear that nuclei with low intensity also have good probabilities.

5   Nuclei Detection

Based on the previously defined probability map, candidate nuclei will be detected. These candidates will correspond to regional maxima in the probability map. In order to overcome the problem of touching nuclei, we will use a similar technique as in [10] to detect the regional maxima. Instead of diffusing the probability map using *gradient vector flow* [14], such as is done in [10], we will use a more computational efficient technique which will be described in the following subsection.

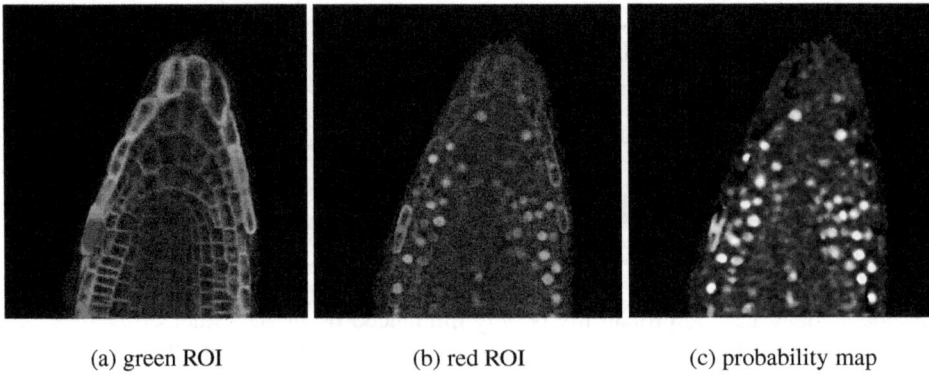

(a) green ROI   (b) red ROI   (c) probability map

Fig. 4. Example of a probability map: based on the green (a) and red (b) micrograph, a probability map is calculated. This probability map denotes how likely it is that a pixel belongs to a cell nucleus.

5.1 Probability Propagation

The goal of this step is to get a map where pixels with high probabilities propagate their probability to surrounding pixels with lower probability. By doing so, we get a map where the probability slowly decreases as it gets further away from pixels with high probability. This can be achieved using the following two scan line algorithm:

1. Scan the map row by row, starting from top to bottom
2. In each row, the pixels are scanned from left to right
3. Replace the pixels by:

$$P(x,y) =$$
$$\max\Big(P(x,y),\ \gamma P(x,y-1),\ \gamma P(x-1,y-1),\ \gamma P(x-1,y),\ \gamma P(x-1,y+1)\Big) \tag{3}$$

Where $\gamma \in [0,1]$ is a weighting factor, which determines the speed at which probabilities decrease. Note that all pixels in eq. (3), except $P(x,y)$, are already processed in previous steps, due to the scanning order. So this propagates the probability of a pixel beyond its direct neighbours with lower probability. An optimal γ has to be found manually.

This algorithm propagates probabilities from top to bottom and from left to right of the map. In order to propagate probabilities in the remaining directions the algorithm is repeated in the opposite direction, from the bottom right corner to the upper left corner.

5.2 Nuclei Detection

By applying a watershed algorithm [15] on the new probability map, the map is split into different partitions. Each regional minimum, e.g. a single cell nucleus, results in a separate partition. However, not every partition contains a nucleus. In Fig. 5.(b) an example of such a partitioning is shown, i.e. this is the partitioned probability map belonging to the micrograph in Fig. 5.(a). Based on the maximum probability belonging to a partition, a selection is made of invalid partitions: if the maximum probability of a partition does not exceed a given threshold, the partition is discarded. An optimal threshold can be learned out of the training dataset: the maximum probability is calculated for each ground truth segment, then the minimum of these maxima can be used as a threshold. The result of this simple classification can be seen in Fig. 5.(c) where discarded partitions are coloured red.

6   Nuclei Segmentation

All remaining partitions will be segmented independently. First each partition is filtered using a variance filter, i.e. each pixel is replaced by the variance in a certain window around the pixel. An example of the filtered partitions can be seen in Fig. 5.(d). Then each filtered partition is segmented using the watershed algorithm. This results in a set of segments for each partition. These segments are delineated green in Fig. 5.(e). For each partition a single segments has to be selected as candidate nucleus. The segment containing the pixel with the highest probability of the partition is considered to be the nucleus of the partition. All other segments in the partition are considered background and are discarded. This results in a single small segment for each partition, however, not all these segments are true nuclei, so a final check is done: the average probability of the nucleus segment is calculated, if it does not exceed a certain threshold it is considered to be noise and is discarded. This threshold can be calculated out of a training dataset in a similar way as the threshold in sec. 5.2. in Fig. 5.(f) the final segmentation result is shown.

7   Results and Discussion

In order to quantitatively evaluate the proposed technique and to compare this techniques with others, the ICIAR *Arabidopsis Thaliana Root Cell Segmentation Challenge* provided a set of 10 ground truth images. This ground truth consists of manually delineated nuclei with their corresponding micrographs. For the training of the algorithm, e.g. calculating the thresholds and defining the necessary probability distributions, a training set of 22 ground truth images was provided. As a validation of the proposed method, two groups of measurements will be used. One group are metrics concerning the cell detection, the second group puts a measure on how well detected cells are segmented. We will first discuss the second group since results of this measurement are used for the other group.

For the validations of the segmentation, the Dice coefficient is used. For each segment found by our method, S, a corresponding segment, GT, will be searched in the ground truth, i.e. the ground truth segment with the maximum overlap with S. Then the Dice coefficient between S and GT is defined as:

$$d(S, GT) = \frac{2\,Area(S \wedge GT)}{Area(S) + Area(GT)} \tag{4}$$

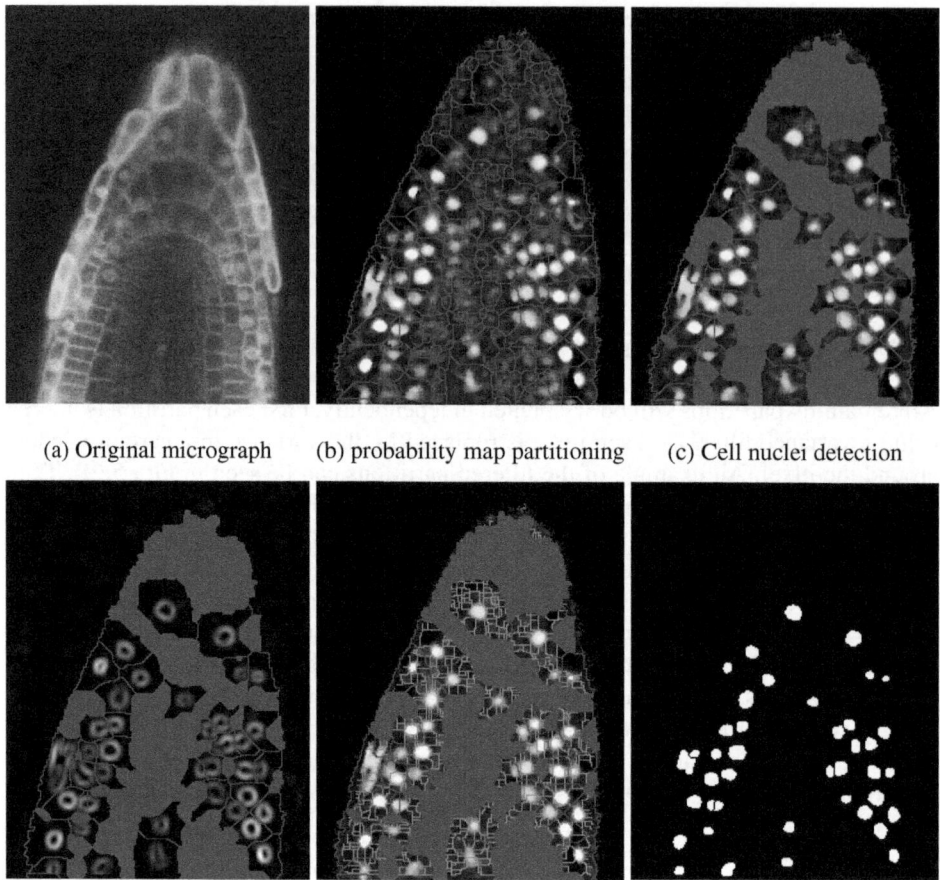

(a) Original micrograph   (b) probability map partitioning   (c) Cell nuclei detection

(d) Variance filter of partitions (e) segmentation of the partitions   (f) Resulting segmentation

Fig. 5. Example of nuclei segmentation

Where $S \wedge GT$ consist of all pixels which both belong to the detected segment as to the ground truth segment. If S and GT are equal, the Dice coefficient is equal to one. The Dice coefficient will approach zero if the overlap is much smaller then area of S and or the area of GT. All detected segments which have no corresponding ground truth segment have a Dice coefficient of zero.

For the validation of cell detection the following measures will be used:

- True positives (TP): each segment having a Dice coefficient higher than 0.5 is considered a TP.
- False positives (FP): all detections which have no corresponding ground truth segment are considered to be FPs
- False negatives (FN): All ground truth segments with no corresponding detections are FN's

- Precision is the ratio of the number of TP's in an image, over the total number of detected cells in that image, i.e. including FP's
- Recall is the ratio of the number of TP's in an image, over the total number of cells in the ground truth image

Table 1. The validation results: column A shows the results for the proposed technique, column B shows the results for the WCIF method

	A	B
average # TP	93.60	71.6
average # FP	8.50	21.90
average # FN	15.40	28.40
average precision	0.86	0.62
average recall	0.83	0.64
average Dice coefficient	0.77	0.62

Table 1 shows in the A column the results of the proposed method. These results are compared to the WCIF method [16], i.e. a method based on Otsu thresholding and watershed for splitting touching nuclei. Note that the first five measurements in Table 1 are the average results over the images, where the last measurement is the average Dice coefficient for all TP's. As can be seen, does the proposed method have significant better results than the WCIF method. One of the main reasons for this significant improvement is the use of a probability map, which reduces the influence of bleed through and noise.

8 Conclusion

We presented a new method for the segmentation of cell nuclei in fluorescent micrographs of Arabidopsis Thaliana roots. This method calculates a probability map expressing how likely it is a pixel belongs to a nucleus. Based on this map, cell nuclei are first detected and then segmented. The method was validated on the *Arabidopsis Thaliana Root Cell Segmentation Challenge* test set, resulting in a precision and recall of respectively 0.85 and 0.83

Acknowledgment

Jonas De Vylder is funded by the Institute for the Promotion of Innovation by Science and Technology in Flanders (IWT).

References

1. Kuijper, A., Zhou, Y.Y., Heise, B.: Clustered cell segmentation - based on iterative voting and the level set method. In: Visapp 2008: Proceedings of the Third International Conference on Computer Vision Theory and Applications, vol. 1, pp. 307–314 (2008)

2. Ray, N., Acton, S.T.: Motion gradient vector flow: An external force for tracking rolling leukocytes with shape and size constrained active contours. IEEE Transactions on Medical Imaging 23(12), 1466–1478 (2004)
3. Ruberto, C., Dempster, A., Khan, S., Jarra, B.: Analysis of infected blood cell images using morphological operators. Image and Vision Computing 20, 133–146 (2002)
4. Vincent, L., Masters, B.: Morphological image-processing and network analysis of cornea endothelial-cell images. Image Algebra and Morphological Image Processing 1769, 212–226 (1992)
5. Carpenter, A., Jones, T., Lamprecht, M.R., Clarke, C., Kang, I., Friman, I., Guertin, D., Chang, J., Lindquist, R., Moffat, J., Colland, P., Sabatini, D.: Cellprofiler: image analysis software for identifying and quantifying cell phenotypes. Genome Biology 7 (2006)
6. Megason, S.G., Fraser, S.E.: Digitizing life at the level of the cell: high-performance laser-scanning microscopy and image analysis for in toto imaging of development. Mechanisms of Development 120(11), 1407–1420 (2003)
7. Unser, M.: The colored revolution of bio-imaging: New opportunities for signal processing. In: Fourteenth European Signal Processing Conference (EUSIPCO'06), Firenze, Italy. September 5-8 (2006); Tutorial
8. Chen, X.W., Zhou, X.B., Wong, S.T.C.: Automated segmentation, classification, and tracking of cancer cell nuclei in time-lapse microscopy. IEEE Transactions on Biomedical Engineering 53(4), 762–766 (2006)
9. Yu, D.G., Pham, T.D., Zhou, X.B., Wong, S.T.C.: Recognition and analysis of cell nuclear phases for high-content screening based on morphological features. Pattern Recognition 42(4), 498–508 (2009)
10. Li, G., Liu, T., Nie, J., Guo, L., Chen, J., Zhu, J., Xia, W., Mara, A., Holley, S., Wong, S.T.C.: Segmentation of touching cell nuclei using gradient flow tracking. Journal of Microscopy-Oxford 231(1), 47–58 (2008)
11. Cloppet, F., Boucher, A.: Segmentation of complex nucleus configurations in biological images. Pattern Recognition Letters (in press, 2010)
12. Marcuzzo, M., Quelhas, P., Mendonca, A.M., Campilho, A.: Evaluation of symmetry enhanced sliding band filter for plant cell nuclei detection in low contrast noisy fluorescent images. In: Kamel, M., Campilho, A. (eds.) ICIAR 2009. LNCS, vol. 5627, pp. 824–831. Springer, Heidelberg (2009)
13. Matlab: Correcting Nonuniform Illumination,
    http://www.mathworks.com/image-videoprocessing/demos.html?
    file=/products/demos/shipping/images/ipexrice.html
14. Xu, C., Prince, J.: Snakes, shapes and gradient vector flow. IEEE Transactions on Image Processing 7, 359–369 (1998)
15. De Bock, J., Philips, W.: Line segment based watershed segmentation. In: Gagalowicz, A., Philips, W. (eds.) MIRAGE 2007. LNCS, vol. 4418, pp. 579–586. Springer, Heidelberg (2007)
16. Wright Cell Imaging Facility: Particle Analysis,
    http://www.uhnresearch.ca/facilities/wcif/imagej/
    particle_analysis.htm

Optical Flow Based Arabidopsis Thaliana Root Meristem Cell Division Detection

Pedro Quelhas[1], Ana Maria Mendonça[1,2], and Aurélio Campilho[1,2]

[1] INEB - Instituto de Engenharia Biomédica
Divisão de Sinal e Imagem, Campus FEUP
[2] Faculdade de Engenharia, Universidade do Porto
Departamento de Engenharia Electrotécnica e Computadores

Abstract. The study of cell division and growth is a fundamental aspect of plant biology research. In this research the Arabidopsis thaliana plant is the most widely studied model plant and research is based on in vivo observation of plant cell development, by time-lapse confocal microscopy. The research herein described is based on a large amount of image data, which must be analyzed to determine meaningful transformation of the cells in the plants.

Most approaches for cell division detection are based on the morphological analysis of the cells' segmentation. However, cells are difficult to segment due to bad image quality in the in vivo images. We describe an approach to automatically search for cell division in the Arabidopsis thaliana root meristem using image registration and optical flow. This approach is based on the difference of speeds of the cell division and growth processes (cell division being a much faster process).

With this approach, we can achieve a detection accuracy of 96.4%.

Keywords: Biology image processing, cell division detection, Arabidopsis Thaliana.

1 Introduction

Cellular division is a fundamental process responsible for originating other cell types in multicellular organisms. In plants, specialized regions, the meristems, concentrate cellular division. Arabidopsis thaliana is a plant with rapid development and with a simple cellular pattern. Due to these characteristics, it is considered a model organism, widely used in plant research. The Arabidopsis root meristem, located at the tip of the root, is responsible for perpetuating this pattern by cellular division [1]. However, the control of the divisions is not completely understood, which motivates in vivo analysis of the Arabidopsis root.

Development biologists studying roots find it difficult to cope with the lack of suitable technology to analyze root growth in vivo [1]. The great amount of data produced leads to the development of image analysis tools to automatically extract useful information, such as identifying cell division and growth. Some of these solutions focus on the analysis of Arabidopsis development. Cell growth is

A. Campilho and M. Kamel (Eds.): ICIAR 2010, Part II, LNCS 6112, pp. 217–226, 2010.

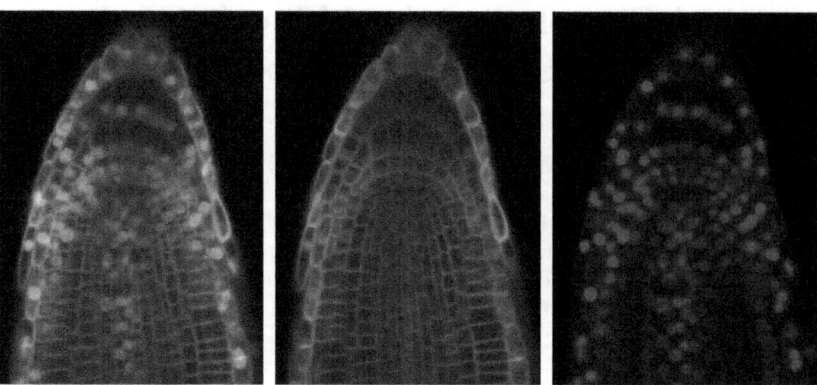

Fig. 1. *In vivo* microscopy image of the Arabidopsis thaliana root. Cell wall and nuclei channels shown in middle and right images respectively.

analyzed using different approaches, such as mathematical models [2] and motion estimation methods [3]. The relation between cell division and elongation in the regulation of organ growth rate is also investigated [4]. These studies show that in vivo imaging of the root is a valuable tool. However, none of them provide an automatic way to study the images at a cellular level.

In most automated cell analysis approaches, the first step is image segmentation, as in approaches described in [5,6]. Individual cell tracking is then performed based on proximity and cell division events are detected due to the mitotic cell morphological changes [7]. To avoid making a decision based on morphological features Yang et. al. proposed a shape independent division detection method based on 2D+time segmentation using level set methods [8]. However, this approach requires a fine time sampling which cannot be obtained in the case of in vivo plant imaging due to bleaching problems arising from excessive sampling. Moreover, segmentation is a difficult problem in computer vision and, in the case of in vivo plant imaging, is made worse by image acquisition process, data variability and noise [9] (Figure 1). These characteristics can often lead to errors in the segmentation process, such as over-segmentation of the cells. This, together with the small number of cell divisions, makes the detection of cell division through segmentation a very difficult task.

Recently, advances have been made in the detection of cell division by analyzing the shape of cell nuclei for mitotic changes [10]. This approach uses local convergence filter [11] for cell detection and an SVM classifier based on morphological features of the detected cells. No segmentation was used. However, problems in cell detection still remained.

In this work, we introduce a novel approach to cell division detection in plants based on optical flow, which does not depend on cell detection. Due to the slow changing overall structure of the root tip's walls, image registration between time-lapse images is possible. Root tip walls do not change rapidly since little elongation occurs in the root tip meristem. Changes in the root tip come from cell division and these changes are relatively fast and localized, being clearly

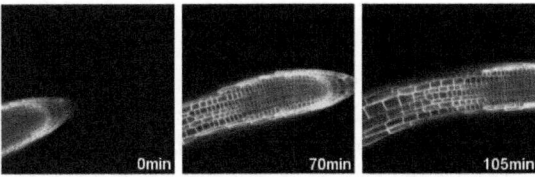

Fig. 2. Uncontrolled time-lapse in vivo imaging (showing only the GFP channel)

visible in the cell nuclei images. Our approach takes advantage of the different rates of evolution (local/global) and based on the optical flow between registered cell nuclei images detects fast changes in the root's morphology which signal a cell division.

The plant cell images used in this paper are confocal fluorescence images, which have separate channels for cell nuclei and cell wall images. Cell walls are marked using Green Fluorescent Protein (GFP) and cell nuclei are marked using Yellow Fluorescent Protein (YFP) [12]. Figure 1 shows an *in vivo* image of the root meristem and each channel separately. Cell walls are shown in the green channel and nuclei are shown in the red channel. However, our approach is not limited to these type of images and could be applied to other images and other types of plant cells where division occurs.

This paper is organized as follows: Section 2 describes the data used in this work and its capturing system, Section 3 describes the approach proposed and Section 4 presents the results of the proposed approach applied to the data. Finally, conclusion is presented in Section 5.

2 Database

For this work we used an *in vivo* image database of Arabidopsis Thaliana roots. This data was obtained using a confocal microscope with a motorized stage, which is automatically controlled to compensate the relative translation introduced by the root's growth. The specific control method used was proposed by Garcia et. al. [12]. The time duration of the experiments ranged between 5 and 13 hours, with images being acquired every 15 minutes.

The reason for the use of microscope stage control is that in *in vivo* experiments, if no compensation is performed, the area under study will grow out of the field of view. Depending of the zoom this can happen in as little as 15 minutes. Also, this type of control is required to releave the biology research from adjusting the microscope stage every few minutes. In Figure 2, we can see an example of images acquired without motion compensation, in this case the area under study leaves the frame in little more than an hour.

While the control method used applies estimation techniques to keep the root in view it does not completely eliminate motion between frames. Furthermore, it does not compensate rotation as the microscope stage is only capable of translation. As such, the resulting time-lapse images retain some residual motion which we will need to eliminate for the success of our approach.

3   Methodology

Our approach is divided into three main steps: time-lapse image registration, time-lapse image pairwise optical flow estimation and root local morphology change detection based on the extracted optical flow.

As mentioned in the previous section the images used in this work have both cell wall and cell nuclei information, stored in the green and red channels respectively. We will base cell division detection on the nuclei image channel, as nuclei suffer clear change during division. As cell walls are more stable the cell wall channel will be used for image registration of time-lapse frames. While some channel bleed through can occur in this type of data, it was not significative for our methodology. In the next subsections we will describe each step of our approach.

3.1   Image Registration

Image registration in time-lapse videos deals with the registration of motion throughout the complete image sequence. However, as we require only that there is no motion between each frame pair, we only register images in pairs ignoring all previous motion information. This allows us to safeguard against error propagation throughout the image sequence.

Given the pair of images constituted by the current image I_t and the next image I_{t+1}, we want to correct their relative motion and obtain an estimation for the next image $\hat{I}_{t+1}$ which has no global motion relative to I_t. As mentioned before, we choose to base such estimation on cell walls as they have more structural information, higher contrast and may be considered to be rigid at the time scale used in the experiments analyzed.

Assuming a rigid root we limit motion between frames to rotation and translation. This type of transformation is defined by equation 1, more details in [13].

$$\begin{bmatrix} x' \\ y' \end{bmatrix} = \begin{bmatrix} \cos(\theta) & -\sin(\theta) \\ \sin(\theta) & \cos(\theta) \end{bmatrix} * \begin{bmatrix} x \\ y \end{bmatrix} + \begin{bmatrix} t_x \\ t_y \end{bmatrix} \tag{1}$$

where (x, y) are the coordinates for a plant root location in frame t and (x', y') are the coordinates the same plant root location coordinates for frame $t + 1$. To compensate for the unknown motion between frames we need to estimate the motion parameters: (t_x, t_y, θ).

Motion estimation between images is a vast area with many possible solutions of varying complexity. In the case of our application we chose to perform motion estimation based on local interest point correspondences using the DOG local interest point detector and SIFT local point descriptors proposed by David Lowe [14]. While simpler approaches may have produced similar results, we chose this combination of local interest point detectors and descriptors due to their robust nature to make our approach applicable to a wider range of experimental conditions.

The local interest point based motion estimation between images used in our approach has three steps: interest point detection and descriptor extraction, point matching between images, and rigid motion parameter estimation:

Interest point detection and descriptor extraction: Given an images I we extract local interests points using the Difference of Gaussian (DOG) interest point detector [14]. For each image we obtain a set of interest points:

$$\mathcal{L}(I) \mapsto P = \{p_i, i = 1, \ldots, n\}, \tag{2}$$

where $\mathcal{L}$ is the DOG interest point detector, P is the set of resulting n local interest points $p_i = (x_i, y_i)$. For each point in set P we extract a SIFT descriptor [14]. This descriptor captures the local texture and structure so that it is possible to distinguish between local points and match similar points.

Point matching: Based on the Euclidian distance between SIFT descriptors we can obtain putative correspondences between point sets P_t and P_{t+1} from time lapse frames I_t and I_{t+1}:

$$C_t(p_i, p_j) = \begin{cases} 1 \; if \; D_E(p_i, p_j) < D_E(p_i, p_{j'}) \forall j' \neq j \\ 0 \; otherwise \end{cases} \tag{3}$$

where D_E is the Euclidian distance, i and j are indexes for point sets P_t and P_{t+1} respectively.

Motion parameter estimation: The set of correspondence C_t can contain a large percentage of incorrect matches. As such to estimate the motion parameters (t_x, t_y, θ) a robust estimation method is required. In our approach we applied a RANSAC estimator [13], based on C_t, to obtain the motion parameters (tx, ty, θ). The RANSAC algorithm is based on the exact solution to equation 1, using random sample correspondences from C_t. This results in a tentative solution for the motion parameters: $\tilde{tx}, \tilde{ty}, \tilde{\theta}$. Using this solution, we project points P_t from image I_t to obtain P'_t. We then compute the distance between the projected points P'_t and the corresponding points P_{t+1}, according to C_t. If the coordinate distance between projected and putative matched points is within a threshold ($15 pixels in our experiments$), we classify the correspondence as an inlier of the tentative solution. This process is repeated by selecting other random correspondence samples and obtaining other tentative solutions for the motion parameters. After a number of iterations the solution with the largest number of inliers is kept as the estimate for the motion parameters: $\hat{tx}, \hat{ty}, \hat{\theta}$. Applying the inverse of the estimated motion parameters $(\hat{tx}, \hat{ty}, \hat{\theta})$ we can transform image I_{t+1} and by doing so obtain a pair of images that is registered, which retain only small amounts of residual rigid motion.

3.2   Optical Flow Estimation

After performing rigid motion registration between the time-lapse images based on the cell wall channel we estimate the optical flow of the cell nuclei channel.

Optical flow deals with the estimation of the displacement field between two images, giving the correspondences between each individual pixels from each image. Contrary to the previously applied image registration the root is free to have non-rigid motion between frames.

To obtain the time-lapse images pairwise optical flow we use the high accuracy estimation method presented by Brox et. al. [15]. This method combines three spatiotemporal assumptions: a brightness constancy assumption, a gradient constancy assumption, and a discontinuity-preserving smoothness constraint. The brightness and gradient constancy are expressed as:

$$I(x, y, t) = I(x + u, y + v, t + 1) \tag{4}$$

and,

$$\nabla I(x, y, t) = \nabla I(x + u, y + v, t + 1), \tag{5}$$

where (u, v) is the searched displacement vector between an image at time t and another image at time $t + 1$. Let $\mathbf{x} := (x, y, t)$ and $w := (u, v, 1)$ the data energy to be minimized in the estimation of the optical flow can be written as:

$$E_{Data}(u, v) = \int_{\Omega} \Psi(|I(\mathbf{x} + w) - I(\mathbf{x})|^2 + \gamma|\nabla I(\mathbf{x} + w) - \nabla I(\mathbf{x})|^2) \, dx \tag{6}$$

where $\Psi = \sqrt{s^2 + \epsilon^2}$ which is a modified L^1 norm and γ is the weight between both assumptions. By using a large value for γ we ignore changes in grey value and base optical flow estimation mostly in image gradient information.

Finally, a smoothness term has to describe the model assumption of a piecewise smooth flow field. This is achieved by penalizing the total variation of the flow field, which can be expressed as:

$$E_{Smooth}(u, v) = \int_{\Omega} \Psi(|\nabla u|^2 + |\nabla v|^2) \, dx \tag{7}$$

with $\Psi = \sqrt{s^2 + \epsilon^2}$ as in the E_{Data} term. The spatial gradient $\nabla := (\partial x, \partial y)$ indicates that a spatial smoothness assumption is involved.

The total energy is the weighted sum between both energy terms:

$$E(u, v) = E_{Data} + \alpha E_{Smooth} \tag{8}$$

with some regularization parameter $\alpha > 0$. Higher α values cause the method to ignore local optical flow variations. In the case of our approach we used the parameters $\alpha = 20$ and $\gamma = 150$ in order to trust mostly the image gradient and have little smoothing of the displacements field [15].

Next a coarse to fine approach based on warping techniques is used to find the displacement field given by u and v that minimizes the energy $E(u, v)$. The scale of this estimation regulates the magnitude allowed for the displacement field. As cell motion is in the order of 5 to 10 pixels we chose 4 levels of coarse to fine estimation.

While division could be analyzed through frame subtraction, the results would be more easily corrupted by local greylevel variations.

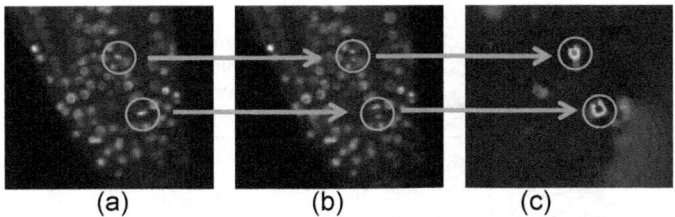

Fig. 3. Optical flow resulting from plant root time-lapse images when cell division occurred: (a) image at time t, (b) image at time $t + 1$, (c) magnitude of the estimated optical flow. Cell divisions occurring are highlighted within green circles.

3.3   Optical Flow Segmentation

Given the estimated optical flow (Figure 3), we can find the locations where cell division is most likely to have occurred. Since most global (rigid) motion was eliminated by the image registration step, any large magnitude displacement in the optical flow should signal a cell division. However, the root is a 3D object and it does not move only in the viewed 2D confocal plane. As such, sometimes cell nuclei can appear or disappear from the image plane even if the cell walls remain visible. As such, some optical flow disturbances are not due to cell division. Figure 4 shows some examples of high magnitude optical flow disturbances which were not caused by cell division, these can cause false cell division detections.

To detect possible cell divisions in a robust way we do not rely on a single threshold but analyze the areas of strong optical flow disturbance at several thresholds. We take 6 different threshold level between $4/12$ and $9/12$ of the full range of values of the optical flow magnitude in the displacement field and obtain 6 different binary images. Segmented regions within those images are grouped based on overlap across thresholds and those that appear in more than 2 threshold images are retained as possible cell divisions. A final check is made that the area of such regions is not too large (900 pixel squared) and the area of the segmented regions does not suffer large area variations between threshold level (variance bellow 1000). This approach is based on the idea that cell division causes an isolated peak in flow disturbance.

4   Results and Discussion

To validate our methodology we used images from two time-lapse biology experiments, comprising a total of 49 images. The time-lapse fluorescence microscopy sequences were recorded using a temporal resolution of one image per 15 minutes. The dataset contained a total of 4412 cells and 35 of those cells where annotated as undergoing division during the time-lapse capture.

Table 1 shows the cell division detection results for our approach, when applied to the dataset. The results show that our method has a good accuracy and recall, while precision has a low value. However, it has to be taken into account that

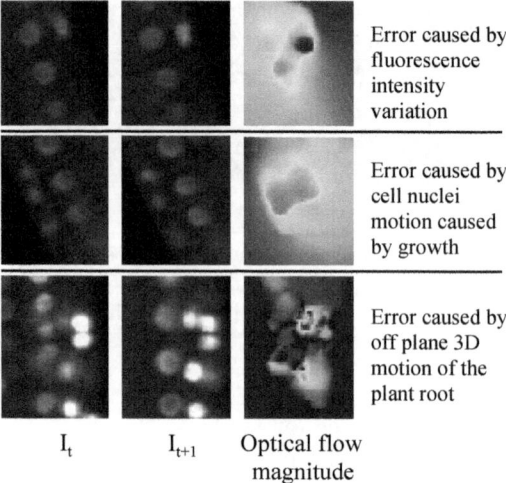

			Error caused by fluorescence intensity variation
			Error caused by cell nuclei motion caused by growth
			Error caused by off plane 3D motion of the plant root
I_t	I_{t+1}	Optical flow magnitude	

Fig. 4. Examples of root regions where nuclei changed appearance between frames causing peaks in optical flow which did not originate in cell division

Table 1. Detection performance results of our approach for cell division detection

	Classification results
Accuracy(%)	96.4
Precision(%)	14.9
Recall(%)	74.3

the dataset has 4412 cells from which only 0.8% are dividing. As such, false positives are always to be expected. Overall the results are good and present an improvement over shape based approaches [10].

During the experiments presented here we found that the main source of error is the time-lapse interval excessive duration. Based on the results from our experiments we believe that by using a time-lapse interval shorter than 15 minutes the proposed approach would be more efficient. The second problem found is the 3D of-plane motion of the plant root. This is a more complex problem as it would required a more complex microscope stage compensation method and 3D root acquisition.

5   Conclusion

We introduced an automatic segmentation free approach to detect cell division in *in vivo* time-lapse sequences of growing Arabidopsis thaliana. This method is based on the analysis of the optical flow between time-lapse images and detects the localized distortion caused by the, relatively rapid, cell nuclei division event. Results show an improvement over previously proposed methods with higher detection accuracy(96.4). Being segmentation free means that this method can

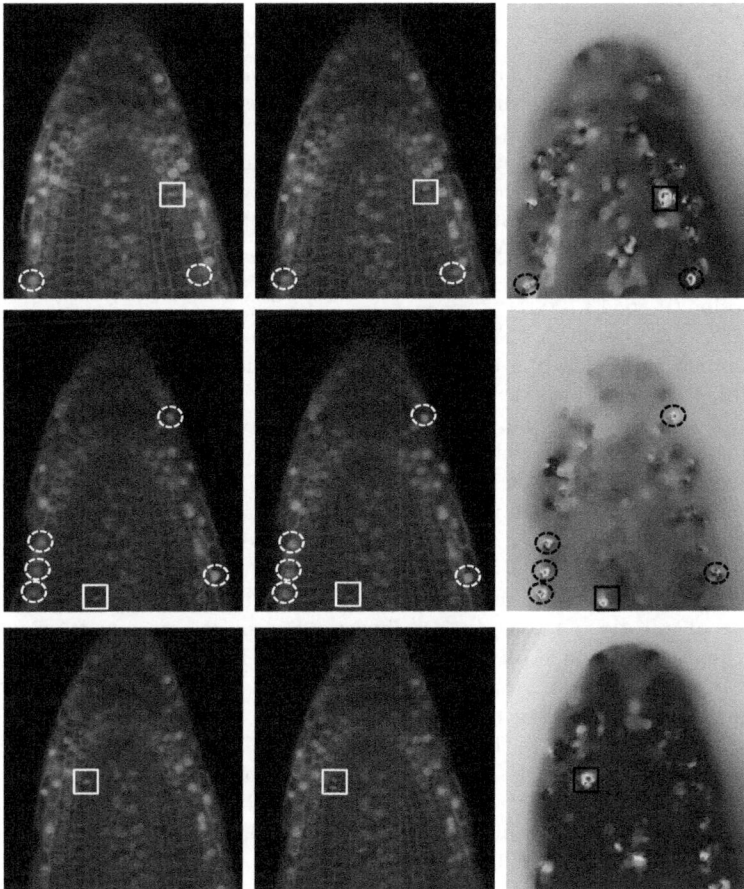

Fig. 5. Cell division results overlayed on current (left) and next image (middle) as well as on the resulting optical flow (right). False detections show in dashed circles and correctly detected divisions shown in solid squares.

be used even in the presence of high noise levels and low contrast. While the test images were only of the root tip, we believe that as long as the plant images can be properly registered this approach will work with the shoot meriterms and with other image types.

As future work we expect that combining both optical flow and cell morphology analysis approaches will produce improved results leading to the performance levels necessary for a laboratory prototype implementation.

Acknowledgements

The authors acknowledge the funding of Fundação para a Ciência e Tecnologia, under contract ERA-PG/0007/2006.

References

1. Campilho, A., Garcia, B., Toorn, H., Wijk, H., Campilho, A., Scheres, B.: Time-lapse analysis of stem-cell divisions in the arabidopsis thaliana root meristem. The Plant Journal 48, 619–627 (2006)
2. Iwamoto, A., Satoh, D., Furutani, M., Maruyama, S., Ohba, H., Sugiyama, M.: Insight into the basis of root growth in arabidopsis thaliana provided by a simple mathematical model. J. Plant Res. 119, 85–93 (2006)
3. Roberts, T., Mckenna, S., Wuyts, N., Valentine, T., Bengough, A.: Performance of low-level motion estimation methods for confocal microscopy of plant cells in vivo. In: Motion'07, p. 13 (2007)
4. Beemster, G., Baskin, T.: Analysis of cell division and elongation underlying the developmental acceleration of root growth in arabidopsis thaliana. Plant Physiology 116, 1515–1526 (1998)
5. Chen, X., Zhou, X., Wong, S.T.C.: Automated segmentation, classification, and tracking of cancer cell nuclei in time-lapse microscopy. IEEE Trans. on Biomedical Engineering 53(4), 762–766 (2006)
6. Mao, K.Z., Zhao, P., Tan, P.: Supervised learning-based cell image segmentation for p53 immunohistochemistry. IEEE Trans. on Biomedical Engineering 53(6), 1153–1163 (2006)
7. Harder, N., M-Bermudez, F., Godinez, W., Ellenberg, J., Rohr, K.: Automated analysis of mitotic cell nuclei in 3rd fluorescence microscopy image sequences. In: Workshop on Bio-Image Inf.: Bio. Imag., Comp. Vision and Data Mining (2008)
8. Yang, F., Mackey, M.A., Ianzini, F., Gallardo, G., Sonka, M.: Cell segmentation, tracking, and mitosis detection using temporal context. In: Medical Image Computing and Computer-Assisted Intervention (MICCAI), pp. 302–309 (2005)
9. Marcuzzo, M., Quelhas, P., Campilho, A., Mendonça, A.M., Campilho, A.: Automated arabidopsis plant root cell segmentation based on svm classification and region merging. Comp. in biology and medicine 39, 785–793 (2009)
10. Marcuzzo, M., Guichard, T., Quelhas, P., Maria Mendonça, A., Campilho, A.: Cell division detection on the arabidopsis thaliana root. In: Araujo, H., Mendonça, A.M., Pinho, A.J., Torres, M.I. (eds.) IbPRIA 2009. LNCS, vol. 5524, pp. 168–175. Springer, Heidelberg (2009)
11. Pereira, C.S., Mendonça, A.M., Campilho, A.: Evaluation of contrast enhancement filters for lung nodule detection. In: Kamel, M.S., Campilho, A. (eds.) ICIAR 2007. LNCS, vol. 4633, pp. 878–888. Springer, Heidelberg (2007)
12. Garcia, B., Campilho, A., Scheres, B., Campilho, A.: Automatic tracking of arabidopsis thaliana root meristem in confocal microscopy. In: Campilho, A.C., Kamel, M.S. (eds.) ICIAR 2004. LNCS, vol. 3212, pp. 166–174. Springer, Heidelberg (2004)
13. Hartley, R., Zisserman, A.: Multiple View Geometry in Computer Vision. Cambridge University Press, Cambridge (2003)
14. Lowe, D.G.: Distinctive image features from scale-invariant keypoints. International Journal of Computer Vision 60, 91–110 (2004)
15. Brox, T., Bruhn, A., Papenberg, N., Weickert, J.: High accuracy optical flow estimation based on a theory for warping. In: Pajdla, T., Matas, J(G.) (eds.) ECCV 2004. LNCS, vol. 3024, pp. 25–36. Springer, Heidelberg (2004)

The West Pomeranian University of Technology Ear Database – A Tool for Testing Biometric Algorithms

Dariusz Frejlichowski and Natalia Tyszkiewicz

Faculty of Computer Science and Information Technology,
West Pomeranian University of Technology, Zolnierska 49, 71-210, Szczecin, Poland
{dfrejlichowski,ntyszkiewicz}@wi.zut.edu.pl

Abstract. The biometric identification of persons became a very important problem nowadays. Various aspects of criminality, and lately terrorism, forced us to increase the financial outlays allocated for safety measures based on biometrics. Fingerprints and face are the most popular biometric features. However, new modalities are still desirable (moreover, nowadays the usage of multimodal systems becomes very popular). An ear seems to be a very interesting one. The auricle has complex and stable structure and is as distinguishable for various persons as the face. During the development of biometric algorithms their proper and reliable testing is one of the most important aspects. For that purpose a test database can be very helpful. There are many test bases with faces, but ear databases are very rare. Moreover, the existing ones are usually strongly limited. Therefore, in this paper a new ear images database (the West Pomeranian University of Technology Ear Database) is presented for usage in various scientific applications, e.g. for testing biometric algorithms.

Keywords: Ear biometrics, ear recognition, ear database.

1 Introduction to Ear Biometrics

The identification of human individuals became the key problem of our times and concerns many aspects of life and science. An increasing competition in the world market and development of technological thought caused necessity of data security, whereas a wave of terrorism and criminality contributed to intensified protection of the society. In areas of finances, health care, transport, entertainment, communication as well as governmental organizations, automatic identification issue gains more and more importance. Secure systems appear to be an indispensable element of our life. A verification of a person, who has access to specific buildings or data is counted among routine procedures that prevent abusing and state fundamental protection ([1-4]).

The usage of biometry became an essential component of effective person identification as biometric features of the human being can not be lost, copied or the same for different individuals. Biometric characteristics can be divided into behavioral (e.g. signature, keystroke dynamics) and physical (related to the shape of the body, [5]). The second group leads to the passive identification methods that do not require user's active participation in the process of identification (demanded data can be even

A. Campilho and M. Kamel (Eds.): ICIAR 2010, Part II, LNCS 6112, pp. 227–234, 2010.

acquired without person's knowledge). Systems used for monitoring of crowded public places like airports or sport centers, are ones of the most desired applications. Usually, a face as a source feature is used for that purpose. In studies devoted to biometrics an ear is often compared to it, because similar recognition methods and equipment (camera or camcorder) are applied to it ([1,6]).

1.1 Brief History of an Ear as a Biometric Feature

The genesis of scientific interest in potential usage of human ear took place in 1854, when for the first time patrimonial features of an auricle were noticed by A. Joux. In 1890 primary application of ears for personal identification was presented and advocated by the French criminologist A. Bertillon, who pointed out ears' rich, stable and time independent character. Later, in the beginning of the 20th century fundamental assumptions for application of ear in human identification were published by R. Imhofer, who also asserted that in the set of 500 ears only four characteristic features were required to prove their uniqueness ([7]).

In 1910 the lecture about possibilities of exploiting ears in human identification by J. H. Evans has become a scientific sensation. It was presented for the members of The Forensic Medicine Association. In the same year for the first time the person identification based on ear photos was used by the court of Liverpool. Since early 50's of the last century ear-prints have played a significant role in forensic science, where they have been used quite extensively ([8]).

Application of advanced computer techniques to the discussed problem in the 90's of the last century made possible the automatic identification of individuals basing on pattern recognition and image analysis methods. Almost decade ago Burge et al. ([9]) were amongst the first scientists exploring an ear as a biometric. They were using graph matching techniques on a Voronoi diagram of curves extracted from the Canny edge map. Moreno et al. ([10]) presented two approaches employing neural network classifiers based on feature points (recognition rate of 43% was reported) and ear morphology (recognition rate - RR of 83%). Hurley et al. ([11,12]) used a "force field" feature extraction based on simulated potential energy fields. This method achieved a 99.2% RR on a dataset of 252 images selected from the XM2VTS face database. Chang et al. ([13]) used Principal Components Analysis (PCA) and reported performance not significantly different for both face and ear.

Some of approaches use geometrical parameters, e.g. Mu et al. (85% RR, [14]), Choras (100% RR for a dataset of non-occluded ear images, [15,16]). Yan et al.([17,18]) used Iterative Closest Point (ICP) method for recognition performed on a set of 3D ear images and reported 97.8% RR. Bhanu et al. presented recognition system based on ICP and a local surface descriptor (also using 3D images) for which rate of 90.4% has been reported ([19]). After testing combinations of 2D-PCA, 3D-PCA and 3D edges on database of 203 images, Yan et al. concluded that better result can be achieved after a fusion of the all three ([20]). Akkermans et al. ([21]) developed a biometric system based on measurements of acoustic transfer function of an ear by projecting a sound wave and recording it reflected signal. They achieved equal 1-5% error rates, depending on the device used for measuring. Description of many other approaches for ear recognition can be found in [22].

1.2 Biometric Test Databases

The databases for testing the effectiveness of recognition methods are required, regardless of the chosen biometric feature. Such a repository of biometric templates seems to be one of the most important elements determining correctness and efficiency of an identification system. Since the face is the most popular and explored biometrics, face databases are the most numerous and rich collections of images, as opposed to ear databases, which are very small in numbers and usually have strong limitations (e.g. lighting, background conditions). Three exemplary ear databases exist: owned by the University of Technology and Life Sciences ([6,16]), the Notre Dame University ([23]) and the one based on profiles of the XM2VTS face database ([24]). Generally, the research material was collected in the above cases only at maternal universities, where datasets were created, what caused unintentional limitation to similar individuals (students). Lack of age differences in a population are followed usually by another restriction in the form of controlled environment in which photos where taken (e.g. uniform background or lighting conditions). Moreover, no outdoors photos have been taken. The above assumptions make the recognition less difficult, however they have significant impact on reliability of tested methods as they do not reflect real-life conditions. Detailed specification of ear in the presence of occlusion is another factor that effects recognition results, but this problem has been omitted in all mentioned cases. Moreover, the single image description contains usually only basic information reduced to ID, sex and possibly the lighting conditions.

Analysis of the state of the art convinced us that a new database of ear images avoiding the listed limitations is desirable. That is why our goal was to construct a representative and diverse database (regarding age, sex, occlusion etc.). Full concept and presumptions of the West Pomeranian University of Technology Ear Database (WPUTED) were stated in our previous article ([25]). Here the database is presented in details. We assume that it can be used in all scientific applications, which can be very helpful in testing of new biometric algorithms. The database is available at *http://ksm.wi.zut.edu.pl/wputedb/*. Conditions of its usage and some other information are also provided there.

2 The Characteristic of the West Pomeranian University of Technology Ear Database

The priority of created database was the elimination of problems concerning existing biometric ear images databases. One of the most important assumptions was the selection of suitable data that would be enriched by additional factors and information (disregarded in other collections) so that it could provide wide range of cases of ear images for their recognition and identification.

The collection contains 2071 ear images of 501 individuals (254 women and 247 men) of various age. The more detailed description (e.g. angles, ages) are provided in table 1. At least two photos per ear (profile and half-profile) for each subject were taken. Therefore usually four ear images for every person were collected. However, in some cases more images were acquired (see table 2). In order to make further testing of appropriate identification methods easier, the ear images were extracted and stored separately.

Table 1. Overview of the conditions characterizing the ear database

Attribute	Range
No. of subjects	501 (254 female and 247 male)
No. of photos	2071
Photos per ear	Min. 2 + doubles
Poses	90 deg., 75 deg.
Age	0-20 (25%)
	21-35 (48%)
	36-50 (15%)
	50 and above (12%)

Table 2. Number of ear images per subject in the database

No. of photos per subject	No. of individuals	Repeated sessions
4	451	No
5	40	Yes
6	6	Yes
7	1	Yes
8	3	Yes

Table 3. Overview of the conditions characterizing the database

Distortion	Percentage of images	Total percentage	
One earring	24.75		Major occlusion
More than one earring	2.60	27.35	
Occlusion caused by hair	27.15		
Significant occlusion caused by hair	5.99	33.14	
Glasses	14.37		Minor occlusion
Dirt, dust	0.60		
Hat	2.79		
Headband/ Headscarf	1.00	19.56	
Hearing aid	0.40		
Head phones	0.40		

For the first time in the preparation of the ear database some of pictures were taken outdoors (15,6% photos) and some were taken in the dark (2%).

Occlusions within an area of an ear seem to be the most important problem in human identification, therefore those cases became especially desirable when collecting the images. Ear deformations were recorded for 80% of photos. They appear mostly in form of hair covering (166 subjects), earrings (147 subjects) and in minority as glasses, headdresses, noticeable dirt, dust, birth-marks, ear-pads etc. The whole range of occurred auricle deformations are presented in table 3. On the other hand, a 20% of acquired images are free of any auricle occlusions. Another problem is the appearance of additional artifacts, which happened in 8% of photos and was usually caused by motion blur effect.

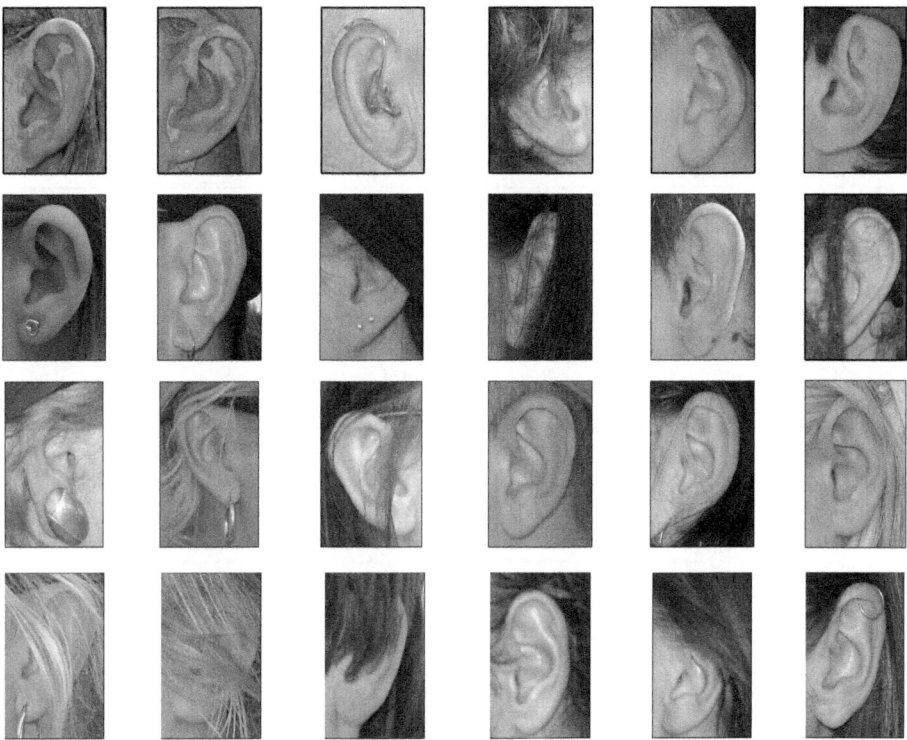

Fig. 1. Some examples from the WPUT Ear Database

The WPUTEDB is organized in one-directory structure, in which a filename of each image has encoded information that includes conditions, distortions of an image as well as basic data about a subject. The strict description of single image is crucial regarding the possible dissimilarities in performance of various recognition algorithms in the presence of variable ear features. Moreover, the precise specification makes opportunity of creating a proper set fulfilling requirements of a specific method (e.g. distinguishing the set of images without hear coverings). Rich information encoding within the filename makes easier strict determination of parameters that significantly impact the recognition results. After prior analysis of ear deformations four categories were proposed in order to systematize information about an ear image: personal data (ID, sex, race, age group), technical data (image position, angle, side of chosen ear), external conditions (factors describing environment like lighting and background conditions), occlusions (all significant deformations of shape and appearance of ear).

232   D. Frejlichowski and N. Tyszkiewicz

Table 4. A coding description for a single ear image

	Pos.	Attribute	Description
personal data	1-3	ID number	number from 0 to 999
	4	Sex	F – female; M – male
	5	Age	0 – 0-20 years; 1 – 21-35 years; 2 – 36-50 years; 3 – 51 years and above
	6	Skin Color	W – white; B – black A – yellow; O – other
technical data	7	Image category	0 – original photo; 1 – extracted ear
	8	Side	L – left ear; R – right ear
	9	Angle	0 – 90 degrees (profile); 1 – 75 degrees (half-profile); 2 – unknown / other
	10	Image orientation	V – vertical; H – horizontal
	11	Distance	0 – less than 1m; 1 – about 1-1,5m; 2 - higher
external factors	12	Place	I – indoor; O – outdoor
	13	Natural lighting conditions	0 – daylight; 1 – dark
	14	Background	0 – uniform; 1 – heterogeneous 2 – other people are visible on a photo
occlusion	15	Hair	0 – none; 1 – covered; 2 – strongly covered
	16	Earrings	0 – none; 1 – one; 2 – more than one
	17	Head covering	0 – none; 1 – hat/cap; 2 – band/scarf
	18	Glasses	0 – none; 1 – present
	19	Dirt	0 – none; 1 – significantly covered; 2 – local dirt
	20	Marks	0 – none; 1 – mole; 2 – ear burst; 3 – tattoo
	21	Special features	0 – none; 1 – hearing aid; 2 – ear pads; 3 - other
other	22	Artifacts	0 – none; 1 – underexposure; 2 – overexposure 3 – motion blur; 4 – options 1 & 3; 5 – options 2 & 3
	23	Face covered	0 – no; 1 – yes; 2 – cut
	24	No. of session	Numbered from 0 to 9

To encode the most important information within a filename, alphanumeric characters as well as their order in a string were used. The meaning and coding rules for a single image description are presented in table 4. There are some possible cases which did not appear in the database so far (e.g. tattoos or ear burst), but may show up during its future development.

3 An Example of Encoded Filename and Its Interpretation

In this section filename coding for a sample image from WPUTEDB and its description is provided (position numbers within the filename string are put in braces). The image (see fig. 2) presents left ear {8} of white {6} female {4} individual no. 039

{1-3} of age within 21-35 years {5}. This is extracted ear image {7} from its original form of full women profile {9} which was taken indoors {12} in poor natural lighting conditions (in dark {13}) in horizontal orientation{10} from distance lower than 1m {11}. The background of the image is non-homogeneous, because other people were visible on the initial image before ear extraction {14}. Only one earring appears {16}. Other occlusions in form of hair {15}, head covering {17}, glasses {18}, dirt {19}, ear marks {20}, hearing aid or earphones {21} have not been reported. This is a first session for this subject {24}.

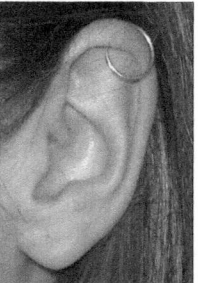

0	3	9	F	1	W	0	L	0	H	0	I	1	2	0	1	0	0	0	0	0	0	0	0	.jpg
1	2	3	4	5	6	7	8	9	10	11	12	13	14	15	16	17	18	19	20	21	22	23	24	

Fig. 2. Example filename encoding for an image from the WPUTEDB

The biometric features hold the most promising solution and the future direction of human identification. An ear for which recognition methods have shown encouraging progress in the last decade, appears as one of the most unique and exquisite data source for the purpose of identity verification. To support those algorithms, reliable and extensive database provided with crucial information about each ear is required. More demanding sets with more complex cases of ear images (e.g. in terms of occlusions, the environment, the number of samples per subjects, changes over time, image description etc.) are not guaranteed by existing databases especially due to population selection, no-age variance and environment choice (so far, none of photos were taken outdoors).

Therefore in this paper a database was presented, which can be used in scientific applications, e.g. for testing biometric algorithms. The database can be downloaded from *http://ksm.wi.zut.edu.pl/wputedb/*. Some other information about the West Pomeranian University of Technology Ear Database is also provided there.

References

1. Bobcow, A., Dabrowski, M.: Biometric Access Control. Pomiary Automatyka Kontrola 53(4), 87–90 (2007) (in Polish)
2. Jain, A.K., Pankanti, S.: A Touch of Money. IEEE Spectrum 43(7), 22–27 (2006)
3. Jain, A.K., Ross, A., Prabhakar, S.: An Introduction to Biometric Recognition. IEEE Transactions on Circuit and Systems for Video Technology, Special Issue on Image and Video-Based Biometrics 14(1), 4–20 (2004)
4. Jain, A.K., Pankanti, S.: Biometrics: A Tool for Information Security. IEEE Transactions on Information Forensics and Security 1(2), 125–143 (2006)

5. Kukharev, G., Kuzminski, A.: Biometric Techniques Part I - Face Recognition Methods. Szczecin University of Technology Press, Szczecin (2003) (in Polish)
6. Choras, M.: Ear Biometrics Based on Geometrical Feature Extraction. Electronic Letters on Computer Vision and Image Analysis 5(3), 84–95 (2005)
7. Kasprzak, K.: Criminalistic Otoscopy. University of Warmia and Mazury Press, Olsztyn (2003) (in Polish)
8. Holyst, B.: Criminalistics. LexisNexis Press, Warsaw (2004) (in Polish)
9. Burge, M., Burger, W.: Ear Biometrics for Computer Vision. In: Proc. of the 21st Workshop of the Austrian Association for Pattern Recognition, pp. 275–282 (1997)
10. Moreno, B., Sanches, A.: On the use of outer ear images for personal identification in security applications. In: Proc. of the IEEE 33rd Annual Intl. Conf. on Security Technology, pp. 469–476 (1999)
11. Hurley, D., Nixon, M., Carter, J.: Automatic Ear Recognition by Force Field Transformations. In: IEEE Colloquium on Visual Biometrics, vol. 5, pp. 7/1–7/5 (2000)
12. Hurley, D., Nixon, M., Carter, J.: Force Field Energy Functionals for Image Feature Extraction. Image and Vision computing 20(5-6), 311–317 (2002)
13. Chang, K., Bowyer, K., Sakar, S., Victor, B.: Comparison and combination of ear and face images in appearance-based biometrics. IEEE Transactions on Pattern Analysis and Machine Intelligence 25(9), 1160–1165 (2003)
14. Mu, Z., Yuan, L., Xu, Z., Xi, D., Qi, S.: Shape and structural feature based ear recognition. In: Li, S.Z., Lai, J.-H., Tan, T., Feng, G.-C., Wang, Y. (eds.) SINOBIOMETRICS 2004. LNCS, vol. 3338, pp. 663–670. Springer, Heidelberg (2004)
15. Choras, M.: Biometric Methods of Person Identification Basing on Ear Images. Biuletyn Informacyjny Techniki Komputerowe, January 2004, 59–69 (2004) (in Polish)
16. Choras, M.: Ear Biometrics - Methods of Feature Extraction Basing on Geometrical Parameters. Przeglad Elektrotechniczny 82(12), 5–10 (2006) (in Polish)
17. Yan, P., Bowyer, K.: A Fast Algorithm for ICP-Based 3D Shape Biometrics. Computer Vision and Image Understanding 107(3), 195–202 (2007)
18. Yan, P., Bowyer, K., Chang, K.J.: ICP-Based Approaches for 3D Ear Recognition. In: Jain, A.K., Ratha, N.K. (eds.) Biometric Technology for Human Identification II, Proc. of the SPIE, vol. 5779, pp. 282–291 (2005)
19. Bhanu, B., Chen, H.: Human Ear Recognition in 3D. In: Proc. of the Workshop on Multimodal User Authentication, pp. 91–98 (2003)
20. Yan, P., Bowyer, K.: 2D and 3D ear recognition. In: Biometric Consortium Conference 2004, Arlington, Virginia (2004)
21. Akkermans, T.H.M., Kevenaar, T.A.M., Schobben, D.W.E.: Acoustic Ear Recognition. In: Zhang, D., Jain, A.K. (eds.) ICB 2005. LNCS, vol. 3832, pp. 697–705. Springer, Heidelberg (2005)
22. Hurley, D., Arbab-Zavar, B., Nixon, M.S.: The Ear as a Biometric. In: Proc. of EUSIPCO 2007, Poznan, Poland, pp. 25–29 (2007)
23. The Notre Dame University Database, http://www.nd.edu/~cvrl/
24. XM2VTSDB Database, http://www.ee.surrey.ac.uk/CVSSP/xm2vtsdb/
25. Frejlichowski, D., Tyszkiewicz, N.: The Database of Digital Ears Images for Testing of Biometric Systems. In: Borzemski, L., Grzech, A., Swiatek, J., Wilimowska, Z. (eds.) Information Systems Architecture and Technology. Web Information Systems: Models, Concepts & Challenges, pp. 217–227. Wroclaw University of Technology Press, Wroclaw (2008)

Associating Minutiae between Distorted Fingerprints Using Minimal Spanning Tree

En Zhu[1,2], Edwin Hancock[2], Peng Ren[2], Jianping Yin[1], and Jianming Zhang[3]

[1] School of Computer Science, National University of Defense Technology,
Changsha 410073, China
[2] Department of Computer Science, University of York, Heslington, York, YO10 5DD, UK
[3] Department of Computer Science, Hunan City University, Yiyang 413000, China
enzhu@nudt.edu.cn

Abstract. This paper proposes to associate minutiae between distorted fingerprints by locating two matched minimal spanning trees, and then apply affine
transformation to one of the two images. This process is iterated until correspondences are stable. A radial basis function –based warping model is established from the final correspondences, and is applied to nonlinearly align the
two images. The experimental results show that the proposed method leads to
an improved matching performance.

Keywords: Fingerprint matching, minutiae correspondence, minimal spanning
tree.

1 Introduction

Fingerprint recognition is still a challenging problem, due to the difficulties of extracting features from low quality images, and the difficulties of matching low quality or
nonlinearly distorted images. This paper discusses the problems posed by the matching of nonlinearly distorted fingerprint images. Fig. 1 (a) and (b) show examples of
distorted images of the same fingerprint from FVC2004 [1].

In order to deal with distortions of fingerprint images, Watson [2] uses multiple
images of a fingerprint to establish a composite distortion-tolerant filter to improve
fingerprint correlation matching performance. Kovacs-Vajna [3] uses triangular
matching to cope with the deformation of fingerprint images. Senior [4] copes with
the distortion by normalizing the ridge distance and accordingly removing distortion.

Some methods such as the reported in [5],[6] use a bounding box to develop a distortion tolerant method. The representative bounding box -based method in [5] rigidly
aligns (rotation and translation) images and uses bounding box to detect correspondences. We refer to this method as RIGID model. For distorted images, especially
heavily distorted images, this method gives large number of false correspondences.
Fig. 1 (c) shows the rigid alignment result for the two distorted images, Fig. 1 (a) and
(b), under a pair of reference minutiae. The regions far away from the reference minutiae are not well aligned.

There have been a number of attempts to use thin plate spline to match distorted
fingerprint images. For instance, Bazen [7] and Ross [8] use a thin-plate spline (TPS)

A. Campilho and M. Kamel (Eds.): ICIAR 2010, Part II, LNCS 6112, pp. 235–245, 2010.
© Springer-Verlag Berlin Heidelberg 2010

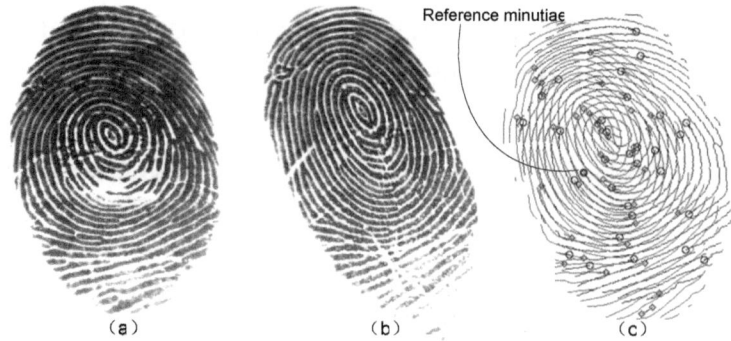

Fig. 1. Distorted fingerprints and their rigid alignment. (a) Template image. (b) Query image. (c) Rigid alignment of the skeletonised images. Regions around the reference minutiae are well aligned, while regions far away from the reference are not.

function [9] to match minutiae. Ross [8] use the thin-plate spline function to establish an average distortion model from multiple template fingerprints so as to improve the system performance at the identification stage. Bazen [7] uses the TPS model to warp fingerprint images from initial correspondences and new correspondences are identified between the warped images. This process is iterated until correspondences no longer change. The TPS model is effective for matching distorted fingerprints. It aims to find new correspondences and needs correct initial correspondences to establish warping model that can be used to detect attached correspondences. Compact distribution or false results in initial correspondences or in middle stages may well establish false warping model and subsequently lead to the successive false correspondences. The TPS model is a specific case of image registration based on radial basis functions (RBF). Liang [10] divides the fingerprint into two regions, namely the rigid region (inner region) and the nonrigid region (outer region). They then applies a rigid transformation and multiquadric RBF function to the two regions respectively.

In order to match distorted fingerprints, in this paper, we propose to find correspondences using minimal spanning tree. The motivation and the intuitions understanding the proposed method are explained in the following section.

2   Motivations and the Proposed Method

The challenge of matching two distorted fingerprints is how to associate minutiae between the two images, i.e. locating minutiae correspondences. Using a bounding box is effective if only minor distortions are present. For nonlinear or large distortions as illustrated in Fig. 1 (a) and (b), a bounding box will fail to locate true correspondences and produce false correspondences. Fig. 2 (a) shows the correspondences obtained using a bounding box (distance is equal to or less than 15 pixels, and the direction difference is equal to or less than 10 degree) after rigidly aligning the two images using the reference of template minutia 19 and query minutia 20. Denote by $c(u,v)$ a correspondence between the template minutiae u and the query minutiae v, and denote by $r(i,j)$ the reference minutiae pair consisting of the template minutia i and the query minutia j. The correct correspondences between the two given images should be:

$c(3,1)$, $c(4,2)$, $c(5,3)$, $c(6,5)$, $c(7,11)$, $c(8,10)$, $c(9,7)$, $c(10,15)$, $c(11,13)$, $c(12,9)$, $c(13,14)$, $c(16,18)$, $c(17,17)$, $c(19,20)$, $c(20,23)$, $c(21,19)$, $c(22,21)$, $c(23,24)$, $c(24,26)$, $c(25,25)$, $c(26,28)$, $c(27,29)$, $c(29,32)$, and $c(30,31)$, giving 24 correspondences in total. Under the reference of $r(19,20)$ as show in Fig. 2 (a), 9 correspondence minutiae pairs are located, of which two, $c(9,9)$ and $c(27,30)$, are false correspondences. From Fig. 1 (c) it is clear that $c(9,9)$ and $c(27,30)$ are close. The false correspondences $c(9,9)$ and $c(27,30)$ are both far away from the reference $r(19,20)$. We turn to another reference minutiae pair $r(26,28)$ as shown in Fig. 2 (b), where $c(27,30)$ is no longer a correspondence, and a $c(27,29)$ is a correspondence. If $c(19,20)$ and $c(26,28)$ are true correspondences, then $c(27,29)$ under $r(26,28)$ should be more reliable than $c(27,30)$ under $r(19,20)$, since $c(27,29)$ is closer to $r(26,28)$ than $c(27,30)$ to $r(19,20)$. From Fig. 2 (a), it is also clear that $c(29,32)$ is not a correspondence under $r(19,20)$, while they are in correspondences under $r(26,28)$ in Fig. 2 (b). Therefore, if we properly combine the correspondences under the two reference pairs, it may be possible to both remove false correspondences and to locate more true correspondences. In general, each pair of reference minutiae will detect some true correspondences together with some false correspondences and miss some true correspondences. The located correspondences far away from the reference are likely to be unreliable, and the missed correspondences are also in general far away from the reference. If a template minutia is in correspondence to a different query minutia under a different reference, the correspondence with the closer reference minutiae pair is likely to be more reliable. These observations motivate us to consider locating correspondences using a minimal spanning tree (MST).

We abstract the relationships between minutiae using a graph, in which vertices are minutiae. Given a pair of reference minutiae $r(u,v)$, if $c(i,j)$ are in correspondence, then (u,i) is a directed edge in the template graph, and (v,j) is the corresponding directed edge in the query graph. Corresponding minutiae pairs under a given reference pair can be taken as new reference pairs, and in turn, new correspondences will be found and new edges can be added to the template graph and query graph. The two graphs are implicit (not constructed explicitly), and are implied by the correspondences under each pair of reference minutiae. Our idea is to find two matched minimal spanning trees from the template graph and query graph respectively. The minimal spanning tree is located using Prim algorithm. In order to decrease false match rate, additional constraints will be imposed on the spanning tree, i.e. the intersection between tree edges is forbidden, which can reduce false correspondences for genuine matching and decrease correspondences for impostor matching.

There are two main steps for the proposed method:

(1) Locate correspondences under each pair of reference minutiae using rigid alignment.

(2) Select a pair of reference minutiae with the largest number of correspondences. Commencing from the selected reference minutiae, which are considered as the roots of the minimal spanning tree to be located, locate two matched minimal spanning trees. The vertices from the two trees are then the minutiae correspondences.

With the correspondences produced by step (2), a transformation, for instance a rigid affine or a non-rigid warping, can be applied to the template minutiae. The two steps can then be interleaved and iterated to convergence until the correspondences are stable. Using the final correspondences, a warping model based on radial basis function is established to warp the template image for alignment.

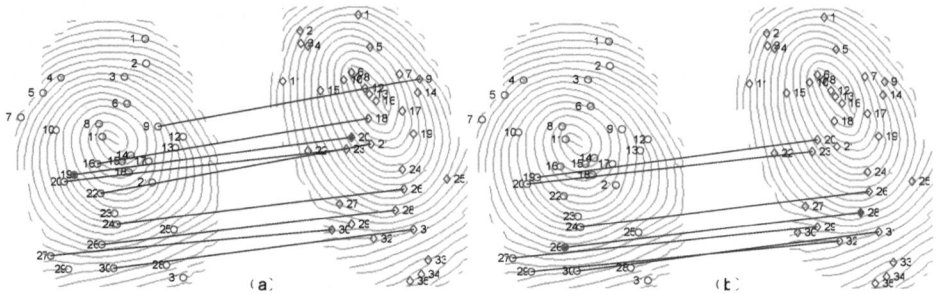

Fig. 2. Different correspondence sets are produced using rigid alignment under different reference minutiae pairs. (a) Correspondences under the reference of $r(19,20)$ (alignment is shown in Fig. 1 (c)). (b) Correspondences under reference of $r(26,28)$.

3 Rigid Transformation and Initial Correspondences Estimation

Minimal spanning tree requires initial correspondences under each pair of reference minutiae pair, so that an implicit template graph and query graph can be established using the correspondence relations. Let $\{t_1, t_2, ..., t_m\}$ be the template minutiae set, and $\{q_1, q_2, ..., q_n\}$ be the query minutiae set. Each template minutiae t_i is represented by a triple $(x_i^t, y_i^t, \theta_i^t)$, where (x_i^t, y_i^t) are the minutiae coordinates and θ_i^t is the minutiae direction. Similarly a query minutia q_i is represented by $(x_i^q, y_i^q, \theta_i^q)$. Hereinafter, we use $r(v, w)$ to denote a pair of reference minutiae, t_v and q_w. Given a pair of reference minutiae $r(v, w)$, we can translate and rotate the minutiae to allow the reference minutia to overlap the origin centering, and the reference minutia direction to point along the positive x axis. Therefore, in order to locate correspondences under each pair of reference minutiae, the transformation will be executed only $m + n$ times, m for the template and n for the query. In this case, $(x_i^t, y_i^t, \theta_i^t)$ is transformed to $(x_i^{t'}, y_i^{t'}, \theta_i^{t'})$, and $(x_i^q, y_i^q, \theta_i^q)$ is transformed to $(x_i^{q'}, y_i^{q'}, \theta_i^{q'})$ as follows:

$$\begin{pmatrix} x_i^{t'} \\ y_i^{t'} \\ 1 \end{pmatrix} = \begin{pmatrix} \cos(\theta_v^t) & \sin(\theta_v^t) & 0 \\ -\sin(\theta_v^t) & \cos(\theta_v^t) & 0 \\ 0 & 0 & 1 \end{pmatrix} \begin{pmatrix} 1 & 0 & -x_v^t \\ 0 & 1 & -y_v^t \\ 0 & 0 & 1 \end{pmatrix} \begin{pmatrix} x_i^t \\ y_i^t \\ 1 \end{pmatrix} \tag{1}$$

$$\theta_i^{t'} = \theta_i^t - \theta_v^t \tag{2}$$

$$\begin{pmatrix} x_i^{q'} \\ y_i^{q'} \\ 1 \end{pmatrix} = \begin{pmatrix} \cos(\theta_w^q) & \sin(\theta_w^q) & 0 \\ -\sin(\theta_w^q) & \cos(\theta_w^q) & 0 \\ 0 & 0 & 1 \end{pmatrix} \begin{pmatrix} 1 & 0 & -x_w^q \\ 0 & 1 & -y_w^q \\ 0 & 0 & 1 \end{pmatrix} \begin{pmatrix} x_i^q \\ y_i^q \\ 1 \end{pmatrix} \tag{3}$$

$$\theta_i^{q'} = \theta_i^q - \theta_w^q \tag{4}$$

The rigid transformation brings the two images into alignment, and a bounding box can be used to detect correspondences. We use $c(i, j) | r(v, w)$ to denote that t_i is in correspondence to q_j under the reference $r(v, w)$. Define $corres_{v,w}(i)$ as follows:

$$corres_{v,w}(i) = \begin{cases} j & \text{if } c(i,j) \mid r(v,w) \\ 0 & \text{if } t_i \text{ has no correspondence under } r(v,w) \end{cases} \qquad (5)$$

Let $score(v,w)$ be the number of correspondences under the reference minutiae pair $r(v,w)$. It is possible that $corres_{v,w}(i) = corres_{v,w}(j) = k > 0, i \neq j$. In this case, there is a conflict between t_i and t_j. Before locating the minimal spanning tree, we resolve this conflict in the following manner: If $score(i,k) \geq score(j,k)$, then we set $corres_{v,w}(j) = 0$ and decrease $score(j,k)$ by 1, else set $corres_{v,w}(i) = 0$ and decrease $score(i,k)$ by 1. Finally, t_{root_t} and q_{root_q} will be selected as the roots of the two minimal spanning trees, where

$$(root_t, root_q) = \arg\max_{v,w} score(v,w) \qquad (6)$$

Fig. 2 (a) and (b) show the correspondences for the images in Fig. 1 (a) and (b) under $r(19,20)$ and $r(26,28)$ respectively. Here $score(19,20)=9$ and $score(26,28)=7$, and $score(19,20)=9$ is the maximum score among all the reference minutiae pairs. Therefore $(root_t, root_q)=(19,20)$ in this example.

4 Locate Minimal Spanning Tree

The root minutiae t_{root_t} and q_{root_q} are acquired from the initial correspondences described in Section 3, and the initial correspondences are represented by $corres_{v,w}(i)$ $(1 \leq v \leq m, 1 \leq w \leq n, 1 \leq i \leq m)$. In this section we describe the process of locating the two matched minimal spanning trees (MST), namely the template MST and query MST. Let $MST_T = (V_T, E_T)$ and $MST_Q = (V_Q, E_Q)$ be the template MST and query MST respectively. The vertex sets V_T and V_Q index the template minutiae and the query minutiae. The edge sets of the trees are $E_T \subseteq V_T \times V_T$ and $E_Q \subseteq V_Q \times V_Q$. MST_T is found by locating the minimal spanning tree, and MST_Q is generated using the correspondences between V_Q and V_T. Let $d(i,j)$ be the distance between t_i and t_j. The distance matrix $(d(i,j))$ will be used to locate the minimal edges representing the arrangement of minutiae in the first pattern. Using $corres_{MST}(i)$ to record the minutiae correspondences between MST_T and MST_Q, $corres_{MST}(i)=0$ indicates that t_i has no corresponding query minutia, and $j=corres_{MST}(i)>0$ indicates that t_i corresponds to t_j. At the outset of the process, $corres_{MST}(root_t)=root_q$, and $corres_{MST}(i)=0$ for $1 \leq i \leq m, i \neq root_t$.

Supposing that function Intersect(e,f) ($e, f \in E_T$, or $e, f \in E_Q$) returns *true* if the edge e intersects with the edge f, else returns *false*, MST_Corres (Algorithm 1) locates the matched minimal spanning trees. In step (2.4) of the algorithm, if there is a $k \in V_T$, satisfying the condition $corres_{MST}(k)=j$, then conflict occurs between t_k and t_i, since t_i attempts to be in correspondence with q_j which has already been associated with t_k. In this case, we cease attempting to associate t_i to q_j. An example is provided in Fig. 2 where $c(27,30)$ and $c(27,29)$ are in conflict with each other (different query minutiae are associated to the same template minutiae). Step (2.4) also requires that the newly generated edges of the tree do not intersect with any existing edges, because not allowing intersection can decrease the number of correspondence for impostor matching and decrease false correspondence for genuine matching. In step (2.4), $E_T = E_T + \{(u,i)\}$ means that a new edge (u,i) is appended to the template MST, and t_u is the parent minutia of t_i. Similarly, $E_Q = E_Q + \{(v,j)\}$ means that a new edge (v,j) is appended to the query MST, and q_v is the parent minutia of q_j.

Algorithm 1. MST_Corres

Input: $d(i,j)$ ($1 \le i \le m$, $1 \le j \le n$), $(root_t, root_q)$, and $corres_{v,w}(i)$ ($1 \le v \le m$, $1 \le w \le n$, $1 \le i \le m$).

Output: $MST_T = (V_T, E_T)$, $MST_Q = (V_Q, E_Q)$, and $corres_{MST}(i)$ ($1 \le i \le m$).

1: Initialization: $M_T = \{1,2,...,m\}$, $M_Q = \{1,2,...,n\}$, $V_T = \{root_t\}$, $V_Q = \{root_q\}$, $E_T = \varnothing$, $E_Q = \varnothing$, $corres_{MST}(root_t) = root_q$, and $corres_{MST}(i) = 0$ ($1 \le i \le m, i \ne root_t$).

2: While ($V_T \ne M_T$), do:

2.1: $(u,i) = \arg\min_{k,l}\{d(k,l) | k \in V_T, l \in M_T - V_T\}$.

2.2: If $d(u,i) = \infty$, return.

2.3: $v = corres_{MST}(u)$, $j = corres_{u,v}(i)$, $e_1 = (u,i)$, $e_2 = (v,j)$.

2.4: If $j > 0$, and $\forall k \in V_T$ $corres_{MST}(k) \ne j$, and $\forall e' \in E_T$ $\neg$Intersect(e_1,e'), and $\forall e'' \in E_Q$ $\neg$Intersect(e_2,e''), then $V_T = V_T + \{i\}$, $V_Q = V_Q + \{j\}$, $E_T = E_T + \{(u,i)\}$, $E_Q = E_Q + \{(v,j)\}$, $corres_{MST}(i) = j$, else $d(u,i) = d(i,u) = \infty$.

Fig. 3 (a) provides an example of locating matched minimal spanning trees for the images in Fig. 1 (a) and (b). In Fig. 3 (a), 12 correspondence pairs are located, namely $c(5,3)$, $c(9,14)$, $c(11,13)$, $c(16,18)$, $c(19,20)$, $c(20,23)$, $c(22,21)$, $c(24,26)$, $c(26,28)$, $c(27,29)$, $c(29,32)$, and $c(30,32)$, among which there is one false correspondence $c(9,14)$. In Fig. 2 (a), 9 correspondences are located, in which there are 2 false correspondences, $c(9,14)$ and $c(27,30)$. The false match $c(27,30)$ in Fig. 2 (a) is removed in Fig. 3 (a) and is replaced with $c(27,29)$. In Fig. 3 (a), the parent node of template minutia 27 is minutia 29 in the minimal spanning tree, therefore $c(27,30)$ is produced under reference $r(29,32)$, and the correspondence $c(27,29)|r(29,32)$ is more reliable than $c(27,30)|r(19,20)$, since template minutiae 27 is much closer to minutia 29 than to minutiae 19. Compared with Fig. 2 (a), Fig. 3 (a) gives 3 additional correspondence pairs, $c(5,3)$, $c(11,13)$, and $c(29,32)$, and replaces a false correspondence $c(27,30)$ with the true correspondence $c(27,29)$. In this example, the algorithm MST_Corres can locate a greater number of true correspondences and can also decrease the number of false correspondences.

We observe that many potential correspondences are missed. The missed correspondences can be located after affine transforming the template image. The affine transformation of the template image from $(x_i^t, y_i^t, \theta_i^t)$ to $(x_i^{t'}, y_i^{t'}, \theta_i^{t'})$ is given by

$$\begin{pmatrix} x_i^{t'} \\ y_i^{t'} \end{pmatrix} = \begin{pmatrix} a_1 \\ b_1 \end{pmatrix} + \begin{pmatrix} a_2 & a_3 \\ b_2 & b_3 \end{pmatrix} \begin{pmatrix} x_i^t \\ y_i^t \end{pmatrix} \tag{7}$$

$$\theta_i^{t'} = angle\left(\begin{pmatrix} \cos(\theta_i^t) \\ \sin(\theta_i^t) \end{pmatrix}^T \begin{pmatrix} a_2 & a_3 \\ b_2 & b_3 \end{pmatrix}^T \right) \tag{8}$$

where a_1, a_2, a_3, b_1, b_2 and b_3 are the least square parameters for the correspondences produced by the minimal spanning tree, and $angle((x,y))$ returns the direction (in the interval $[0,2\pi)$) of vector (x,y). This transformation brings the template minutiae (template image) and the query minutiae (query image) into alignment

(Fig. 4 (a)) better than rigid alignment (Fig. 1 (c)). After the transformation, $corres_{v,w}(i)$ ($1 \leq v \leq m$, $1 \leq w \leq n$, $1 \leq i \leq m$) is re-estimated, and the algorithm MST_Corres is applied again to locate two new minimal spanning trees. Fig. 3 (b) shows the result of applying this procedure to the result shown in Fig. 3 (a). In Fig. 3 (b), 20 correspondences are located, and each of them is correct. This procedure can be interleaved and iterated until correspondences are stable. In fact, we find that increasing the number of iterations does not improve correspondences, because MST is stable for significant heavy distortions. The MST_Corres procedure is therefore iterated just two times in the experiments.

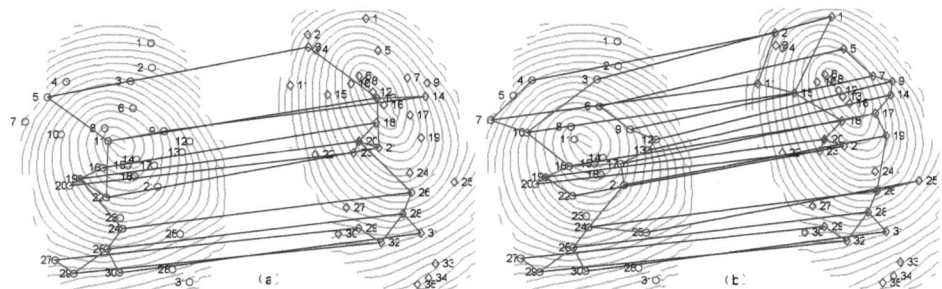

Fig. 3. Minutiae correspondences by finding minimal spanning tree. (a) Minimal spanning tree based on initial correspondence. (b) Minimal spanning tree after affine-transforming the template (the transformed result is shown in Fig. 4.)

Based on the final correspondences, we can use a warping-transformation based on the radial basis function (RBF) to better align the template minutiae (template image) and query minutiae (query image). Suppose that there are C final correspondences, and in which the coordinates of template minutiae are (x_i, y_i) ($1 \leq i \leq C$), and the corresponding query minutiae are (x_i', y_i') ($1 \leq i \leq C$). The RBF model is as follows:

$$\begin{pmatrix} G_{1,1} & \cdots & G_{1,C} & 1 & x_1 & y_1 \\ \vdots & \ddots & \vdots & \vdots & \vdots & \vdots \\ G_{C,1} & \cdots & G_{C,C} & 1 & x_C & y_C \\ 1 & \cdots & 1 & 0 & 0 & 0 \\ x_1 & \cdots & x_C & 0 & 0 & 0 \\ y_1 & \cdots & y_C & 0 & 0 & 0 \end{pmatrix} \begin{pmatrix} w_{a,1} & w_{b,1} \\ \vdots & \vdots \\ w_{a,C} & w_{b,C} \\ a_1 & b_1 \\ a_2 & b_2 \\ a_3 & b_3 \end{pmatrix} = \begin{pmatrix} x_1' & y_1' \\ \vdots & \vdots \\ x_C' & y_C' \\ 0 & 0 \\ 0 & 0 \\ 0 & 0 \end{pmatrix} \quad (9)$$

$$G_{i,j} = G\left(\|(x_i - x_j, y_i - y_j)\|\right) \quad (10)$$

where $w_{a,i}$, $w_{b,i}$ ($1 \leq i \leq C$), a_k and b_k ($1 \leq k \leq 3$) are parameters acquired by solving equation (9), and $G(r)$ is a radial basis function, and for thin-plate spline [9],

$$G(r) = \begin{cases} \lambda & r = 0 \\ r^2 \log(r) & r > 0 \end{cases} \quad (11)$$

where λ is a positive constant for approximating warping transformation. When transforming the image or the minutiae, the coordinates (x, y) will be transformed to

$$(x', y')^T = A \cdot (1, x, y)^T + \sum_{i=1}^{C} \left[(w_{a,i}, w_{b,i})^T \cdot G \left(h(| (x - x_i, y - y_i) |) \right) \right] \tag{12}$$

$$h(r) = \begin{cases} \varepsilon & r = 0 \\ r & r > 0 \end{cases} \tag{13}$$

where ε is a small positive constant for approximately transforming the landmark minutiae. After applying the warping-transformation to the template image based on the final correspondences of Fig. 3 (b), we obtain the alignment results shown in Fig. 4 (b).

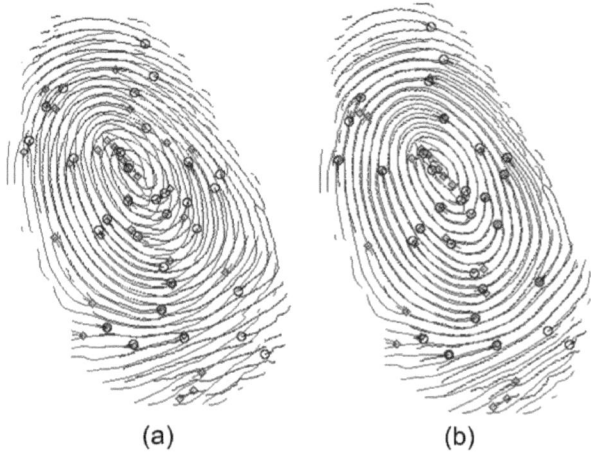

(a)   (b)

Fig. 4. Alignment of images from Fig. 1. (a) and (b). (a) Using affine transformation based on correspondences produced by the first application of MST_Corres. (b) Using warping-transformation based on the final correspondences.

5   Experimental Results

Four genuine matchings using images in Fig. 5 are tested for correspondences comparison between the proposed MST model and the TPS model used in [7] and the RIGID model used in [5]. The four matchings are (1)MATCH_1: image_1 against image_2, (2) MATCH_2: image_2 against image_3, (3) MATCH_3: image_3 against image_4, (4) MATCH_4: image_5 against image_6. Minutiae extraction method is from [11],[12],[13]. The correspondences estimation results are shown in Fig. 6, which tells us that the MST model improves the correspondences results compared with the RIGID model and TPS model. The three models are tested on FVC2004DB1_A (100 fingers, 8 samples for each finger) for matching performance comparison. We feed 3 values into a BPNN network (with 3 input nodes, 3 hidden nodes, and 1 output node) to estimate the matching score. The 3 values are (1) C, the number of correspondences, (2) n_1, the number of template minutiae in the overlapping area, (3) n_2, the number of query minutiae in the overlapping area. The ROC curves of them are shown in Fig. 7, which tells us that the MST model performs better than the TPS model and RIGID model. The matching time of the MST model is similar to the TPS, about 30 ms on Intel Core 2 U7600 1.2GHz.

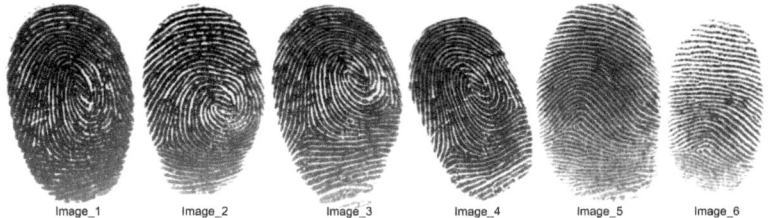

Fig. 5. Experimental images for correspondences estimation

Fig. 6. Examples of correspondences

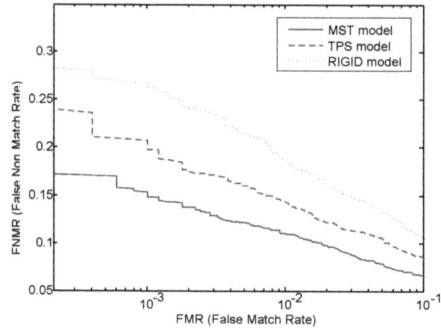

Fig. 7. ROC curves on FVC2004DB1_A

6 Conclusion

We have proposed to use minimal spanning tree (MST) to locate minutiae correspondences, in which each correspondence is determined under the reference of their parent minutiae. Each determined correspondence is more reliable, since they are close (shortest edge) to their parent minutiae. MST can find more true correspondences, and decrease the number of false correspondences, and it is effective for matching heavily distorted fingerprints.

Acknowledgments. This work was supported by the National Natural Science Foundation of China (Grant No. 60603015; 60970034), the Foundation for the Author of National Excellent Doctoral Dissertation (Grant No. 2007B4), and the Foundation for the Author of Hunan Provincial Excellent Doctoral Dissertation).

References

1. Maio, D., Maltoni, D., Cappelli, D., et al.: FVC 2004: Third Fingerprint Verification Competition. In: Zhang, D., Jain, A.K. (eds.) ICBA 2004. LNCS, vol. 3072, pp. 1–7. Springer, Heidelberg (2004)
2. Watson, C.I., Grother, P.J., Casasent, D.P.: Distortion-Tolerant Filter for Elastic-Distorted Fingerprint Matching. In: Proc. SPIE Optical Pattern Recognition, pp. 166–174 (2000)
3. Kovacs-Vajna, Z.M.: A Fingerprint Verification System Based on Triangular Matching and Dynamic Time Warping. IEEE Trans. Pattern Analysis and Machine Intelligence 22(11), 1266–1276 (2000)
4. Senior, A., Bolle, R.: Improved Fingerprint Matching by Distortion Removal. IEICE Trans. Information and Systems 84(7), 166–174 (2001)
5. Jain, A.K., Hong, L., Bolle, R.: On-line Fingerprint Verification. IEEE Trans. Pattern Analysis and Machine Intelligence 19, 302–314 (1997)
6. Zhu, E., Yin, J.P., Zhang, G.M.: Fingerprint Matching Based on Global Alignment of Multiple Reference Minutiae. Pattern Recgonition 38(10), 1685–1694 (2005)
7. Bazen, A.M., Gerez, S.H.: Fingerprint Matching by Thin-Plate Spline Modelling of Elastic Deformations. Pattern Recognition 36, 1859–1867 (2003)

8. Ross, A., Dass, S., Jain, A.: Fingerprint Warping Using Ridge Curve Correspondences. IEEE Trans. Pattern Analysis and Machine Intelligence 28(1), 19–30 (2006)

9. Bookstein, F.: Principal Warps: Thin-Plate Splines and the Decomposition of Deformations. IEEE Trans. Pattern Analysis and Machine Intelligence 11(6), 567–585 (1989)

10. Liang, X., Xiong, N., Yang, L.T.: A Compensation Scheme of Fingerprint Distortion Using Combined Radial Basis Function Model for Ubiquitous Services. Computer Communication 31, 4360–4366 (2008)

11. Zhu, E., Yin, J.P., Hu, C.F., Zhang, G.M.: A Systematic Method for Fingerprint Ridge Orientation Estimation And Image Segmentation. Pattern Recognition 39(8), 1452–1472 (2006)

12. Zhu, E., Yin, J., Zhang, G., Hu, C.: A Gabor filter based fingerprint enhancement scheme using average frequency. International Journal of Pattern Recognition and Artificial Intelligence 20(3), 417–429

13. Zhu, E.: Research on the Technologies of Feature Extraction and Recognition for Low Quality Fingerprint Image. PhD dissertation, National University of Defense Technology, China (2005)

Application of Wave Atoms Decomposition and Extreme Learning Machine for Fingerprint Classification

Abdul A. Mohammed, Q.M. Jonathan Wu, and Maher A. Sid-Ahmed

Department of Electrical Engineering,
University of Windsor, Ontario, Canada
{mohammea,jwu,ahmed}@uwindsor.ca

Abstract. Law enforcement, border security and forensic applications are some of the areas where fingerprint classification plays an important role. A new technique based on wave atoms decomposition and bidirectional two-dimensional principal component analysis (B2DPCA) using extreme learning machine (ELM) for fast and accurate fingerprint image classification is proposed. The foremost contribution of this paper is application of two dimensional wave atoms decomposition on original fingerprint images to obtain sparse and efficient coefficients. Secondly, distinctive feature sets are extracted through dimensionality reduction using B2DPCA. ELM eliminates limitations of classical training paradigm; trains data at a considerably faster speed due to its simplified structure and efficient processing. Our algorithm combines optimization of B2DPCA and the speed of ELM to obtain a superior and efficient algorithm for fingerprint classification. Experimental results on twelve distinct fingerprint datasets validate the superiority of our proposed method.

1 Introduction

Biometric verification has received considerable attention during the last decade due to increased demand for automatic person categorization. Automated classification of an individual based on fingerprints is preferred since they are less vulnerable to be copied, stolen and lost [1] and due to their uniqueness and stability [2,3]. Fingerprint detection is a technology that has been widely adopted for personal identification in many areas such as criminal investigation, access control and internet authentication.

Fingerprint classification algorithms are classified into two categories; local and global. Local feature based methods include correlation, minutiae and ridge feature based matching algorithms. Global features are obtained using mathematical transforms; a classifier compares energies of different fingerprints and classifies them based on the global trends. Correlation-based techniques utilize gray level information of an image and take into account all dimensional attributes of a fingerprint, thereby providing enough image resolution. These techniques have been successfully applied for fingerprint classification [5] but they suffer from higher computational cost. Minutiae based techniques [7] extract minutiae from two fingerprints and store them as sets of points in a two dimensional plane and execute matching by generating an alignment between the template and the input minutiae set that result in maximum

A. Campilho and M. Kamel (Eds.): ICIAR 2010, Part II, LNCS 6112, pp. 246–255, 2010.

pairings. In low quality fingerprint images minutiae extraction becomes extremely difficult and thus ridge patterns [4] are reliably extracted for classification.

Researchers have also used fast Fourier transforms (FFT) and multi-resolution analysis tools that extract global features from fingerprint images for classification. Fitz and Green [8] used a hexagonal fast Fourier transform (FFT) to transform fingerprint images into frequency domain and employed a "wedge-ring detector" to extract features. A fingerprint classifier based on wavelet transform and probabilistic neural network is proposed in [9]. Wilson et al. [10] developed a Federal Bureau of Investigation (FBI) fingerprint classification standard that incorporates a massively parallel neural network structure. Other neural network classification schemes, using self organizing feature map, fuzzy neural networks, radial basis function neural network (RBFNN) and ellipsoidal basis function neural networks (EBFNN) have also been proposed [11].

In this work, we present a fast and accurate fingerprint classification algorithm that extracts sparse fingerprint representation using wave atoms decomposition; these coefficients are dimensionally reduced using bidirectional two-dimensional principal component analysis (B2DPCA). An extreme learning machine (ELM) classifier, based on a fast single hidden layer feedforward neural network (SLFNN), is trained and tested using dimensionally reduced extracted features. The proposed classification algorithm requires less human interventions and can run at thousand folds faster learning speed than conventional neural networks. ELM determines network parameters analytically, avoids trivial human intervention and makes it efficient for online applications.

The remainder of the paper is divided into 5 sections. Section 2 discusses wave atoms decomposition, followed by a discussion of B2DPCA in section 3. ELM algorithm for classification is discussed in section 4 and the proposed method is described in section 5. Experimental results are discussed in section 6.

2   Wave Atoms Decomposition

Wave atoms [12] are the most recent mathematical transforms for harmonic computational analysis. They are a variant of 2D wavelet packets that retain an isotropic aspect ratio. Wave atoms encompass a sharp frequency localization that cannot be achieved using a filter bank based on wavelet packets and offer a significantly sparser expansion for oscillatory functions than wavelets, curvelets and Gabor atoms. Wave atoms capture coherence of a pattern across and along oscillations whereas curvelets capture coherence only along the oscillations. Wave atoms precisely interpolate between Gabor atoms and directional wavelets in the sense that the period of oscillations of each wave packet (wavelength) is related to the size of essential support via parabolic scaling i.e. wavelength is directly proportional to square of the diameter.

Two distinct parameters α; indexing multiscale nature, and β representing directional selectivity are adequate for indexing all known forms of wave packet architectures namely wavelets, Gabor, ridgelets, curvelets and wave atoms. The triangle formed by wavelets, curvelets and wave atoms, as shown in the Fig. 1, indicates the wave packet families for which sparsity is preserved under transformation. Wave atoms are defined for $\alpha=\beta=1/2$, where α indexes the multiscale nature of the transform,

from $\alpha = 0$ (uniform) to $\alpha = 1$ (dyadic). β measures the wave packet's directional selectivity (0 and 1 indicate best and poor selectivity respectively). Wave atoms represent a class of wavelet packets where directionality is sacrificed at the expense of
preserving sparsity of oscillatory patterns under smooth diffeomorphisms. Essential
support of wave packet in space (left) and in frequency (right) is shown in Fig. 2.

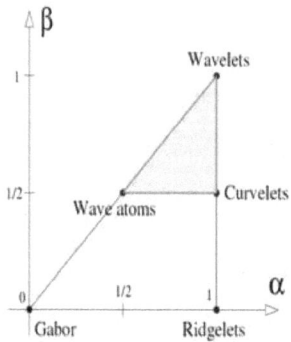

Fig. 1. Comparison of different wave packets architectures with respect to multiscale nature
and directional selectivity [12]

2.1 1D Discrete Wave Atoms

Wave atoms are constructed from tensor products of adequately chosen 1D wave
packets. Let $\psi_{m,n}^{j}(x)$ represent a one-dimensional family of wave packets, where
$j, m \geq 0$, and $n \in Z$, centered in frequency around $\pm \omega_{j,m} = \pm \pi 2^{j} m$, with $C_{1}2^{j} \leq m \leq C_{2}2^{j}$ and
centered in space around $x_{j,n} = 2^{-j}n$. Dyadic scaled and translated versions of $\hat{\psi}_{m}^{0}$ are
combined in the frequency domain and the basis function is written as:

$$\psi_{m,n}^{j}(x) = \psi_{m}^{j}(x - 2^{-j}n) = 2^{j/2}\psi_{m}^{0}(2^{j}x - n). \qquad (1)$$

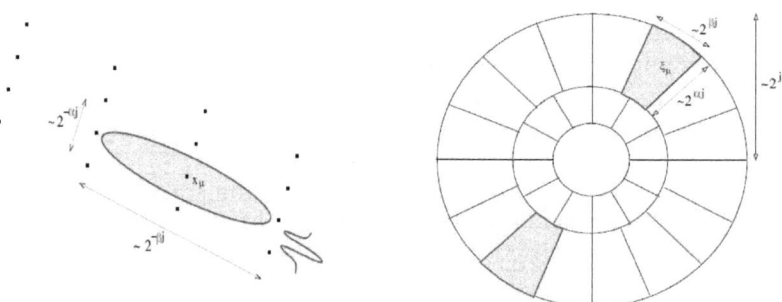

Fig. 2. Wave atoms tiling in space and frequency [12]

The coefficients $c_{j,m,n}$, for each wave $w_{j,m,n}$, are obtained as decimated convolution at scale 2^{-j}. Input sample u is discretized at $xk=kh$, $h=1/N$, $k=1,\ldots N$, and discrete coefficients $c_{j,m,n}^D$ are computed using a reduced inverse FFT inside an interval of size $2^{j+1}\pi$ centered about origin:

$$c_{j,m,n}^D = \sum_{k=2\pi(-2^j/2+1:2^j/2)} e^{i2^{-j}nk} \sum_{p\in2\pi Z} \overline{\hat{\psi}_m^j(k+2^j p)}\,\hat{u}(k+2^j p). \tag{2}$$

A simple wrapping technique is used for the implementation of discrete wavelet packets and the steps involved are:

1. Perform an FFT of size N on the samples of $u(k)$.
2. For each pair (j,m), wrap the product $\hat{\psi}_m^j\hat{u}$ by periodicity inside the interval $[-2^j\pi, 2^j\pi]$ and perform an inverse FFT of size 2^j to obtain $c_{j,m,n}^D$.
3. Repeat step 2 for all pairs (j,m).

2.2   2D Discrete Wave Atoms

A two-dimensional orthonormal basis function with 4 bumps in frequency plane is formed by individually taking products of 1D wave packets. 2D wave atoms are indexed by $\mu=(j,m,n)$, where $m=(m_1,m_2)$ and $n=(n_1,n_2)$. Construction is not a simple tensor product since there is only one scale subscript j. This is similar to the non-standard or multi-resolution analysis wavelet bases where the point is to enforce same scale in both directions in order to retain an isotropic aspect ratio. In 2D eq. (1) is modified accordingly.

$$\varphi_\mu^+(x_1,x_2) = \psi_{m1}^j(x_1 - 2^{-j}n_1)\,\psi_{m2}^j(x_2 - 2^{-j}n_2). \tag{3}$$

Combination of (3) and its Hilbert transform provides basis functions with two bumps in the frequency plane, symmetric with respect to the origin and thus directional wave packets oscillate in a single direction.

$$\varphi_\mu^{(1)} = \frac{\varphi_\mu^+ + \varphi_\mu^-}{2}, \quad \varphi_\mu^{(2)} = \frac{\varphi_\mu^+ - \varphi_\mu^-}{2} \tag{4}$$

$\varphi_\mu^{(1)}$ and $\varphi_\mu^{(2)}$ together form the wave atoms frame and are jointly denoted by φ_μ. Wave atoms algorithm is based on the apparent generalization of the 1D wrapping strategy to two dimensions.

3   Bidirectional Two Dimensional Principal Component Analysis

Principal Component Analysis (PCA) is a data representation technique widely used in pattern recognition and compression schemes. In the past researchers used PCA and bunch graph matching techniques for enhanced representation of face images. PCA cannot capture even a simple variance unless it is explicitly accounted in the training data. In [13] Yang et al. proposed two dimensional PCA for image representation. As opposed to PCA, 2DPCA is based on 2D image matrices rather than 1D vector so the image matrix does not need to be vectorized prior to feature extraction. Instead an image covariance matrix is computed directly using the original image matrices.

Let X denote a q dimensional unitary column vector. To project a $p \times q$ image matrix A to X; linear transformation $Y=AX$ is used which results in a p dimensional projected vector Y. The total scatter of the projected samples is characterized by the trace of the covariance matrix i.e. matrix of the projected feature vectors, $j(X)=tr(S_x)$, where $tr( )$ represents the trace of S_x, and S_x denotes covariance matrix of the projected features.

$$S_x = E(Y - E(Y))(Y - E(Y))^T = E[(A - EA)X][(A - EA)X]^T. \qquad (5)$$

$$tr(S_x) = X^T[E(A - EA)^T(A - EA)]X. \qquad (6)$$

$G_t = E[(A - EA)^T(A - EA)]$ is $q \times q$ nonnegative image covariance matrix. If there are M training samples, the jth image sample is denoted by $p \times q$ matrix A_j.

$$G_t = \frac{1}{M}\sum_{j=1}^{M}(A_j - \overline{A})^T(A_j - \overline{A}). \qquad (7)$$

$$J(X) = X^T G_t X, \qquad (8)$$

where $\overline{A}$ represents an average image of all the training samples. Above criterion is called the generalized total scatter criterion. The unitary vector X that maximizes the criterion $j(X)$ is called the optimal projection axes. An optimal value represents a collection of d orthonormal eigen vectors $X_1, X_2, \ldots X_d$ of G_t corresponding to d largest eigen values. A limitation of 2DPCA based dimension reduction is the processing of higher number of coefficients since it works along row directions only. Zhang and Zhou [14] proposed $(2D)^2$ PCA based on the assumption that training sample images are zero mean, and image covariance matrix can be computed from the outer product of row/column vectors of images.

4   Extreme Learning Machine

Feedforward neural networks (FNNs) are widely used in classification techniques due to their approximation capabilities for non-linear mappings. Slow learning speed of FNNs is a major bottleneck encountered, since input weights and hidden layer biases are updated using a parameter tuning approach such as gradient descent algorithm. Huang et al. [16] proposed an extremely fast learning algorithm called ELM for training a SLFNN. ELM randomly assigns input weights and hidden layer biases if the hidden layer activation function is infinitely differentiable. In ELM a learning paradigm is converted to a simple linear system whose output weights are analytically determined through a generalized inverse operation of the hidden layer weight matrices.

An N dimension random distinct sample (x_i, t_i) where $x_i = [x_{i1}, x_{i2}, \ldots x_{in}]^T \in \mathfrak{R}^n$ and $t_i = [t_{i1}, t_{i2}, \ldots t_{im}]^T \in \mathfrak{R}^m$, ELM with L hidden nodes and an activation function $g(x)$ is modeled as:

$$\sum_{i=1}^{L} \beta_i g_i(x_j) = \sum_{i=1}^{L} \beta_i g_i(w_i.x_j + b_i) = o_j, \; j = \{1,2,...N\}, \tag{9}$$

Where $w_i=[w_{i1},w_{i2},....w_{in}]^T$, $\beta_i=[\beta_{i1}, \beta_{i2},....\beta_{iL}]^T$ represent weight vectors connecting input nodes to an ith hidden node and from the ith hidden node to all output nodes. b_i indicates threshold for i^{th} hidden node whereas $w_i.x_j$ represents an inner product of w_i and x_j. An ELM can reliably approximate N samples with zero error.

$$\sum_{i=1}^{L} \beta_i g_i(w_i.x_j + b_i) = t_j, \; j = \{1,2,...N\}. \tag{10}$$

Eq. (10) is modified as $H\beta=T$, $H=(w_1,...,w_L,b_1,...,b_L,x_1,...,x_N)$, such that i^{th} column of H is the output of ith hidden node with respect to inputs $x_1,x_2,....x_N$. If the activation function $g(x)$ is infinitely differentiable, it is proved that the number of hidden nodes are such that $L<<N$. Training of SLFNN requires minimization of an error function E.

$$E = \sum_{j=1}^{N} \left(\sum_{i=1}^{L} \beta_i g(w_i x_j + b_i) - t_j \right)^2 = E = \|H\beta - T\|. \tag{11}$$

H is determined using gradient descent and the weights w_i, β_i and bias parameters b_i are tuned iteratively with a learning rate ρ. A small value of ρ causes the learning algorithm to converge slowly whereas a higher rate leads to instability and divergence to local minima. To avoid these limitations, ELM incorporates a minimum norm least-square solution, and instead of tuning the entire network parameters a random allocation of input weights and hidden layer biases help to analytically determine the hidden layer output matrix H and curtail the problem to a least-square solution of $H\beta=T$. H is a non-square matrix, the norm least-square solution of the above linear system becomes $\beta=H^*T$, where H^* is the moore-penrose generalized inverse of H. The above relationship holds for a non-square matrix H whereas the solution is straightforward for $N=L$. An infinitely small training error is achieved using the above model since it represents a least-square solution of the linear system.

$$\left\| H \hat{\beta} - T \right\| = \left\| HH^*T - T \right\| = \min_{\beta} \| H\beta - T \|. \tag{12}$$

5   Proposed Fingerprint Classification Algorithm

The proposed classification scheme is independent of fingerprint patterns and is based on individual features and the number of trained fingerprint classes. Table 1 consists of detailed steps that demonstrate our proposed technique. Our system classifies fingerprint images into one of the trained classes; therefore, only one verification process is required per image. Our proposed scheme deals with classification of fingerprint images using ELM design and utilizes dimensionally reduced feature vectors obtained from wave atoms decomposition. Wave atoms decomposition is used for sparse representation of fingerprint images since they belong to a category of images that oscillate smoothly in varying directions. Discrete 2D wave atoms decomposition is applied on the original fingerprint image to efficiently capture coherence patterns along and across the oscillations. Fingerprint images are digitized using 256 gray levels therefore a transformation in color space is not required. Dimension of fingerprint images within each database were reduced to *64×64* prior to wave atoms decomposition. Image resizing was the only pre-processing performed on all datasets to minimize computations and to guarantee uniformity with other methods used for comparison. An orthonormal basis is used instead of a tight frame since each basis function oscillates in two distinct directions instead of one. This orthobasis variant property is important in applications where redundancy is undesired.

In addition to the aforementioned alterations there were no further changes made to the images as it may lead to image degradation. We randomly divide image database into two sets namely training set and testing set. All images within each database have the same dimension, i.e. *R×C*. Similar image sizes support the assembly of equal sized wave atoms coefficients and feature vector extraction with identical level of global content. 2D wave atoms decomposition of every image is computed and coefficients are saved as initial feature matrix. Wave atoms decomposition is a relatively new technique for multiresolution analysis that offers significantly sparser expansion, for oscillatory functions, than other fixed standard representations like wavelets, curvelets and Gabor atoms.

Table 1. Outline of our Proposed Classification Scheme

INPUT: Randomly divide image database into two subsets TR_i and TE_j where i={1,2,…,n} and j={1,2,…,m} representing training and test image sets respectively.
OUTPUT: Classifier - f(x)
1. Resize fingerprint images from all databases to RxC.
2. Compute the wave atoms decomposition of each training and test images and extract feature sets. Each feature set is of dimension RxC. (Refer to section 2 for details of wave atoms decomposition)
3. Calculate image covariance matrix of test and train images to obtain intermediate featue matrix. $$G_{iR} = \frac{1}{n}\sum_{i=1}^{n}(A_i - \overline{A_R})^T(A_i - \overline{A_R})$$ $$G_{iE} = \frac{1}{m}\sum_{j=1}^{m}(A_j - \overline{A_E})^T(A_j - \overline{A_E})$$
4. Evaluate the maximizing criteria *J(X)* for train and test images. $$J(X_R) = X_R^T G_{iR} X_R$$ $$J(X_E) = X_E^T G_{iE} X_E$$
5. Repeat steps 3-4 on the transposed intermediate feature matrix to obtain B2DPCA based feature vectors, f_p of size UxV.
6. Train Extreme Learning Machine (ELM) classifier: Generate set of 2DPCA based feature vectors (vectorized feature vectors obtained in previous step) for training.
7. Classify images with test feature vectors using ELM trained in step 6.

Application of ELM based classification on original wave atoms coefficients is computationally expensive due to higher dimensionality of data originating from large image datasets. Outliers and irrelevant image points being included into classification task can also affect the performance of our algorithm; hence B2DPCA is employed to reduce dimensionality of initial feature vectors. Features are extracted by computing 2DPCA of initial feature matrix along image rows, called as intermediate feature matrix. 2DPCA is again applied on the transposed intermediate feature matrix along its rows so as to generate a final feature matrix. Application of 2DPCA using the modified approach retains better structure and correlation information amongst neighboring pixel coefficients. Dimensionally reduced wave atoms coefficients are vectorized into a $U \times V$ dimension vector, final feature vector, where $U \times V \ll R \times C$.

B2DPCA based feature vectors better retain the global structure of input space and facilitate accurate classification with lower computational complexity, diminished outliers and irrelevant information. ELM is trained using labeled B2DPCA feature vectors and classified using the trained network.

6   Results and Discussion

Extensive experiments were performed using 3 standard and distinctive collections of fingerprint datasets; FVC2000, FVC2002 and FVC2004 [2] to test the practicality of our proposed method. Each dataset consists of four diverse databases generated using various fingerprint acquisition techniques. Each database contains 8 fingerprints of each of the 100 distinctive subjects.

All images were resized to 64×64 in our experiments and 5 images from each database were used as prototypes and the remaining 3 for testing to ensure consistency with other methods used for comparison. Experiments were also performed on original fingerprint image without resizing and consistently better results were obtained since detailed fingerprint information is incorporated at the expense of large feature vectors. Both the testing and training sets of images are decomposed using 2D wave atoms transform using an orthonormal basis function and dimensionally reduced through application of B2DPCA. Dimensionally reduced features are vectorized and classification is performed by using ELM. The above process was repeated 10 times for all the databases and averaged results of few experiments are documented in the paper. The classification accuracy for Db1 database from FVC2000, FVC2002 and FVC2004 is compared with wavelet transform (WT) based RBFNN and EBFNN fingerprint classification algorithms. Results, obtained with the proposed method (only 6 principal components are used for consistency with other methods), are compared with the classification accuracy reported in [11] using WT-2DPCA-RBFNN and WT-2DPCA-EBFNN.

Table 2. Fingerprint classification rates (%) for different techniques

Database	WT-2DPCA-RBFNN	WT-2DPCA-EBFNN	Proposed Method
FVC 2000	91	91	93.25
FVC 2002	87	87	92.63
FVC 2004	86.5	87	89.62

We conclude from the results in Table 2 that our proposed fingerprint classification algorithm performs significantly better than the wavelet based RBFNN and EBFNN fingerprint classification algorithms. In addition to the improved classification accuracy, our proposed ELM based classifier performs training and testing thousands folds faster than conventional neural network based classification algorithms [15].

From Fig. 3 it is evident that several factors influence classification accuracy, namely, fingerprint acquisition techniques, climatic and environmental conditions and most notably the number of principal components. Dataset Db4 from each of the databases is generated using a synthetic fingerprint generator; consequently the effects of environment and other irrepressible conditions are trifling and are substantiated by improved classification accuracy at low principal components.

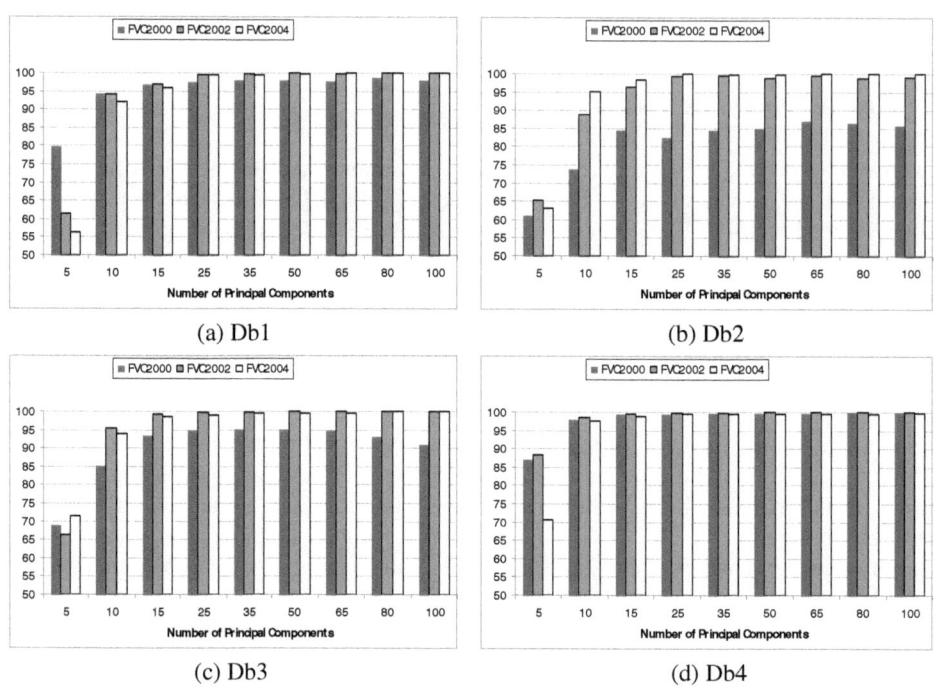

(a) Db1   (b) Db2

(c) Db3   (d) Db4

Fig. 3. Classification accuracy vs. number of principal components

7   Conclusion

An original supervised fingerprint classification algorithm for multiclass categorization based on wave atoms decomposition and bidirectional two-dimensional principal component is proposed. Improvements in classification accuracy validate the fact that wave atoms multiresolution analysis offers significantly sparser expansion, for oscillatory functions, than other fixed standard representations like wavelets, curvelets and Gabor atoms. The proposed classifier is capable of handling marginal fingerprint orientations, illumination variations, moderate pressure changes against the sensor surface and climatic conditions. The algorithm combines the strengths of both B2DPCA and ELM; creates distinctive and improved feature set, an efficient and fast algorithm for fingerprint classification. The proposed fingerprint classification algorithm is independent of the number of prototypes used for training and or testing and is also free of the amount of hidden neurons used for classification, unlike traditional neighborhood based classifiers whose accuracy is greatly affected by the number of prototypes and neighborhood size. Law enforcement, multimedia, and data mining related applications can benefit from our proposed classification scheme.

References

1. Jain, A.K., Ross, A., Pankanti, S.: Biometrics: A tool for information security. IEEE Transaction on Information Forensics Security 1(2), 125–143 (2006)
2. Maltoni, D., Maio, D., Jain, A.K., Prabhakar, S.: Handbook of fingerprint recognition. Springer, New York (2003)
3. Yager, N., Amin, A.: Fingerprint classification: a review. Pattern Analysis and Applications 7(1), 77–93 (2004)
4. Jain, A., Chen, Y., Demitrius, M.: Pores and Ridges: Fingerprint matching using Level 3 features. In: Proc. 18th International Conference on Pattern Recognition, vol. 4, pp. 477–480 (2006)
5. Maio, D., Maltoni, D., Cappelli, R., Wayman, J., Jain, A.K.: FVC 2004: Third fingerprint verification competition. In: Zhang, D., Jain, A.K. (eds.) ICBA 2004. LNCS, vol. 3072, pp. 1–7. Springer, Heidelberg (2004)
6. Bazen, A., Gerez, S.: Systematic methods for the computation of the direction fields and singular points of fingerprints. IEEE Transaction on Pattern Analysis and Machine Intelligence 24(7), 905–919 (2002)
7. Bhanu, B., Tan, X.: Fingerprint indexing based on novel features of minutiae triplets. IEEE Transactions on Pattern Analysis and Machine Intelligence 25(5), 616–622 (2003)
8. Fitz, A., Green, R.: Fingerprint classification using a hexagonal fast Fourier transform. Pattern Recognition 29(10), 1587–1597 (1996)
9. Seokwon, L., Boohee, N.: Fingerprint Recognition using Wavelet Transform and Probabilistic Neural Network. In: Proc. of International Joint Conference on Neural Networks, vol. 5, pp. 3276–3279 (1999)
10. Wilson, C., Candela, G., Grother, P., Watson, C., Wilkinson, R.: Massively parallel neural network fingerprint classification system. National Institute of Standards and Technology; NISTIR 4880 (1992)
11. Luo, J., Lin, S., Lei, M., Ni, J.: Application of dimensionality reduction analysis to fingerprint recognition. In: ISCID, vol. 2, pp. 102–105 (2008)
12. Demanet, L., Ying, L.: Wave Atoms and Sparsity of Oscillatory Patterns. Applied and Computational Harmonic Analysis 23(3), 368–387 (2007)
13. Yang, J., Zhang, D., Frangi, A., Yang, J.: Two-dimensional PCA: a new approach to appearance based face representation and recognition. IEEE Transaction on Pattern Analysis and Machine Intelligence 26(1), 131–137 (2004)
14. Zhang, D., Zhou, Z.: $(2D)^2$ PCA: Two-directional two-dimensional PCA for efficient face representation and recognition. Neurocomputing 69(1), 224–231 (2005)
15. Pankanti, S., Prabhakar, S., Jain, A.K.: On the Individuality of Fingerprints. IEEE Transactions on Pattern Analysis and Machine Intelligence 24(8), 1010–1025 (2002)
16. Huang, G., Zhu, Q., Siew, C.: Extreme learning machine: Theory and applications. Neurocomputing 70(1), 489–501 (2006)
17. Ross, A., Jain, A.K., Reisman, J.: A Hybrid Fingerprint Matcher. Pattern Recognition 36(7), 1661–1673 (2003)

Unideal Iris Segmentation Using Region-Based Active Contour Model

Kaushik Roy[1], Prabir Bhattacharya[2], and Ching Y. Suen[1]

[1] Department of Computer Science and Software Engineering,
Concordia University, Montreal, QC, Canada H3G 1M8
[2] Department of Computer Science, College of Engineering and Applied Sciences
University of Cincinnati, Cincinnati, OH 45221-0030
{kaush_ro,prabir,suen}@encs.concordia.ca

Abstract. Robust segmentation of an iris image plays an important role in iris recognition. Most state-of-the-art iris segmentation algorithms focus on the processing of the ideal iris images that are captured in a controlled environment. In this paper, we process the unideal iris images that are acquired in an unconstrained situation and are affected severely by gaze deviation, eyelids and eyelashes occlusion, non uniform intensity, motion blur, reflections, etc. The novelty of this research effort is that we apply the modified Chan-Vese curve evolution scheme, which extracts the intensity information in local regions at a controllable scale, to find the pupil and iris boundaries accurately. A data fitting energy is defined in terms of a contour and two fitting functions that locally approximate the image intensities on the two sides of the contour. This energy is then incorporated into a variational level set formulation with a regularization term. Due to the kernel function used in energy functional, the extracted intensity information of the local regions is deployed to guide the motion of the contour, which thereby assists the curve evolution scheme to cope with the intensity inhomogeneity that occurs in the same region. The contours represented by the proposed variational level set method may break and merge naturally during evolution, and thus, the topological changes are handled automatically. The verification performance of the proposed scheme is validated using the UBIRIS Version 2, the ICE 2005, and the WVU unideal datasets.

Keywords: Iris recognition, iris segmentation, region-based active contour, Chan-Vese curve evolution, intensity inhomogeneity.

1 Introduction

The current stress on security and surveillance has resulted in a rapid development of automated personal identification systems based on biometrics [1]. Recently, the iris recognition is in the limelight for many high security biometrics applications [2, 3]. The exact segmentation of the iris plays perhaps the most important role in iris recognition [4, 5]. The main task of the segmentation routine is to localize the inner/outer boundary from the iris. Apart from the proper localization of the iris structure, the segmentation scheme should also identify the eyelid and eyelash occlusions and

A. Campilho and M. Kamel (Eds.): ICIAR 2010, Part II, LNCS 6112, pp. 256–265, 2010.

detect the other noisy regions such as reflections. The localization error may result in lower recognition performance due to incorrect encoding of the textural content of the iris [6, 7]. For iris segmentation, most of the researchers assume that the iris is circular or elliptical. However, in the case of unideal iris images, which are captured in an uncontrolled environment, iris may appear as noncircular or nonelliptical [3]. Also, in the iris images where the eyes are not properly opened, highly occluded regions cannot be extracted, and thus, the segmentation performance is deteriorated [6]. The iris images may also be affected by the intensity inhomogeneity, deviated gaze, nonlinear deformations, pupil dilation, head rotation, motion blur, reflections, non-uniform intensity, low image contrast, camera angles and diffusion, and presence of eyelids and eyelashes. Recently, several researchers proposed different unideal iris recognition schemes. In [2], inner and outer boundaries were detected in terms of active contours based on the discrete Fourier series expansions of the contour data. In [3], two approaches were proposed in which the first approach compensated for the off angle gaze direction, and the second approach used an angular deformation calibration model. In [6, 7], curve evolution approaches were applied based on geometrics active contours to segment the non-frontal iris images. To localize the unideal iris images accurately, we proposed the level set based curve evolution approaches using the edge-stopping function and the energy minimization algorithm in [8], and in [9], we deployed a variational level set based curve evolution scheme, which uses significantly larger time step for numerically solving the evolution *partial differential equation* (PDE). The segmentation approaches proposed in [6-8] consume huge computational time due to costly reinitialization process. In [2, 7], curves evolve from the previously obtained pupil boundary to the outer boundary, which in turn, slows down the segmentation process. The parametric active contours based iris segmentation scheme may terminate at certain local minima such as the specular reflections, the thick radial fibers in iris or the crypts in ciliary region [2]. The active contours with an edge stopping function as a halting criteria proposed in [6, 7, 8] may fail to detect the outer boundary accurately if the iris is separated from the sclera region by relatively a smooth boundary. Furthermore, the intensity inhomogeneity often occurs in the unideal iris images due to reflections, motion blur, luminosity, etc. Most of the current unideal iris segmentation schemes based on active contours models proposed in [2, 6-9] tend to rely on intensity homogeneity in each of the regions to be segmented, and furthermore, most of the unideal iris localization algorithms [2, 6-8] consume huge computational time due to expensive curve evolution approach. This impedes the traditional level set based iris recognition systems to be deployed in real-time scenario. Addressing the above problems, we apply a modified Chan-Vese curve evolution scheme proposed in [10], which extracts the intensity information in local regions at a controllable scale, to find the pupil and iris boundaries accurately. A data fitting energy is defined in terms of a contour and two fitting functions that locally approximate the image intensities on the two sides of the contour [11]. This energy is then incorporated into a variational level set formulation with a regularization term. Due to the kernel function used in energy functional, the extracted intensity information of the local regions is deployed to guide the motion of the contour, which thereby assists the curve evolution scheme to cope with the intensity inhomogeneity that occurs in the same region [10]. In addition, the level set regularization term is used to ensure the accurate computation and to avoid expensive reinitialization of the evolving curve. The contours represented by the variational level set may break and merge naturally during evolution, and thus, the topological changes are handled automatically. Prior to applying the curve evolution approach using the active contours, we deploy Direct Least Square (DLS) based elliptical fitting to obtain an initial approximation of the pupil and the iris boundaries [8, 9].

2   Unideal Iris Segmentation

The segmentation of the unideal iris image is a difficult task because of the noncircular shape of the pupil and the iris, and the shape differs depending on the image acquisition techniques [3]. We divide the iris segmentation process into two steps. In the first step, we use an elliptical model to approximate the inner (pupil) and outer (iris) boundaries of the iris, and then, we apply the region-based active contour model to find the exact inner and outer boundaries of the iris based on the approximated boundaries obtained in the previous step.

Before applying the curve evolution approach, we deploy DLS based elliptical fitting to approximate the pupil boundary. However, the accuracy of the ellipse fitting process degrades in the presence of the outliers such as eyelashes. Therefore, we apply a morphological operation, namely, the opening to an input image to suppress the interference from the eyelashes. DLS based elliptical fitting returns five parameters $(p_1, p_2, r_1, r_2, \varphi_1)$: the horizontal and vertical coordinates of the pupil center (p_1, p_2), the length of the major and minor axes (r_1, r_2), and the orientation of the ellipse φ_1. To approximate the outer boundary, we apply the DLS based elliptical fitting scheme again, and obtain five parameters $(I_1, I_2, R_1, R_2, \varphi_2)$: the horizontal and vertical coordinates of the iris center (I_1, I_2), the length of the major and minor axes (R_1, R_2), and the orientation of the ellipse φ_2. This method, thus, provides the rough estimation of iris and pupil boundaries.

Based on the approximation of the inner and outer boundaries, the curve is evolved using the modified Chan-Vese functional [10, 11] for accurate segmentation of the pupil and iris regions. In the following paragraphs, we briefly discuss the segmentation process based on active contour approach [10].

In our proposed curve evolution method, the following energy functional is deployed [10]:

$$F\left(\varphi, f_1, f_2\right) = E\left(\varphi, f_1, f_2\right) + \mu P(\varphi) \tag{1}$$

The term $E\left(\varphi, f_1, f_2\right)$ in (1) can be defined as:

$$E\left(\varphi, f_1, f_2\right) = \textstyle\sum_{i=1}^{2} \lambda_i \int \left(\int G_\sigma(x-y)|I(y) - f_i(x)|^2 N_i\left(\varphi(y)\right) dy\right) dx + \\ v \int |\nabla H\left(\varphi(x)\right)| dx \tag{2}$$

where the level set function φ represents the closed contour C in the image domain Ω, and this closed contour separates Ω into two regions: $\Omega_1 = outside\ (C)$ and $\Omega_2 = inside\ (C)$. λ_i are positive constants, and the functions $f_i(x)$ are the values that approximate image intensities outside and inside the closed contour C. In this research effort, level set function φ takes the positive and negative values outside and inside C, respectively. In (2), H is called the Heaviside function and $N_1(\varphi) = H(\varphi)$, $N_2(\varphi) = 1 - H(\varphi)$. The intensities $I(y)$, which are effectively involved in the above energy term, are in a local region centered at the point x, whose size can be controlled by the Gaussian kernel function, $G_\sigma = \frac{1}{(2\pi)^{n/2}\sigma^n} e^{-|u|^2/2\sigma^2}$ with a scale parameter $\sigma > 0$. The last term $\int |\nabla H(\varphi(x))| \, dx$ in (2) computes the length of the zero level contour of φ. The length of the zero level contour can be equivalently defined by the integral $\int \delta(\varphi)|\nabla\varphi| dx$ with the Dirac delta function δ. The Heaviside function H can be approximated as follows:

$$H_\epsilon(x) = \frac{1}{2}\left[1 + \frac{2}{\pi}\arctan\left(\frac{x}{\epsilon}\right)\right] \tag{3}$$

The derivative of H_ϵ is $\delta_\epsilon(x) = H'_\epsilon(x) = \frac{1}{\pi}\frac{\epsilon}{\epsilon^2+x^2}$. Now if we replace H in (2) with H_ϵ, the energy functional E in (2) can be approximated by:

$$E_\epsilon(\varphi, f_1, f_2) = \sum_{i=1}^{2}\lambda_i \int\left(\int G_\sigma(x-y)|I(y) - f_i(x)|^2 \, N_i^\epsilon(\varphi(y))\, dy\right) dx + \nu \int |\nabla H_\epsilon(\varphi(x))| \, dx \tag{4}$$

where $N_i^\epsilon(\varphi) = H_\epsilon(\varphi)$ and $N_i^\epsilon(\varphi) = 1 - H_\epsilon(\varphi)$.

The level set regularized term $\mu P(\varphi)$, $(\mu > 0)$ in (1) which is used for accurate computation and stable level set evolution can be defined as $P(\varphi) = \int \frac{1}{2}(|\nabla\varphi(x)| - 1)^2 dx$, and this term measures the deviation of the function φ from a signed distance function. Now we minimize the energy functional $F(\varphi, f_1, f_2)$ with respect to φ using the standard gradient descent method by solving the gradient flow equation as follows:

$$\frac{\partial\varphi}{\partial t} = -\delta_\epsilon(\varphi)(\lambda_1 e_1 - \lambda_2 e_2) + \nu\delta_\epsilon(\varphi)div\left(\frac{\nabla\varphi}{|\nabla\varphi|}\right) + \mu\left(\nabla^2\varphi - div\left(\frac{\nabla\varphi}{|\nabla\varphi|}\right)\right) \tag{5}$$

where e_1 and e_2 are functions which can be expressed as:

$$e_i(x) = \int G_\sigma(y-x)|I(x) - f_i(y)| dy, \quad i = 1, 2 \tag{6}$$

The above (5) is the required active contour model. The term $-\delta_\epsilon(\varphi)(\lambda_1 e_1 - \lambda_2 e_2)$ is responsible for driving the active contour toward the iris/pupil boundaries. The second term $\nu\delta_\epsilon(\varphi)div\left(\frac{\nabla\varphi}{|\nabla\varphi|}\right)$ has a length shortening or smoothing impact on the zero level contour which is useful to maintain the regularity of the contour. The third term $\mu\left(\nabla^2\varphi - div\left(\frac{\nabla\varphi}{|\nabla\varphi|}\right)\right)$ is denoted as level set regularization term, which is used to maintain the regularity of the level set function. In order to estimate the exact boundary of the pupil, we initialize the active contour φ to the approximated pupil boundary, and evolve the curve in the narrow band of ± 10 pixels. We evolve the curve from outside the approximated inner boundary to remove the effect of reflections. Similarly, for computing outer boundary, the active contour φ is initialized to the estimated iris boundary, and the optimal estimation of the iris boundary is computed by evolving the curve in a narrow band of ± 20 pixels. In this case, the curve is evolved from inside the approximated iris boundary to reduce the effects of the eyelids and the eyelashes. Fig. 1 (b, c) shows the segmentation results.

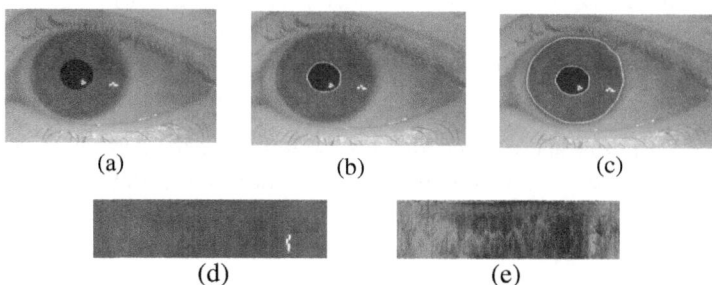

(a)   (b)   (c)

(d)   (e)

Fig. 1. (a) Original image from WVU dataset, (b) Pupil Detection, (c) Iris detection, (d) Normalized image, and (e) Enhanced image

Besides reflections, eyelid occlusion, and camera noise, the iris image data may be corrupted by the occlusion of the eyelashes [2]. We deploy one dimensional Gabor filters and variance of intensity to isolate the eyelashes as used in our previous works [8, 12]. We unwrap the iris region to a normalized rectangular block with a fixed dimension of size 64   512 by adopting the circle fitting strategy proposed in [7]. Since the normalized iris image has relatively low contrast and may have non-uniform intensity values due to the position of the light sources, a local intensity based histogram equalization technique is applied to enhance the contrast of the normalized iris image within a small image block of size 20   20. Fig. 1(d, e) shows the unwrapped image and the effect of contrast enhancement, respectively. In this paper, Daubechies Wavelet Transform (DBWT) is used to extract the distinctive features set form normalized and enhanced iris image [9], and for iris pattern matching, we use Hausdorff distance [8].

3   Performance Evaluation

The extensive experimentation is conducted on three datasets, namely, ICE 2005 [13], WVU Unideal [14], and UBIRIS Version 2 [15]. The ICE 2005 [13] contains 2953 images corresponding to 244 classes. The ICE database consists of left and right iris images for experimentation (1528 left iris images from 120 classes and 1425 right iris images from 124 classes). The WVU unideal [14] iris dataset has a total of 800 iris images from 200 different persons. This dataset consists of iris images that are captured off axis under different lighting conditions, and contains the iris images of varying quality ranging from very high quality images with prominent textural details to very low quality images with non linear deformation, reflections, and occlusions. The UBIRIS [15] dataset contains 2410 iris images from 241 classes. The iris images are captured in two sessions. The iris images captured in the first session represent the good quality images whereas the images in the second session have the irregularities with respect to focus, intensity, and reflection. In order to perform an extensive experimentation and to validate our proposed approach, we generate a non-homogeneous dataset by combining the above three datasets, and this dataset consists of 6163 images corresponding to 685 classes. The selected parameters values to find the inner

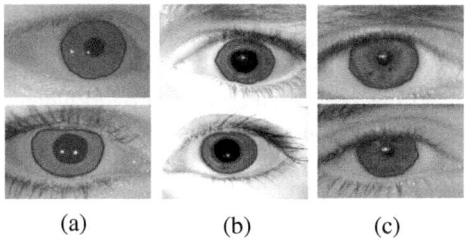

(a)   (b)   (c)

Fig. 2. Segmentation results on datasets (a) WVU, (b) ICE, and (c) UBIRIS

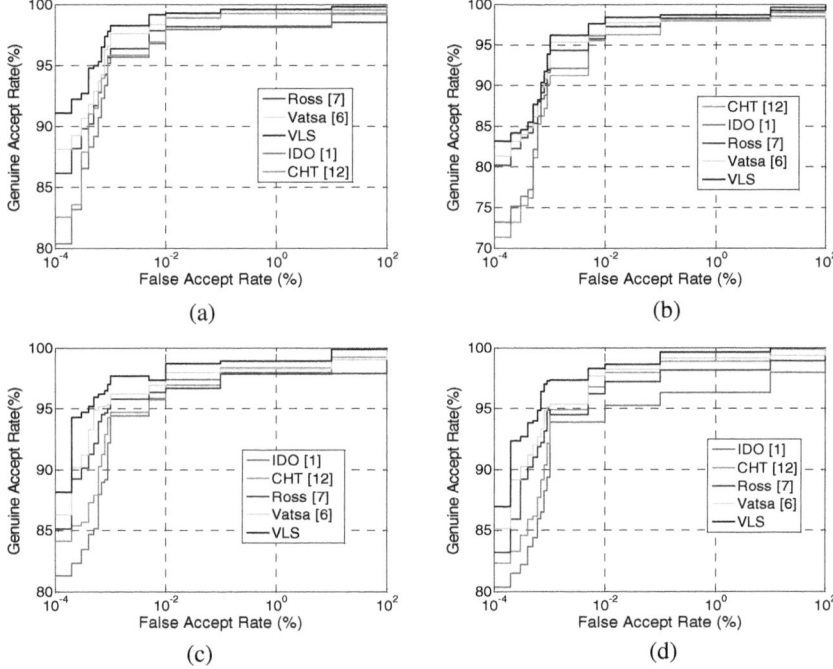

(a)   (b)

(c)   (d)

Fig. 3. ROC curves show the comparison of different existing segmentation techniques on (a) ICE, (b) WVU, (c) UBIRIS, and (d) Combined datasets

and outer boundaries using region based active contour algorithm are $\sigma = 2.0, \lambda_1 = 0.5, \lambda_2 = 1.0, \mu = 1, v = 200.0$, and time step $\Delta t = 0.4$. Fig 2 shows the segmentation results on the three datasets, and we find form this figure that our segmentation scheme performs well even if the intensity inhomogeneity occurs in the iris and the pupil regions. In order to exhibit the effectiveness of our segmentation approach, we compare the region-based active contour model using variational level set formulation (VLS) with integro-differential operator (IDO) proposed by Daugman [1], the Canny edge detection and Hough transform (CHT) based approach applied in our early work [12], and the active contour based localization approaches proposed by Vatsa *et al.* [6]

and Shah and Ross [7] on three datasets. For comparison purpose, we only implement the segmentation approaches proposed in [6, 7], and for feature extraction and matching, we apply our proposed algorithms to each of those schemes. ROC curves in Fig. 3 show that the matching performance is improved when the region-based active contours are used for segmentation with variational level set formulation. The proposed segmentation scheme shows better performance than the traditional active contour based methods reported by Vatsa *et al.* [6] and Shah and Ross [7], and the reason seems to be the advantage of taking into account the image inhomogeneity, which often occurs in the iris region, over the traditional level set based methods. Another advantage of the proposed segmentation algorithm is that this technique performs well even when the iris/sclera boundary is separated by a quite blurry boundary. Also, the proposed variational level set based algorithm speeds up the curve evolution process to a great extent since the used regularization term avoids the costly reinitialization process. Moreover, a morphological operation is applied to an input image to restrain the interference from the eyelashes prior to deploying the DLS based elliptical fitting. The DLS fitting approach provides a reasonable approximation of the inner and the outer boundaries. Furthermore, the level set curve is evolved over a narrowband and, this process, thus, reduces the overall segmentation time. The GAR at a fixed FAR of 0.001% is (a) 96.28% in WVU, (b) 98.23% in ICE, and (c) 97.29% in UBIRIS datasets. The GAR on the combined dataset at the fixed FAR of 0.001% is 97.43%. However, the proposed level set based segmentation algorithm fails to perform on some images of UBIRIS dataset due to huge occlusion as shown in Fig. 4. Fig. 5 also exhibits that only DLS based elliptical fitting strategy may fail to detect the outer boundary accurately; however, it provides an initial estimation of the inner and outer boundaries for the final segmentation using active contour model.

Fig. 4. Sample of iris images from UBIRIS Version 2 dataset [15] on which the proposed segmentation scheme fails to detect the iris/pupil boundary

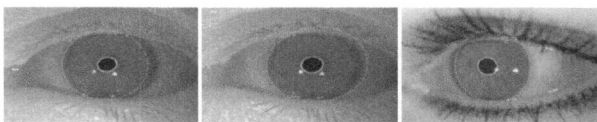

Fig. 5. Sample of iris images from WVU dataset [14] on which the DLS elliptical

Furthermore, we compare the performance of the proposed algorithm with the other existing iris recognition algorithms. We implement the well known iris recognition algorithms proposed by Daugman [1] and Ma *et al.* [4, 5], and compare our approach with those methods on the combined dataset. Fig. 6 exhibits the ROC curve of the proposed algorithm based on level set curve evolution approach on the non-homogeneous combined dataset. ROC curves of the approaches demonstrated in [1, 4, 5] are also plotted for comparison, and it is observed from this figure that the proposed algorithm achieves higher GAR with a very low EER of 0.42% for the combined dataset. It means that the proposed algorithm achieves higher discriminating capabilities than the approaches reported in [1, 4, 5]. Moreover, the approaches proposed in [1, 4, 5] were not adjusted specifically for the unideal situation. The proposed approach based on the region-based active contour algorithm obtains a higher GAR of 97.49 % at the fixed FAR of 0.001% on the combined dataset that contains the iris images with the irregularities due to motion blur, off angle gaze deviation, diffusion, and other real-world problems. Therefore, ROC curves in Fig 6 reveal the effectiveness of the proposed scheme in an unideal situation. Furthermore, we compare the proposed region-based active contour model with three of our most recent works reported in [8, 9, 12]. The schemes proposed in [8, 9] represent the unideal iris recognition algorithms whereas the method demonstrated in [12] recognizes the iris images in an ideal situation. The ROC curves in Fig. 7 clearly demonstrate that our proposed approach outperforms the other methods since region-based curve evolution scheme takes into account the intensity inhomogeneity.

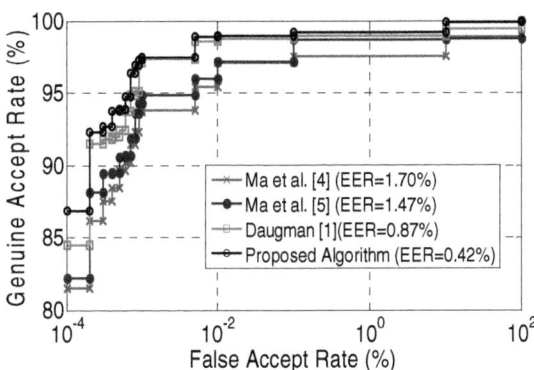

Fig. 6. Comparison with existing iris recognition schemes on the combined dataset

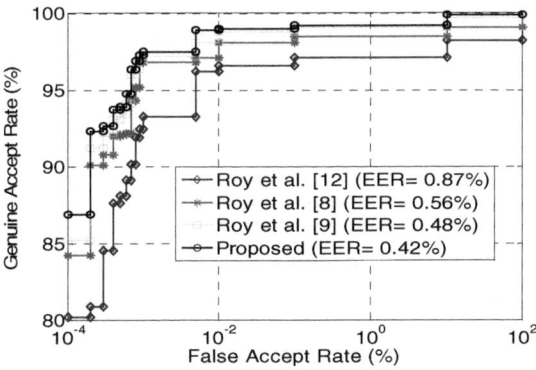

Fig. 7. Comparison with our previous iris recognition schemes on the combined dataset

A new technology development project for iris recognition, namely, the Iris Challenge Evaluation (ICE), has been conducted by the National Institute of Standards and Technology (NIST), USA [13]. We further compare our results with the results summarized by NIST [12] for the ICE 2005 dataset. In ICE 2005, NIST conducted two different experiments: Experiment 1 (right iris and right iris comparison): 12,214 genuine attempts and 1,002,386 imposter attempts. Experiment 2 (left iris and left iris comparison): 14,653 genuine attempts and 1,151,975 imposter attempts. In these experiments, the genuine and imposter matching scores should be evaluated for all possible combinations [13]. In our experiments, we strictly follow the instructions of the fully automatic test on ICE 2005 dataset [13]. In our experiment, ROC curve is used to evaluate the matching performance. Fig. 8 shows the ROC curves of the proposed approach on the ICE 2005 dataset. The performance of the participants is evaluated by verification rate at FAR=0.1%. For the experiment 1, we observe from the Fig 8 that the proposed scheme achieves the GAR of 99.60% whereas the ROC curve for the experiment 2 demonstrates the GAR of 98.80%, and this is encouraging. From the existing iris literatures, it is observed that the segmentation error is very common to any dataset, and it degrades the overall recognition performance substantially. However, the segmentation approach described in section 2 works well for most of the cases even with the iris images of deviated gazes and weak iris boundaries. The DLS based elliptical fitting provides an initial estimate of the inner and outer boundaries, and the variational level set approach localizes the iris and pupil regions accurately based on that initial estimation.

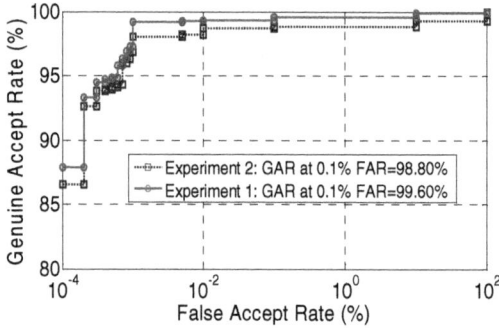

Fig. 8. ROC curves on ICE 2005 dataset. The experimentation was conducted according to instructions provided by NIST [13].

4 Conclusions

The accurate segmentation of the iris plays an important role in iris recognition. In this research effort, we have achieved three performance goals. First, the accurate segmentation of the iris/pupil regions from the degraded eye images that are affected by severe gaze deviation, diffusion, non linear deformation, low intensity, poor acquisition process, eyelid and eyelash occlusions and small opening of eyes. Second, the proposed localization scheme based on region-oriented active contour model addresses the issue of processing the iris images where the inner and outer boundaries are not exactly circular, elliptical and concentric. Third, the intensity inhomogeneity often occurs in iris images and may cause considerable difficulties in iris/pupil segmentation. Our proposed algorithm provides a better performance than the existing unideal iris recognition algorithms when the iris images suffer from intensity inhomogeneity. The proposed localization scheme is validated on the ICE 2005, the WVU unideal, the UBIRIS version 2, and the nonhomogeneous combined datasets with an encouraging performance.

Acknowledgements

We have used the iris dataset "Iris Challenge Evaluation" (ICE) [13] owned by the University of Notre Dame, USA. We have also used the WVU [14] and the UBIRIS Version 2 [15] datasets owned by the West Virginia University, USA and the department of computer science, University of Beira Interior, Portugal, respectively. This work is funded by NSERC, Canada and Concordia Institute for Information Systems Engineering (CIISE), Concordia University, Canada.

References

1. Daugman, J.: How iris recognition works. IEEE Transaction on Circuits, Systems and Video Technology 14(1), 1–17 (2003)
2. Daugman, J.: New methods in iris recognition. IEEE Transactions on Systems, Man, and Cybernetics-Part B 37(5), 1167–1175 (2007)
3. Schuckers, S.A.C., Schmid, N.A., Abhyankar, A., Dorairaj, V., Boyce, C.K., Hornak, L.A.: On techniques for angle compensation in nonideal iris recognition. IEEE Transactions on Systems, Man, and Cybernetics-Part B 37(5), 1176–1190 (2007)
4. Ma, L., Tan, T., Wang, Y., Zhang, D.: Personal identification based on iris texture analysis. IEEE Trans. Pattern Anal. Machine Intell. 25(12), 1519–1533 (2003)
5. Ma, L., Tan, T., Wang, Y., Zhang, D.: Efficient iris recognition by characterizing key local variations. IEEE Trans. Image Process. 13(6), 739–750 (2004)
6. Vatsa, M., Singh, R., Noore, A.: Improving iris recognition performance using segmentation, quality enhancement, match score fusion, and indexing. IEEE Transactions on Systems, Man, and Cybernetics-Part B 38(4), 1021–1035 (2008)
7. Shah, S., Ross, A.: Iris Segmentation Using Geodesic Active Contours. IEEE Trans. Info. Forensic and Security 4(14), 824–836 (2009)
8. Roy, K., Bhattacharya, P.: Nonideal iris recognition using level set approach and coalitional game theory. In: Fritz, M., Schiele, B., Piater, J.H. (eds.) ICVS 2009. LNCS, vol. 5815, pp. 394–402. Springer, Heidelberg (2009)
9. Roy, K., Bhattacharya, P.: Variational level set method and game theory applied for nonideal iris recognition. In: Proc. Int. Conf. on Image Process (ICIP), pp. 2721–2724 (2009)
10. Li, C., Kao, C.Y., Gore, J.C., Ding, Z.: Implicit active contours driven by local binary fitting energy. In: Proc. Int. Conf. on Comp. Vis. and Pattern Recog., pp. 1–7 (2007)
11. Chan, T., Vese, L.: Active contours without edges. IEEE Transaction on Image Process. 10(2), 266–277 (2001)
12. Roy, K., Bhattacharya, P.: Adaptive Asymmetrical SVM and Genetic Algorithms Based Iris Recognition. In: Proc. Int. Conf. on Pattrn. Recog (ICPR), pp. 1–4 (2008)
13. Iris Challenge Evaluation (ICE) dataset found, http://iris.nist.gov/ICE/
14. Iris Dataset obtained from West Virginia University (WVU), http://www.wvu.edu/
15. UBIRIS dataset obtained from department of computer science, University of Beira Interior, Portugal, http://iris.di.ubi.pt/

Secure Iris Recognition Based on Local Intensity Variations*

Christian Rathgeb and Andreas Uhl

University of Salzburg, Department of Computer Sciences, A-5020 Salzburg, Austria
{crathgeb,uhl}@cosy.sbg.ac.at

Abstract. In this paper we propose a fast and efficient iris recognition algorithm which makes use of local intensity variations in iris textures. The presented system provides fully revocable biometric templates suppressing any loss of recognition performance.

1 Introduction

Over the past years plenty of biometric traits have been established to be suitable for personal identification [1,2], iris being one of the most reliable [3]. Several iris recognition algorithms have been proposed throughout literature, reporting impressive recognition rates of over 99% and EERs below 1% on diverse datasets. However, iris recognition algorithms are still left to be improved with respect to computational performance as well as template protection, which has recently become an important issue [4,5]. Elapsed time during matching becomes relevant if huge databases are introduced whereas template protection guards users from identity theft. Fig. 1 shows a diagram of a generic iris recognition system.

The contribution of this work is the proposal of a new, computationally fast iris recognition algorithm providing practical recognition rates. By examining local intensity variations in preprocessed iris textures, features are extracted. We demonstrate the efficiency of our algorithm through recognition rates as well as comparing time measurements to a well-established algorithm. Furthermore, fully revocable templates are generated, meeting demands of high security applications. Revocable templates are created without the loss of recognition performance, while in many schemes, degradation of accuracy is observed [6].

This paper is organized as follows: first related work regarding iris recognition is summarized (Sect. 2). Subsequently the proposed system is described in detail (Sect. 3) and experimental results are given (Sect. 4). The security of our algorithm is discussed and a technique for providing secure revocable templates is proposed (Sect. 5). Finally, a conclusion is given (Sect. 6).

2 Related Work

Pioneer work in iris recognition was proposed by Daugman [7,8]. Daugman's algorithm forms the basis of today's commercially used iris recognition systems.

---

* This work has been supported by the Austrian Science Fund, project no. L554-N15.

A. Campilho and M. Kamel (Eds.): ICIAR 2010, Part II, LNCS 6112, pp. 266–275, 2010.

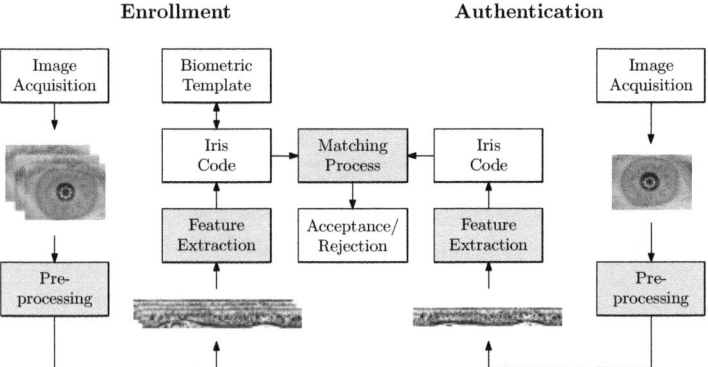

Fig. 1. Iris Recognition System: the common operation mode of enrollment and authentication in an iris recognition system

Within Daugman's approach each point of a preprocessed iris texture is treated as center of a 2D Gabor wavelet. For each of these wavelets the coefficients are generated out of which two bits are extracted resulting in an iris-code of a total number of 2048 bits. The matching process is performed using the Hamming distance as metric, comparing the number of mismatching bits against a threshold reaching an almost perfect recognition rate. A different approach to that presented by Daugman was proposed by Wildes [9]. Here an isotropic bandpass decomposition derived from application of Laplacian of Gaussian filters is applied to the preprocessed image data at multiple scales. That is, filtered images are realized as a Laplacian pyramid to generate the biometric template. In the matching process normalized correlation between acquired samples and stored template is calculated. Since these two first algorithms several approaches have been proposed suggesting several different filters to be used in the feature extraction step. Ma *et al.* [10] as well as Masek [11] examine 1D intensity signals applying a dyadic wavelet transform and a Log-Gabor filter, respectively. Chenhong and Zhaoyang [12] and Chou *et al.* [13] convolve iris images with a Laplacian-of-Gaussian filter. Ko *et al.* [14] apply cumulative sum based change analysis where iris textures are divided into cells out of which mean gray scale values are calculated and furthermore, upward and downward slopes of grayscale values are detected.

Approaches to template protection regarding iris biometrics have been proposed in so-called *Biometric Cryptosystems* [5]. Davida *et al.* [15] were the first to create a so-called "private template scheme" in which a hashed value of preprocessed iris codes and user specific attributes serves as a cryptographic key. By introducing error correcting check bits the scheme is capable of regenerating the hash at the time of authentication. Jules and Wattenberg [16] introduced a novel cryptographic primitive termed "fuzzy commitment scheme" which they suggest to be used in biometric cryptosystems. The key idea is to bind a cryptographic key prepared with error correcting codes with biometric data in a secure template. Additionally, a hash of the key is stored together with the template.

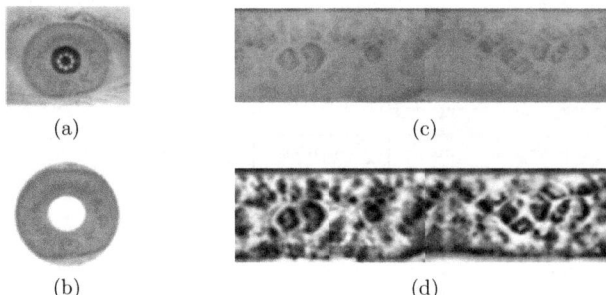

(a)   (c)

(b)   (d)

Fig. 2. Preprocessing: (a) an image of a person's eye is acquired (b) the iris is located and extracted (c) the iris ring is unwrapped to create a normalized iris texture (d) Gaussian blur and CLAHE contrast enhancement technique are applied to obtain a well distributed image

During authentication biometric data which is "close enough" (to some specified metric) to that captured during enrollment is able to reconstruct the key with the use or error correction decoding. The resulting key is then hashed and tested against the previously stored hash. Several systems applying the above concepts have been proposed. Although the main target of these schemes is biometric key management these techniques provide template protection as well [4]. Focusing on reported performance, in general security is increased at the cost of recognition rates. Additionally, iris-based cryptosystems are mostly based on existing iris recognition algorithms performing non-trivial feature extraction. Thus, performance with respect to runtime remains an issue.

In summary, most iris recognition systems exhibit high recognition rates, while these lag the requirement of providing secure biometric templates. Additionally, some algorithms are rather slow due to complex feature extraction techniques. Template protection schemes, such as biometric cryptosystems, provide secure templates, yet, security is mostly achieved at the cost of recognition performance.

3   System Architecture

3.1   Preprocessing

Preprocessing corresponds to the approach presented by Daugman [8]. Having detected the pupil of an eye, the inner and outer boundary of the iris are approximated. Subsequently, pixels of the resulting iris ring are mapped from polar coordinates to cartesian coordinates to generate a normalized rectangular iris texture. Due to the fact that the top and bottom of the iris are often hidden by eyelashes or eyelids, these parts of the iris are discarded (315° to 45° and 135° to 225°). To obtain a smooth image a Gaussian blur is applied to the resulting iris texture. To enhance the contrast we use an advanced contrast enhancement technique called CLAHE (Contrast Limited Adaptive Histogram Equalization) [17]. Compared to other contrast enhancement algorithms, for example histogram equalization, this algorithm operates on local image regions.

For this purpose the image is subdivided into image tiles (so-called contextual regions) and the contrast is enhanced within each of these regions. To avoid artifacts between two adjacent tiles an interpolation algorithm is employed. In Fig. 2 the entire preprocessing procedure is illustrated.

3.2   Feature Extraction

The applied feature extraction technique represents the fundamental part of our system. By tracing intensity variations in horizontal stripes of distinct height of preprocessed iris textures, so-called "pixel-paths" are extracted. We found that these paths are suitable to identify users.

First of all, the preprocessed iris texture of a person i, I_i (in form of a rectangle), is divided into n different horizontal texture stripes

$$I_i \rightarrow \{I_{i1}, I_{i2}, ..., I_{in}\} \tag{1}$$

of height h pixels (needless to say n depends on the size of h). Each texture strip is of dimension $l \times h$, where l denotes the length of preprocessed iris textures.

In the next step two pixel-paths, representing light and dark intensity variations are created for each texture strip I_{ij}. We define these paths as,

$$P_{Lij} := \{p_{Lij0}, p_{Lij1}, ..., p_{Lijl}\} \tag{2}$$
$$P_{Dij} := \{p_{Dij0}, p_{Dij1}, ..., p_{Dijl}\} \tag{3}$$

To calculate the value of elements p_{Lijk} and p_{Dijk} of each of these paths the first element of each path is defined as

$$p_{Lij0} \leftarrow h/2, \qquad p_{Dij0} \leftarrow h/2 \tag{4}$$

In other words, each path starts at the leftmost center of the according strip. Elements p_{Lij1} and p_{Dij1} are then calculated by examining the three directly neighboring pixel values of p_{Lij0} and p_{Dij0} ($p_{Lij0} = p_{Dij0}$) in next pixel column. Then p_{Lij1} is set to the y-value of the maximum and p_{Dij1} is set to the y-value of the minimum of these three values (maxima and minima correspond to brightest and darkest grayscale values of pixels). Thus, we define the values of p_{Lijk} and p_{Dijk} recursively such that,

$$L := MAX \begin{pmatrix} I_{ij}[k+1, p_{Lijk} - 1], \\ I_{ij}[k+1, p_{Lijk}], \\ I_{ij}[k+1, p_{Lijk} + 1] \end{pmatrix} \tag{5}$$

$$D := MIN \begin{pmatrix} I_{ij}[k+1, p_{Dijk} - 1], \\ I_{ij}[k+1, p_{Dijk}], \\ I_{ij}[k+1, p_{Dijk} + 1] \end{pmatrix} \tag{6}$$

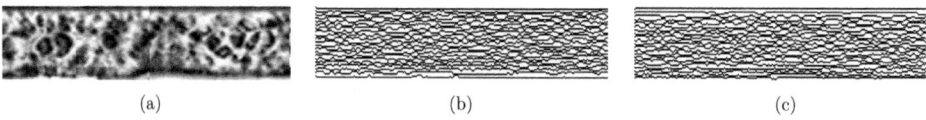

(a)   (b)   (c)

Fig. 3. Feature Extraction: (a) the preprocessed iris texture (b)-(c) two pixel-paths are extracted for each texture strip respresenting light and dark intensity variations for a height of $h = 3$ (notice that the paths of light and dark intensity are not necessarily complementary)

where $I_{ij}[x, y]$ represents the pixel value of the jth texture strip at coordinates $[x, y]$ and the MAX and MIN functions derive the according points. The values of p_{Lijk+1} and p_{Dijk+1} are then set to the y-values of L and D:

$$p_{Lijk+1} \leftarrow L_y \tag{7}$$

$$p_{Dijk+1} \leftarrow D_y \tag{8}$$

This means, $p_{Lijk} \in \{0, 1, ..., h-1\}$ and $p_{Dijk} \in \{0, 1, ..., h-1\}$. In the case p_L or p_D reach the top or bottom of the texture strip, only the according two directly neighboring pixel values are taken into account. An example for constructing light and dark intensity paths is shown in Fig. 3.

To complete the feature extraction extracted paths are further smoothed, which means small peaks are discarded. For this purpose a threshold t is defined and variations of y-position of pixel-paths occurring within a range of t pixels are discarded in order to smooth the whole path. An example of smoothing pixel-paths is illustrated in Fig. 4. The top and the bottom strip are discared since we found that those stripes normally do not carry useful information.

As a result of the described feature extraction procedure extracted paths are stored for the ith user. The size of the generated template depends on the size of the preprocessed iris texture as well as parameter h. For a number of n stripes $2 \times n \times l$ elements out of the set $\{0, 1, ..., h - 1\}$ form the biometric template. At the time of enrollment, where a user i registers with the system, feature extraction is performed for a single iris image and a biometric template T_i is stored. In Sect. 5 we will discuss how to secure these templates.

The feature extraction is based on simple comparisons, thus, no complex calculations are required. With respect to systems where computational simplicity and runtime of feature extraction are issues (for example, smart-card based verification systems) the proposed feature extraction method provides fast computation based on simple comparisons.

3.3   Template Matching

For the ith user, the feature extraction generates a template, denoted by T_i. This template consists of $2 \times n \times l$ integer values which correspond to y-positions of elements of extracted pixel-paths for light and dark intensity variations. To calculate the similarity between two templates T_j and T_k the square of differences of all elements of T_j and T_k are summed up in a matching value M such that

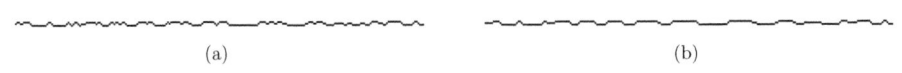

(a)   (b)

Fig. 4. Smoothing Pixel-Paths: (a) the original pixel-path resulting out of a texture stripe (b) the smoothed pixel-path where t is set to 4. That is, all peaks in the y-direction lying within a range smaller than 4 are discarded.

$$M_{jk} := \sum_{m=0}^{N} (abs(T_{jm} - T_{km}))^2 \qquad (9)$$

where N is set to $2 \times n \times l$. By definition small differences increase the matching value slightly, large differences increase the matching value significantly. A small matching value indicates high similarity between templates and vice versa. Depending on the chosen size of h the highest possible match value varies. An appropriate threshold has to be set up according to intra-class and inter-class distributions of genuine and non-genuine users.

In comparison to existing approaches aiming at extracting distinct binary iris-codes which are matched by comparing the Hamming distances of two iris-codes against a predefined threshold (for example, [8,9,11,10]), the proposed matching process lags performance and, thus, is expected to be inappropriate for template matching on large-scale databases. To overcome this restriction we introduce a more efficient way of matching templates for an appropriate height h in Sect. 4.2.

4   Experimental Results

Experiments are carried out using the CASIAv3-Interval [18] iris database, which comprises iris images over two-hundred different persons, where on average about 6 iris images are available per person. As a result of the preprocessing procedure iris textures of 256×64 pixels are extracted ($\Rightarrow l = 256$). A total number of $2 \times n \times 256$ integers are extracted and stored as biometric template. For each person a single iris image is processed in the enrollment step. Additionally, a circular shift of ten pixels to the left and to the right is implemented in order to provide rotation invariance for small head tilts.

4.1   Recognition Performance

In our experiments best results were obtained for stripes of height $h = 3$ pixels. For a height of $h = 3$ we get $64/3 = 21$ stripes where the top and bottom strip are discarded, resulting in a total number of 19 stripes. Each strip consists of 256 integers in the range $[0, 3)$. That is, an iris-code of $256 \times 19 \times 2 = 9728$ codewords out of the set $\{0, 1, 2\}$ is stored in the template. Experimental results for several different values of h are summarized in Table 1, where for all values of h best results were achieved with a threshold of $t = 4$.

The false rejection rate (FRR) and false acceptance rate (FAR) for a height of $h = 3$ and a threshold of $t = 4$ are plotted in Fig. 5. For zero FAR a FRR of

Table 1. Performance measurements for the proposed systems according to different values of h for a threshold of $t = 4$ and recognition rates of existing algorithms

Height (Pixels) / Algorithm	FRR(%) @ FAR = 0	EER (%)
2	3.829	2.227
3	1.978	1.016
4	3.959	2.128
5	5.817	2.992
Masek [11]	3.952	2.477
Ma *et al.* [10]	1.817	1.073
Ko *et al.* [14]	20.531	4.738

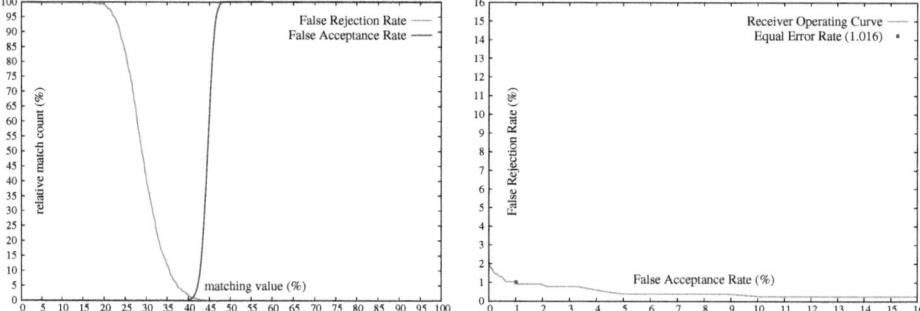

Fig. 5. The false rejection rate and the false acceptance rate of the proposed algorithm for a height of $h = 3$ and a threshold of $t = 4$

Fig. 6. The receiver operating curve and the equal error rate of the proposed algorithm for a height of $h = 3$ and a threshold of $t = 4$

1.978% is obtained. The according receiver operating curve is plotted in Fig. 6 resulting in an EER of 1.016%. Compared to our own implementations of existing iris recognition algorithms (see Table 1), these are satisfying results with respect to the simplicity of the proposed feature extraction method.

4.2   Computational Performance

The above described system was implemented in C and tested on a 1.3 GHz Linux system. As mentioned earlier the feature extraction method is based on simple comparisons between grayscale values. In detail, the maximum and minimum of three numbers are calculated using three comparisons. That is, for a height of $h = 3$ a total number of $256 \times 19 \times 3 = 14592$ comparisons are necessary. Measuring the runtime of the feature extraction method for a single preprocessed iris texture an average processing time of 0.0344 seconds was obtained. To emphasize the performance of the feature extraction method we compare the computational performance to our C implementation of the algorithm of Ma *et al.* [10]. In the algorithm of Ma, a 1-D wavelet transform is applied to ten 1-D intensity signals of average grayscale values of pixel blocks in the preprocessed iris texture. Detected minima and maxima serve as features where sequences of 1 and 0 are assigned to the iris-code until new maxima or minima are found. This whole process is applied to two subbands extracting a total number of 10240 bits where the Hamming distance is applied as similarity metric. As experimental results of our implementation of this algorithm we achieved a FRR of 3.821% for zero FAR and a EER of 1.401% for the whole CASIAv3-Interval [18] iris database (circular shifts are implemented as well; we do not consider any bit-masking information). We measured the runtime of the feature extraction of this algorithm on the same system resulting in an average processing time of 0.1345 seconds for a single feature extraction. On average, the proposed feature extraction method is three times faster than that of Ma.

While most iris recognition systems compare iris-codes by calculating the Hamming distance between these, the matching procedure of our algorithm involves the calculation of a matching value which is expected to be much slower. Since we obtained best results for a height of $h = 3$ we are able to gain performance. To retrieve binary iris-codes we encode elements of calculated pixel paths with a Gray code:

$$0 \leftarrow 00, \qquad 1 \leftarrow 01, \qquad 2 \leftarrow 11 \qquad (10)$$

By applying this encoding, calculating the Hamming distance between two iris-code generates the same results as the previously described matching. For a height of $h = 3$ the matching process is now computationally efficient as well. Compared to the algorithm of Ma which extracts 10240 bits we extract a total number of $2 \times 9728 = 19456$ bits. However, calculating the Hamming distance for larger bitstreams does not drastically decrease performance. For the algorithm of Ma we measured an avergage time of 0.0137 seconds and for the proposed we now achieve a average time of 0.0193 seconds for the matching of two templates. The time for calculating the Hamming distance between two bitstreams of twice the length of those generated by the algorithm of Ma takes only slightly longer, due to system overhead.

5   Cancellable Templates

Recently template security has become an important issue [4]. If biometric templates are stolen or compromised these can not be modified ex post and, thus, become useless. Ratha et al. [6] introduced the concept of cancellable biometrics. The idea of cancellable biometrics consists of intentional, repeatable distortion on a biometric signal based on a chosen transform where the matching process is performed in transformed space. Recovering of original biometric template data becomes infeasible. If the transformed biometric data is compromised the transform function is changed, that is, the biometric template is updated.

In order to provide cancellable biometric templates we suggest a permutation of extracted paths following the idea of line permutation as proposed by [19]. For example, for a height of $h = 3$ we calculate a total number of 38 paths.

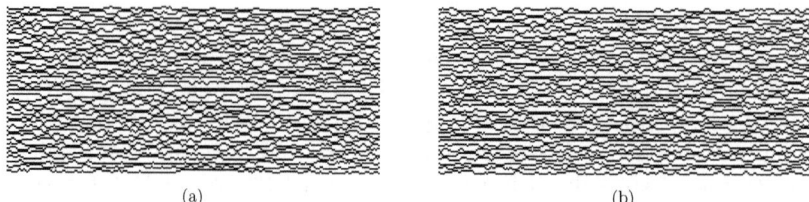

(a)   (b)

Fig. 7. Secure Template Creation: (a) paths calculated during feature extraction (b) secure template defined by a distinct permutation

Paths are permutated according to some user or application specific permutation, where a total number of $38! \simeq 5.23 \cdot 10^{44}$ different permutations are possible. By permutating paths, reconstructing original iris images becomes highly complicated. Guessing a specific permutation is assumed to be computationally infeasible ($38! \gg 2^{128}$). Thus, high security regarding template protection is provided. In Fig. 7 a sample invertible permutation is illustrated. If non-invertible permutations are applied in our system, performance decrease is expected as pointed out in [19]. In case a specific permutation is compromised, an imposter may reconstruct the original order of paths. However, reconstructing the original iris texture from iris codes is not possible, since the feature extraction is non-invertible by definition.

By introducing a two-factor authentication scheme permutations are integrated in the system, where secret permutations represent the second factor. For example, user-specific permutations could be stored on smart-cards, so that permutations are applied after feature extraction and permutated templates are matched against stored templates, previously permuted during enrollment. In comparison to template encryption our system is capable of performing the matching procedure in the encrypted (permuted) domain. Furthermore, compared to approaches to cancellable iris biometrics which operate in the image domain [20], the proposed system does not suffer from performance degradation if invertible permutations are applied. This is one important aspect of the presented approach since security applications must not require a decryption of encrypted templates prior to matching [4].

6   Conclusion

In this work we presented a new, computationally efficient iris recognition algorithm. Besides providing practical recognition rates we demonstrate that the proposed algorithm is suitable for generating secure and fully revocable biometric templates.

Acknowledgements

We would like to thank Michael Liedlgruber for providing us with an implementation of the CLAHE contrast enhancement technique.

References

1. Jain, A.K., Ross, A., Prabhakar, S.: An introduction to biometric recognition. IEEE Trans. on Circuits and Systems for Video Technology 14, 4–20 (2004)
2. Jain, A.K., Flynn, P.J., Ross, A.A.: Handbook of Biometrics. Springer, Heidelberg (2008)
3. Bowyer, K., Hollingsworth, K., Flynn, P.: Image understanding for iris biometrics: a survey. Computer Vision and Image Understanding 110, 281–307 (2008)
4. Jain, A.K., Nandakumar, K., Nagar, A.: Biometric template security. EURASIP J. Adv. Signal Process., 1–17 (2008)
5. Uludag, U., Pankanti, S., Prabhakar, S., Jain, A.K.: Biometric cryptosystems: issues and challenges. Proceedings of the IEEE 92(6), 948–960 (2004)
6. Ratha, N.K., Connell, J.H., Bolle, R.M.: Enhancing security and privacy in biometrics-based authentication systems. IBM Systems Journal 40, 614–634 (2001)
7. Daugman, J.: High confidence visual recognition of persons by a test of statistical independence. IEEE Transactions on Pattern Analysis and Machine Intelligence 15(11), 1148–1161 (1993)
8. Daugman, J.: How Iris Recognition Works. IEEE Trans. CSVT 14(1), 21–30 (2004)
9. Wildes, R.P.: Iris recognition: an emerging biometric technology. Proceedings of the IEEE 85, 1348–1363 (1997)
10. Ma, L., Tan, T., Wang, Y., Zhang, D.: Efficient Iris Recognition by Characterizing Key Local Variations. IEEE Transactions on Image Processing 13(6), 739–750 (2004)
11. Masek, L.: Recognition of Human Iris Patterns for Biometric Identification, Master's thesis, University of Western Australia (2003)
12. Chenhong, L., Zhaoyang, L.: Effcient iris recognition by computing discriminable textons. In: Interantional Conference on Neural Networks and Brain, vol. 2, pp. 1164–1167 (2005)
13. Chou, C.T., Shih, S.W., Chen, W.S., Cheng, V.W.: Iris recognition with multiscale edge-type matching. In: Interantional Conference on Pattern Recognition, pp. 545–548 (2006)
14. Ko, J.G., Gil, Y.H., Yoo, J.H.: Iris Recognition using Cumulative SUM based Change Analysis. In: Intelligent Signal Processing and Communications, IS-PACS'06, pp. 275–278 (2006)
15. Davida, G., Frankel, Y., Matt, B.: On enabling secure applications through off-line biometric identification. In: Proc. of IEEE, Symp. on Security and Privacy, pp. 148–157 (1998)
16. Juels, A., Wattenberg, M.: A fuzzy commitment scheme. In: Sixth ACM Conference on Computer and Communications Security, pp. 28–36 (1999)
17. Zuiverveld, K.: Contrast Limited Adaptive Histogram Equalization. In: Graphics Gems IV, pp. 474–485. Morgan Kaufmann, San Francisco (1994)
18. The Center of Biometrics and Security Research: CASIA Iris Image Database, http://www.sinobiometrics.com
19. Zuo, J., Ratha, N.K., Connel, J.H.: Cancelable iris biometric. In: Proceedings of the 19th International Conference on Pattern Recognition (ICPR'08), pp. 1–4 (2008)
20. Hämmerle-Uhl, J., Pschernig, E., Uhl, A.: Cancelable iris biometrics using block remapping and image warping. In: Samarati, P., Yung, M., Martinelli, F., Ardagna, C.A. (eds.) ISC 2009. LNCS, vol. 5735, pp. 135–142. Springer, Heidelberg (2009)

Transforming Rectangular and Polar Iris Images to Enable Cancelable Biometrics*

Peter Färberböck, Jutta Hämmerle-Uhl, Dominik Kaaser,
Elias Pschernig, and Andreas Uhl**

Department of Computer Sciences, Salzburg University, Austria
uhl@cosy.sbg.ac.at

Abstract. To enable *cancelable biometrics*, we apply two classical transformations, *block re-mapping* and *texture warping*, in two variants to iris image data: first, the transformations are applied to rectangular iris imagery prior to iris detection and iris texture unwrapping, and second, the transformations are applied to polar iris images after the generation of the corresponding iris texture patch. The CASIA V3 Iris database is used and experimental results on the matching performance and key sensitivity of a popular iris recognition method employing the cancelable transforms are given.

1 Introduction

The use of biometrics comes with different problems as compared to conventional authentication methods, like passwords or ID cards. As biometric features are specific to an individual person, they cannot be changed (or not often, as one person for example has only ten fingerprints and two iris patterns available). So where a password can simply be changed or an e-card invalidated, this is not possible with biometrics. In the same way, it is not possible to use different keys for different applications - for example if one wants to use a different key for the bank account and for access to the workplace computer.

A possible approach to cope with this problem are cancelable biometrics [1], which apply a key-dependent transformation to the captured biometric signals in order to achieve revocability of biometric templates. The transformation must be non-invertible so that the original data cannot be reconstructed from the stored transformed version even in case the key used for the transformation is compromised. At the same time matching still has to be possible with the transformed version of the biometric data.

The classical approach for cancelable biometrics [1] proposes to apply transformations in the image domain of sample data prior to feature extraction. This has the advantage that existing recognition algorithms can be used unmodified

---

* This work has been partially supported by the Austrian Science Fund, project no. L554-N15.
** Corresponding author.

A. Campilho and M. Kamel (Eds.): ICIAR 2010, Part II, LNCS 6112, pp. 276–286, 2010.

for the later feature extraction and matching stages. However, the obvious problem of possible difficulties in extracting features from transformed data needs to be considered. Alternative solutions are to apply transformations to biometric template data or to adopt a key-dependent feature extraction process including a non-invertible stage.

In iris recognition, several ideas for cancelable systems have been discussed already [2,3,4]. In this work we focus on applying transformations to iris enrolment and sample image data before the feature extraction and template generation stages. Two types of these data exist. "Rectangular iris images" are pictures of the entire eye with the surrounding region including eye lids, while "polar iris images" consist of iris texture data only and are the result of iris detection and iris texture unwrapping to polar coordinates being applied to rectangular iris images.

When applying a transform to rectangular iris data we have the advantage that after having acquired the image, the transformation can be conducted immediately, eventually even integrated into the acquisition process. This is especially beneficial when viewing the biometric system as a "black box" we cannot trust – transformation is separated here from any further processing and specifically from the recognition process itself. A possible disadvantage are potential difficulties when detecting and unwrapping the iris texture from transformed image data.

When applying the transformations to polar iris image data (as being suggested in [5,4]), only feature extraction itself can be influenced by the transformed data, however, the downside is that the unprotected sample data is subject to iris detection and texture unwrapping which can be seen as a part of the matching process (in this case, this process deals with data subject to privacy constraints). While both approaches have their respective advantages and disadvantages, we will investigate how their actual application interferes with recognition accuracy and impacts on system security.

In section 2, we will investigate how the two classical transforms originally proposed for cancelable biometrics (block re-mapping / permutations and grid morphing / image warping [1]) can be applied to the two types of iris image data. The iris recognition system we use is described in section 3 while section 4 presents our experimental results.

2 Cancelable Iris Recognition

2.1 Transforming Rectangular Iris Images

Running a block permutation or grid morphing on the rectangular iris image would render the image useless for any further processing, especially iris texture segmentation would of course fail. Iris detection and texture unwrapping as implemented for almost all iris recognition techniques can only be successful if the circular nature of the iris texture boundaries (inner pupil and outer sclera boundary) is preserved. In order to accomplish this, we transform the image to polar coordinates using the center of the pupil as origin (see Fig. 1.a).

(a) Polar coordinates   (b) Polar permutation and result in image space

Fig. 1. Block-permutation in polar coordinates

In this polar space, permutations can be applied as follows. As depicted in Fig. 1.a, the image data is cut into vertical stripes which can be subsequently permuted. To reduce the impact of high frequency block boundaries, we use a blurred edge overlay, which basically softens the edges before it overlays a block onto another block. By using the same block multiple times and carefully dropping other blocks out, it is possible to achieve a non-invertible block re-mapping (see Fig. 4.a). An example of a block permutation together with the corresponding reconstruction in the image domain can be seen in Fig. 1.b.

In addition to block re-mapping, a composition of arbitrary functions can be used as a morphing function, still the pupil borders need to remain intact. Therefore, the morphing needs to either leave these borders unchanged or apply translations only in direction of the axis which represents the angle in polar space. For simplicity and mere testing if morphing would gain any further advantages we apply a sine function with amplitude and frequency as parameters. An example can be seen in Fig. 2.a – in the left image, the distortion is applied along the x-axis, which preserves the pupil borders. The right image shows a possible distortion parallel to the y-axis, which cannot be used, as it distorts the pupil border as can be seen in Fig. 2.b.

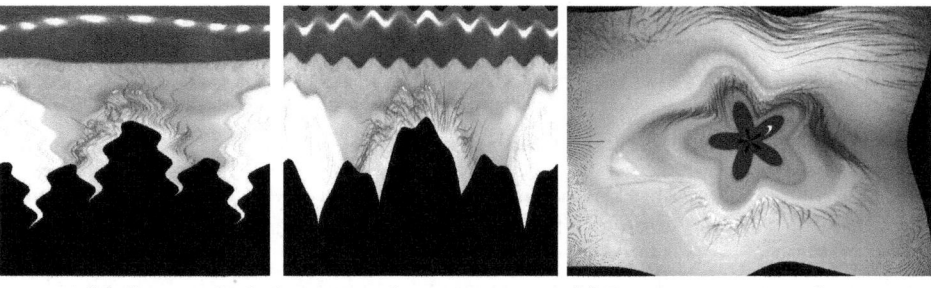

(a) Sine applied along x- and y-axis   (b) Image space: sine along y-axis

Fig. 2. Sinoid distortions in polar coordinates

A combination of sinoid distortion and block permutation in polar coordinates can be seen in Fig. 3.a which demonstrates that such a combination is feasible in principle. Contrasting to this, Fig. 3.b reveals that in some cases even the pupil border is distorted (caused by inaccuracies in finding the pupil center and by the fact that those boundaries often do not exactly correspond to circles) and that eye lashes seem to get scattered over the entire image with obvious impact on iris texture segmentation.

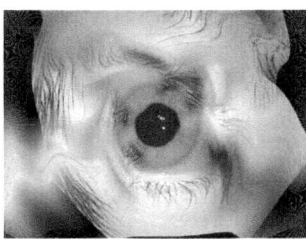
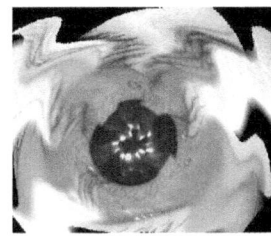
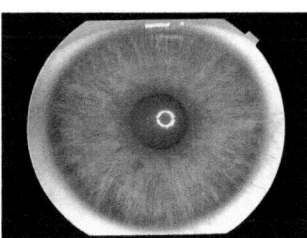

(a) Permutation plus sinoid distortion

(b)   Corrupted   pupil border

(c) Optimal image

Fig. 3. Examples in image domain

Fig. 3.c shows an example image (taken from a specific iris image database[1] also used in rating iris recognition schemes [6]) which better fits our approach as there are no occlusions and as almost the entire image is covered by iris texture. However, typical iris image data do not look like this.

2.2   Transforming Polar Iris Images

For this approach, first iris detection and texture unwrapping is conducted. Subsequently, transformations are applied to the polar iris image data. While line-based transformations have been also considered [4], we employ block-based transformations as being applied in earlier work in context with a different iris recognition scheme [5].

Block Re-mapping. For block re-mapping, each block of the target texture is mapped to a block from the source texture. An example of such a mapping scheme is shown in Fig. 4.a.

As stated in [1], using a re-mapping of blocks instead of a permutation should be preferred for the application of cancelable biometrics, as it is not reversible. Source blocks which are not part of the mapping are not contained in the transformed texture at all, and therefore it is impossible to reconstruct the complete original for an attacker, even in case the key defining the mapping gets compromised. For a discussion on key-space size of this approach please refer to [5].

---

[1] www.inf.upol.cz/iris/

Image Warping. Another transformation we applied to textures is a distortion called mesh warping [7]. In this approach the texture is re-mapped according to a distorted grid mesh laid over it. A key is used to specify one particular distortion, by offsetting each vertex in the original mesh by some amount. This is done by starting with a regular grid placed over the texture, in which the vertices are then randomly displaced using the key as seed to a pseudo random number generator.

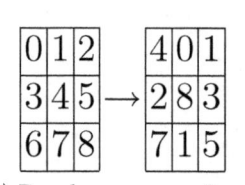

(a) Random re-mapping of blocks

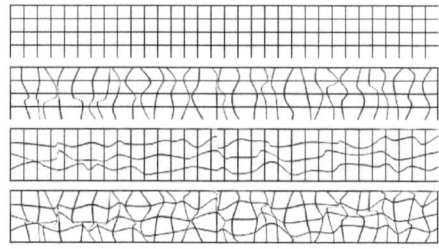

(b) Warping a regular grid

Fig. 4. Re-mapping and Image Warping

The transformation distorts the texture by sampling each pixel in the target texture from the corresponding area in the source texture, so that each vertex of the source mesh is placed to its translated position in the target mesh, interpolating pixels inside grid cells accordingly. In the version we used, this works in two passes, distorting rows along the offset of vertical splines through the mesh vertices, and then columns along offsets of horizontal splines. In the case of miniaturisation, a box-filter is applied to rows and columns, and linear interpolation is used in case of magnification. An illustration for the two passes is shown in Fig. 4.b. Due to the interpolation strategies applied, the transformation is non-reversible as the original data may not be exactly recovered even if the warping parameters are known. The effect of non-revertability is more pronounced of course in the case of miniaturisation. For a discussion on the size of the key-space, please refer to [5].

3   Iris Recognition

Many iris recognition methods follow a quite common scheme [8], close to the well known and commercially most successful approach by Daugman [9]. After image acquisition, in a first step the iris texture is localised and extracted. From this texture, discriminative features are derived, which then can be used for comparison.

We extract iris texture from rectangular iris images as a first step. In our approach (following e.g. Ma et al. [10]) we assume the texture to be the area between the two almost concentric circles of the pupil and the outer iris. These two circles are found by contrast adjustment, followed by Canny edge detection and Hough transformation. After the circles are detected, unwrapping along polar coordinates is done to obtain a rectangular texture of the iris. In our case, we always resample the texture to a size of 512x64 pixels.

Working now only on these texture patches, we divide the data into N stripes to obtain N one-dimensional signals, each one averaged from the pixels of M adjacent rows. We used $N = 10$ and $M = 5$ for our 512x64 pixel textures (only the 50 rows close to the pupil are used from the 64 rows, as suggested in [10]). For feature extraction, we use a custom C implementation similar to Libor Masek's Matlab implementation[2] of a 1-D version of the Daugman iris recognition algorithm. A row-wise convolution with a 1-D complex Log-Gabor filter is performed on the texture pixels. The phase angle of the resulting complex value for each pixel is discretized into 2 bits. Those 2 bit of phase information are used to generate a binary code, which therefore is 512x20 bit.

Once bitcodes are obtained, matching can be performed using Hamming distance as distance measure. If bit codes are obtained from textures of different eyes, the Hamming distance is expected to lie around 0.5. For codes computed from the same eye but different samples, the distance is smaller. For matching to work well, two more refinements are made. The first one aims to improve rotation invariance. To overcome the problem of slightly rotated iris images, we match each bit code multiple times, shifted a few bits left or right. In our results, we use 33 comparisons, from shifting -16 pixel positions to +16 pixel positions. The other improvement is to ignore parts of the binary code we can recognise as not belonging to the iris. This includes parts of the iris cut off the image, or parts hidden by the lids, which our implementation tries to detect approximately.

4   Experiments

4.1   Experimental Setup

For testing, we used the *Interval* dataset out of the *CASIA Iris V3 database*, consisting of 2653 images in 396 classes (i.e. persons). In the first test focused on matching performance (Test 1), we assigned a random key to each class, then calculated the Hamming distance of resulting bit-codes between any two images (3517878 iris comparisons, 9008 of which are intra-class comparisons). If irises from the same class match worse after transformation, or irises from different classes match better after transformation, this shows in the match results as increased false non-match rate (FNMR) and increased false match rate (FMR), respectively. We plot the resulting FNMR against FMR as receiver operating characteristics (ROC) curves, and also indicate the equal error rate (EER) where FNMR and FMR are closest to each other. This is then used as indication in how far matching performance is influenced by the transformation applied.

However, even when we can see no degradation in matching performance in Test 1, this does not mean the key-dependent transformations actually increase security. For example, applying the same key to each class could lead to good

---

[2] http://www.csse.uwa.edu.au/~pk/studentprojects/libor/sourcecode.html

results in the first test but does not influence security at all. Therefore, we performed a second test (Test 2) to evaluate how discernible one transformation is from another, when they result from different keys (i.e. key-sensitivity is investigated). For this purpose, an iris class is copied multiple times, and each such identical class is then assigned a random key as before. If the key-dependent transformations do not lead to sufficiently distinct features, in this case we will observe higher FMR because features of different classes will match. For this second test, we used the first 20 classes with at least 10 samples out of the *Interval* dataset, and created 50 random keys for each to have a roughly similar number of comparisons as compared to the first test (2495000 iris comparisons, 45000 of which are intra-class).

4.2   Experimental Results

Transforming Rectangular Iris Images. The results for this approach are based on a subset of the described data only and we restrict our attention to vertical block permutation only (no additional morphing is used). In Fig. 5 we display the Hamming distances found when conducting Test 1.

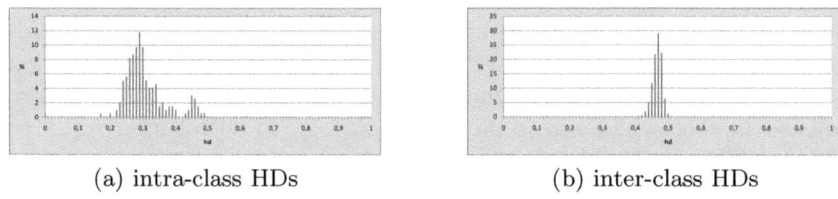

(a) intra-class HDs   (b) inter-class HDs

Fig. 5. Hamming distances when transformations are applied to rectangular iris images

While the inter-class distances are concentrated between 0.4 and 0.5 as desired (which would enable to apply a threshold at e.g. 0.4 as being suggested for the used iris recognition system), we observe a bi-modal distribution for the intra-class distances. While the first peak centred about 0.3 is the "desired" one, we face a second peak similarly distributed as the inter-class distances. This second peak results from failed iris texture segmentation attempts for data as shown in Fig. 3.b, where especially the degraded pupil boundary severely impacts iris texture extraction. With these highly overlapping distributions, sensible recognition results cannot be obtained.

Therefore, we omit further results for added image morphing since a further degradation can be observed. As a consequence, although highly desirable from a conceptual and security point of view, we have to abandon the idea of applying the discussed transforms to the rectangular iris image data directly and do not provide recognition results for the complete dataset.

Transforming Polar Iris Images. Figure 6.a shows the matching results using different block sizes for block re-mapping applied to polar iris images in Test 1. For

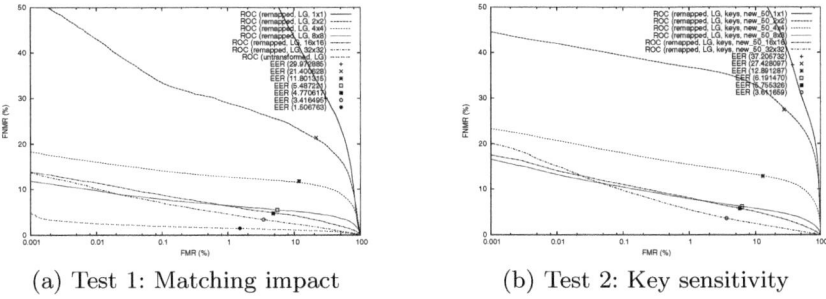

(a) Test 1: Matching impact   (b) Test 2: Key sensitivity

Fig. 6. ROC curves with block re-mapping using different block-sizes

Table 1. EERs with block re-mapping for Test 1 (recognition accuracy impact) with rectangular block sizes

size(pixel)	56x7	64x8	73x9	85x10	102x12	128x16
blocks	81	64	49	36	25	16
EER (%)	3.2	3.7	3.1	3.3	4.6	4.8

comparison, also the ROC curve for matching without any transformation applied is included (which is the bottom-most curve), with a resulting EER of 1.5%.

When permuting blocks of size 32x32 (note that only 32 such blocks fit into the used 512x64 pixel textures), 16x16 and 8x8 pixels, the EER is 3.3%, 4.8% and 5.5.%. Even smaller block sizes increase EER to over 10%. Note that this is different as compared to applying the algorithm of Ma et al. [10] for feature extraction, where 4x4 pixel blocks yielded the highest EER and smaller block exhibited decreasing error rates [5]. Instead of quadratic blocks, blocks also can be rectangles. Table 1 compares the match results when using different rectangular grid sizes for the block re-mapping, from fitting 81 blocks of 56x7 pixels to the 512x64 texture, down to fitting 16 blocks of 128x16 pixel. Re-mapping these rectangular blocks results in EERs from 3.2% to 4.8%. We see that even for the best settings, EER is clearly worse compared to the original algorithm without permutations being applied.

Table 2. EERs with block re-mapping for Test 2

size(pixel)	56x7	64x8	73x9	85x10	102x12	128x16
blocks	81	64	49	36	25	16
EER (%)	3.1	3.5	2.8	2.6	4.2	5.1

Figure 6.b shows again the result of using the same block sizes as Figure 6.a, but now using the same images but only different keys for each class (Test 2), to get an indication on the sensitivity of the keys, as described earlier. Note that now, a FMR of 1% means that 1% of comparisons of images from different classes

resulted in a wrong match. As all classes use the same images, it means despite having different keys and therefore different block re-mappings, they were still close enough to match. Block sizes 32x32, 16x16 and 8x8 have an EER of 3.6%, 5.8% and 6.2%, with smaller block sizes over 10%. The EER for big rectangular block sizes is shown in Table 2, going up to 5.1% in the case of only 16 blocks. The best error rates are obtained when using 73x9 and 85x10 pixel blocks, with an EER of 3.1%/3.3% for matching and an EER of 2.8%/2.6% when only using different keys.

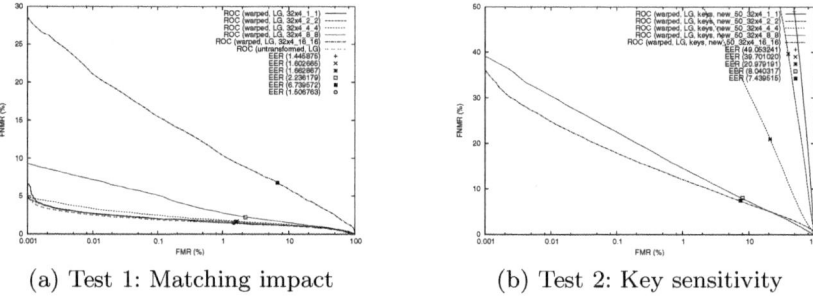

(a) Test 1: Matching impact   (b) Test 2: Key sensitivity

Fig. 7. ROC curves for warping a fixed 32x4 grid by different amounts

When using the mesh warping transformation, the size of the used mesh as well as the range of random offsets available can be adjusted. Results on Test 1 are shown in Figure 7.a, for different warp offsets and a fixed grid of 16x16 pixel blocks with 128 vertices (note the expression 32×4 in the legend of the figure designates the number of 16x16 pixel blocks that can be accommodated in our texture patch). In the transformed grid, varying ranges for the horizontal and vertical pixel offsets of each mesh vertex are used. In the case of the largest offset where warped blocks can overlap the strong distortions result in less information in the features and the resulting EER is 6.7%. Using offsets in the range of 8x8 pixel, the EER is 2.2%, and for even smaller distortions the EER gets close to the untransformed case. However, when using identical pictures in different classes (Test 2), the small transformations do not lead to sufficient difference, as can be seen in figure 7.b, which compares the same transformation parameters. Using different warp amounts for the 16x16 pixel blocks, the EER never goes below 7.4% in the case of 16x16 pixel offsets. Looking at both figures, the best case is the warp amount of using half the block dimension, in this case 8x8 pixels, as it also had a low EER in figure 7.a.

Figure 8.a shows ROC curves of Test 1 for some rectangular block sizes. For normal matching, the worst result is an EER of 2.5% for warping with a mesh of 6x6 nodes, offsetting each one by up to 42x5 pixels. Using the same parameters in figure 8.b, again with only different transformations in classes (Test 2), it can be seen that the FNMR to FMR ratio gets higher when using only few vertices - but the EER for using a mesh of 2x2 vertices still is at 2.1%. The best overall result is using a mesh with 9x9 vertices, which results in an EER of 2.0% for normal matching (Test 1) and 1.8% when comparing keys (Test 2).

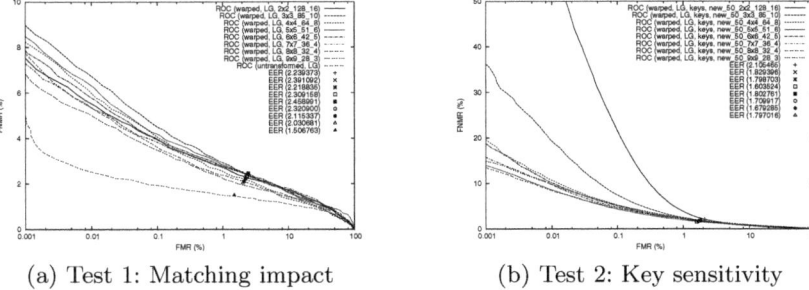

(a) Test 1: Matching impact   (b) Test 2: Key sensitivity

Fig. 8. ROC curves for warping with different grid sizes, with the maximum offset so the grid can not self-overlap

5 Conclusion

We applied the concept of cancelable biometrics to iris recognition by performing two transformations to iris textures. The two transformations, one a simple block re-mapping, the other a mesh deformation, were applied in two variants. The first variant applies the transformations to rectangular iris data as being captured by a sensor, the second variant applies the transformations to polar iris images after extracting the iris texture, but prior to feature extraction.

The strategy to apply transformations to rectangular iris image data was not successful at all due to a high rate of iris segmentation errors caused by problems for iris detection in transformed image data. For this approach, other types of transforms need to be used preserving the circular nature of the iris texture boundaries better. For the transformations being applied to polar iris images, the best parameters found for block re-mapping resulted in an EER of 3.1% instead of 1.5%. For the mesh-warping transformation, our tests resulted in 2.0% EER instead of 1.5% for the best parameters found.

References

1. Ratha, N., Connell, J., Bolle, R.: Enhancing security and privacy in biometrics-based authentication systems. IBM Systems Journal 40(3), 614–634 (2001)
2. Chong, S.C., Jin, A.T.B., Ling, D.N.C.: High security iris verification system based on random secret integration. Computer Vision and Image Understanding 102(2), 169–177 (2006)
3. Chong, S.C., Jin, A.T.B., Ling, D.N.C.: Iris authentication using privatized advanced correlation filter. In: Proceedings of the 1st International IAPR Conference on Biometrics (ICB'06). Volume 4642 of Springer Lecture Notes on Computer Science. (2006) 382–388
4. Zuo, J., Ratha, N.K., Connel, J.H.: Cancelable iris biometric. In: Proceedings of the 19th International Conference on Pattern Recognition 2008 (ICPR'08), pp. 1–4 (2008)
5. Hämmerle-Uhl, J., Pschernig, E., Uhl, A.: Cancelable iris biometrics using block re-mapping and image warping. In: Samarati, P., Yung, M., Martinelli, F., Ardagna, C.A. (eds.) ISC 2009. LNCS, vol. 5735, pp. 135–142. Springer, Heidelberg (2009)

6. Dobeš, M., Machala, L., Tichavský, P., Pospíšil, J.: Human eye iris recognition using the mutual information. Optik 115(9), 399–405 (2006)
7. Wolberg, G.: Image morphing: a survey. The Visual Computer 14(8/9), 360–372 (1998)
8. Bowyer, K., Hollingsworth, K., Flinn, P.: Image understanding for iris biometrics: A survey. Computer Vision and Image Understanding 110(2), 281–307 (2008)
9. Daugman, J.: How iris recognition works. IEEE Transactions on Circiuts and Systems for Video Technology 14(1), 21–30 (2004)
10. Ma, L., Tan, T., Wang, Y., Zhang, D.: Efficient iris recognition by characterizing key local variations. IEEE Transactions on Image Processing 13, 739–750 (2004)

Advances in EEG-Based Biometry

António Ferreira, Carlos Almeida, Pétia Georgieva, Ana Tomé, and Filipe Silva

Department of Electronics, Telecommunications and Informatics/IEETA,
University of Aveiro, Aveiro, Portugal
petia@ua.pt

Abstract. This paper is focused on proving the concept that the EEG signals collected during a perception or mental task can be used for discrimination of individuals. The viability of the EEG-based person identification was successfully tested for a data base of 13 persons. Among various classifiers tested, Support Vector Machine (SVM) with Radial Basis Function (RBF) exhibits the best performance. The problem of static classification that does not take into account the temporal nature of the EEG sequence was considered by an empirical post classifier procedure. The algorithm proposed has an effect of introducing a memory into the classifier without increasing its complexity. Control of a classified access into restricted areas security systems, health disorder identification in medicine, gaining more understanding of the cognitive human brain processes in neuroscience are some of the potential applications of EEG-based biometry.

Keywords: Classification, support-vector machine, biometry, electroencephalogram (EEG).

1 Introduction

The Electroencephalogram (EEG) is an effective non-invasive method to analyze the brain electrical activity. Recently the interest in decoding and interpreting Event-Related Potentials (ERPs) induced by mental or perception tasks is rapidly growing. ERPs are transient components in the EEG generated in response to a stimulus (e.g. presentation of images, motor imagery or mental tasks). The reported ERP decoding success brought new application scenarios as for example the Brain Machine Interface (BMI), Biometrics, Neuro-feedback (NF) treatment. In all these emerging applications, the major challenges with the study of ERPs are: i) Low signal-to-noise ratio (SNR) due to the large and salient background noise in the EEG; ii) Identification of relevant patterns in the ERPs, which is related with more efficient procedures of feature extraction and classification; iii) Identification of targeted brain functional states and the associated experimental protocols (looking for appropriate stimulus, temporal aspects as the number and timing of sessions, adequate feedback to the tested subjects).

With this paper we pretend to tackle the above defined problems in the particular framework of EEG-based biometry. We report our study on proving the concept that EEG data collected following a strictly defined protocol can be reliably used for person identification.

A. Campilho and M. Kamel (Eds.): ICIAR 2010, Part II, LNCS 6112, pp. 287–295, 2010.

There is very little research work published using brain signals as biometric tools to identify individuals, Poulos *et al.* (1999), Paranjape, *et al.* (2001); Palaniappan and Mandic, (2007). Nevertheless, in these studies it was suggested that the brain-wave pattern of every individual is unique and, therefore, the EEG can be used for building personal identification or authentication systems. The identification attempts to establish the identity of a given person out of a closed list of persons (one from many), while the authentication aims to confirm or deny the identity claimed by a person (one to one matching), Marcel and Millan, (2007). The identified person is exposed to a stimulus (usually visual or auditory) for a certain time and the EEG signals coming from a number of electrodes spatially distributed over the subject's scalp are collected and input to the biometry system.

The raw EEG signals are too noisy and variable to be analyzed directly. Therefore, the EEG signals need to go through a sequence of processing steps: i) Data acquisition, storage and format transforming; ii) Filtering (removal of interferences from other unwanted sources, as for example physiological artifacts or baseline electrical trends); iii) Feature extraction and classification; iv) Feedback generation and visualization.

The identification/authentication systems built so far differ basically in filtering and classification components, Palaniappan and Mandic, (2007); Marcel and Millan, (2007). However, our initial study (Ferreira, 2009, Almeida, 2009) has shown that the discrimination process is slightly dependent on the specific filter and classifier. Critical issues related with building an efficient EEG based biometry system are briefly discussed below.

Biometry as a modeling problem. The EEG recordings are unique for each person and the problem of EEG-based biometry can be interpreted as a modeling problem, *i.e.*, design a feature model that belongs to a certain person and design a personal classifier with a respective owner. The trained identification model has to identify the subject from a data base of personal profiles and the authentication system has to confirm or not that the subject being evaluated is who he claims to be.

Stimulus. Study on the type and the duration of the evoked potentials (visual or auditory) that would enhance the identification/authentication capacity. Preliminary tests have demonstrated that the type of the stimulus (for example mental task, motor task, image presentation or a combination of them) is crucial for reliable extraction of personal characteristics. It seems that some mental tasks are more appropriate than others. At the same time, experiments with combination of stimuli appear to be more advantageous for the personal uniqueness of the EEG patterns.

Post-processing. Ongoing research suggests that post-processing techniques on the classifier output as instant error correction and averaging would improve the identification/authentication capacity.

Real-time biometry. Optimization of the evoked potential duration (EPD) in order to implement the paradigm in an on-line scheme. Current study has shown that both two short or too long EPD worsen the biometrical system, Ferreira, (2009). The compromise can be learned by cross validation during the classifier training.

The paper is organized as follows: Section 2 presents the experimental setup for the present study. In section 3 and 4 the main modules of the EEG biometry system are discussed, namely the feature extraction, the classification and the post-processing procedure. In Section 5 the effect of the EPD is analyzed. Finally, the concluding remarks are addressed in section 6.

2 Experimental Setup

Visually Evoked Potential (VEP) signals were extracted from thirteen female subjects (20-28 years old). All participants had normal or corrected to normal vision and no history of neurological or psychiatric illness. Neutral, fearful and disgusting faces of 16 different individuals (8 males and 8 females) were selected, giving a total of 48 different facial stimuli. Images of 16 different house fronts to be superimposed on each of the faces were selected from various internet sources. This resulted in a total of 384 grey-scaled composite images (9.5 cm wide by 14 cm high) of transparently superimposed face and house with equivalent discriminability.

Participants were seated in a dimly lit room, where a computer screen was placed at a viewing distance of approximately 80 cm coupled to a PC equipped with software for the EEG recording. The images were divided into two experimental blocks. In the first, the participants were required to attend to the houses (ignoring the faces) and in the other they were required to attend to the faces (ignoring the houses). The participant's task was to determine, on each trial, if the current house or face (depending on the experimental block) is the same as the one presented on the previous trial. Stimuli were presented in sequence, for 300ms each and were preceded by a fixation cross displayed for 500 ms. The inter-trial interval was 2000 ms.

EEG signals were recorded from 20 electrodes (Fp1, Fp2, F3, F4, C3, C4, P3, P4, O1, O2; F7, F8, T3, T4; P7, P8, Fz, Cz, Pz, Oz) according to the 10/20 International system (see Fig.1). EOG (Electrooculogram - eye movemen) signals were also recorded from electrodes placed just above the left supraorbital ridge (vertical EOG) and on the left outer canthus (horizontal EOG). VEP were calculated off-line averaging segments of 400 points of digitized EEG (12 bit A/D converter, sampling rate 250 Hz). These segments covered 1600ms comprising a pre-stimulus interval of 148 ms (37 samples) and post-stimulus onset interval of 1452 ms. Before processing, EEG was visually inspected and those segments with excessive EOG artifacts were manually eliminated. Only trials with correct responses were included in the data set. The experimental setup was designed by Santos *et al.* (2008) for their study on subject attention and perception using VEP signals.

3 Person Identification

3.1 Feature Extraction

The neuro-engineering theoretical and application studies related with the EEG signals are based on the knowledge that the EEG signals are composed of waves inside the 0-60 Hz frequency band and that different brain activities can be identified based on the recorded oscillations. For example, signals within the delta band (below 4 Hz) correspond to a deep sleep, theta band (4-8 Hz) signals are typical for dreamlike state, alpha frequencies (8-13 Hz) correspond to relaxed state with closed eyes, beta band

(13-30 Hz) are related with waking activity and gamma frequencies (30-50 Hz) are characteristics for mental activities as perception and problem solving. The relationship between the EEG and the brain functions is well documented in Niedermayer and Lopes da Silva, (1999).

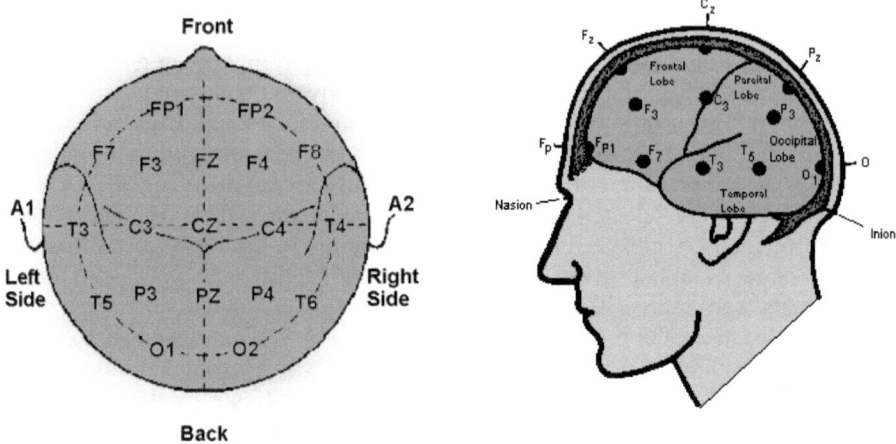

Fig. 1. Spatial location of the EEG electrodes over the frontal, central and parietal areas

For the present study the gamma-band spectral power of the VEP signals was computed by the Welch's periodogram method. The temporal segments, over which one value of the spectral power matrix is computed, correspond to one trial (around 1600 ms), *i.e.*, the samples collected during one image presentation. The normalized gamma-band spectral power for each channel was computed. It is a ratio of the spectral power of each channel and the total gamma-band spectral power of all channels. The level of perception and memory access among individuals are different and this reflects in significant difference between the gamma-band spectral power ratios of the subjects which is the key for the VEP based individuals identification.

3.2 Classifiers

Multiclass Support Vector Machine (SVM). Two strategies of training multiple binary classifiers for classification of the VEP spectral power ratios were implemented, Tan (2006): i) Support Vector Machine - One Against Other (SVM_OAO) and ii) Support Vector Machine - One Against All (SVM_OAA). Each strategy creates a set of binary classifiers that are afterwards combined to output the final labeling. Linear or nonlinear functions are comparatively tested as the SVM feature space mapping functions. Radial Basis Function (RBF) is selected for the nonlinear SVM case. The SVM-OAO creates P(P-1)/2 binary classifiers where P is the number of the persons identified. The classification principle is the max-wins voting strategy, in which every classifier assigns the instance to one of the two classes, the class with most votes determines the instance classification. The SVM-OAA creates P binary classifiers with the classification principle - the winner-takes-all and the binary classifier with the highest output function assigns the class.

Two training scenarios were considered:

- **Scenario 1:** The classifier is trained with data set coming from one experimental block (subject has to attend to the faces ignoring houses) and tested with data from the other experimental block (subject has to attend to the houses and ignore the faces).
- **Scenario 2:** The classifier is trained with data coming from both experimental blocks and tested with unseen data from the same blocks.

3.3 Principal Component Analysis (PCA)

A possible way to increase the signal to noise ratio is to accompany the feature extraction step with the principal component analysis (PCA). For the case considered, the PCA was designed first to extract only principal components of the normalized gamma-band spectral power (the feature space) that accumulates 95% of the signal energy (this is equivalent to feature space reduction). Then, it follows a step to reconstruct the feature space with the same dimensionality. The performance of both SVM classifiers was evaluated with or without PCA processing in the framework of the two scenarios. The results, summarized in Table 1 and Table 2, suggest that while the PCA is aimed at capturing the main EEG patterns, the individual specificity is lost and the classification accuracy is worsen. A possible interpretation is that the energy in the 30-50 Hz band of the original data set is already attenuated due to an embedded filtering process of the EEG acquisition apparatus. The PCA processing additionally reduces the VEP power spectral density and, therefore, all classifiers studied exhibit worse generalization performance (Table 1).

Table 1. Average classification error with PCA feature selection

With PCA		Classifier	1st PP step	2nd PP step	3rd PP step	4th PP step	5th PP step
SVM_ OAO (One Against One)	Linear (Scenario 1)	65,94	63,10	60,01	59,71	59,79	59,61
	Linear (Scenario 2)	56,42	51,58	48,05	47,12	46,25	45,57
	Nonlinear (Scenario 1)	44,53	37,26	31,24	27,95	26,19	24,07
	Nonlinear (Scenario 2)	**36,43**	**28,00**	**22,08**	**19,01**	**17,41**	**14,49**
SVM_ OAA (One Against All)	Linear (Scenario 1)	58,65	54,24	50,64	49,88	49,34	48,60
	Linear (Scenario 2)	59,79	56,55	54,42	53,42	52,36	51,24
	Nonlinear (Scenario 1)	43,78	36,76	31,12	28,03	24,93	23,33
	Nonlinear (Scenario 2)	**35,99**	**27,60**	**21,24**	**18,67**	**16,44**	**15,17**

Table 2. Average classification error without PCA feature selection

Without PCA		Classifier	1st PP step	2nd PP step	3rd PP step	4th PP step	5th PP step
SVM_OAO (One Against One)	Linear (Scenario 1)	38,21	35,36	33,43	31,89	31,63	30,37
	Linear (Scenario 2)	29,98	24,88	23,19	23,55	22,77	21,54
	Nonlinear (Scenario 1)	26,42	20,31	17,42	16,87	15,97	14,95
	Nonlinear (Scenario 2)	**15,67**	**10,16**	**8,32**	**6,95**	**5,54**	**5,10**
SVM_OAA (One Against All)	Linear (Scenario 1)	30,57	25,02	23,56	22,58	21,27	20,26
	Linear (Scenario 2)	26,84	21,17	17,87	16,45	14,52	13,71
	Nonlinear (Scenario 1)	26,99	21,54	18,32	16,70	15,16	14,49
	Nonlinear (Scenario 2)	**17,43**	**12,05**	**9,78**	**8,49**	**6,96**	**6,62**

4   Post Processing (PP) Procedure

Both classifiers perform a static (memoryless) classification that does not consider explicitly the temporal nature of the VEP signals. Time accounting classifiers, as for example Recurrent Neural Networks (NNs), Time Lag NNs or Reservoir Computing, have the disadvantage to require complex training procedures that not always converge.

In order to keep low complexity of the biometrical system, we propose here an empirical way to introduce memory into the classifiers. During a post processing (PP) procedure, a moving window of a sequence of n past classifier outputs (personal labels) is isolated and following a predefined strategy the labels are corrected. For example, during the first PP step a window of the last three labels is defined (n=3) and, in case the first and the last labels are the same but different from the central one, this label is corrected to be equal to the others. The window dimension of the second PP step is increased with one (n=4). If the first and the last elements have the same label, but the two central elements are different from each other and from the lateral elements they are corrected. It was observed that increasing the dimensionality of the moving window (third PP step with n=5; fourth PP step with n=6; fifth PP step with n=7) the overall performance of both classifiers improved. The strategy of each next step is to increase the number of central elements and to correct them in case they are different from the equal lateral elements of the moving (with one sample) window. After the fifth PP step the performance started to decrease, therefore five PP steps were subsequently implemented in the EEG-based biometry system (see Table 1 and Table 2 above).

In Fig.2 an example of classifier response for 5 classes with a sequence of 10 samples per class is depicted. Though the classifier recognizes in general the different persons correctly some of the responses are incorrect and the aim of the PP procedure is to correct these wrong guesses. The incorrect responses of the classifier decrease after each subsequent PP step.

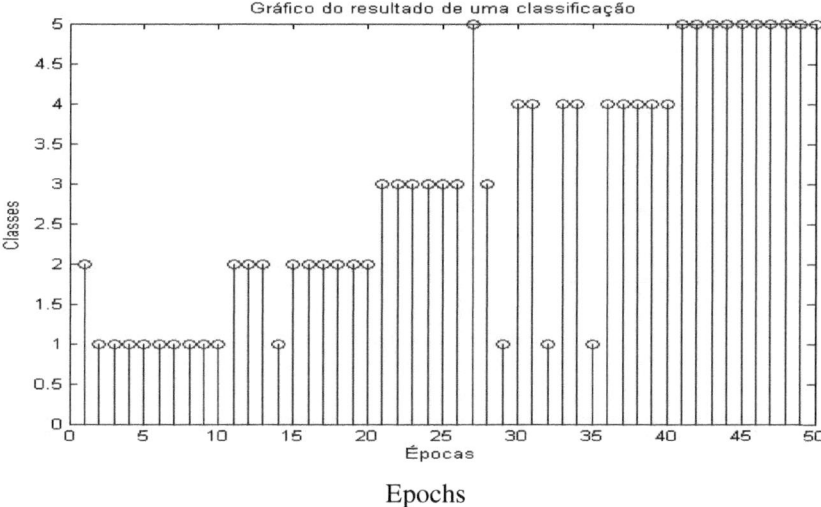

Epochs

Fig. 2. Example of classifier response for 5 classes with a sequence of 10 samples per class

5   Evoked Potential Duration

The effect of the Evoked Potential Duration (EPD) was particularly studied since it defines the viability of the biometry system. If the identified person has to be exposed too long time to a stimulus in order to be identified, it would make the system not quite practical and difficult to realize in real time. Therefore, the length of the ERP time series required for person identification needs to be reasonably short. The results of this study are summarized in Figs. 3-5 where the average classification error (ACE) is depicted as a function of the training segment length (N^o of trails). This analysis was done for the two studied SVM classifiers: SVM_OAO (Fig.3), SVM_OAA (Fig.4) and confirmed also for the k-Nearest Neighbor (k-NN) basic classifier (Fig.5) with k=3 and k=5. Note that for all classifiers there is a number of trails for which the ACE is minimized and longer time exposure does not suggest better person's discrimination. These results are averaged over the total number of identified subjects (13 persons) and an interval of 25-30 trials is determined as the optimal duration. Each trial corresponds to 400 samples with duration of about 1.5 s. Subsequently, 40-45 s. is going to be the expected time for stimulus expose before the classifier identify one person with the highest probability to make a correct guess. Though the conclusions go beyond of what can be analytically proved, the intuition behind is that too long time exposure to visual stimuli leads to accommodation and tiredness, thus the personal specificity encoded in the ERPs is vanishing and the classifier error increases.

294   A. Ferreira et al.

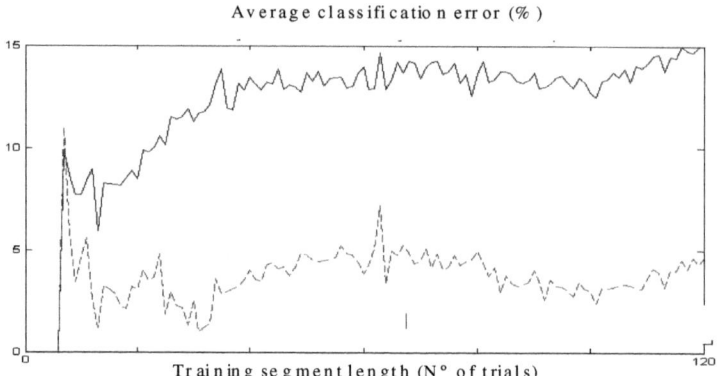

Fig. 3. SVM_OAO: ACE without PP (bold line) & after the 5th PP step (dashed line)

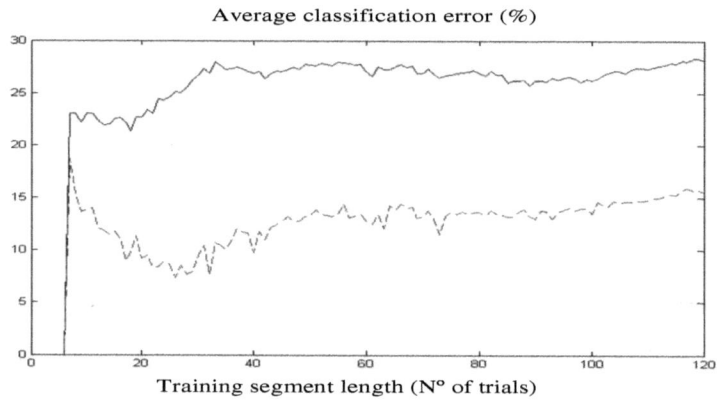

Fig. 4. SVM_OAA: ACE without PP (bold line) & after the 5th PP step (dashed line)

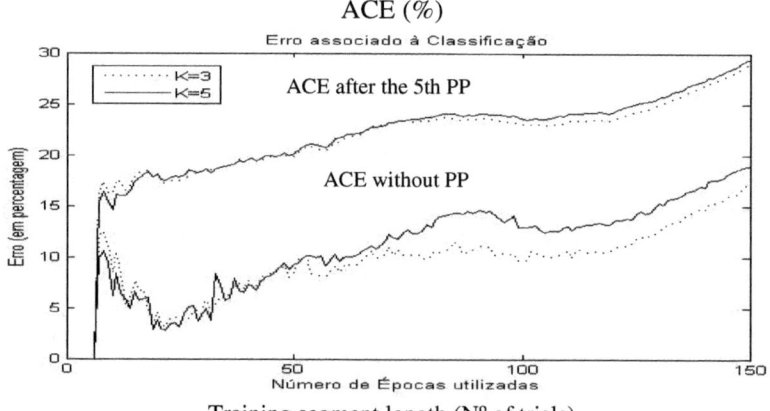

Fig. 5. k-NN: ACE without PP (bold (K=5) and dashed (K=3) lines below) & after the 5th PP step (bold (K=5) and dashed (K=3) lines above)

6 Concluding Remarks

This paper tested the feasibility of the EEG-based person identification. Although the results are only for 13 person subject pool, it does provide evidence of stability and uniqueness in the EEG shapes across persons. However, the classification accuracy of the EEG biometry currently cannot compete with the conventional biometrics (such as fingerprint, iris or palm recognition systems) and in general the EEG person identification modality can be seen just as a supplement ("a second opinion"). However, our long term goal is to use the principles of EEG-based biometry to detect abnormal scenarios, i.e. scenarios where a person is not acting as it would normally do in similar circumstances. Cognitive functions, such as attention, learning, visual and audio perception and memory, are critical for many human activities (for example driving) and they trigger numerous brain activities. Assuming that those brain activities follow a pattern for each person in normal circumstances (*reference pattern*), they are likely to change when the person is stressed, fatigued (physically, visually or mentally), or under the influence of several substances (alcohol, stimulants, drugs, etc.) (*deviation pattern*). In this context the EEG-based biometry would be particularly effective in health care applications, where it could be used not only to verify a patient's identity in medical records, prior to drug administration or other medical procedures but also to detect early in advance abnormal physiological or mental states of the patient.

In all, we expect several potential applications to emerge in the future. Control of the classified access into restricted areas security systems, illnesses or health disorder identification in medicine, gaining more understanding of the cognitive human brain processes in neuroscience are among the most appealing.

References

1. Ferreira, A.J.C.: EEG-based personal identification, Master thesis, University of Aveiro (2009) (in Portuguese)
2. Almeida, C.M.A.: EEG-based personal authentication, Master thesis, University of Aveiro (2009) (in Portuguese)
3. Marcel, S., del Millán, J.R.: Person authentication using brainwaves (EEG) and maximum a posteriori model adaptation. IEEE Transactions on Pattern Analysis and Machine Intelligence 29(4), 743–752 (2007)
4. Niedermeyer, E., Lopes da Silva, F.: Electroencephalography. Lippincott Williams and Wilkins (1999)
5. Palaniappan, R., Mandic, D.P.: Biometrics from Brain Electrical Activity: A Machine Learning Approach. IEEE Transactions on Pattern Analysis and Machine Intelligence 29(4) (2007)
6. Paranjape, R.B., Mahovsky, J., Benedicenti, L., Koles, Z.: The Electroencephalogram as a Biometric. In: Proc. CCECE, vol. 2, pp. 1363–1366 (2001)
7. Poulos, M., Rangoussi, M., Chrissikopoulos, V., Evangelou, A.: Person identification based on parametric processing of the EEG. In: Proc. IEEE ICECS, vol. 1, pp. 283–286 (1999)
8. Santos, I.M., Iglesias, J., Olivares, E.I., Young, A.W.: Differential effects of object-based attention on evoked potentials to fearful and disgusted faces. Neuropsychologia 46(5), 1468–1479 (2008)
9. Tan, P.-N., Steinbach, M., Kumar, V.: Introduction to Data Mining (2006)

Two-Factor Authentication or How to Potentially Counterfeit Experimental Results in Biometric Systems*

Christian Rathgeb and Andreas Uhl

University of Salzburg, Department of Computer Sciences, A-5020 Salzburg, Austria
{crathgeb,uhl}@cosy.sbg.ac.at

Abstract. Two-factor authentication has been introduced in order to enhance security in authentication systems. Different factors have been introduced, which are combined for means of controlling access. The increasing demand for high security applications has led to a growing interest in biometrics. As a result several two-factor authentication systems are designed to include biometric authentication.

In this work the risk of result distortion during performance evaluations of two-factor authentication systems including biometrics is pointed out. Existing approaches to two-factor biometric authentication systems are analyzed. Based on iris biometrics a case study is presented, which demonstrates the trap of untruly increasing recognition rates by introducing a second authentication factor to a biometric authentication system. Consequently, several requirements for performance evaluations of two-factor biometric authentication systems are stated.

1 Introduction

Reliable personal recognition is required by a wide variety of access control systems. Examples of these systems include ATMs, laptops and cellular phones [1]. If these systems fail to meet the demands of reliable and robust authentication potential imposters may gain access to these systems. In order to enhance the security of access control systems two factor authentication (T-FA) has been introduced, wherein two factors are combined in order to authenticate a user. The key idea of T-FA is to sum up the security of two factors. These factors include, passwords, representing "something you know", or physical tokens, such as smart-cards, representing "something you have". Additionally, biometric traits are applied, respresenting "something you are".

However, several problems may occur when introducing biometric authentication to T-FA systems. Performance gain with respect to recognition rates is often achieved due to the assumption of unrealistic preconditions. Resulting performance distortions may not be recognized at first sight, yet, these could lead to serious security vulnerabilities. In order to shed light on the use of biometrics as additional factors in T-FA schemes we demonstrate a way of how to untruly

---

* This work has been supported by the Austrian Science Fund, project no. L554-N15.

A. Campilho and M. Kamel (Eds.): ICIAR 2010, Part II, LNCS 6112, pp. 296–305, 2010.

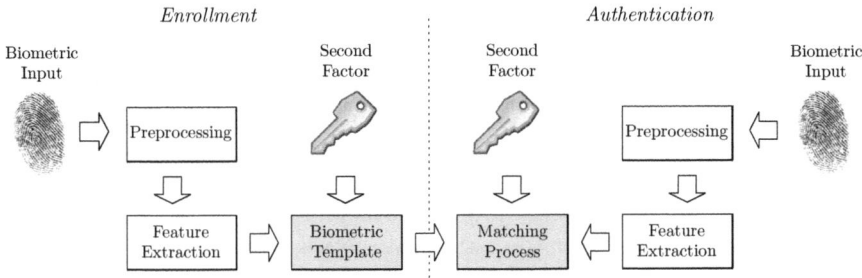

Fig. 1. The basic operation mode of a two-factor biometric authentication system. During enrollment and authentication additional factors are presented to the system.

improve recognition rates in an biometric recognition system by introducing a token-based T-FA scheme. As a consequence, we manifest requirements with respect to performance measurements in T-FA systems including biometrics.

The remainder of this paper is organized as follows: in Sect. 2 we first summarize the fundamentals of T-FA and biometric systems. Performance evaluation in biometric verification systems is described and existing T-FA systems involving biometric authentication are analyzed. In Sect. 3 we propose a case study in which we apply T-FA to an iris-biometric verification system. Experimental results are analyzed and a summary of T-FA and biometric verification systems is given in Sect. 4. In conclusion, requirements for performance evaluations in such systems are stated.

2   T-FA and Biometric Systems

An authentication factor is a piece of information used to authenticate or verify the identity of a user. In a T-FA system two different factors are combined in order to authenticate a user. It is claimed that T-FA generally delivers a higher level of authentication assurance compared to using just one factor. Three basic classes of factors can be distinguished: personal factors, such as user-defined passwords, physical factors, such as smart-cards [2] or human factors, such as biometric traits [3]. Combining two factors from two different classes yields T-FA where each factor is applied independently.

A common example of T-FA systems are ATMs, where physical factors are combined with personal factors. T-FA only applies to systems which use factors of different classes – authentication schemes based on, for example, two biometric modalities are referred to as multi-modal authentication [4]. Besides known vulnerabilities of T-FA schemes [5], such as "man-in-the-middle" attacks, where an imposter does not need to be in possession of any physical factor, we will focus on the risk of false performance evaluation of T-FA schemes involving biometrics as human factor. Fig. 1 shows the basic operation mode of a T-FA system including biometric authentication. At the time of enrollment biometric traits and a second personal or physical factor are presented. During authentication this factor is presented again in order to achieve successful authentication.

With respect to biometric authentication two different modes are distinguished: verification and identification [1]. Since we aim at analyzing T-FA schemes including biometrics we will only focus on verification, since the presentation of an additional personal or physical token represents an identity claim per se. Due to the variance in biometric measurements biometric systems do not provide perfect matching, as it is easily implemented in password or PIN-based authentication systems [3]. Thus, a fuzzy matching is performed where decision-thresholds are set up in order to distinguish between genuine and non-genuine users, respectively. Hence, several magnitudes define the performance of a biometric system. Widely used measures include False Rejection Rate (FRR), False Acceptance Rate (FAR) and Equal Error Rate (EER).

2.1   Biometric T-FA Systems

Obviously, T-FA increases the security of biometric authentication systems since potential imposters have to compromise the second factors in order to gain access to the system as a first attack stage. However, if a T-FA scheme is constructed where one factor is represented by a biometric trait additional factors are either personal or physical tokens. Both of these factors require a perfect matching. This means a wrong PIN or a wrong smart-card would result in a rejection of the imposters. If each imposter would be in posession of the correct second factor of the account he wants to gain access to (i.e. the second factor has been compromised), the overall recognition rate of system is expected to remain the same. Applying a sequential check of both factors (regardless of the order), the recognition rate is equal to a system which only performs biometric authentication. That is, additional factors become meaningless with respect to recognition performance since these are potentially compromised by all imposters. By analogy, if no imposter would be in posession of the correct second factor of the account he wants to gain access to, the overall recognition rate is of course expected to increase. This is because imposters which may have tricked the biometric authentication system are rejected at the time the additional factor is checked. Throughout literature several approaches have been proposed where T-FA is introduced to biometric systems. In any case, authors claim to introduce personal or physical tokens in order to enhance the security of the system. However, in some cases it is doubtful if the proposed system can maintain recognition rates without the use of a second factor (i.e. in case the second factor has been compromised). In order to underline the problem of evaluating the performance of biometric T-FA schemes in terms of recognition performance, we will discuss several approaches which we found questionable regarding reported performance results. Hence, we do not cover all approaches to biometric T-FA schemes, but only a small selection to emphasize that potential incorrect performance evaluations should be considered an important issue.

The introduction of biometrics to generic cryptographic systems resulted in a new field of research, named biometric cryptosystems [6]. Most approaches which can be subsumed under the category of biometric cryptosystems aim at extracting cryptographic keys out of biometric data. Performance evaluations are adopted such that correctly generated keys are equivalent to successful authentication and vice versa. Due to biometric variance a widespread usage of helper data, for example error correction codes, has proven to be worthwhile. However, in several approaches the application of helper data conceals the actual performance rates. In the following, we provide three examples for that.

Teoh et al. [7,8] introduced a technique to generate cryptographic hashes out of face biometrics which they refer to as "BioHashing". Like in generic T-FA schemes, in the BioHashing approach random numbers are associated with each user. These user-specific random numbers are used as seed in order to generate biometric hashes. These random numbers, which represent the second factor, have to be presented to the system in addition to biometric data at authentication. The authors report almost perfect performance rates for the generation of biometric hashes. In order to expose the true performance of BioHashing, Kong et al. [9] presented an implementation of FaceHashing. It was found that the reported performance was achieved under the hidden assumption that random numbers are never lost, stolen, or compromised. This assumption does not hold in general. Physical tokens can be stolen or duplicated while personal tokens can be easily guessed or broken, for example by dictionary attacks [10]. In order to associate cryptographic with biometric data Reddy et al. [11] proposed a so-called "fuzzy vault scheme" [12] based on iris biometrics. The authors achieve T-FA by embedding an additional layer of security, namely a password. With this password the generated vault as well as the secret key is encrypted. In experiments the security of a fuzzy vault scheme which exhibits a FRR of 8% and a FAR of 0.03% is increased, where a total number of 100 templates are used. As result of the hardening scheme the FRR increases to 9.8% due to misclassification of a few minutiae. At the same time the FAR decreases to 0.0%. It is claimed that this is due to the fact that minutiae are distributed more randomly. If this was the case for the use of one single password (identical for all users), this could be integrated into the original algorithm to increase performance. However, if passwords are compromised the systems' security decreases to that of an ordinary fuzzy vault scheme which indicates that the FAR of 0.0% was calculated under unrealistic preconditions. In recent work Jassim et al. [13] proposed a method of improving the performance of PCA based face recognition system. The authors introduce random projections based on the Gram-Schmidt process which are used to map biometric features onto secret personalized domains. For this purpose a secret permutation matrix is assigned to each user, which represents a second factor. This means, each user is in possession of a unique matrix. In experimental results accumulations of genuine users remain the same while inter-class distances increase. While the original PCA based system reveals an EER of 17% an EER of 0.2% is reported if random projections are applied for each user. The authors do not consider the case where the same permutation matrix is assigned to each user. Therefore, again the results are achieved under the assumption, that the second factor has not been compromised.

Ratha et al. [14] introduced the concept of "cancellable biometrics". Biometric data can be compromised and therefore become useless because it can not be modified ex post. The idea of cancellable biometrics consists of intentional, repeatable distortion of a biometric signal based on transforms where matching is performed in the transformed space. Thus, if a potential imposter is able to steal the stored template recovering of original biometric data becomes infeasible. In contrast to biometric recognition systems or biometric cryptosystems the scope of cancellable biometrics is template security. By definition, a T-FA system is constructed since user-specific transformations are applied. That is, different transformations represent second factors which are used to secure biometric templates. All of the approaches presented in Section 2.1 can be seen as systems which provide cancellable biometrics, since random numbers, passwords or transformations which are applied to biometric features can be updated easily. With respect to recognition performance approaches to cancellable biometrics aim at maintaining the performance of the original underlying systems (in general loss of performance is expected). This means, approaches to cancellable biometrics in which better performance as compared to the original algorithm is reported should be examined carefully. For example, the BioHashing approach of Teoh et al. [7,8] was extended to be used as cancellable biometrics. As pointed out by Cheung et al. [15], experimental results were obtained under the unpractical assumptions stated above.

3   T-FA and Iris Recognition: A Case Study

In this section we describe an existing iris recognition system which we apply to construct a T-FA system. First we will consider the performance of the biometric system. Subsequently, we will construct a generic T-FA scheme by introducing user-specific bit streams as second factor.

3.1   Iris Recognition System

In order to apply biometric authentication we use our own implementation of the algorithm of Ma et al. [16]. In their approach the iris texture is treated as a kind of transient signal which is processed using a 1-D wavelet transform. The local sharp variation points, which denote important properties of transient signals, are recorded as features. We always extract an iris texture from eye images as a first step. We assume the texture to be the area between the two almost concentric circles of the pupil and the outer iris. These two circles are found by contrast adjustment, followed by Canny edge detection and Hough transformation. After the circles are detected, unwrapping along polar coordinates is done to obtain a rectangular texture of the iris. In our case, we always resample the texture to a size of 512x64 pixels.

The texture is subsequently divided into N stripes to obtain N one-dimensional signals, each one averaged from the pixels of M adjacent rows. We used $N = 10$ and $M = 5$ for our 512x64 pixel textures (only the 50 rows close to the pupil are used from the 64 rows, as suggested in [16]). A dyadic wavelet transform

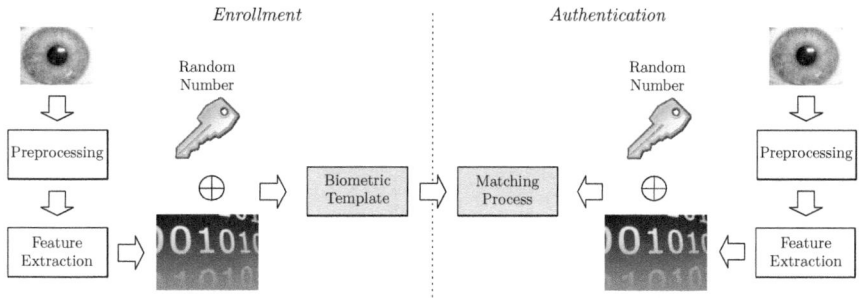

Fig. 2. T-FA scheme: random numbers are introduced with which iris codes are sequentially XORed during enrollment. At authentication the biometric template is XORed with another random number and the result is matched against an extracted iris code.

is then performed on each of the resulting 10 signals, and two fixed subbands are selected from each transform. This leads to a total of 20 subbands. In each subband we then locate all local minima and maxima above some threshold, and write a bitcode alternating between 0 and 1 at each extreme point. Using 512 bits per signal, the final code is then 512x20 bit.

Once bitcodes are obtained, matching can be performed on them and Hamming distance lends itself as a very simple distance measure. For matching to work well, we compensate for eye tilt by shifting the bit-masks during matching by three pixels in each direction.

3.2   Two-Factor Iris Recognition System

In algorithm described above, a users iris serves as the only authentication factor. In order to construct a T-FA system a second factor has to be introduced. Therefore we simply apply random numbers which are associated with specific users. These random numbers can be stored on a smart-card, representing a physical factor. Additionally, we choose rather short random bit streams, hence, these are easily remembered representing personal factors as well. At this point it is important that the application of random bit streams yields a generic approach to T-FA, since these just represent a user-specific secret.

During enrollment for each user a randomly generated bitstream is generated. The iris code of a user, which is extracted during enrollment is sequentially XORed with the random number in order to generate a secure template. That is, the stored iris code is protected by the random bit stream, similar to the approach presented by Zuo *et al.* [17]. If a user wants to authenticate with the system an appropriate random number has to be presented firstly. Subsequently, the stored template is sequentially XORed with this random number and the resulting iris code is matched against the one extracted from the presented iris image. Hence, a T-FA system is realized by simply introducing random numbers which are associated with users by sequentially XORing these with iris codes. The operation mode of the whole system is illustrated in Fig. 2.

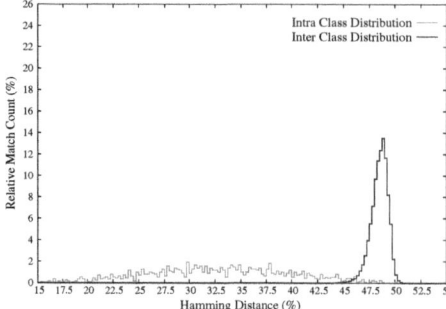

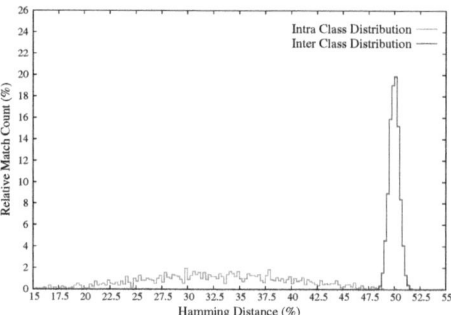

Fig. 3. Intra-class and inter-class distribution of the algorithm of Ma

Fig. 4. Intra-class and inter-class distribution of the algorithm of Ma using T-FA

3.3   Experimental Results

Experiments are carried out using the CASIAv3-Interval iris database[1], a widely used test set of iris images of over two hundred persons. The database comprises iris images of size 320×280 pixels out of which normalized iris textures of 512×64 pixels are extracted in the preprocessing step as described earlier. Applying our implementation of the feature extraction of Ma *et al.* to preprocessed iris textures, a total number of 10240 bits are extracted per iris image. Matching is performed by calculating the Hamming distance between extracted iris codes, where a circular shift of three pixels to the left and right is implemented in order to provide some degree of rotation invariance. In Fig. 3 the distribution of the intra-class distance and the inter-class distance are plotted. Fig. 5 shows the FRR and the FAR resulting in an EER of 1.76%. For a threshold of 42% (in terms of correct bits), a FRR of 5.61% and zero FAR is achieved. In other words, the system will in general reject 5.61% of all genuine users while no imposters are untruly accepted. We are aware that these results are worse than those reported by Ma. *et al.*, however, the absolute performance of the algorithm is not the topic of this work. Thus, our implementation serves its purpose.

For the construction of a T-FA system we introduce random numbers consiting of 8 bits. Performance is measured in the same way as in the iris recognition system applying the same test set, however, now users have to present biometric data which has to pass the fuzzy match of the recognition system as well as a random number which is sequentially XORed with stored templates. Since genuine users are in possession of correct random numbers the construction of the T-FA system does not effect the intra-class distribution. Therefore, calculated Hamming distances between genuine iris codes remain the same as can be seen in Fig. 4. If we make the assumption that imposters are not in possesion of valid random numbers the performance of the whole system is increased. This means we calculate the inter-class distribution applying the random numbers to users which were assigned to them during enrollment. In other words, users claim the

---

[1] The Center of Biometrics and Security Research: CASIA Iris Image Database, http://www.sinobiometrics.com

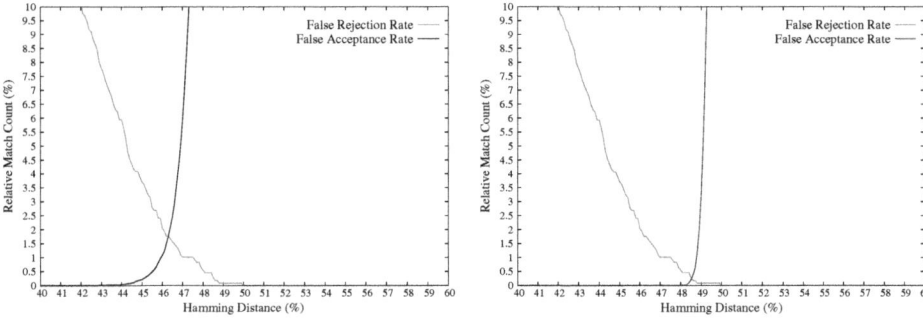

Fig. 5. FRR and FAR of the algorithm of **Fig. 6.** FRR and FAR of the algorithm of
Ma   Ma using T-FA

identity of other users but present their own biometric data as well as their
own random number (a randomly chosen number could be used as well). The
distribution of the resulting inter-class distances is also plotted in Fig. 4 and the
FRR and the FAR are plotted in Fig. 6 resulting in a EER of 0.25%. It becomes
clear that now inter-class distances accumulates around 50% instead of around
47% in the original system. This is because iris codes are now XORed with
potentially different random numbers resulting in almost random bitstreams.
Since binary iris codes are extracted, the Hamming distance between random
bit streams is expected to be 0.5. This means we can now increase the threshold
and result in overall (virtual) performance gain of the system. For an increased
threshold of 48% correct bits we now achieve a FRR of 0.43% and zero FAR
resulting in an EER of 0.25% which is about 1.5% better than the performance
of the original iris recognition system (see Fig. 6).

3.4   Analysis

In the above presented biometric system two authentication factors are combined
by sequentially XORing iris codes with 8-bit random numbers. Based on the
assumption that random numbers are never compromised we increase inter-class
distances and are able to gain performance by increasing the threshold which is
used to separate genuine and non-genuine users.

The problem of the presented scenario is that additional factors are considered to never be stolen, lost, shared or duplicated where in practice the opposite
is true. The assumption that imposters would try to infiltrate the system by
presenting some random personal or physical factor is rather absurd. Additional
factors such as passwords or PIN must not be considered secure since these are
easily compromised [3]. As we demonstrated inter-class distributions increase.
In case the decision-threshold is increased according to the new inter-class distribution the biometric system becomes more tolerant. This is because access to
the system is even eased if potential imposters are in possession of a valid second
factor which is a realistic scenario. In case a potential imposter is in possesion
of a valid random number the recognition rate degrades to that of the original

biometric system. In this case, a threshold of 48% correct bits would yield a FAR of 24.21%, that is, the system becomes highly vulnerable. Thus, performance only holds if random numbers are never compromised. Having adjusted the system decision threshold to the virtual performance as determined under unrealistic preconditions, the accuracy of the system is actually severely degraded, in case the second factor is compromised.

4   Summary and Conclusion

In all of the above discussed T-FA schemes additional factors (random numbers, passwords or permutation matrices) are considered to never be stolen, lost, shared or duplicated. If this would be the case the introduction of biometrics becomes meaningless since the system could rely on these random numbers or passwords without any risk. That is, authentication could be performed just by presenting appropriate random numbers or passwords. In case inter-class distributions are calculated under these assumptions for performance evaluations, the FAR of the system is kept artificially low. Hence, thresholds can be adapted to generate better results like in our presented scheme. That is, the biometric system is set to be more tolerant since inter-class distances become larger. However, if imposters are in possession of valid random numbers or passwords, T-FA systems become highly vulnerable as has been shown.

We conclude that it is incorrect and also severely misleading to claim that T-FA does increase the recognition performance of an biometric authentication system. In practice, security may be enhanced since two factors are necessary to achieve successful authentication, yet it is essential that the recognition performance and the corresponding decision parameters remain the same as compared to the employment of the "pure" biometric system. In all of the presented systems claimed performance is achieved through the unpractical assumption that non-genuine users are not in possession of valid second factors. If performance evaluations are carried out like in the above presented systems the true performance of the underlying biometric system is concealed.

4.1   Requirements for Performance Evaluations of T-FA Systems

From the analysis of all the above presented approaches and our case study several requirements to performance evaluations regarding T-FA schemes including biometric authentication can be derived:

1. It is required that in experiments, especially when calculating inter-class distances, any type of personal or physical token has to be considered compromised. Focusing on biometric systems T-FA must not be interpreted as a way to increase the recognition performance of a system.
2. If any sort of helper data is introduced, especially in biometric cryptosystems, this helper data must be considered compromised during experimental results if this helper data is not dependent on biometric data only.
3. The scenario where potential imposters are in possession of additional second factors must not be ignored since physical or personal factors are easily

compromised in general (decision thresholds have to be set up according to this scenario).

4. The security provided by introducing second factors to biometric systems must not and cannot be measured in terms of FRR or FAR.

References

1. Jain, A.K., Ross, A., Prabhakar, S.: An introduction to biometric recognition. IEEE Trans. on Circuits and Systems for Video Technology 14, 4–20 (2004)
2. Chien, H., Jan, J., Tseng, Y.: An efficient and practical solution to remote authentication: Smart card. Computer and Security 21, 372–375 (2002)
3. Jain, A.K., Flynn, P.J., Ross, A.A.: Handbook of Biometrics. Springer, Heidelberg (2008)
4. Jain, A.K., Nandakumar, K., Ross, A.A.: Handbook of Multibiometrics. Springer, Heidelberg (2006)
5. Schneier, B.: Two-factor authentication: too little, too late. ACM Commun. 48(4), 136 (2005)
6. Uludag, U., Pankanti, S., Prabhakar, S., Jain, A.K.: Biometric cryptosystems: issues and challenges. Proceedings of the IEEE 92(6), 948–960 (2004)
7. Teoh, A.B.J., Ngo, D.C.L., Goh, A.: Biohashing: two factor authentication featuring fingerprint data and tokenised random number. Pattern Recognition 37, 2245–2255 (2004)
8. Teoh, A.B.J., Ngo, D.C.L., Goh, A.: Biometric Hash: High-Confidence Face Recognition. IEEE Transactions on Circuits and Systems for Video Technology 16(6), 771–775 (2006)
9. Kong, A., Cheunga, K.H., Zhanga, D., Kamelb, M., Youa, J.: An analysis of Bio-Hashing and its variants. Pattern Recognition 39, 1359–1368 (2006)
10. Klein, D.V.: Foiling the cracker: a survey of, and improvements to, password security. In: Proceedings of the 2nd USENIX Workshop Security, pp. 5–14 (1990)
11. Reddy, E., Babu, I.: Performance of Iris Based Hard Fuzzy Vault. IJCSNS International Journal of Computer Science and Network Security 8(1), 297–304 (2008)
12. Juels, A., Sudan, M.: A fuzzy vault scheme. In: Proc. 2002 IEEE International Symp. on Information Theory, p. 408 (2002)
13. Jassim, S., Al-Assam, H., Sellahewa, H.: Improving performance and security of biometrics using efficient and stable random projection techniques. In: Proceedings of the 6th International Symposium on Image and Signal Processing and Analysis, ISPA '09, pp. 556–561 (2009)
14. Ratha, N.K., Connell, J.H., Bolle, R.M.: Enhancing security and privacy in biometrics-based authentication systems. IBM Systems Journal 40, 614–634 (2001)
15. Cheung, K.H., Kong, A.Z.D., Kamel, M., You, J., Lam, T.H.W.: An Analysis on Accuracy of Cancelable Biometrics based on BioHashing. In: Khosla, R., Howlett, R.J., Jain, L.C. (eds.) KES 2005. LNCS (LNAI), vol. 3683, pp. 1168–1172. Springer, Heidelberg (2005)
16. Ma, L., Tan, T., Wang, Y., Zhang, D.: Efficient Iris Recogntion by Characterizing Key Local Variations. IEEE Transactions on Image Processing 13(6), 739–750 (2004)
17. Zuo, J., Ratha, N.K., Connel, J.H.: Cancelable iris biometric. In: Proceedings of the 19th International Conference on Pattern Recognition 2008 (ICPR'08), pp. 1–4 (2008)

Automated Detection of Sand Dunes on Mars

Lourenço Bandeira[1], Jorge S. Marques[2], and José Saraiva[1] and Pedro Pina[1]

[1] CERENA, Instituto Superior Técnico, Av. Rovisco Pais, 1049–001 Lisboa, Portugal
[2] ISR, Instituto Superior Técnico, Av. Rovisco Pais, 1049–001 Lisboa, Portugal
lpcbandeira@ist.utl.pt, jsm@isr.ist.utl.pt, jose.saraiva@ist.utl.pt,
ppina@ist.utl.pt

Abstract. In this paper we show that the detection of dune fields on images of the surface of Mars, however varied they are, can be achieved through the application of an automated methodology. The procedure is based on the extraction of local information from images after they are organized according to a regular grid which defines cells, in turn aggregated into larger regions (blocks) that constitute the detection units. A set of gradient features is extracted and tested with Boosting and Support Vector Machine classifiers. A detection rate of 98.7% was obtained for a 5-fold cross validation on a set of images captured by the Mars Orbital Camera on board the Mars Global Surveyor probe.

Keywords: Gradient and HOG features, SVM and Boosting classifiers, Mars.

1 Introduction

Dunes are the most frequent aeolian features on the Martian surface, and their study contributes to the understanding of the interactions between the atmosphere and the surface of the planet, of the way the climate has evolved along the history of Mars and of how it works currently [1,2]. Dunes on Mars were first observed in the early 1970s on Mariner 9 images, but only the largest kilometric fields were detected. In the late 1990s, with the orbital mission of the Mars Global Surveyor probe (MGS), equipped with a higher spatial resolution camera, many more dune fields were resolved and it was confirmed that the shapes visible showed many similarities with those occurring on Earth [3]. Recently, a group of planetary scientists created the Mars Dune Consortium (http://www.mars-dunes.org) whose stated intention is to produce a catalogue containing all dune fields identifiable on the surface of Mars [4]. The results of their program of search and delineation of dune fields, which has been performed manually, are available online in a geographical database, the MGD^3-Mars Global Digital Dune Database [5]. This database only contains, at the moment of writing, information about the area between latitudes 65°N and 65°S, in which dunes cover an area of approximately 70,000 km^2. A rough estimation of the total area covered by dune fields on Mars gives about 120,000 km^2 in the southern hemisphere and about 680,000 km^2 in the northern [6]; thus, more than 90% of the Martian dune

A. Campilho and M. Kamel (Eds.): ICIAR 2010, Part II, LNCS 6112, pp. 306–315, 2010.

fields have yet to be mapped. Furthermore, the remotely sensed images used so far to construct the database still leave outside this venture the smallest dune fields.

Besides this huge quantity of dunes still unmapped, the detection of changes on the characteristics of these dynamic aeolian features is another issue that could also benefit a lot if an automated method were available to delineate them on remotely sensed images at different scales and moments in time. In the last years, some techniques have begun to be implemented to automatically detect structures on planetary surfaces but, so far, only the field of impact crater studies has achieved some maturity. A large variety of methodologies have been developed and tested, using the most recent and powerful tools, with steadily improving performances [7,8,9,10,11,12,13,14]. On the contrary, there is as yet no automated approach to deal with the identification of sand dunes. There are some applications dealing with temporal change detection or measuring the height of dunes, but they are restricted to geographically confined case studies.

Thus, the objective of this paper is to test the adequacy of recent and up-to-date machine learning methodologies for the detection of aeolian dunes on remotely sensed images of Mars. This work is partly inspired on some previous strategies and algorithmic sequences used for automated crater detection. For the purpose now considered, we have selected two types of features that work best in the extraction of the directional and periodic characteristics of the dunes (gradient and histogram-of-oriented-gradients features), and which were both used on Boosting and Support Vector Machine classifiers to indicate if a given region of the image contains dunes. The performance of those methods is evaluated with a set of high spatial resolution images acquired by the MOC camera of the MGS probe which represent the diversity of Martian dune types.

2 Formulation of the Problem

2.1 Dune Types

A geological classification scheme of sand dunes was proposed by McKee [15] for terrestrial examples, mostly based on field work. It considers the different shapes that exist and relates them to specific environments of deposition and the factors acting upon it. The dunes so far identified on the Martian surface have been classified according to that scheme, and although most of them fit into the main types there are some undefined morphologies not known to occur on Earth [4]. On Fig. 1, we present some examples of the predominant Martian dune type (barchan dunes). From this, it becomes clear the multitude of factors that affect the visual aspect of dune fields - constituents, size, shape and density, association to seasonal advance and withdrawal of ice cover, angle of illumination, just to name some - and that must be tackled by any automated approach designed to detect their presence on an image. Thus, the nature and varied characteristics of occurrence of sand dunes on images of the Martian surface demand a learning strategy, able to adapt itself to distinct situations.

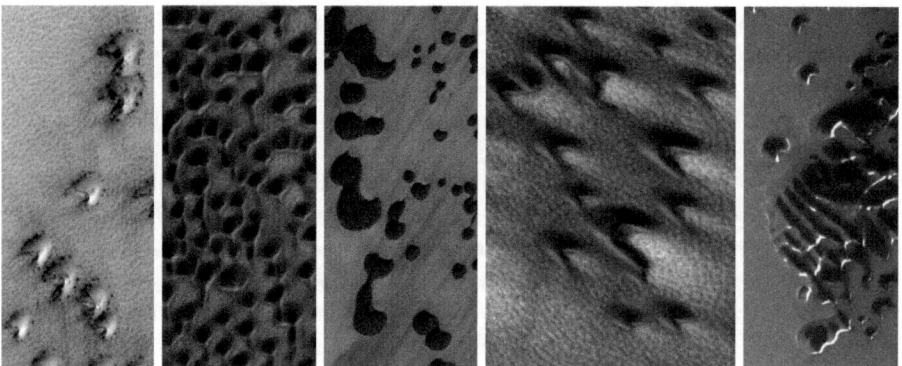

Fig. 1. Diversity of Martian sand dunes on portions of MGS/MOC images (from left to right): R18-01906, S01-00739, S01-00925, E23-00005 and R22-00101. Each image covers an area of 1500x3000 m^2 [image credits: MSSS/NASA/JPL].

2.2 Image Analysis

The procedure adopted for the identification of the dunes is based on the analysis of the local information of the image along a regular grid. For that purpose, an image is divided into cells (Fig. 2a) from which given features will be extracted. To increase the invariance to specific factors such as illumination and shadowing, an aggregation of the local features is performed within larger regions, blocks constituted by 3×3 cells, which are the detection windows (Fig. 2b). The displacement of the block along the entire image grid is performed with an overlapping between adjacent blocks equal to one cell side (Fig. 2c).

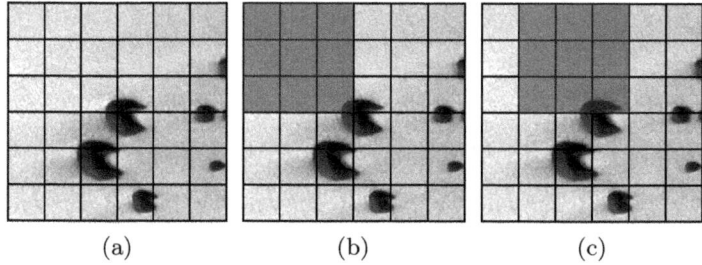

(a)   (b)   (c)

Fig. 2. Tiling an image in (a) cells and (b) blocks (3×3 cells, in red); (c) Block displacement with overlapping. This region corresponds to a sample of image E02-01086 [image credits: MSSS/NASA/JPL].

3 Features and Classifiers Used

Important advances were achieved in the last decade in the field of computer vision, both in the type of features used to characterize objects and in the recognition methods which are needed to learn and classify the objects present in the images. From those recent contributions we selected the features we consider to be among the most appropriate to detect the patterns presented by sand dunes. We considered features based on the image gradient $g(x) \in \mathbb{R}^2$ computed at each image point x. The gradient vector is characterized by its amplitude $|g(x)|$ and phase $\phi(x)$. These features are grouped into the following four sets:

- **HP(9):** This corresponds to the features introduced by Dalal and Triggs [16] in their face detection problem, the histogram-of-oriented gradients or HOG features. They intend to capture the characteristic edge structure of the local shape of the dune, with a controlled degree of invariance to local geometric and radiometric factors. To obtain them, it is necessary to compute the weighted histogram for each cell which results from the multiplication of the gradient phase by its magnitude for each contributing point. Therefore, the histogram value associated to the kth cell C^k is:

$$h_i^k = \sum_{x \in C^k} |g(x)|.b_i(\phi(x)).$$ (1)

where $b_i(\phi) = \begin{cases} 1, & \text{if } \phi \in i^{th} \text{ bin} \\ 0, & \text{otherwise} \end{cases}$

For our problem, we defined an angular interval of 20° for the computation of the directional histograms, so we have a total of 81 features per block (9 histogram bins/cell × 9 cells). The phase histograms were not normalized.

- **HPM(9):** This consists of a modified version of the HOG features, by using separately the phase histogram:

$$h_i^k = \sum_{x \in C^k} b_i(\phi(x)).$$ (2)

and the magnitude histogram:

$$\tilde{h}_i^k = \sum_{x \in C^k} \tilde{b}_i(|g(x)|).$$ (3)

where $\tilde{b}_i(|g|) = \begin{cases} 1, & \text{if } |g| \in i^{th} \text{ bin} \\ 0, & \text{otherwise} \end{cases}$

For the phase, with the same angular interval of 20°, we have 81 features (9 histogram bins/cell × 9 cells), and for the magnitude, considering 11 bins (resulting from a 4-unit interval between a minimum of 0 and a maximum of 40), we have 99 features (11 histogram bins/cell × 9 cells) Thus, for this set, a total of 180 features per block are obtained.

- **HP:** This refers to the histograms of the gradient phase for each image block B:

$$h_i = \sum_{x \in B} b_i(\phi(x)).$$ (4)

in the same 9 bins (angular interval of 20°). Thus, 9 features (9 histogram bins/block × 1 block) are used in this situation. In this case, the phase votes were not weighted by the gradient magnitude.

- **HPM:** It consists of using the histograms of phase, as defined previously for HP, and of magnitude of the gradient on each block separately. The histogram amplitude is given by:

$$\tilde{h}_i = \sum_{x \in B} \tilde{b}_i(|g(x)|). \tag{5}$$

Thus, it gives 9 features for the phase (1 histogram × 9 bins of 20° of angular interval) and 11 features for the magnitude (the same bins of the HP features). For this set, a total of 20 features are extracted.

The size of each cell is the same for all images and is equal to 40 × 40 pixels. In order to have the features varying between 0 and 1, a normalization step was performed globally for each image and for each individual feature. For the classification of the blocks we used two of the most advanced and powerful classifiers that have already proven their ability in dealing with a variety of classification problems, namely in remotely sensed imagery of the Earth and other planetary surfaces: Boosting and Support Vector Machines (SVM).

Boosting algorithms achieve remarkable results by combining a large number of weak classifiers, using weighted majority vote [17]. They are also able to performe feature selection i.e., to select a subset of informative features for a given problem. This can be done by assuming that each weak classifier depends on a single feature [18]. The application of boosting algorithms in object recognition lead to excellent results in, for instance, face [18] and impact craters [12] problems.

SVM are kernel methods that use an implicit transformation to a higher dimensional space in order to achieve good separability by means of a linear classifier in the new space [19]. The hyperplane used for separation in the higher dimensional space is chosen in such a way that the so-called margin (the distance to the closest samples in each class) is maximized. The samples determining the margin are called the support vectors. Different transformation kernels, such as Gaussian, polynomial, linear and circular can be used, yielding different classifiers.

4 Results

4.1 Dataset

To test our approaches, we have selected a set of 20 remotely sensed images captured by the Mars Orbiter Camera N/A (narrow angle mode) of the Mars Global Surveyor probe. Those images are from different locations on the planet, cover a total area of about 1320 km² and are representative of the diversity of barchan dunes, one of the most common types on Mars that we have chosen to test these features and classifiers. These single-band images with 256 grey levels have a spatial resolution between 3.22 and 6.79 metres/pixel. Their dimension is variable, but a typical value is of about 1000 columns per 6000 lines.

For each image we constructed ground-truth information, by manually delineating the dunes therein contained (examples are shown in Fig. 3a and Fig. 4a), indicating the 'dune' and 'not-dune' regions). The tiling of the ground-truth into cells was then performed. In this process, only the cells containing more than 30% of dune area were considered as 'dune', whereas the cells with less than 10% of dune area were considered as 'not-dune'; the cells with dune areas comprised in the interval 10-30% were not considered (examples in Fig. 3b and Fig. 4b).

4.2   Evaluation

Every classifier was tested with each of the four sets of features using a 5-fold cross-validation, i.e., the total number of image blocks was divided into five subsets of the same size: four of them were used for training, the remaining one was used for testing. This procedure was repeated five times, so that each subset was used once for testing.

For the tests with the SVM classifier we have used the freely distributed package SVMLight [20]. Several kernels were exploited, but among those with higher performances we chose the linear kernel since it is the most simple. The performance of each classifier with each one of the 4 sets of features is evaluated through the computation of the probabilities of false negatives ($p_{FN} = FN/(FN+TP)$), false positives ($p_{FP} = FP/(FP+FN)$) and of a global error ($p_{error} = p_N.p_{FP} + p_P.p_{FN}$), where FN stands for the number of false negative blocks, TN the number of true negative blocks, FP the number of false positive blocks, TP the number of true positive blocks, N the total number of negative blocks and P the total number of positive blocks.

The classification output is illustrated in Fig. 3c and Fig. 4c with two distinct MOC images (R17-00333 and S01-00925). The overall performances obtained for all images are synthesized in Table 1. Globally, the values achieved are very good, with the majority of situations (6 out of 8) presenting probabilities of error below 0.024: these refer to the features HP(9), HPM(9) and HPM, with both classifiers. The exception is given by the HP features which, both for Boosting and SVM classifiers, attain a probability of error of 0.347 and 0.436. This means that the phase of the gradient is not, by itself, a discriminative feature.

Table 1. Performance of the two classifiers for the detection of Martian dunes

Features	Boosting			SVM		
	p_{FN}	p_{FP}	p_{error}	p_{FN}	p_{FP}	p_{error}
HP(9)	0.019	0.032	0.023	0.0408	0.007	0.024
HPM(9)	0.017	0.016	0.017	0.0362	0.007	0.022
HP	0.353	0.330	0.347	0.6431	0.228	0.436
HPM	0.013	0.014	**0.013**	0.0341	0.004	**0.019**

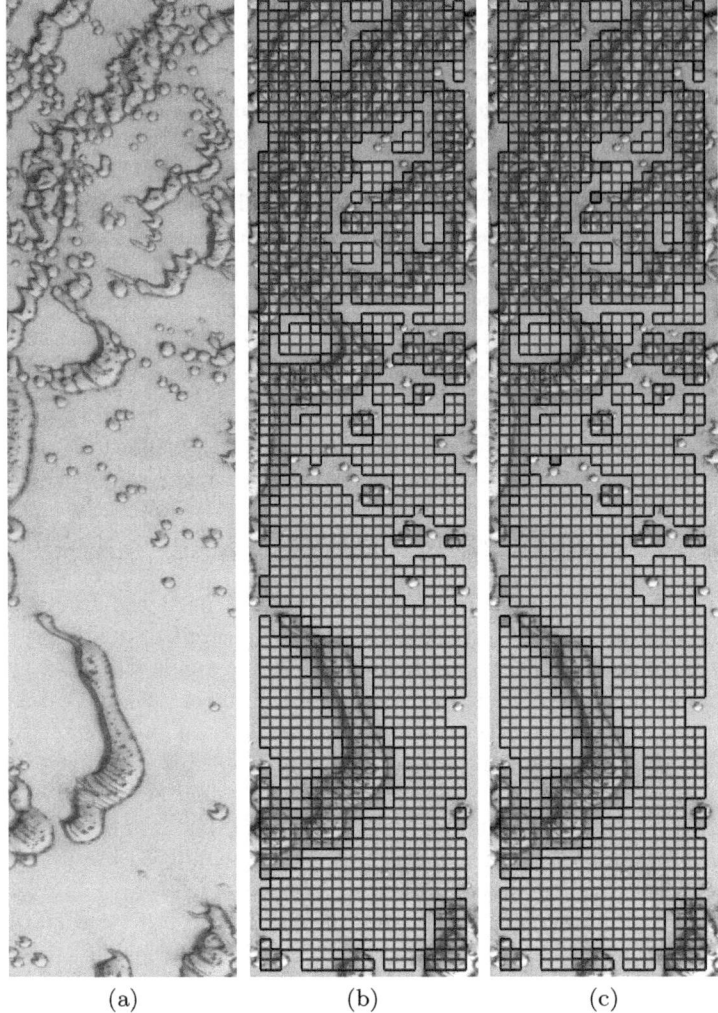

(a)   (b)   (c)

Fig. 3. First example of dune classification on part of the image R17-00333 (TP in green, TN in yellow, FN in red, FP in blue): (a) input image with overlapping of manual ground-truth; (b) ground-truth tiling in cells; (c) output of SVM classifier with HPM features [image credits: MSSS/NASA/JPL].

Although the best performances of each classifier are achieved with the same HPM features (0.013 for Boosting and 0.019 for SVM), their difference is not relevant when compared to the values obtained with HP(9) and HPM(9) features. There is some concordance in these results, since both classifiers perform excellently for the same sets of features (HP(9), HPM(9) and HPM) and both have a weak performance when using the HP features. The histogram of magnitude seems to be the most discriminative feature and no advantage is observed in this problem by splitting the image block into 9 cells.

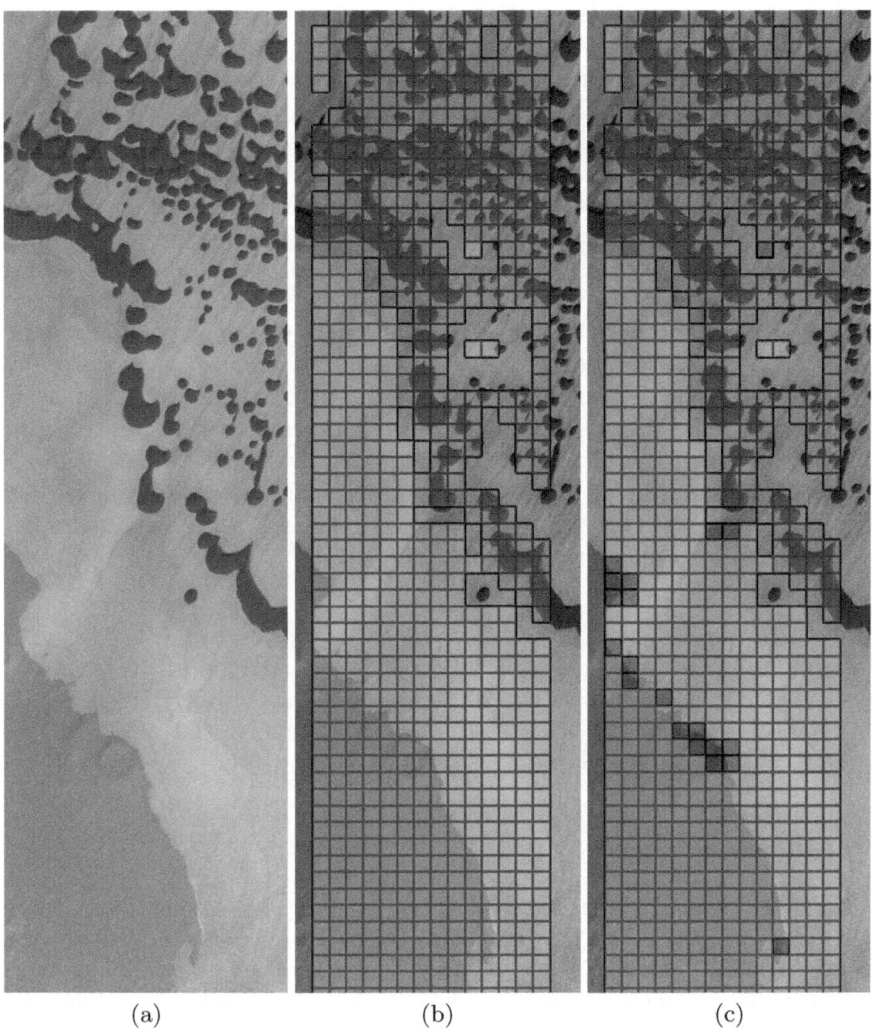

(a)   (b)   (c)

Fig. 4. Second example of dune classification on part of the image S01-00925 (TP in green, TN in yellow, FN in red, FP in blue): (a) input image with overlapping of manual ground-truth; (b) ground-truth tiling in cells; (c) output of Boosting classifier with HP(9) features [image credits: MSSS/NASA/JPL].

5 Conclusions

The major conclusion put forth in this paper is that the adequacy of automated methods for dealing with the diversity of sand dunes on the Martian surface was verified, as correct detections with significant performances were achieved. The values obtained in this first experiment indicate that the key factor resides on the selection of the features that are more adequate for describing the characteristics of these aeolian structures. In particular, the amplitude of the gradient has proved to be the most informative feature.

Although a set of powerful features and classifiers were successfully used on representative samples of the large diversity of Martian dune fields, we must remind that this is only a preliminary study. We have dealt with dune fields composed by individuals of different sizes, shapes and densities in distinct illumination conditions, but we are aware that many more different situations will have to be faced, namely considering the scale and the diversity of the Martian landscape where many other geomorphological features can and will sometimes be present. Nonetheless, we believe that the adaptive and learning nature of the methods we are using will be able to deal with those different circumstances.

In future work we intend to greatly expand the datasets by incorporating images of every type of Martian dunes and testing on them the approaches we have employed here; we will also test additional types of features and classifiers. Moreover, and with the ultimate goal of making available a robust tool to be used in the cartography of Martian dunes at a planetary scale, we also intend to automatically classify the Martian dunes according to the scheme used in the classification of analogue terrestrial structures [15].

Acknowledgments. This work was partially supported by FCT Portugal under the pluriannual funding attributed to CERENA/IST and ISR/IST and the project PTDC/CTE-SPA/099041/2008. L. Bandeira (SFRH/BD/40395/2007) and J. Saraiva (SFRH/BD/37735/2007) acknowledge financial support by FCT Portugal.

References

1. Greeley, R., Kuzmin, R.O., Haberle, R.M.: Aeolian processes and their effects on understanding the chronology of Mars. Space Science Reviews 96, 393–404 (2001)
2. Wilson, S., Zimbelman, J.: Latitude-dependent nature and physical characteristics of transverse aeolian ridges on Mars. Journal of Geophysical Research-Planets 109, E10003 (2004)
3. Edgett, K.S., Malin, M.C.: New views of Mars eolian activity, materials, and surface properties: three vignettes from the Mars Global Surveyor Mars Orbiter Camera. Planetary and Space Science 50, 151–155 (2000)
4. Hayward, R.K., Mullins, K., Fenton, L., Hare, T., Titus, T., Bourke, M., Colaprete, A., Christensen, P.: Mars global digital dune database and initial science results. Journal of Geophysical Research-Planets 112, E1107 (2007)
5. Hayward, R.K., Mullins, K.F., Fenton, L.K., Titus, T.N., Tanaka, K.L., Bourke, M.C., Colaprete, A., Hare, T.M., Christensen, P.R.: Mars global digital dune database (MGD3): User's guide. In: Planetary Dunes Workshop: A Record of Climate Change, abs. 7013, Alamogordo, NM (2008)
6. Hayward, R.K., Mullins, K.F., Fenton, L.K., Titus, T.N., Bourke, M.C., Colaprete, T., Hare, T., Christensen, P.R.: Mars digital dune database: Progress and application. In: Lunar and Planetary Science XXXVIII, abs. 1360, Houston, TX (2007)

7. Barata, T., Alves, E.I., Saraiva, J., Pina, P.: Automatic recognition of impact craters on the surface of Mars. In: Campilho, A.C., Kamel, M.S. (eds.) ICIAR 2004. LNCS, vol. 3212, pp. 489–496. Springer, Heidelberg (2004)
8. Bue, B.D., Stepinski, T.F.: Machine detection of Martian impact craters from digital topography data. IEEE Transactions on Geoscience and Remote Sensing 45(1), 265–274 (2007)
9. Bandeira, L., Saraiva, J., Pina, P.: Impact crater recognition on Mars based on a probability volume created by template matching. IEEE Transactions on Geoscience and Remote Sensing 45(12), 4008–4015 (2007)
10. Salamunićcar, G., Lončarić, S.: Open framework for objective evaluation of crater detection algorithms with first test-field subsystem based on MOLA data. Advances in Space Research 42(1), 6–19 (2008)
11. Salamunićcar, G., Lončarić, S.: GT-57633 catalogue of Martian impact craters developed for evaluation of crater detection algorithms. Planetary and Space Science 56(15), 1992–2008 (2008)
12. Martins, R., Pina, P., Marques, J.S., Silveira, M.: Crater detection by a boosting approach. IEEE Geoscience and Remote Sensing Letters 6(1), 127–131 (2009)
13. Urbach, E.R., Stepinski, T.F.: Automatic detection of sub-km craters in high resolution planetary images. Planetary and Space Science 57(7), 880–887 (2009)
14. Stepinski, T.F., Mendenhall, M.P., Bue, B.D.: Machine cataloging of impact craters on Mars. Icarus 203(1), 77–87 (2009)
15. McKee, E.D.: Introduction to a study of global sand seas. In: McKee, E.D. (ed.) A study of global sand seas, pp. 1–19. University Press of the Pacific, Honolulu (1979)
16. Dalal, N., Triggs, B.: Histograms of oriented gradients for human detection. In: CVPR 2005-Computer Vision and Pattern Recognition Conference, vol. 1, pp. 886–893. IEEE Press, New York (2005)
17. Schapire, R.E., Freund, Y., Bartlett, P., Lee, W.S.: Boosting the margin: a new explanation for the effectiveness of voting methods. Annals of Statistics 26(5), 1651–1686 (1998)
18. Viola, P., Jones, M.: Robust real-time face detection. International Journal of Computer Vision 57(2), 137–154 (2004)
19. Vapnik, V.N.: The nature of statistical learning theory. Springer, Berlin (1995)
20. Joachims, T.: Estimating the generalization performance of an SVM efficiently. In: Proceedings of the Seventeenth International Conference on Machine Learning, pp. 431–438. Morgan Kaufmann Publishers Inc., San Francisco (2000)

Directional Gaze Analysis in Webcam Video Sequences

V. Vivero, N. Barreira, M.G. Penedo, D. Cabrero, and B. Remeseiro

Department of Computer Science, University of A Coruña (Spain)
v.viverogarcia@gmail.com, {nbarreira,mgpenedo,cabrero,bremeseiro}@udc.es

Abstract. The analysis of the gaze direction has many applications. Most of the proposed techniques need special devices to estimate the gaze direction, but, in practice, the high cost of these devices prevents a widespread use. For this reason, the research in this field is currently focused on the development of techniques that work with low-cost devices. In this paper, we present a novel approach to perform a directional gaze analysis from webcam video sequences. This approach is based on well-known segmentation and pattern recognition techniques. It is fully automatic since it does not need user interaction and it can be applied in real time. We also present preliminary results that prove the efficiency and accuracy of the proposed methodology.

1 Introduction

Eye gaze is one of the most studied facial features due to its practical applications. The detection of sleepiness in drivers, the improvement of the usability in web sites, and the use of the gaze as a human-computer interface are some examples that highlight the importance of the research in this field.

Most of the current methodologies need specific devices or several light sources to follow the eye gaze. On one hand, several works use special cameras to gather eye information [1,2]. Even though much research has been devoted to the development of lightweight devices [3], these approaches are uncomfortable for the final users. On the other hand, other techniques [4,5,6] use infrared illumination to make the pupil tracking easier. Nevertheless, in spite of their accuracy, the cost and complexity of the hardware makes difficult the proliferation of systems based on these devices.

In this sense, many efforts have been focused on the development of low-cost approaches, mainly based on webcams, to analyze the eye gaze. Lin et al. [7] presented an eye tracking system to control the mouse. Their approach is based on a K-nearest neighbor classifier and an adaptive skin model to locate the face and, after that, to segment the eyes using color information. Torricelli et al. [8] used the Hough transform to extract the eyes from the video frames. The eye areas were the input to a neural network that estimated the gaze direction. Valenti et al. [9] proposed a method that uses a scale space framework to locate the eyes and a linear mapping method to estimate the eye gaze. As a consequence, this approach needs some user calibration. Also, the OpenGazer project [10] is an

A. Campilho and M. Kamel (Eds.): ICIAR 2010, Part II, LNCS 6112, pp. 316–324, 2010.

open source gaze tracking application that needs some user interaction. First, the user selects several relevant feature points on the input video. The input frames and the selected feature points are used to train a Gaussian process that represents the mapping between the image of an eye and the position on the screen. After that, the system is ready to perform the gaze tracking using the Viola-Jones algorithm [11] for face and eye detection and the Gaussian process for gaze prediction. The calibration process is performed for each video sequence and needs some user interaction (e.g. the user should keep the head still).

In this paper, we propose a methodology to perform a basic directional gaze analysis in a real time scenario using inexpensive technology, such as a webcam. In our methodology, we receive the input video from the webcam and we search for the area where the eyes are located. Then, we use threshold techniques to characterize the skin and, consequently, the sclera and pupil. Finally, we apply directional patterns in order to decide the gaze direction in each frame. Our approach is fully automatic since it does not need further training, calibration, user interaction, or special lighting conditions.

This paper is organized as follows. Section 2 describes the proposed methodology. Section 3 shows the preliminary results obtained with our approach. Finally, Section 4 presents the conclusions and future work.

2   Methodology

The goal of this work is the development of a generic methodology to perform a directional analysis of the eye gaze. There were three main requirements in the design of this methodology. First, it should process data in real time in order to be used in real scenarios. Second, it should work with low-cost non-intrusive acquisition devices, such as standard webcams. Third, it should avoid interaction from the final users in order to simplify its operation. To this end, we propose a methodology with three stages, as Fig. 1 shows. First, we extract the image area where the eyes are located from the webcam video frames. Then, we identify each eye within the interest region and, finally, we analyze the gaze direction in both eyes. This section explains in detail these three stages.

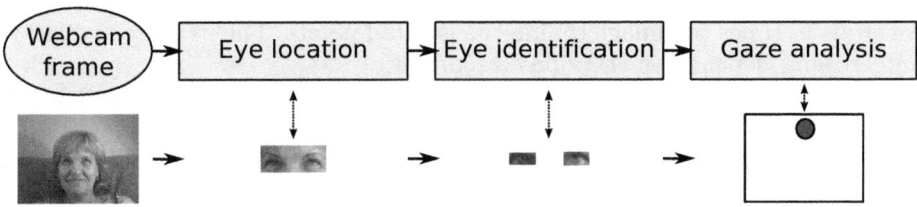

Fig. 1. Main stages of the proposed methodology for directional gaze analysis

2.1 Eye Location

The eye location task is performed by means of the Viola-Jones algorithm [11]. It is based on the combination of a *cascade* of classifiers that explore the image in multiple scales and locations. In the training stage, simple features, similar to Haar basis functions, are located. After that, an Ada-Boost learning algorithm selects the best features and trains the classifiers with them. A strong classifier is finally built from a collection of weak classifiers. In the detection stage, a series of classifiers are applied to every subwindow. The background areas are eliminated by the initial classifiers with very little processing and only the interesting areas are analyzed in detail by subsequent classifiers. Thus, this algorithm is very efficient and can be used in real-time applications. Also, this algorithm was successfully applied in several face detection and eye location applications [10].

2.2 Eye Identification

Once the Viola-Jones algorithm locates the eye region within the frame, next step is the identification of both right and left eyes. This task is very complex since the eye areas are very small and are affected by the lighting conditions. Some approaches use threshold techniques to identify the sclera within the image region [7]. However, the results depend on the lighting conditions since the light sources can generate white spots in the skin or shadows in the eye region. Other approaches find circles in the image in order to locate the iris [8]. But, depending on the gaze direction or the head pose, the iris is not always a circle.

In this work, we propose a novel approach. Since we want to segment the eyes from the rest of the face in the eye region, we try to discard the skin pixels. To this end, we convert all the pixels in the eye region to the TSL (*Tint, Saturation, Lightness*) color space [12]. This color space is very robust for the segmentation of skin pixels under different lighting conditions since it isolates the usual skin colors from the rest of the color values. Thus, a simple threshold technique identifies the skin pixels. Results of the skin segmentation using the TSL color space can be seen in Fig. 2.

Since the quality of the segmentation decreases with poor lighting conditions, as Fig. 3 shows, we apply morphological operators in order to improve the eye segmentation. First, we apply n dilations to the segmented image and, finally, we erode m times the image to link the isolated points. The number of dilations and erosions depends on the video resolution.

The last step in this stage is a region growing algorithm to put together each eye region. Two seeds are selected to start the growing procedure, one seed in each eye. Note that the eyes are usually located in the same position within the extracted eye region. If we draw a horizontal line that splits the eye region down the middle, the line will go through both eyes. Moreover, if we draw a vertical line that divides the eye region into equal portions, each eye will be located on each side of the line. If the input eye regions do not fulfill these conditions and, as a consequence, a correct seed is not found, the current frame is discarded from subsequent analysis.

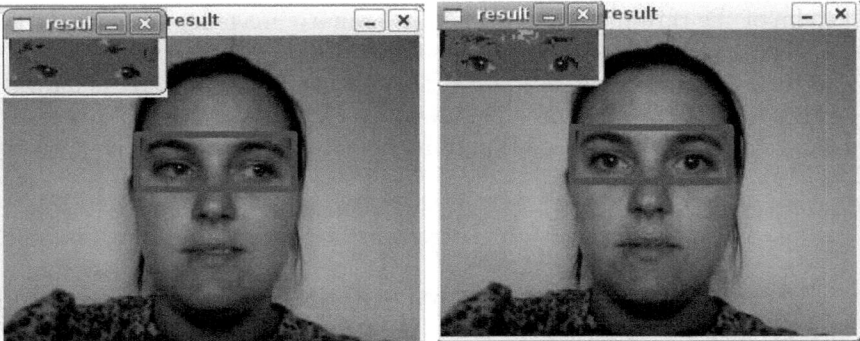

Fig. 2. Examples of skin segmentation using the TSL color space. The red pixels in the subwindows point out the skin pixels. The rectangles show the eye region detected by the Viola-Jones algorithm.

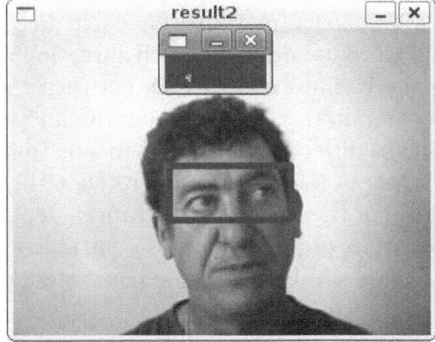

Fig. 3. The eye segmentation gets complicated in different lighting conditions. In this example, the video frames are super-exposed.

Once the seeds are selected, the region growing algorithm puts together all the non-skin pixels. We consider several criteria as the stop conditions of the algorithm. Besides the classic criterion of similarity between each pixel and its neighboring pixels, we take into account the size and the shape of the regions. If the final regions does not fulfill these criteria, the region growing algorithm is applied again from different seeds.

2.3   Gaze Analysis

Once both eyes are segmented, we analyze the gaze direction in each eye separately and we combine the results in order to increase the accuracy of the final decision. This analysis includes two steps. First, the input images are thresholded in order to obtain binary images. Then, the gaze direction is defined by means of a pattern matching technique.

The aim of the thresholding is obtaining a binary image for each eye. In these images, the black pixels should correspond to the iris and the pupil while the white pixels represent the sclera. Since the lighting conditions are unknown, we apply an adaptive thresholding. This way, we compute a threshold α for each pixel using the intensity information of its neighborhood as follows:

$$\alpha = \frac{\sum_{i=x-\frac{\tau}{2}}^{x+\frac{\tau}{2}} \sum_{j=y-\frac{\tau}{2}}^{y+\frac{\tau}{2}} EyeImage(i,j)}{\tau^2} - \varsigma \tag{1}$$

where $EyeImage(i,j)$ is the intensity value of the eye input image at the position (i,j), τ is the window size, and ς is a correction factor. The correction factor prevents the pixel being marked as background in a region with constant intensity. Note that the window size affects, not only the segmentation quality (bigger windows are more affected by lighting conditions), but also the computation times.

Once the binary image is computed, we compare the position of the black pixels (pupil and iris) with four directional patterns (North, South, East, and West) and we obtain a similarity measure for each direction. Figure 4 shows the directional patterns we have defined. Each directional pattern is scaled to the eye image size and the Hamming distances between the eye image and each directional pattern are computed. The Hamming distance counts the number of pixels in which both images differ. A high value means that the eye is looking at the opposite direction whereas a low value means that the eye is looking at this direction. Since the values of the Hamming distances depend on the image size, we normalize the distances in order to obtain a suitable similarity measure. To this end, the distance D in each direction a is computed as follows:

$$D_a = \frac{HD_{a'} * \beta}{HD_a} \tag{2}$$

where HD_a and $HD_{a'}$ represent the Hamming distance between the image and two patterns in opposite directions $(a, a') = \{(N, S), (S, N), (W, E), (E, W)\}$. β weights the distances and makes the thresholding easier. When the gaze is centered, $HD_a \approx HD_{a'}$ so that $D_a \approx \beta$ and $D_{a'} \approx \beta$. This way, we can say that values similar to β represent the center direction.

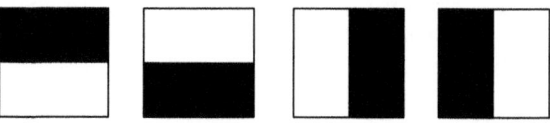

Fig. 4. Directional patterns. From left to right: North, South, East, and West patterns.

The highest D_a value points out the gaze direction. However, we can also determine *how much* the user is looking in a specific direction. Moreover, the similarity between D_a and β represents how centered is the gaze.

Finally, we compare the directions obtained for both right and left eyes. If both directions coincide, we set the gaze direction for the current frame. Otherwise, the gaze direction remains unknown. Figure 5 shows two frames of a video sequence where the gaze direction is correctly detected.

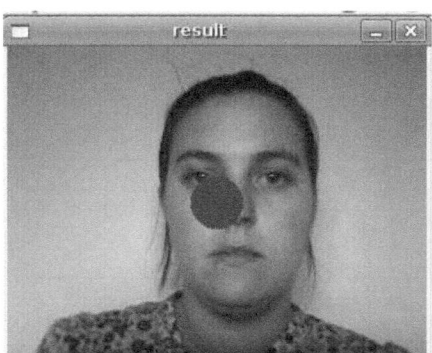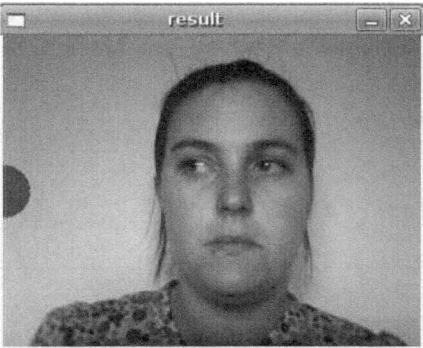

Fig. 5. Directional analysis. The red dots point out the gaze direction in each frame.

Since this methodology comprises several stages, a wrong processing in any stage implies that the gaze direction will be unknown in the current frame. However, changes in the gaze direction involve several frames so that we include a delay parameter δ in the methodology. If the gaze direction is lost in a frame, we keep the direction of the previous frame for δ frames. Nevertheless, δ should be set to a small value in order to prevent errors in case of abrupt eye movements. The results show that this parameter improves the performance of the proposed methodology.

3   Results

We apply the proposed methodology to several webcam video sequences in order to prove the suitability of our approach. Table 1 shows the number of frames in each video sequence. All the videos were captured at 10 fps with a *Sony Visual Communications VGP-VCC1* camera and with a resolution of 640x480 pixels. In order to validate our approach, we tested, not only the complete system, but also each single stage of the proposed methodology.

In the *eye location* stage, the training set for the Viola-Jones algorithm consisted of 10 eye regions and 667 non-eye regions extracted from the *Faces94* database [13]. This database contains images of both men and women with different facial expressions and no lighting variations. To test the trained algorithm, we select 236 frames at random from the video sequences. Table 2 shows the statistical results of this stage in the set of frames randomly selected. We assess the quality of our approach in terms of sensitivity since it measures the proportion

Table 1. Number of frames in each video sequence used for the validation of the proposed methodology

Sequence	1	2	3	4
Number of Frames	113	105	109	85

Table 2. Statistical analysis of the methodology in the test set. The sensitivity measures the proportion of gaze directions correctly identified.

Stage	Test set	TP	FN	FP	Sensitivity
Eye location	236 random frames	234	2	13	0.99
Eye identification	Sequence 1	86	23	4	0.79
	Sequence 2	90	15	0	0.86
	Sequence 3	106	1	2	0.99
	Sequence 4	69	10	6	0.87
Gaze analysis	Sequence 1	81	3	2	0.96
	Sequence 2	69	16	5	0.81
	Sequence 3	101	3	2	0.97
	Sequence 4	65	3	1	0.96
Complete system	Sequence 1	100	7	6	0.93
	Sequence 2	75	25	5	0.75
	Sequence 3	105	3	1	0.95
	Sequence 4	72	10	3	0.88

of gaze directions correctly identified. In this sense, the high sensitivity value (0.99) proves the accuracy of the Viola-Jones algorithm and the suitability of the selected train set.

Table 2 also shows the results of the *eye identification* and *gaze analysis* stages. First, we apply the *eye identification* stage to the whole video sequences. Even though the sensitivity values are not very high in some cases, the number of false positives is low. Then, we only apply the *gaze analysis* stage to those frames correctly segmented in the *eye identification* stage. In this case, the sensitivity values are very high. The *gaze analysis* stage has several parameters. In the adaptive thresholding, the parameters τ and ς were empirically set to 7 and -5, respectively. Bigger windows are affected by the lighting conditions and slow down the thresholding process whereas smaller windows produce irregular eye boundaries. In the directional analysis, the central direction is represented by $\beta = 1200$. Finally, the delay parameter δ was set to 3 frames.

The *complete system* row in Table 2 shows the results of the whole methodology in the test sequences. Note how the delay parameter δ allows the correction of a high number of undetected gaze directions. Even though Sequence 2 presents a high number of false negatives due to lighting conditions, the sensitivity values are high.

The methodology was implemented using the OpenCV library in GNU/Linux and the tests were performed in an Intel Core 2 Duo at 2.40 GHz with 2GB of RAM. Figure 6 shows the computation times for several video sequences.

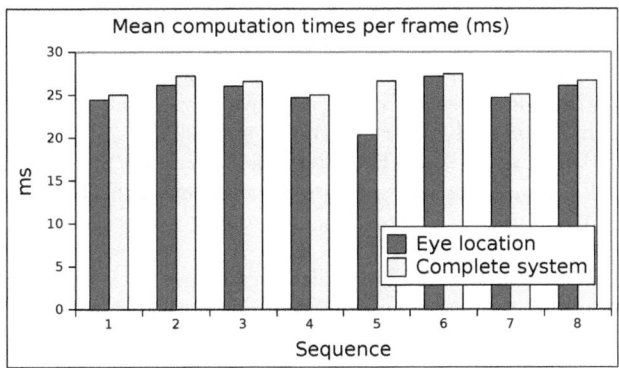

Fig. 6. Mean computation times per frame in several video sequences. The bottle-neck of the system is the *eye localization* stage.

The mean computation time is lower than 26 ms so that this methodology can process 38 fps. The *eye localization* stage is the most time-consuming stage since it analyzes the whole image. The other stages work with small eye regions.

4   Conclusions

This work presents a novel approach for the directional gaze analysis in video sequences using a low-cost acquisition device. The proposed methodology is fully automatic and does not need user interaction. First, we apply a Viola-Jones algorithm to identify the eye region. This algorithm requires a previous training stage to distinguish the regions of interest. This stage is performed once using a public face database and its results can be applied to any video sequence. After that, we distinguish between skin and eye pixels using a thresholding technique in the TSL color space. The result of this stage is the segmentation of both right and left eyes. The gaze direction is given by a pattern matching between directional patterns and the binary representation of each eye. The methodology is very fast and can be applied in real time applications. The sensitivity of the proposed methodology is greater than 0.75 under usual lighting conditions.

Future work in this field includes the study of the lighting conditions in order to increase the accuracy of the segmentation. In this sense, the application of histogram techniques or the analysis of the color components could improve the selection of suitable threshold values. Also, the directional analysis can be extended using more directional patterns or taking into account all the directional similarity measures.

Acknowledgments. This paper has been partly funded by Ministerio de Ciencia y Tecnología–Instituto de Salud Carlos III (Spain) through the grant contract PI08/90420 and FEDER funds.

References

1. Pelz, J.B., Canosa, R., Babcock, J.S., Kucharczyk, D., Silver, A., Konno, D.: Portable eyetracking: A study of natural eye movements. In: Proceeding of the SPIE: Human Vision and Electronic Imaging (2000)
2. Ryan, W.J., Duchowski, A.T., Birchfield, S.T.: Limbus/pupil switching for wearable eye tracking under variable lighting conditions. In: ETRA '08: Proceedings of the 2008 symposium on Eye tracking research & applications, pp. 61–64. ACM, New York (2008)
3. Babcock, J.S., Pelz, J.B.: Building a lightweight eyetracking headgear. In: ETRA '04: Proceedings of the 2004 symposium on Eye tracking research & applications, pp. 109–114. ACM, New York (2004)
4. Pérez, A., Córdoba, M.L., García, A., Méndez, R., Muñoz, M.L., Pedraza, J.L., Sánchez, F.: A precise eye-gaze detection and tracking system. In: WSCG (2003)
5. Chen, J., Tong, Y., Gray, W., Ji, Q.: A robust 3d eye gaze tracking system using noise reduction. In: ETRA '08: Proceedings of the 2008 symposium on Eye tracking research & applications, pp. 189–196. ACM, New York (2008)
6. Ohno, T., Mukawa, N., Yoshikawa, A.: Freegaze: a gaze tracking system for everyday gaze interaction. In: ETRA '02: Proceedings of the 2002 symposium on Eye tracking research & applications, pp. 125–132. ACM, New York (2002)
7. Lin, Y.P., Chao, Y.P., Lin, C.C., Chen, J.H.: Webcam mouse using face and eye tracking in various illumination environments. In: 27th Annual International Conference of the Engineering in Medicine and Biology Society, IEEE-EMBS 2005, January 2005, pp. 3738–3741 (2005)
8. Torricelli, D., Conforto, S., Schmid, M., D'Alessio, T.: A neural-based remote eye gaze tracker under natural head motion. Computer Methods and Programs in Biomedicine 92(1), 66–78 (2008)
9. Valenti, R., Staiano, J., Sebe, N., Gevers, T.: Webcam-based visual gaze estimation. In: Foggia, P., Sansone, C., Vento, M. (eds.) Image Analysis and Processing – ICIAP 2009. LNCS, vol. 5716, pp. 662–671. Springer, Heidelberg (2009)
10. Zielinski, P.: Opengazer: open-source gaze tracker for ordinary webcams, http://www.inference.phy.cam.ac.uk/opengazer/
11. Viola, P., Jones, M.: Rapid object detection using a boosted cascade of simple features. In: Proc. CVPR, vol. 1, pp. 511–518 (2001)
12. Terrillon, J.C., David, M., Akamatsu, S.: Automatic detection of human faces in natural scene images by use of a skin color model and of invariant moments. In: FG '98: Proceedings of the 3rd. International Conference on Face & Gesture Recognition, Washington, DC, USA, p. 112. IEEE Computer Society, Los Alamitos (1998)
13. Spacek, L.: Collection of facial images: Faces94, http://cswww.essex.ac.uk/mv/allfaces/faces94.html

Novelty Detection on Metallic Surfaces by GMM Learning in Gabor Space

Yigitcan Savran and Bilge Gunsel

Multimedia Signal Processing and Pattern Recognition Lab., Dept. of Electronics and
Comm. Eng, Istanbul Technical University, 34469 Maslak, Istanbul, Turkey
gunselb@itu.edu.tr

Abstract. Defect detection on painted metallic surfaces is a challenging task in
inspection due to the varying illuminative and reflective structure of the surface.
This paper proposes a novelty detection scheme that models the defect-free surfaces by using Gaussian Mixture Models (GMMs) trained in Gabor space. It is
shown that training using the texture representations obtained by Gabor filtering
takes the advantage of multiscale analysis while reducing the computational
complexity. Test results reported on defected metallic surfaces including pinhole, crater, hav, dust, scratch, and mound type of abnormalities demonstrate
the superiority of developed integrated system with respect to the stand alone
Gabor filtering as well as the spatial domain GMM classification.

Keywords: Novelty Detection, Gaussian Mixtures, Gabor Filters.

1 Introduction

Computer based novelty detection on painted metallic surfaces is a challenging task
due to the structure of the surface. Painted metallic surfaces have varying brightness
and highly reflective features which cause a complex and difficult structure in modeling and novelty detection. In addition, the color of the painted surface and whether
paint includes small metallic particles is another issue to be considered.

Providing a multiscale approach, Gabor filtering is a frequently used technique in
novelty detection for its orientation and frequency selective features. In [1], Jain et all.
developed an inspection technique to grade the painting quality of automotive finishes
by using Gabor filters. In this work, it is shown that under appropriate lighting, Gabor
features can provide sufficient textural information about painted metallic surfaces.
Main difficulty in using Gabor filtering is it requires a precise tuning the filter parameters according to the texture under inspection. Another successfully applied
method in novelty detection with aperiodic and complex textures is Gaussian mixture
models (GMMs). In [2], the complex and aperiodic form of ceramic surfaces is modeled by the GMMs using multiscale gray level as well as chromatic images to improve the modeling performance of GMMs. The defect detection accuracy reported in
[2] reaches to 92.7% when gray level images of ceramic tiles are classified. In this
paper, in order to model complex and difficult structure of the painted metallic surfaces, we propose a GMM classifier trained in Gabor space. It is shown that being a
multiscale filtering scheme, Gabor filtering can provide an efficient feature set that is

A. Campilho and M. Kamel (Eds.): ICIAR 2010, Part II, LNCS 6112, pp. 325–334, 2010.

capable of modeling textural structure of metallic surfaces. It is also shown that GMMs trained in Gabor space is capable of reflecting aperiodic textural structure of metallic paintings that yields a powerful classification task which is highly robust to lighting variations as well as color changes arise from complex form of the surface.

In Section 2, GMM modeling of textural features in Gabor space is presented. Novelty detection using GMMs is formulated in Section 3. Section 4 reports the test results obtained on the metallic surfaces of car paint finishes. In Section 5 we summarize conclusions and give some directions for the future work.

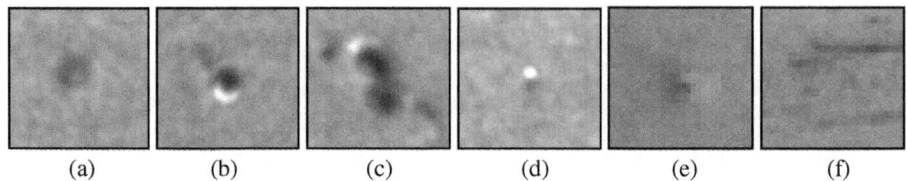

(a)   (b)   (c)   (d)   (e)   (f)

Fig. 1. Painting defects acquired by a digital camera zoomed on a metallic gray finish: (a) crater, (b,c) Hav, (d) mound , (e) bump, (f) scratch

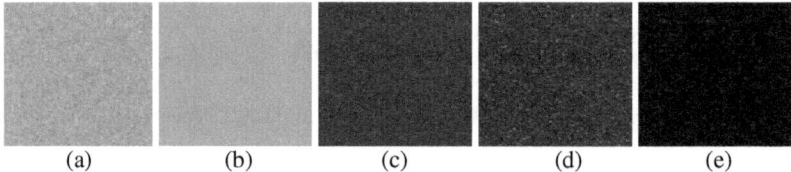

(a)   (b)   (c)   (d)   (e)

Fig. 2. Gray level images of painted metallic surfaces with different colors: (a) metallic gray, (b) white, (c) navy, (d) metallic nacreous black, (e) metallic black.

2 Modeling the Painted Surfaces by GMMs in Gabor Space

We propose a novelty detection method that integrates Gabor filtering with the Gaussian Mixture Modeling. As it can be seen from Fig. 1, defects on painted automotive surfaces may have varying size, direction and depth. This requires a multiscale, multi orientation inspection. Therefore we propose extraction of features which represent the surface texture in Gabor space. In the literature, Gabor filtering has been widely used because it provides a multiresolution feature extraction scheme and offers a wavelet like transformation [3, 4]. However, in our case Gabor filtering itself is not adequate to detect the novelties accurately. This is because of the nonstructural and highly variable nature of metallic surface textures that may cause several false alarms (Fig.2). To eliminate this drawback, unlike the existing works, we propose GMM modeling of Gabor features for automatic novelty detection. Details of the method are presented in the following subsections.

2.1 Extraction of Textural Features by Gabor Filtering

As accepted a good model of 2D receptive profiles of mammalian cortical simple cells, Gabor filters provide optimal localization in space and frequency domains and offer orientation selectivity and spatial locality [4, 5, 6, 7]. In this work these features of Gabor filters are exploited in modeling the background texture of painted metallic surfaces for defect detection.

The Gabor filtered output image at the spatial coordinate (x,y), $O_{m,v}(x,y)$, can be expressed as convolution of an input image I with a Gabor filter family $\psi_{m,v}$ as it is shown in Eq.(1).

$$O_{m,v}(x, y) = I(x, y) * \psi_{m,v}(x, y) \tag{1}$$

The Gabor filter family is formulated by Eq.(2) [4].

$$\psi_{m,v}(x, y) = \frac{\left\| k_{m,v} \right\|^2}{\sigma^2} e^{-\frac{\left\| k_{m,v} \right\|^2 (x^2+y^2)}{2\sigma^2}} \left[e^{(ik_{m,v})} - e^{-\frac{\sigma^2}{2}} \right] \tag{2}$$

As it can be seen from Eq.(2), Gabor function is a sinusoidal function modulated with a Gaussian. The first multiplicative term is the Gaussian function which localizes the oscillation while the second term in the bracket refers to the oscillatory part. Here, m and v determine respectively the orientation and scale of the kernel. σ refers to the ratio of the Gaussian window width to wavelength. ‖.‖ denotes the norm operator and $k_{m,v}$ is the wave vector defined in Eq.(3) where $\phi_m = \pi m / 4$, $m \in \{0,...,3\}$ refers to the orientation of the kernel

$$k_{m,v} = k_v e^{i\phi_m} = k_v \left(\cos \phi_m, \sin \phi_m \right) \quad . \tag{3}$$

It is shown that k_v, the central frequency can be formulated as in Eq.(4), where k_{max} is the maximum frequency of the kernel and is set to $\pi/2$, f is the frequency spacing between kernels and v determines the scale [4].

$$k_v = \left\| k_{m,v} \right\| = \frac{k_{max}}{f^v} \tag{4}$$

We propose a feature extraction scheme that fuse the information extracted at different orientations of a scale by Gabor filtering. Figure 3(a) illustrates the proposed scheme at one scale for four orientations. Assuming that each orientation as a different channel, in each channel $d \times d$ patches are collected as training samples and each training sample is filtered with a Gabor filter yielding a $1 \times d^2$ dimensional feature vector. Concatenating these vectors together from each channel, the training sample set $Z = \{Z_i\}_{i=1}^{P}$ is created and the $4 \times d^2$ dimensional feature sets at different scales are calculated correspondingly, where P is the number of patches.

Conventionally an appropriate Gabor filter set that represents the characteristic of a background texture, i.e., painted metallic surface images in our case, is designed. In novelty detection, the inspection system filters the observed surfaces by using the designed filter and the multiscale multi orientation filter outputs are integrated either

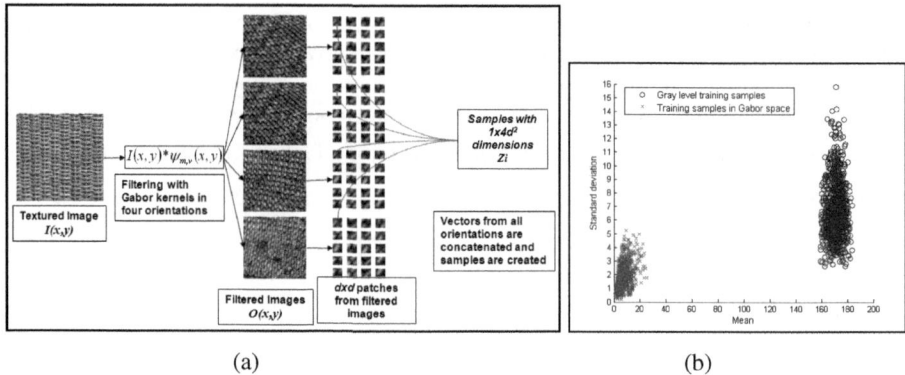

(a)   (b)

Fig. 3. (a) Proposed feature extraction scheme. (b) Standard deviation versus mean plot of features extracted for 3x3 patches taken from gray level image of a metallic gray painted surface and its Gabor filtered image. Number of training samples is 1000. Blue circles represents gray level image patches where red crosses represents Gabor filtered image patches.

by a simple thresholding or by another fusion logic to detect defected pixels. Fusion of multiscale multi orientation information is a difficult task in Gabor filtering. In contrast we propose a feature level integration scheme that models the background texture by Gaussian mixtures. In our scheme the training samples extracted by Gabor filtering are fed into the GMM module. Hence the fusion of Gabor filtered outputs problem is converted to a GMM learning problem that can be expressed as a GMM parameter estimation scheme. Fig.3(b) illustrates the standard deviation versus mean plot of features extracted for 3x3 patches taken from gray level image of a painted metallic gray car surface and its Gabor filtered image. Blue circles and red crosses respectively denote the features extracted in spatial gray level domain and filtered Gabor image domain. As it is seen, characteristic of textural features has a Gaussian form in both domains, however Gabor features provide compactness in terms of mean and standard deviation that leads us to model the background by GMMs in Gabor space.

2.2 Metallic Surface Modelling by GMMs

The proposed inspection system is trained for each color of automotive finish and then novelty detection is achieved as abnormality detection. Abnormalities may arise from several no complete or inappropriate processing steps performed by automotive painting mechanism. Therefore defects may vary in terms of their size, orientation, etc. Basically our system is capable of detecting pinhole, crater, hav, dust, scratch, and mound type of abnormalities that constitute most frequent cases.

We propose modeling the defect free painted metallic surface of a specific color by a K component GMM. Let μ_k and w_k denote respectively the mean vector and the covariance matrix of a mixture component k, where $k=1..K$. Hence the parameter vector of the Gaussian mixture model with K components is represented as $M = \{m_k\}_{k=1}^{K}$ where $m_k = \{\mu_k, w_k\}$ denotes the feature vector of component k. The training set $Z = \{Z_i\}_{i=1}^{P}$ constructed by Gabor filtered defect free painted metallic surface samples is used for the estimation of M where P is the total number of training samples. Note that Z_i is one point in the sample space and space dimension is determined by the patch size d. We use four orientations in Gabor filtering thus the space dimension is $4d^2$. Under the assumption that α_k, the prior probability of the kth Gaussian mixture component, is also unknown, the probability of a training sample Z_i belongs to the mixture component k is modeled by Eq.(5),

$$p(Z_i \mid \psi_k) = N\left(Z_i; \alpha_k, \mu_k, \omega_k\right), \tag{5}$$

where $\psi = \{\alpha_k, \mu_k, \omega_k\}_{k=1}^{K}$ denotes the GMM parameter vector. Note that sum of the prior probabilities is $\sum_{k=1}^{K} \alpha_k = 1$. As all mixture components are unknown the density function of Z_i given the parameter set ψ, can be calculated by the conditional probability given by Eq.(6).

$$p(Z_i \mid \psi) = \sum_{k=1}^{K} p\left(Z_i \mid m_k\right)\alpha_k \tag{6}$$

Then the log-likelihood that needs to be optimized for the entire training data set Z, is formulated by Eq.(7).

$$\log p(Z \mid \psi) = \sum_{i=1}^{P} \log\left(\sum_{k=1}^{K} p\left(Z_i \mid m_k\right)\alpha_k\right) \tag{7}$$

The maximum likelihood estimation of the Gaussian mixture parameters $\hat{\psi}$ given in Eq.(8) is calculated using Expectation Maximization (EM) algorithm.

$$\hat{\Psi} = \arg\max_{\psi} \log p(Z \mid \psi) \tag{8}$$

In the learning of GMM parameters, conventional EM algorithm steps are called recursively. In the first step, initial ψ values are assigned randomly. Then in every t. step $\psi^{(t)}$ is updated as given in Eq.(10), where the likelihood of kth mixture component given Z_i is calculated according to Eq.(9).

$$p\left(m_k \mid Z_i, \psi^{(t)}\right) = \frac{p\left(Z_i \mid m_k, \psi^{(t)}\right)\alpha_k}{\sum_{k=1}^{K} p\left(Z_i \mid m_k, \psi^{(t)}\right)\alpha_k} \tag{9}$$

$$\alpha_k^{(t)} = \frac{1}{P}\sum_{i=1}^{P} p\left(m_k^{t-1} \mid Z_i, \psi_k^{(t-1)}\right)$$

$$\mu_k^{(t)} = \frac{\displaystyle\sum_{i=1}^{P} Z_i\, p\left(m_k \mid Z_i, \psi_k^{(t)}\right)}{\displaystyle\sum_{i=1}^{P} p\left(m_k \mid Z_i, \psi_k^{(t)}\right)} \qquad (10)$$

$$w_k^{(t)} = \frac{\displaystyle\sum_{i=1}^{P} \left(Z_i - \mu_k^{(t)}\right)\left(Z_i - \mu_k^{(t)}\right)^T p\left(m_k \mid Z_i, \psi^{(t)}\right)}{\displaystyle\sum_{i=1}^{P} p\left(m_k \mid Z_i, \psi^{(t)}\right)}$$

Until the estimated parameter values are stabilized the Expectation and Maximization steps are called recursively and the Gaussian mixture parameters are updated. The GMM parameters of defect free paintings learned in Gabor space are stored for automatic novelty detection.

3 Automatic Novelty Detection

In the modeling step, Gaussian mixture parameters representing the metallic surfaces are learned in Gabor space. Automatic novelty detection is performed maximization of the likelihood ratio for each pixel. This is achieved by first filtering the observed test image with the Gabor filters used in the learning phase. For each pixel, test sample patches are created from the filtered test images as explained in sub section 2.2. For each pixel, a *dxd* patch is created from each filtered image and test patches concatenated together. Probability of a pixel belonging to a known defect free metallic surface is calculated by using the likelihood ratio given by Eq.(11).

$$p(T_i \mid \psi) = \sum_{k=1}^{K} p\left(T_i \mid m_k\right)\alpha_k \qquad (11)$$

In Eq.(11), T_i denotes the Gabor feature vector corresponding to the pixel to be classified, $p\left(T_i \mid \psi\right)$ is the probability value which is a similarity measure between that test sample and trained texture. In Eq(11), m_k and α_k respectively denote the mean vector and prior probability of the kth Gaussian mixture component, ψ is the Gaussian mixtures parameter set learned at the training step as it is explained in sub section 2.2.

The test pixel is assigned to one of the known defect free metallic surfaces if the likelihood calculated according to Eq.(11) remains less than a threshold which is specified during the training phase. Otherwise it is labeled as a defected pixel. The same decision criteria is applied through the observed pixels that concludes the novelty detection. In our work the decision threshold is specified by applying k-means clustering on the Gabor filtered defect free surface images while k is set to 2 denoting the defected and defect free cluster labels. Note that it is not easy to apply a similar task in spatial domain because of the high variations of pixel gray levels of metallic surfaces. Another issue that needs to be clarified is the number of defect free parameter sets. Although the same parameter set can be used for similar colors, we have learned the parameter set of each color to increase the performance. This results in a longer training step but does not change the decision criteria.

4 Test Results

Novelty detection performance tests are performed on gray level images acquired from painted metallic surfaces by a digital camera using an inspection lighting system. Performance of the proposed GMM learning in Gabor space method (ND1) is evaluated for different patch sizes at several scales where the number of mixtures k varies from 1 to 10. It is observed that satisfactory performance is achieved for the patch size $d=1$(patch becomes 1 pixel) and $k=2$. Performance achieved by the stand alone Gabor filtering (ND2) and the spatial domain grayscale GMM modeling method introduced in [2] (ND3) are also reported for comparison purposes. For the ND2, Gabor parameters are selected as $f=1.5$, $\sigma=1.5$, $v=2$ and 4 orientations are used. The same Gabor filter set is used for the ND1. For ND3, the best results are obtained with patch size $d=3$ and mixture component number $k=5$. Test results showed that the ND3 is very sensitive to illumination changes hence using smaller patch sizes results in higher false alarm rates.

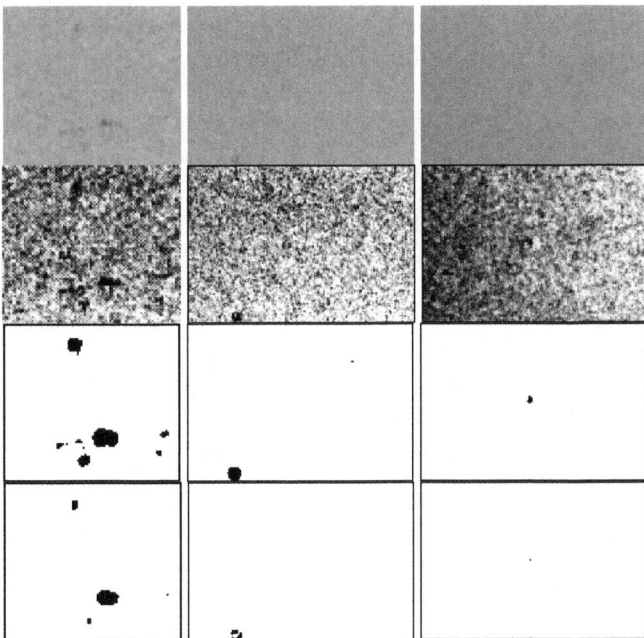

Fig. 4. Test results obtained on metallic gray colored metallic surface. 1. row: Gray level test images, 2.row: Histogram equalized test images. 3. row: Detected abnormalities by the proposed GMM learning in Gabor space (ND1). 4. row: Detected abnormalities by Gabor filtering with simple thresholding (ND2).

In Fig.4 test results obtained by using Gaussian mixture modeling based on Gabor features (ND1) and standalone Gabor filtering (ND2) are illustrated. Method presented in [4] is used for standalone Gabor filtering. Filtered images obtained at one scale and four orientations are fused with an 'OR' operation. First row of Fig.4 illustrates the gray level test images acquired from a metallic gray colored car surface. Second row presents histogram equalized images in order to highlight the defects. Note that histogram equalization amplified the textural components since the contrast is very low. The textural structure of defect free surface is complex hence applying a thresholding method gives high false alarm ratio. Third row and fourth row of Fig.4, respectively illustrate the detected abnormalities by ND1 and ND2. Superiority of the proposed GMM learning in Gabor space (ND1) to standalone Gabor filtering (ND2) on detection of defects with low contrast is clear.

A number of test has been performed to evaluate the performance improvement achieved by GMM learning in Gabor space (ND1) with respect to the spatial domain GMM classification (ND3). Fig. 5(a) illustrates a metallic surface image with a crater type of defect. Histogram equalized defected image is also given (Fig.5(b)) to highlight the defect. Note that histogram equalization amplified the textural components since the contrast is very low. Using the proposed GMM modeling in Gabor space, the defect is detected precisely (Fig.5(c)) where the GMM modeling in gray level space yields a one pixel defect (Fig.5(d)). We run similar routines on the defected image with scratch type of defect (Fig.5(e)). In histogram equalized version of image (Fig.5(f)) defects are seen in a connected form like resulting image obtained by ND1 (Fig.5(g)) where the abnormalities detected by ND3 (Fig.5(h)) are in separated form. In addition, the defect on the left upper side of the image could not be found by gray level GMM due to the reflective structure of the defect.

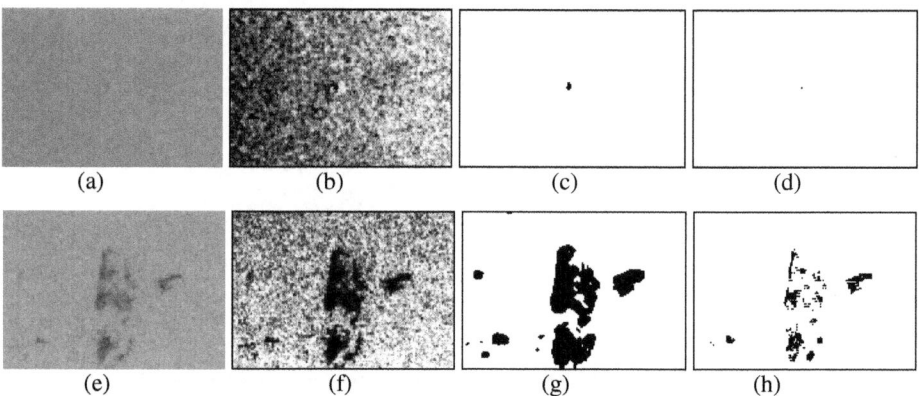

Fig. 5. Abnormalities detected on gray colored metallic surface. a), e) Gray level test images. b), f) Histogram equalized test images. Crater type defect c) detected by our method (ND1), d) detected by ND3. Detected scratch type defect g) by our method (ND1), h) by ND3.

Fig.6(a) illustrates two craters on the test surface. The first one is in the upper right region and the second one is in lower right region. In the histogram equalized image the defect in the upper side could be seen easily (Fig.6(b)). Both defects are detected by GMM learning in Gabor space (ND1) (Fig.6(c)), however GMM in spatial domain

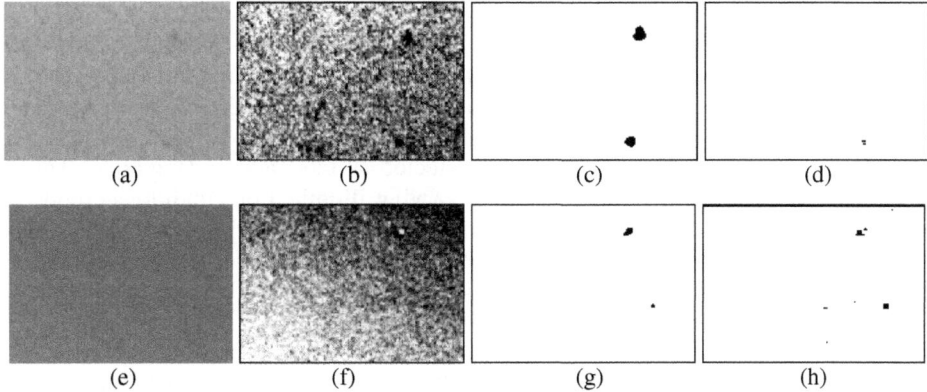

(a)   (b)   (c)   (d)

(e)   (f)   (g)   (h)

Fig. 6. Detected abnormalities on metallic gray colored car surface. a), e) Gray level test images. b), f) Histogram equalized test images. Hav type defect c) detected by our method (ND1), d) detected by ND3. Detected crater type defect g) by our method (ND1), h) by ND3.

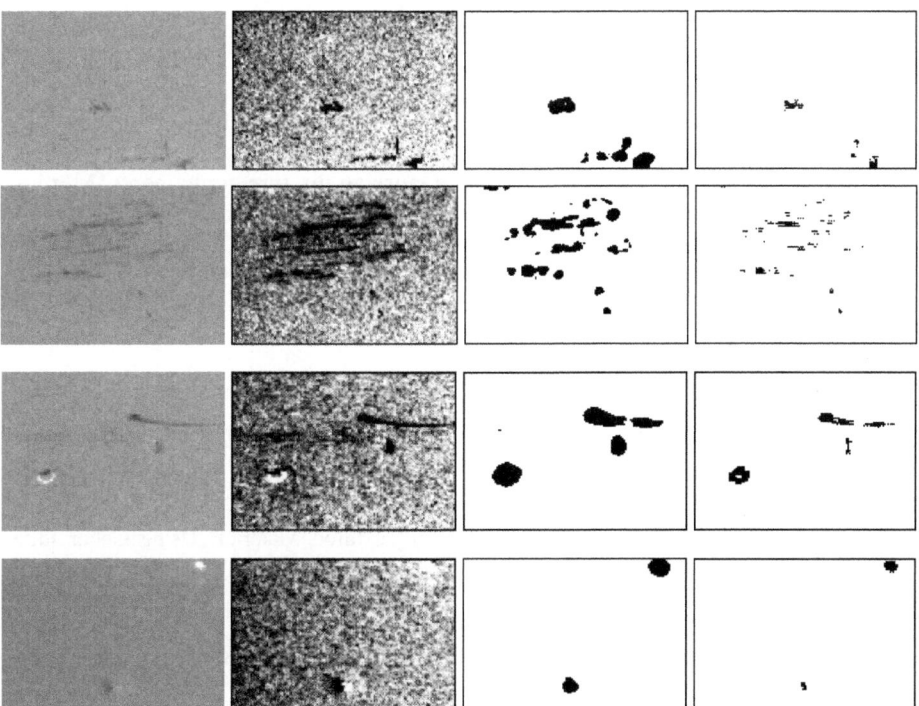

Fig. 7. Test results on metallic gray colored metallic surface. Col 1: Gray level test images, Col 2: Histogram equalized test images. Col.3: Defects detected by our method (ND1). Col 4: Defects detected by by ND3.

(ND3) misses the defect in upper right region (Fig.6(d)). This is because Gabor features are superior to gray level features in detection of defects with low contrast. In Fig.6(e) there is a low illuminated metallic surface patch. Illumination change from lower left corner to upper right corner could be seen in histogram equalized image (Fig.6(f)). Using Gabor features two actual defects are detected (Fig.6(g)), however using gray level image, extra defects are detected as false alarms (Fig.6(h)). This is because Gabor features are more robust to varying illumination conditions compared to gray level features.

Results reported in Fig.7 illustrate superiority of the introduced GMM modeling in Gabor space (ND1) over GMM modeling in spatial domain (ND3) for several type of metallic surface painting defects. In the first column gray level test samples are illustrated and in the second column histogram equalized test samples are given in order to highlight painting defects. Third and fourth columns respectively show the detected abnormalities by ND1 and ND3.

5 Conclusions

We propose a feature level fusion scheme in which the background texture is modeled by Gaussian mixtures that integrate the information extracted at different orientations and scales by Gabor filtering. It is shown that the introduced scheme is capable of correctly detecting common painting defects including pinhole, crater, hav and scratch with a small false alarm ratio. It is also shown that transforming the gray level information into Gabor space provides a more compact representation that yields a modeling capability with a lower number of Gaussian mixture components. Moreover the proposed GMM modeling scheme eliminates the main drawback of standalone Gabor filtering that needs a fine tuning in parameter space.

References

1. Jain, A.K., Farrokhnia, F., Alman, D.: Texture Analysis of Automotive Finishes. In: Proc.Vision '90, pp. 8.1–8.16. Detroit Michigan (1990)
2. Xie, X., Mirmehdi, M.: TEXEMS: Texture Exemplars for Defect Detection on Random Textured Surfaces. IEEE Transactions On Pattern Analysis and Machine Intelligence 29(8), 1454–1464 (2007)
3. Kumar, A., Pang, G.K.H.: Defect Detection in Textured Materials Using Gabor Filters. IEEE Transactions on Industry Applications 38(2), 425–440 (2002)
4. Escofet, J., Navarro, R., Millán, M.S., Pladellorens, J.: Detection of Local Defects in Textile Webs Using Gabor Filters. Opt. Eng. 37, 3140–3149 (1996)
5. Daugman, J.G.: Two-Dimensional Spectral Analysis of Cortical Receptive Field Profiles. Vision Research 20, 847–856 (1980)
6. Liu, C.: Gabor-Based Kernel PCA with Fractional Power Polynomial Models for Face Recognition. IEEE Transactions on Pattern Analysis and Machine Intelligence 26(5), 572–581 (2004)
7. Jain, A.K., Farrokhnia, F.: Unsupervised Texture Segmentation Using Gabor Filters. In: Proc. IEEE Conference on System, Man and Cybernetics, pp. 14–19 (1990)

Digital Instrumentation Calibration Using Computer Vision

Fernando Martín-Rodríguez[1], Esteban Vázquez-Fernández[1,2], Ángel Dacal-Nieto[2], Arno Formella[3], Víctor Álvarez-Valado[2], and Higinio González-Jorge[2]

[1] Communications and Signal Theory Department, University of Vigo
[2] Laboratorio Oficial de Metroloxía de Galicia
[3] Computer Science Department, University of Vigo,
ETSET, C/ Maxwell S/N (Ciudad Universitaria), 36310 Vigo, Spain
fmartin@tsc.uvigo.es, evazquez@lomg.net

Abstract. This paper describes a computer vision system designed to automatically read the displays of digital instrumentation. The system is used in calibration sessions where many measurements have to be made and where we are interested in getting the whole numerical series downloaded on a host computer. Before our system was running, a human operator had to inspect the instruments at the right times (required by the calibration procedure) and to write down all the results. Note that we are speaking of very simple and sometimes old instruments that usually do not provide a digital interface or a removable memory (and if they do, we do not have a standard interface accepted by all the manufacturers). Results show the benefits of this system, obtaining a success rate higher than 99% in display recognition

Keywords: Computer vision, text segmentation, character recognition, digital instrumentation.

1 Introduction

1.1 Purposes of Development

The computer vision system was designed to automatically read digital instrumentation measurements avoiding a time-consuming work and minimizing errors due to tiredness or forgetfulness. The application was first implemented at the "Laboratorio Oficial de Metroloxía de Galicia" (LOMG: www.lomg.es) for digital thermometer calibration (Fig. 1). Nevertheless, our application will be useful with all types of instruments that exhibit a numerical display.

The process starts with a photograph of the instrument displaying a stable measurement. Then we use standard image processing techniques to segment the image characters. Finally, we will see how we recognize the digits with a new approach that combines two different classifiers.

A. Campilho and M. Kamel (Eds.): ICIAR 2010, Part II, LNCS 6112, pp. 335–344, 2010.

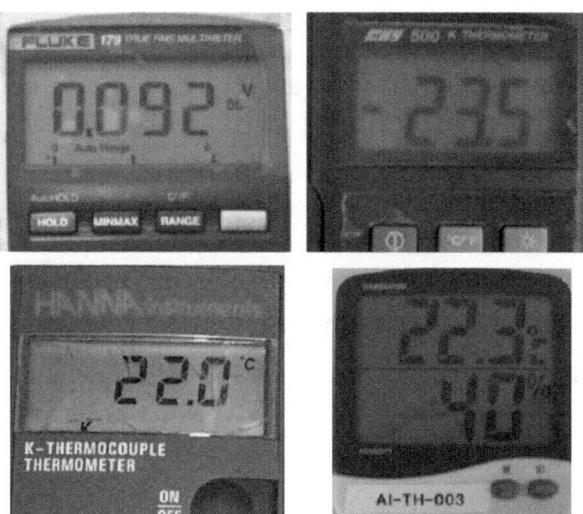

Fig. 1. Examples of different instruments, some of them are showing display defects (bubbles in the first one, stripes in the third one)

1.2 Related Work

Our system is an example of character recognition in scene images like that in [1]. We have taken some ideas from there such as the interpolated threshold in non uniformly illuminated images.

Our system also shares ideas with car plate recognition systems [2], [3]. In such plate recognition systems, plate has to be located automatically but a fixed character font is assumed. In contrast, for digital instrumentation, we have to take into account multiple fonts (Fig. 2) which is, as we will see, a source of problems. As the operator must set up the calibrating experiment, we started with a system where the region of interest (ROI) is user supplied.

Fig. 2. Examples of different displays using different character fonts. Our system deals with all of them.

As we will describe in the following, image preprocessing and segmentation are based on standard image processing techniques adequately adjusted to our problem. The recognition stage combines two methods: first a classical one based on feature extraction followed by 1-NN classification; second an original method especially suited to recognize instrumentation digits (as we will see is somewhat inspired by the classical 7-segment display). The fusion of both recognizers is also an original contribution where we use the second one to correct possible errors of the first one.

2 Image Capturing and Preprocessing

2.1 Image Capturing

Capturing is perhaps the most important part of the whole system. A good capture will make recognition easy while a bad one would make it impossible.

Due to some in-site restrictions at LOMG, we cannot modify the environment illumination. We use a C-Cam BCi4 camera with 1280x1024 resolution (Fig. 3). In most of the cases, we use a 25 mm lens (able to focus from 15 cm to 1 m). However, sometimes the physical conditions oblige to the use of a 75 mm lens with a focal distance from 1.5 m to 10 m.

To help in capturing, we designed a mechanical arrangement that can be used with most instruments to get always the same capturing conditions (Fig. 3).

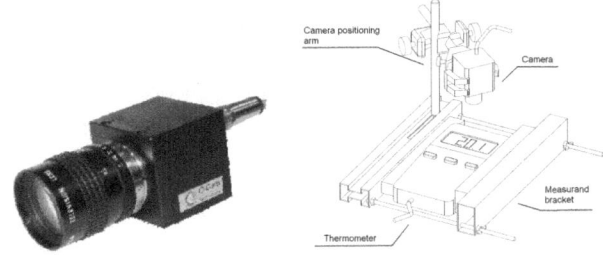

Fig. 3. System camera and mechanical arrangement for stable capture

As the observed instrument is not moving during image capturing (remember we are calibrating the instrument, not using it dairy), we decided to rely on user to extract the ROI (Fig. 4). The region will have to be marked only once for all the series. Afterwards, we developed an automatic extraction system using the methods described in [3]. This new subsystem gets correct location in more than 99% of the test images [4] (Fig. 5) but it is not integrated with the rest of the system yet.

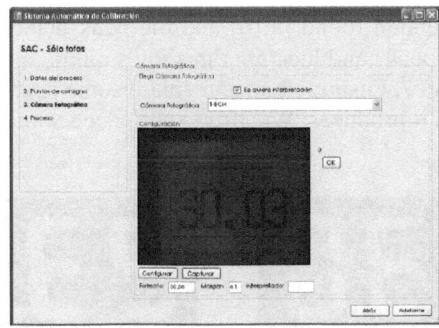

Fig. 4. User selection of Region of Interest (ROI)

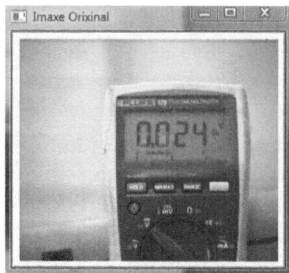

Fig. 5. Automatic location [4]. Still not used at LOMG.

2.2 Binarization

Binarization is the process that converts a grayscale or color image into a binary one with only two levels. We start by converting our colored images to grayscale using the ITU-R BT.709 recommendation (gray=0.2125R+0.7154G+0.0721B). We use this equation because instrument displays are often green or, at least, dominated by the green channel. As expected, the resulting gray level distribution shows a bimodal histogram (two main peaks, see Fig. 6).

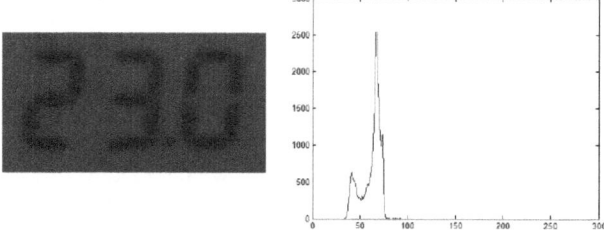

Fig. 6. Grayscale display image and its gray histogram

We use a combination of the well known Otsu method [5], implementing it via an approximate iterative version found in [6] and the peak detection method [7] (based on searching histogram peaks and locating thresholds on the minima between them).

As can be seen in Fig. 7, Otsu method can create some segmentation problems due to the thicker characters it produces. We also experienced some problems with images

Fig. 7. Left: Peak detection method. Right: iterative Otsu method.

with important illumination gradients (Fig. 8). In these cases, a global threshold is not enough. This can be solved splitting the image into sub-images and applying interpolated thresholds [1] (see results in Fig. 9).

Fig. 8. <u>Left</u>: Image with illumination gradient (not very evident, more visible in an equalized image: <u>middle</u>). <u>Right</u>: binarization with a single threshold.

Fig. 9. Images binarized with interpolated thresholds. Left: 8x8 sub-images. Right: 4x3 sub-images.

The final solution consists on applying first the peak detection method and then measuring threshold quality using the histogram area in the threshold neighborhood. If that area is bigger than usual the threshold is considered incorrect and we switch to an interpolated threshold with 12 sub-images (4x3 sub-image grid). We use Otsu threshold on each piece.

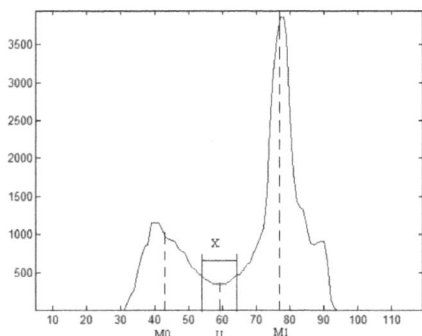

Fig. 10. Threshold quality test based on area in the threshold neighborhood. If this area exceeds 2.5% of total histogram area, threshold is not good.

2.3 Skew Angle Correction

To correct a possible skew angle, we estimate the upper contour of the characters and compute the slope of the resulting straight line (Fig. 11).

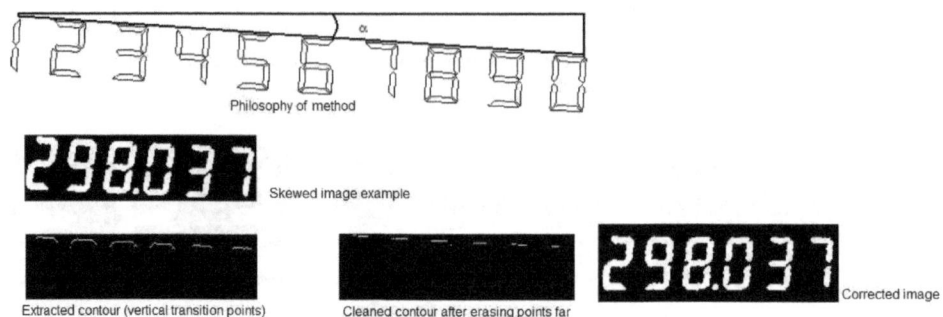

Fig. 11. Skew angle correction

2.4 Extracting the Character Row (Presegmentation)

Extracting the character row is the same as removing the blank lines above and below characters and also to the left and to the right of them. As shown in Fig. 12, this is an easy process using horizontal and vertical image projections.

On each projection, we detect the region of interest beginning and ending by searching for large gradients, first from left to right (top to bottom) and afterwards in the opposite direction.

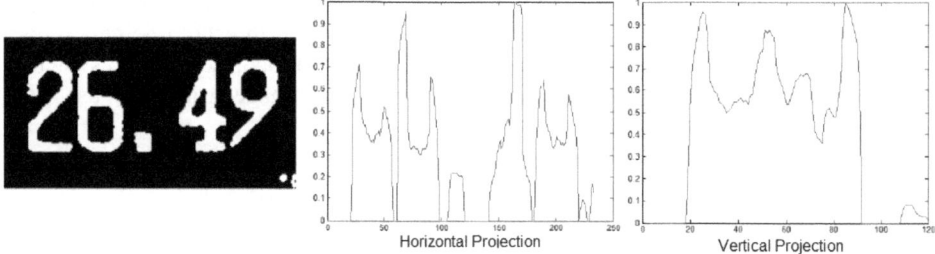

Fig. 12. Binary display image and both projections

3   Character Segmentation

To isolate the different characters in a preprocessed row, we use what we call "enhanced projections". The enhanced horizontal projection of an image is the vector that contains in position i the dot product ($< x, y >= \sum x_j y_j$) between the (i-1)th and (i+1)th column. With this kind of projection, minima that mark character transitions are deeper than with standard projections.

The main procedure consists of searching the horizontal projection from right to left while applying a kind of hysteresis process. The right to left direction is chosen because digit ending is usually more evident than its beginning (most digits end by a vertical line, i.e., a strong projection gradient). We use the word hysteresis because the threshold to detect a character beginning is different (bigger) to the threshold used to detect an ending (Fig. 13).

To detect segmentation errors (linked characters), we compute the aspect ratio of all segmented items (R=height/width). For a ratio R of less than 1.2 we decide that we rather have two characters. In this case, we first compute the local maxima and minima of the projection (using a sliding window procedure as described in [8]). The deeper minimum that is between two peaks is selected as the optimum breaking point.

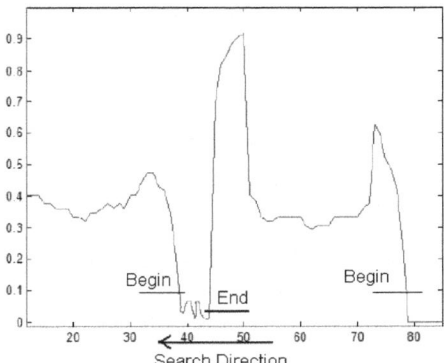

Fig. 13. Detection with hysteresis

4 Digit Recognition

First, we normalize the extracted characters by scaling them to a fixed size of 16x16 points. Note that we do not maintain the aspect ratio. We started by keeping that ratio (getting a character with 16 points height and less than 16 points width) followed by centering with vertical lines [8]. Nevertheless, we discovered that, in the end, we yielded slightly worse results. Distorting input characters is not a problem as long as we also distort the patterns as well.

4.1 Feature Extraction

We extract features in two ways. First, we use horizontal and vertical projections of the individual characters. As different characters may have almost the same projection (Fig. 14), we split each character into two halves (upper and lower) and then we compute 4 vectors: upper horizontal (16 values), upper vertical (8 values), lower horizontal (16 values), and lower vertical (8 values). Final feature vector has 48 values.

Second, we use features based on Kirsch gradients [6], [8]. The Kirsch operator computes a first order derivative (similar to operators from Prewitt, Sobel, Canny... [9]). Our purpose is to compute image components along four directional axes: horizontal, vertical, right diagonal and left diagonal. For example, the horizontal component is computed via a vertical gradient (being always perpendicular to the desired direction).

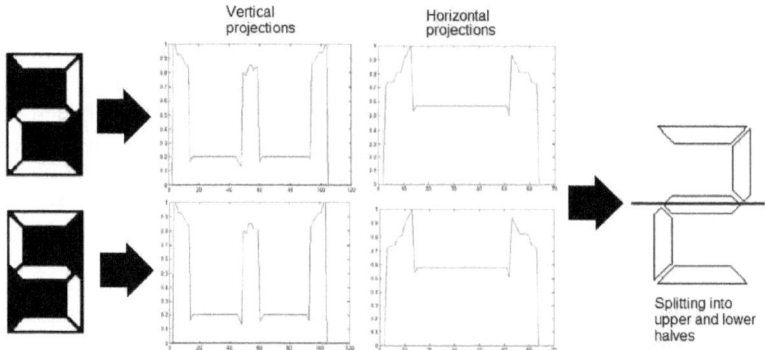

Fig. 14. Characters with almost identical projections

Eventually, we obtain four local feature maps as 16x16 images. Using all of them as a feature vector would result in a 1024 length vector. In [8] it is suggested to decimate the 16x16 images to size 4x4. However, in our particular system we obtained better results leaving the 4 directional components in its original size (Fig. 15).

We combine the two feature vectors, i.e., projections and Kirsch, yielding a total vector of length 1024+48=1072.

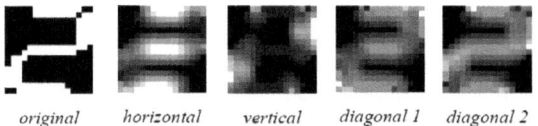

original   horizontal   vertical   diagonal 1   diagonal 2

Fig. 15. Example of directional components

4.2 Classification

We tried various classifiers (like probabilistic neural networks, Gaussian classifiers and k-NN). Best results were for the nearest neighbor algorithm (1-NN). This is not very surprising as it is explained in [10]. This system has to deal with several different character types (7-segment, graphical fonts, skewed, not skewed, etc.). In this multi-font situation, there exists sometimes more variance between the samples of the same character in different fonts than between different characters in the same font (intra-class variance greater than inter-class variance).

As patterns for the 1-NN classifier we chose perfect ones (obtained from the different fonts). We tried to use patterns from segmented input digits but the 1-NN got better results for the artificial, perfect ones.

4.3 Visual Inspection and Fusion with 1-NN Classification

Visual inspection is an intuitive method. We developed it studying the reasoning that people express when they describe how they recognize characters. For example, no matter which font, number **'2'** has always two openings: one in the upper left part and the other in the lower right one.

We implemented a complete recognizer [11] based on a template that defines the regions of interest (Fig. 16). We check whether each region is active or not by majority voting between foreground and background pixels. See that the template is a 7-segment digit with no corners (sharp in some fonts and rounded in others).

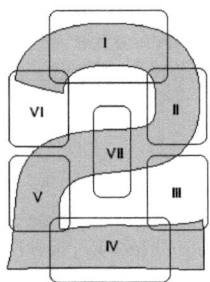

Fig. 16. Regions of interest tested (a 7-segment like template)

To benefit from both classification schemes, we run them in parallel and combine their results in the following manner:

- We run the feature extraction and compute the norm-1 distance to every pattern yielding a distance vector.
- We run the visual inspection algorithm yielding an estimate for the digit to be recognized. This is coded as a binary vector of length 10 where an active bit at a position corresponds to recognition of that class.
- We reduce the distances that correspond to the class that was recognized by visual inspection by 20%, empirical.
- We apply 1-NN and minimum distance wins.

5   Results

Our test set consisted of 16 image sequences, with a total of 448 images. The system obtained the correct values 445 times, id. est: 99.33% recognition rate (measured on display images, not on individual digits). We have tried samples from all accessible fonts: 7-segment, skewed, not skewed, graphic display... In routine work of the LOMG, 7-segment displays (easier to recognize) are the most common ones.

Average execution time is 25 milliseconds per image (Intel Core Duo, 2.53 GHz).

Our results are similar to other found in literature like [12], although in this publication they only consider 7-segment displays.

6   Conclusions and Future Lines

We have designed and implemented a useful system able to read almost any display of digital instrumentation devices.

We have employed standard image processing techniques adapted to this problem. We also designed a hybrid recognizer, which combines a classical classifier with a visual inspection algorithm. Final recognition rate suggests that we have solved the problem despite the intra-class variance due to the presence of multiple fonts.

As future work lines, we emphasize on the following: integrating the automatic ROI location into the industrial system, optimizing the feature vector trying to detect the principal components and, finally, using knowledge from previous images (for instance the font type) when recognizing subsequent images in a sequence.

References

1. Ohya, J., Shio, A., Akamatsu, S.: Recognizing Characters in Scene Images. IEEE Transactions of Pattern Analysis and Machine Intelligence 16(2), 214–220 (1994)
2. Cowell, J.R.: Syntactic Pattern Recognizer for Vehicle Identification Numbers. Image & Vision Computing (1995)
3. Fernández-Hermida, X., et al.: Automatic and Real Time Recognition of V.L.P.'s (Vehicle License Plates). In: Del Bimbo, A. (ed.) ICIAP 1997. LNCS, vol. 1311, pp. 552–559. Springer, Heidelberg (1997)
4. Martín-Rodríguez, F., et al.: Localización de Caracteres en Imágenes de Instrumentación Digital. In: Proceedings of URSI-2009 (National Meeting of the International Scientific Radio Union). Santander, Spain (2009)
5. Otsu, N.: A Threshold Selection Method for Gray Level Histograms. IEEE Transactions on System, Man and Cybernetics (1979)
6. González, R.C., Woods, R.E.: Digital Image Processing, 3rd edn. Prentice Hall, Englewood Cliffs (2008)
7. Martín-Rodríguez, F.: Analysis Tools for Gray Level Histograms. In: Proceedings of SPPRA-2003, Signal Processing Pattern Recognition and Applications. Rhodes, Greece (2002), http://www.iasted.org
8. Proceedings of the IEEE (Special Isue on O.C.R.'s) 80(7) (1992)
9. Jain, A.K.: Fundamentals of Digital Image Processing. Prentice Hall, Englewood Cliffs (1989)
10. Blue, J.L., et al.: Evaluation of Pattern Classifiers for Fingerprint and OCR Applications. Pattern Recognition (Pergamon Press) 27(4), 485–501 (1994)
11. Vázquez-Fernández, E., et al.: Human Visual Perception as a Complementary Method for Digit Recognition. In: Proceedings of VIIP-2009 (Visualization, Imaging and Image Processing). Palma de Mallorca, Spain (2009), http://www.iasted.org
12. Corrêa Alegria, F., Cruz Serra, A.: Automatic Calibration of Analog and Digital Measuring Instruments Using Computer Vision. IEEE Transactions on Intrumentation and Measurement 49(1), 94–99 (2000)

Dynamic Scenes HDRI Acquisition

Anna Tomaszewska and Mateusz Markowski

West Pomeranian University of Technology, Szczecin,
Faculty of Computer Science and Information Technology,
Zolnierska 49, 71-210, Szczecin, Poland
atomaszewska@wi.zut.edu.pl, mmarkowski@wi.zut.edu.pl

Abstract. We present fast, robust and fully automatic method for high dynamic range images acquisition for non-static scenes. The key components of our technique are probability maps calculated from sequences of hand-held photographs. In practice, several basic problems occur with image sequences used for creating HDR images. Firstly, camera movement causes images to misalign, what results in blurred output image. Secondly, objects in the scene are in movement, causing ghost artifacts. In our method we focus on removing such artifacts in order to generate sharp HDR images. We validate our results via HDR VDP and compare them with other known approaches.

1 Introduction

Image composition from input image sequences is a well-known approach. The example is panorama generating from photographs showing different parts of the scene. Another example is HDR image acquisition, where component images depict the same part of the scene. Common problem in both cases is overlapping of parts of input images where ghost artifacts may occur.

Recently, there is tremendous progress in the development and accessibility of high dynamic range (HDR) imaging technology [1]. Its popularity comes from possibility of registration in one image the radiance of real scenes that contain dozen of orders of magnitude. Modern image processing and graphics software becomes HDR enabled. Also HDR digital photography replaces low dynamic range (LDR) technologies. HDR photographs are of much better quality and easier to be processed in a digital darkroom. Unfortunately, HDR cameras are still very expensive and not available for average users. On the other hand, taking HDR photographs seems to be legitimate and crucial. The development of high dynamic range imagery (HDRI) has brought us to verge of arguably the largest change in image display technologies since the transition from black-and-white to color television. Novel capture and displays hardware will soon enable consumers to enjoy the HDR experience in their own homes [2].

The multi-exposure HDR capture technique [3] seems to be a good alternative to HDR cameras, which can be used to create an HDR image from photographs taken with a conventional LDR camera. The technique uses differently exposed photographs to recover the response function of a camera. The main disadvantage

A. Campilho and M. Kamel (Eds.): ICIAR 2010, Part II, LNCS 6112, pp. 345–354, 2010.

Fig. 1. Problem: LDR sequence (top row), HDRI: conventional approach (bottom left), our approach (bottom right)

of those techniques is a necessity of using a tripod. Any movement of a camera causes misalignments between hand-held images what makes blurry image at the end. The second problem is when the objects on the photographed scenes are in movement what causing so-called ghost artifacts. In this paper we propose a technique for HDRI acquisition of non-static scenes (Figure 1). Our application of this technique allows to create correct HDR image based on a simple sequence of three LDR photographs with overlapped ghost regions. In our application we introduce additional modules for image aligning and image de-noising. All functionality is fully automatic. The technique is robust and fast due to GPU-based implementation. We validated them by HDR VDP algorithm.

The paper is organized as follows. In section 2 previous works are discussed. In section 3, the application of our HDR acquisition technique is presented in details. In section 4 we briefly describe a GPU implementation of our method. Section 5 shows and discusses achieved results. In the last section, the paper is concluded and possible future work is suggested.

2   Previous Work

There is a growing demand for HDR image of both static and dynamic scenes. That is why as far as hardware solutions of HDRI acquisition are not easily available, software solutions will be needed. A few approaches have been developed in order to remove ghosts artifacts during HDRI acquisition. The first technique is based on tracking non-static objects by matching their key points in a sequence of images [4]. The method fails for occluded objects or for patterns for which it is not possible to find correct matching. Another approach replaces the whole regions, where ghost artifacts are likely to occur, with reference ones. The regions

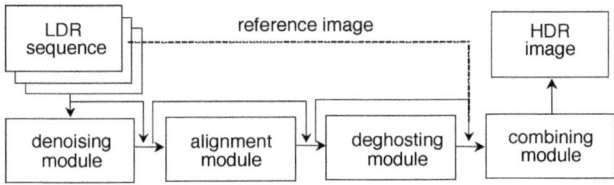

Fig. 2. Image acquisition pipeline

can be selected manually [5][8] or detected automatically [6][7]. Unfortunately, the technique works correctly only when the whole dynamic range of a region can be registered in a single image exposure. In [13] they proposed a histogram based ghost removal method, in which object motion and background change between two exposures are detected using multi-level threshoding of the intensity histogram. A different solution was presented in [9] where iterative propagation of ghost probability was used. The method requires a large number of images in LDR sequence and still background for moving objects. Moreover, it is time consuming and must be computed in many iterations.

Image registration is another problem during acquisition of HDR image. Misalignments between photographs in a sequence can appear due to camera movement (in the case of hand-held photographs) or not careful usage of tripod. There are two basic techniques of image registration: matching key points and checking pixels difference. In the first case the same drawbacks as during ghost removal occurs (matching key-points problem). The solutions based on pixels difference generally give better results. In some software solution for alignment only horizontal and vertical shifts without rotation compensation are considered due to complexity of computations [10].

In the paper we propose modified pixel-based approach for ghosts removing. Deghosting is based on the ghost maps. The ghost maps are calculated using probability of belonging of pixels to background. They depict regions where ghost artifacts are likely to occur, or regions with under- and overexposed pixels. The technique is fast due to GPU based implementation of de-ghosting and alignment modules.

3 Ghost Removal Algorithm

Generating of HDR images of arbitrary static or non-static scene requires introducing a ghost removal component. We developed such module in the GPU based application for HDRI acquisition. The algorithm used in the application has four successive stages: image de-noising, position alignment, ghost removal and HDRI composition. First three of them can be used optionally (Figure 2), however, it is easier to align denoised images or remove ghost from aligned photographs. Therefore each successive stage works better if previous stages are included into the acquisition pipeline.

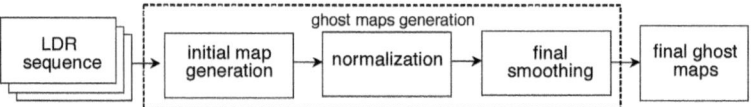

Fig. 3. Ghosts maps generation schema

The image denoising module is based on the wavelet coefficient thresholding method presented in [11]. We optimized the existing implementation of the algorithm to perform it on GPU. The images alignment module compensates camera shifts or rotations. In the first step, an image with middle exposure is chosen as a reference one. Other images in a sequence are aligned to this image. Additionally simple algorithm for removing of misalignments was implemented. We assume, that images taken with auto-bracketing are registered very fast one by one, so misalignments are also limited. Therefore we used a simplified approach to align pictures based on error minimization.

In some situations one of the input images (with the best exposure) must be used as reference image. It happens where only a part of the object is in movement and may be treated as ghost. In this case only a movement effect should be removed but not the whole object. Recognized ghost is replaced by corresponding fragment from reference image.

3.1 Ghost Maps Generation

In order to remove ghosts, for each input LDR image a ghost map is generated. This map shows how much the pixel color influences the pixel color in result image. The module is composed from three states: initial map generation, normalization and final smoothing (Figure 3). As an input to the ghost maps generation we provide normalized image sequences for which maps are calculated. In the first stage, corresponding pixels values for every color channel from LDR sequence are compared. It estimates a probability that a pixel belongs to a still background or to moving object (ghost). The initial ghost map for each LDR image is computed based on (1).

$$G_i(x, y) = \sum_{j=1, j \neq i}^{k} F(P_i(x, y), P_j(x, y)),\qquad(1)$$

where: G_i means ghost map of i-th image, F - deghosting comparison function, P_i, P_j - i-th and j-th normalized LDR images respectively , k - number of images.

$$F(P_i(x, y), P_j(x, y)) = \sum_{c=r,g,b} \frac{1}{1 + \exp\left(-\left(max\left(\frac{P_i(x,y)_c}{P_j(x,y)_c}, \frac{P_j(x,y)_c}{P_i(x,y)_c}\right) - 1.35\right) \cdot 20\right)},$$

$$(2)$$

where: $P_i c$, $P_j c$ - pixel componets c of normalized images i and j respectively.

The deghosting comparison function F is based on sigmoid function (Equation (2)). At development stage function was designed and tuned with registration

Fig. 4. Example LDR images sequence (top raw: $\frac{1}{640}s$, $\frac{1}{160}s$, $\frac{1}{40}s$) and their normalized ghost maps (bottom raw)

error distribution graphs. Even for static scenes captured values are not linearly dependent due to registration errors. That is why distribution graphs were used to determine acceptable and unacceptable limits in color difference. For ghost free images values P_i and P_j should be equal. In ghost regions difference in color is very likely. Function max returns the ratio between brighter and darker value. For each rgb channel such value is transformed by sigmoid function and then sum up to estimate that two pixels match or not.

In the next stage the initial ghost maps must be normalized and the zero value is assigned to over-exposed pixels. Normalized ghost maps are prone to errors. They look noisy due to independent computation of each pixel in ghost maps (see Figure 4). In the next stage the ghost map smoothing is proceeded. It integrates map values with neighbor pixels and is based on dilatation, erosion and convolution. The smoothing is based on dilatation, erosion and convolution. Firstly the (3×3) dilatation is applied. It removes one or two pixels regions recognized as ghost from the ghost maps. After dilation, the (15×15) erosion is applied It fills holes in ghost regions recognized as valid pixels. These regions are additionally extended during dilatation. Finally ghost maps are smoothed based on convolution with (5×5) window kernel The example results after each step are presented in Figure 5 [6].

3.2   HDRI Composition

Generation of HDR image from sequence of LDR pictures depicting dynamic scene, with calibration and ghost removal taken into account, is similar to traditional approach for static scenes [3]. The novelty is using of ghost maps in the final equation, where pixel color is computed according to Equation (3) using weighted function which was tuned experimentally (Equation (4)).

Fig. 5. Ghostmap $i=3$ results after: dilatation – G_i^d (left), erosion – G_i^e (middle), convolution – G_i^c (right)

$$H(x,y) = \frac{\sum_{i=1}^{k} \frac{S_i(x,y)_c}{E_i} W_i(x,y) G_i^c(x,y)}{\sum_{i=1}^{k} W_i(x,y) G_i^c(x,y)}, \tag{3}$$

where: $S_i(x,y)$ is pixel of input LDR image i, E_i is exposure value of image i, $G_i^c(x,y)$ is grayscale pixel of ghost map i and $W_i(x,y)$ means pixel weight of image i according to modified (experimentally) Equation (4). A chart of the weight function for a single color component is shown in Figure 6.

$$W_i(x,y) = \sum_{c=r,g,b} \min \left(\frac{1}{1 + \exp\left(5 - 100 S_i(x,y)_c\right)}, \frac{1}{1 + \exp\left(30 S_i(x,y)_c - 24\right)} \right), \tag{4}$$

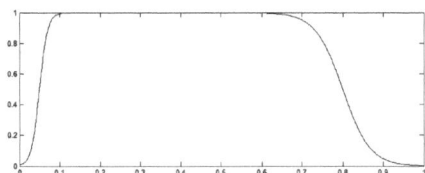

Fig. 6. Weight function graph for single color component

4   Implementation

To speed up calculation we used hardware acceleration. The image is represented by four textures RGBA respectively. In calibration module we implemented aligning algorithm, which enables to calibrate up to 4 images at the same time. This solution was proposed in order to optimal parallel down-sampling of all image channels.

The ghost maps generation module was implemented in GPU. It creates maps in a single rendering pass. Number of input textures to the shader is equal to the number of images in the sequence. Every shader output is RGBA texture where every channel corresponds to the single ghost map. For more than four

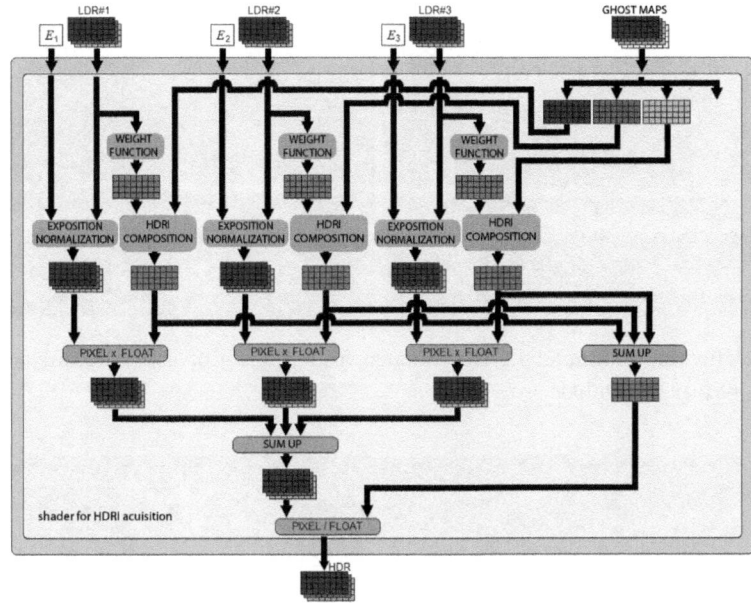

Fig. 7. Detail shader for HDRI acuisition for the three LDR images

images in an image sequence, multi-texture hardware extension (MRT) is used. To improve performance, horizontal and vertical kernels are applied separately in each operation. The HDR image color is computed based on Equation (3) in a single rendering pass. The shader has scalable number of input textures.

Input LDR sequence and ghost maps are combined in single pass shader. Detailed shader schema for three image sequence is depicted in Figure 7.

5 Results

A few example image sequences were used to test quality and performance of the HDRI acquisition application. Both hand-held and tripod sequences with varying number of images and exposure range were assessed. The best results were achieved for exposure difference less or equal to two F-stop. It is compatible with exposure bracketing functionality in typical cameras that allows for such exposure change.

The result of experiment depicted on Figure 8 presents ghosts removal results. In this sequence, moving objects causes ghosts artifacts. Our ghost removal module detects moving objects and removes ghost from a final HDR image.

In the next test we compared result images with and without ghost removal module with the static scene with HDR-VDP algorithm [12] (see Figure 9). HDRI with ghost removal contains much less visible artifacts. It is worth to mention, that differences in HDR-VDP results for our approach come from changes illumination of the scene.

Fig. 8. Deghosting example: LDR sequence from Figure 4, conventional acquisition (left), our approach (right)

Fig. 9. Results for HDR-VDP test for image sequences 8 with static scene (without moving object) with image created (left) without ghost removal module (right) with ghost removal module

We conducted also the test for scene where only a part of the object is treated as ghosts. The best example is a man who waves his arms. His body is still, only arms are moving. In that situation we need to use reference image Figure 10 (right) to avoid removing unwanted parts of image. Without reference image, mans arms would be recognized as ghosts and removed, what is depicted in Figure 10 (left).

Finally we compare our approach with the image technique with existing application for acquisition of a dynamic HDRI scenes: Dynamic Photo HDR, Qtpfsgui and Photomatix. The results of comparison is shown in Figure 11. Our application seems to produce the best images with correctly removed ghosts. Our application requires 11 seconds to align, de-ghost and create a final HDR image based on a sequence of three LDR images of resolution 3039x2014 pixels (GPU textures upload/download 4.55 s, alignment 6.11 s, ghost detection and HDRI composition 0.47 s). The high performance of algorithms computation was achieved due to careful GPU implementation.

Fig. 10. Deghosting comparison: LDR sequence (top), our approach without/with reference image (bottom)

Fig. 11. Deghosting comparison: DynamicPhotoHDR and Qtpfsgui (top), Photomatix and our approach (bottom)

6   Conclusions and Future Work

In the paper we present ghost removal technique and its GPU based implementation. Our approach was applied for HDR image acquisition and validated via HDR-VDP [12] algorithm.

A fully automated tool for HDRI acquisition was presented. It allows to create HDRI images of static and dynamic scenes from hand-held photographs.

In future work we plan to improve the ghost map generation module. We noticed that some ghost removal errors could occur for images with many high dynamic range ghost regions like reflections on a waving water. Performance of alignment module could be also improved, because currently it is 12-times slower module than de-ghosting and final composition modules.

References

1. Reinhard, E., Ward, G., Pattanaik, S., Debevec, P.: High Dynamic Range Imaging: Acquisition, Display and Image-Based Lighting. Morgan Kaufmann Publishers, San Francisco (2005)
2. Akyüz, A.O., Fleming, R.W., Riecke, B.E., Reinhard, E., Bülthoff, H.: Do HDR displays support LDR content? A psychophysical evaluation. ACM Trans. Graph 26(3), 38 (2007)
3. Debevec, P.E., Malik, J.: Recovering High Dynamic Range Radiance Maps from Photographs. In: SIGGRAPH'97, pp. 369–378 (1997)
4. Tomaszewska, A., Mantiuk, R.: Image Registration for Multi-exposure High Dynamic Range Image Acquisition. In: WSCG, Int. Conf. in Central Europe on Computer Graphics, Visualization and Computer Vision, pp. 49–56 (2007)
5. Rota, G.: Qtpfsgui - HDR Imaging Workflow Application (2007), http://qtpfsgui.sourceforge.net
6. Grosch, T.: Fast and Robust High Dynamic Range Image Generation with Camera and Object Movement. In: Vision, Modeling and Visualization, RWTH Aachen, pp. 277–284 (2006)
7. HDRsoft: Photomatix Pro (2003), http://www.hdrsoft.com
8. Mediachance: Dynamic Photo HDR (2008), http://www.mediachance.com
9. Khan, E.A., Akyüz, A.O., Reinhard, E.: Ghost Removal in High Dynamic Range Images. In: IEEE International Conference on Image Procesing, pp. 2005–2008 (2006)
10. Ward Larson, G.: Fast, Robust Image Registration for Compositing High Dynamic Range Photographs from Handheld Exposures. Exponent - Failure Analysis Assoc. (2003)
11. Selesnick, I., Wagner, C.: Double-Density Wavelet Software. Polytechnic University's Brooklyn (2004)
12. Mantiuk, R., Daly, S., Myszkowski, K., Seidel, H.-P.: Predicting Visible Differences in High Dynamic Range Images - Model and its Calibration. In: Human Vision and Electronic Imaging X, IS&T/SPIE's 17th Annual Symposium on Electronic Imaging, vol. 5666, pp. 204–214 (2005)
13. Min, T.-H., Park, R.-H., Chang, S.-K.: Histogram based ghost removal in high dynamic range images. In: IEEE International Conference on Multimedia and Expo., ICME 2009, pp. 530–533 (2009)

Correcting Book Binding Distortion
in Scanned Documents

Rafael Dueire Lins[1], Daniel M. Oliveira[1], Gabriel Torreão[1],
Jian Fan[2], and Marcelo Thielo[3]

[1] Universidade Federal de Pernambuco, Departamento de Eletrônica e Sistemas,
Av. Prof. Luiz Freire, s/n, Cidade Universitária
50740-540, Recife, PE, Brasil
{rdl,daniel.moliveira,gabriel.dsilva}@ufpe.br
[2] Hewlett-Packard Labs, Palo Alto, USA
jian.fan@hp.com
[3] Hewlett-Packard Labs, Porto Alegre, Brazil
marcelo.resende.thielo@hp.com

Abstract. This paper presents an exponential model to correct the distortion
that appears when bound books are digitalized using flatbed scanners. The proposed algorithm infers the height of the book spine for every column in the image and fits an exponential curve. The de-warping is done regarding the scanner
acquisition model and shape binding. This new de-warping method is effective,
fast, accurate and easy to implement.

Keywords: Shape-from-shading, image processing, book binding de-warping.

1 Introduction

Scanners are the device most widely used for document digitalization. Figure 1 (a)
shows only one page scanned at a time, in which the sweeping line is perpendicular to
the book spine. Figure 1(b) depicts the scanning of two pages a time, with the book
spine lying parallel to the scanning line. Examples of images obtained in such scenarios can be seen in Figures 1(c) and 1(e). In both situations, an unpleasant warp appears near the book binding, which is better visualized in the enlarged parts of Figures
1.d and 1.f, respectively. It is important to notice that the two scenarios yield very
different images: the scanner projection is orthographic in the direction of the lamp
movement, but has perspective projection in the direction orthogonal to the lamp
movement [1][11].

This paper proposes a new method to correct both scenarios presented in Fig.1 assuming that the document image has only one page. It is organized as follows: Section
1.1 shows a brief review of the de-warping literature and sketches the proposed method. Section 2 presents the scanner acquisition model and its parameters extraction.
Section 3 explains new distortion correction methods whose results are provided in
Section 4. Conclusions and lines for further work are in Section 5.

A. Campilho and M. Kamel (Eds.): ICIAR 2010, Part II, LNCS 6112, pp. 355–365, 2010.
© Springer-Verlag Berlin Heidelberg 2010

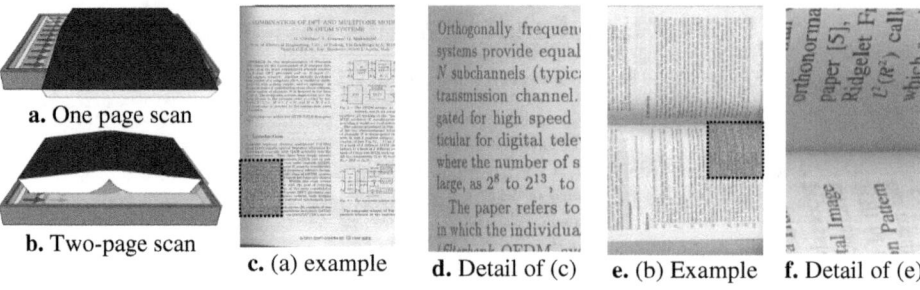

a. One page scan

b. Two-page scan

c. (a) example   **d.** Detail of (c)   **e.** (b) Example   **f.** Detail of (e)

Fig. 1. Book scanning scenarios

1.1 Review of the De-warping Literature

Tan *et al* propose in [1] a de-warping algorithm for scanned document images that reconstructs the volume surface using a shape-from-shading pattern in the scenario of Fig. 1.b. The main disadvantage of this method is the need for several scanner parameters (focus, gain and bias of the photoelectric transformation, etc). Scanner manufacturers seldom provide those parameters, making impossible the application of the algorithm. Ukida and Konishi [2] introduce a warp restoration in scanned images, using the scanner with three different light sources (red, blue and green) at fixed locations. Ukida *et al* [3] propose a shape and color reconstruction method using a scanner equipped with four light sources which are located symmetrically under the scanning line. Most scanners have a single white light source, thus the application of the algorithms in references [2] and [3] may not be possible.

Reference [15] introduces a new method that only asks the user to scan a calibration grid placed at the center of the volume to be scanned. This grid allows inferring the relation between pixel intensity and depth. Such parameters are used to correct the distortion is the other pages of the volume. This method is efficient in a large number of book images, but in a few pages tested the noisy areas had their shading wrongly estimated causing excessive widening of the de-warped area.

There are several works which focus on estimating the 3D-model using the shading pattern of photo objects. Two surveys can be found in references [4] and [5]. Most of those solutions are time intensive, besides not yielding good de-warping results [6].

Another approach, recently proposed by Pintus *et al* in [10] modifies the scanner to gather two images at the same time to build a 3D model by stereo vision 3D reconstruction. This method claims for a customized scanner.

There are also several other methods for removing such distortion for documents captured with portable digital cameras. Most algorithms try to find document text-lines [8][9], assuming that they are parallel in the original document. Those algorithms get the envelope of each line and attempt to "straighten" them by moving font cases, not correcting the narrowing of the fonts in the distorted region. Such algorithms are unsuitable for scanned documents, because as shown in Fig. 1.c, images from scanned books often present binding distortion but lines remain parallel to each other. This happens due to the orthographic/perspective projection of scanners as explained earlier on, therefore most portable camera de-warping algorithms in the literature cannot be used.

The algorithm proposed herein borrows some ideas of [15], by using the document shading pattern to estimate its depth using the relation obtained in section 2. The depth is fitted onto an exponential curve as shown in section 3.

2   The Scanner Acquisition Model

To validate this scanner projection model, a square grid was placed on a flat slope [2] as illustrated in Fig. 2.a. Using HP ScanJet 5300c scanner with 300dpi, Fig. 2.b was acquired with $\Psi=90°$ and $\theta=30°$ (1.c scenario); Fig. 2.d with $\Psi=0°$ and $\theta=30°$ (1.e scenario). The height of the squares is constant, independently of the paper depth (orthographic projection), whereas the square "widths" are narrower and closer to the focal axis. The paper becomes darker with increasing distance, as shown in Fig. 2.c.

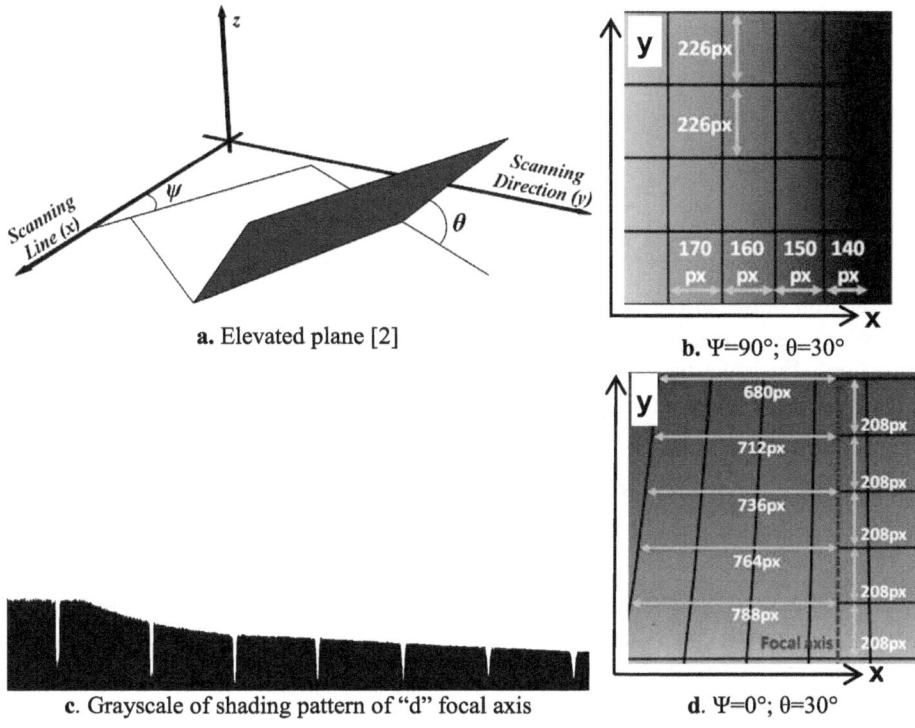

a. Elevated plane [2]

b. $\Psi=90°$; $\theta=30°$

c. Grayscale of shading pattern of "d" focal axis

d. $\Psi=0°$; $\theta=30°$

Fig. 2. Square grid for validating the projection model

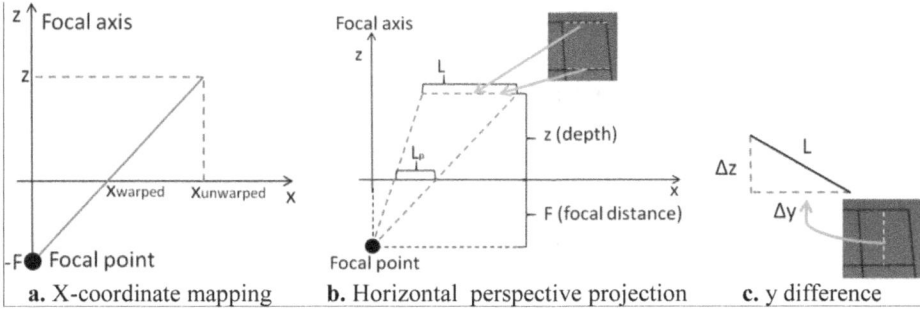

a. X-coordinate mapping   **b.** Horizontal perspective projection   **c.** y difference

Fig. 3. Warped/unwarped (flat) projection relations

From Fig. 3.a it is possible to infer warped/flat x values relation in equation (1) with triangle relations, where F_a is the focal axis x-coordinate. Fig. 3 also shows Fig. 2.d distorted square. Using triangular relations, one gets eq. (2). Let L_{p1} and L_{p2} be the values of the upper and lower projected square sides. The difference between them yields to (3). The value for Δz is calculated using Pythagoras theorem (Fig. 3.b). Finally, the focal distance can be obtained with (4). As there is more than one square available, the focal distance is calculated for each square and the median of the values obtained is chosen for use.

$$x_{warped} = F_a + F \times (x_{unwarped} - F_a)/(z(y) + F) . \tag{1}$$

$$\frac{L}{L_p} = \frac{z+F}{F} . \tag{2}$$

$$L \times \left(\frac{1}{L_{p1}} - \frac{1}{L_{p2}}\right) = \frac{\Delta z}{F}; \ \ where \ \Delta z = z_1 - z_2 . \tag{3}$$

$$F = \frac{1}{L} \times \sqrt{L^2 - \Delta y^2} \times \left|\frac{1}{L_{p1}} - \frac{1}{L_{p2}}\right|^{-1} . \tag{4}$$

The relation between the paper depth and shading is assumed to be linear (see Fig. 2.c), where zero depth means that the paper is on the scanner flatbed (region without distortion). With the focal distance, it is possible to infer the distance to the flatbed (z-value) using equation (2).

- Let I_0 and I_{flat} be the square grid whitest paper shading values in the region around the corner. I_0 is measured from the warped horizontal side, I_{flat} is measured from the flat region.
- Let b and b_{flat} be as above, but measured from the warped document, respectively.

Thus, the shading/depth relation is expressed in (5). One may observe that for $b = b_{flat}$ the depth equals 0, that means that it is on the flatbed. For the HP ScanJet 5300c, the values found for 300dpi were: 2,894.37 pixels (24.51cm using eq. (6)) for F, 233/255 for I_{flat}, 174/255 for I_0 and 270.1412 pixels for Z_0.

$$d(b) = Z_0 \times (1 - b/b_{flat})/(1 - I_0/I_{flat}) . \tag{5}$$

$$pixelsToCm(v) = (v/DPI) \times 2.54 . \tag{6}$$

3 The De-warping Procedure

To better understand the shape of the book binding distortion, 8 pictures were taken with different warp levels (Fig. 4.a shows one of them). Sample points were fitted into an exponential curve as described by eq. (7) (Fig. 4.b shows plotting for 4.a) with a mean error of less than 0.08 cm for all 8 images. Thus, eq. (7) models binding with acceptable accuracy.

$$z(x) = Z_0 \times e^{\lambda x} . \tag{7}$$

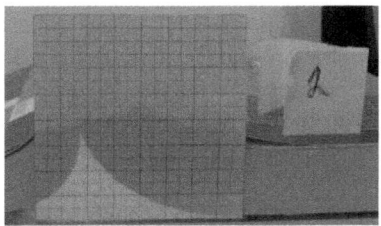
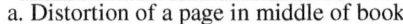
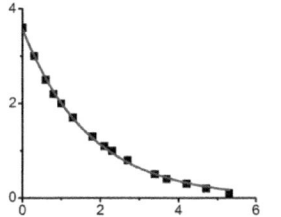

a. Distortion of a page in middle of book   b. Plot and exponential fitting of (a)

Fig. 4. Validation of book binding modeled as an exponential curve

To correct the book binding distortion, the depth is estimated and fitted onto the exponential curve represented by equation (7). One simple and effective method is presented in [14] using equations (8) and (9). De-warping is done by the line integral of (7) which is presented by equations (10) and (11). The next sections show how to obtain the value of the depth for both scenarios in Fig. 1.

$$Z_0 = e^{D_0}; \; D_0 = \frac{\sum_{i=1}^{n} \ln z_i \sum_{i=1}^{n} x_i^2 - \sum_{i=1}^{n} x_i \sum_{i=1}^{n} x_i \ln z_i}{n \sum_{i=1}^{n} x_i^2 - \left(\sum_{i=1}^{n} x_i\right)^2} \tag{8}$$

$$\lambda = \frac{n \sum_{i=1}^{n} x_i \ln z_i - \sum_{i=1}^{n} x_i \sum_{i=1}^{n} \ln z_i}{n \sum_{i=1}^{n} x_i^2 - \left(\sum_{i=1}^{n} x_i\right)^2} \tag{9}$$

$$s(x) = \frac{1}{\lambda}\left(\sqrt{(\lambda z(x))^2 + 1} - \ln\left|\frac{\sqrt{(\lambda z(x))^2 + 1} + \lambda z(x)}{\lambda z(x)}\right|\right) \tag{10}$$

$$x_d(x) = s(x) - s(0) \tag{11}$$

3.1 Shape-from-Shading

Assuming that an image was obtained in the scenario depicted in Fig. 1.a, and the projection to be pure orthographic due to small flatbed distances, the proposed procedure for de-warping is summarized as:

1. For each column i of the image find the mode of the intensity of the pixels in the highest 10-percentile in the histogram of the pixels in the column. This value b_i is taken as representative of the intensity of the paper column.
2. Identify the shading value of a flat region (b_{flat}) by getting the most frequent value of b_i.
3. The exponential fitting is done to obtain I and λ with (i, z_i) (where $z_i = d(b_i)$) using equations (8) and (9) [14]. The values for z_i that are too close to 0 have a greater influence than others as the logarithm goes to minus infinite as the input approaches zero [14]. The curve fitting is done with $|b_i - b_{flat}| \geq 20/255$.
4. Apply the de-warping procedure using the line integral of $z(x)$ as described in equations (10) and (11).

The inverse function for equation (10) cannot be calculated analytically. The direct mapping is done (from source x to target x) with values for the orphan x columns in the target, calculated by linear interpolation of the closest columns. Figure 5 shows the proposed procedure applied to Figure 1.c, with the shading pattern at (a) and the detail of the final result in (b). The advantages of proposed method are:

- The fitting procedure detects the deformation side (left or right) automatically as λ can be positive or negative
- There is no need to detect when the warped region starts. The exponential has low derivative values on the opposite side of the deformation, thus it acts as a constant straight line in this area. Thus equation (7) can be applied to every value of x.

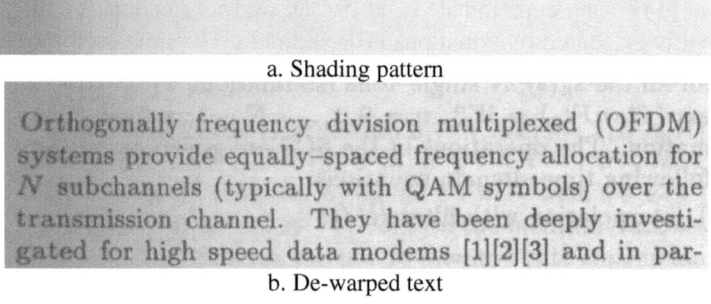

a. Shading pattern

Orthogonally frequency division multiplexed (OFDM) systems provide equally–spaced frequency allocation for N subchannels (typically with QAM symbols) over the transmission channel. They have been deeply investigated for high speed data modems [1][2][3] and in par-

b. De-warped text

Fig. 5. Proposed shape-from-shading applied to Figure 1.c

3.2 Text-Line Depth Extraction

Looking at the central part of the book image in Figure 1.e one may observe that closer to the book spine the top page looks brighter than the bottom page. Thus, at the same depth one finds two different luminance values, therefore shape-from-shading cannot be used to estimate the distance to the scanner flatbed in the scenario of Figure 1.b. As one may observe, the lines are curved in this case, thus text-line segmentation can be used to get depth values. There are several text-line segmentation for black-and-white warped documents such as [8][9]. For the sake of space, the method used herein is not presented. As segmentation requires a monochromatic image as input, image binarization [7] is performed. One assumes that the text-lines are already segmented and the baseline envelope for each of them is found. The following procedure is executed:

1. Un-warped straight lines are estimated by doing a linear regression with text-line central letters, as illustrated in Figures 6.a and 6.b in gray.
2. For each point of the baseline envelope there is a corresponding point on the straight line that represents where it should be if there were no warp. For the given point, depth (z value) can be obtained using eq. (2), where L and L_p are the distance to the focal axis for "un-warped" and warped points, respectively, with the focal axis assumed to be the vertical middle line (e.g. Fig. 6.b);

3. The exponential curve fitting is done with equations (8) and (9) [14] using (y_i, z_i) of baseline points. Not all baseline points can be used, thus they are restricted to:
   a. As a point gets closer to the focal axis L and L_p gets smaller and more prone to errors, thus the curve fitting only uses points far away from the focal axis.
   b. Higher depth values than a given threshold (see shape-from-shading step 3);
4. Use the line integral of $z(y)$ as described in equations (10) and (11) to obtain the de-warped y value, and use equation (1) to obtain the warped x-value for every point.

In step 1, the central letters is in a region where y value is between 25% and 75% of the image height. In step 3.a, the baseline points far from the focal axis are the ones with x-coordinate less than 25% or greater than 75% of the image height. In step 3.b, the low depth values are the ones greater than 0.169 cm (or 20 pixels for 300 dpi using equation (6)). There is no analytical inverse for equation (10), thus orphan line points in the resulting image are calculated by bilinear interpolation of the closest points. Figure 6.d shows details of processing Figure 1.e. This method also has the same advantages as the previous one.

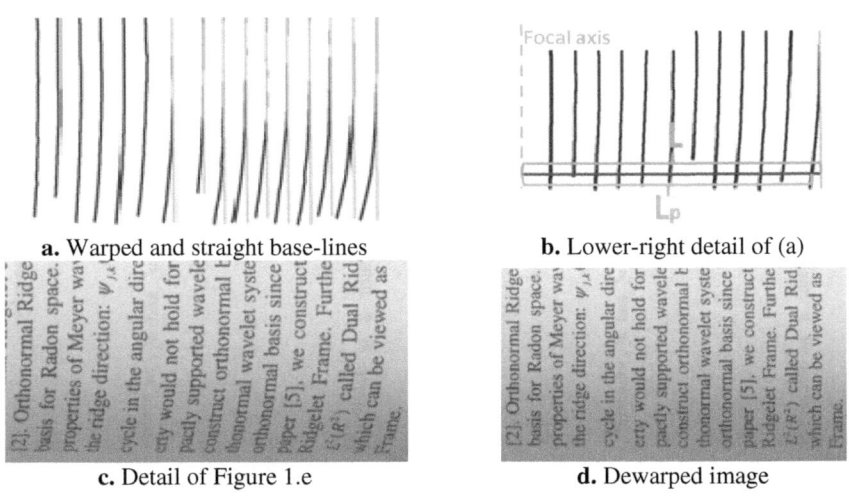

a. Warped and straight base-lines   **b.** Lower-right detail of (a)

c. Detail of Figure 1.e   **d.** Dewarped image

Fig. 6. Text-line depth extraction on Figure 1.e

4   Results

To validate the proposed method 20 different pages were scanned in the scenario of Figure 1.a and 15 in the scenario depicted in Figure 1.b. Some examples can be seen in Table 1 with the original and de-warped images on the left and the right hand sides, respectively. The first row shows a page with text lines with repeated letters in order to compare the result of the width restoration of Figure 1.a. The only limitation perceived is when binding is small and changes rapidly; the last row of Table 1 shows one of such cases. The next sections show pre- and post processing steps to improve the method proposed herein.

Table 1. Dewarping results: original (left) with dewarped (right)

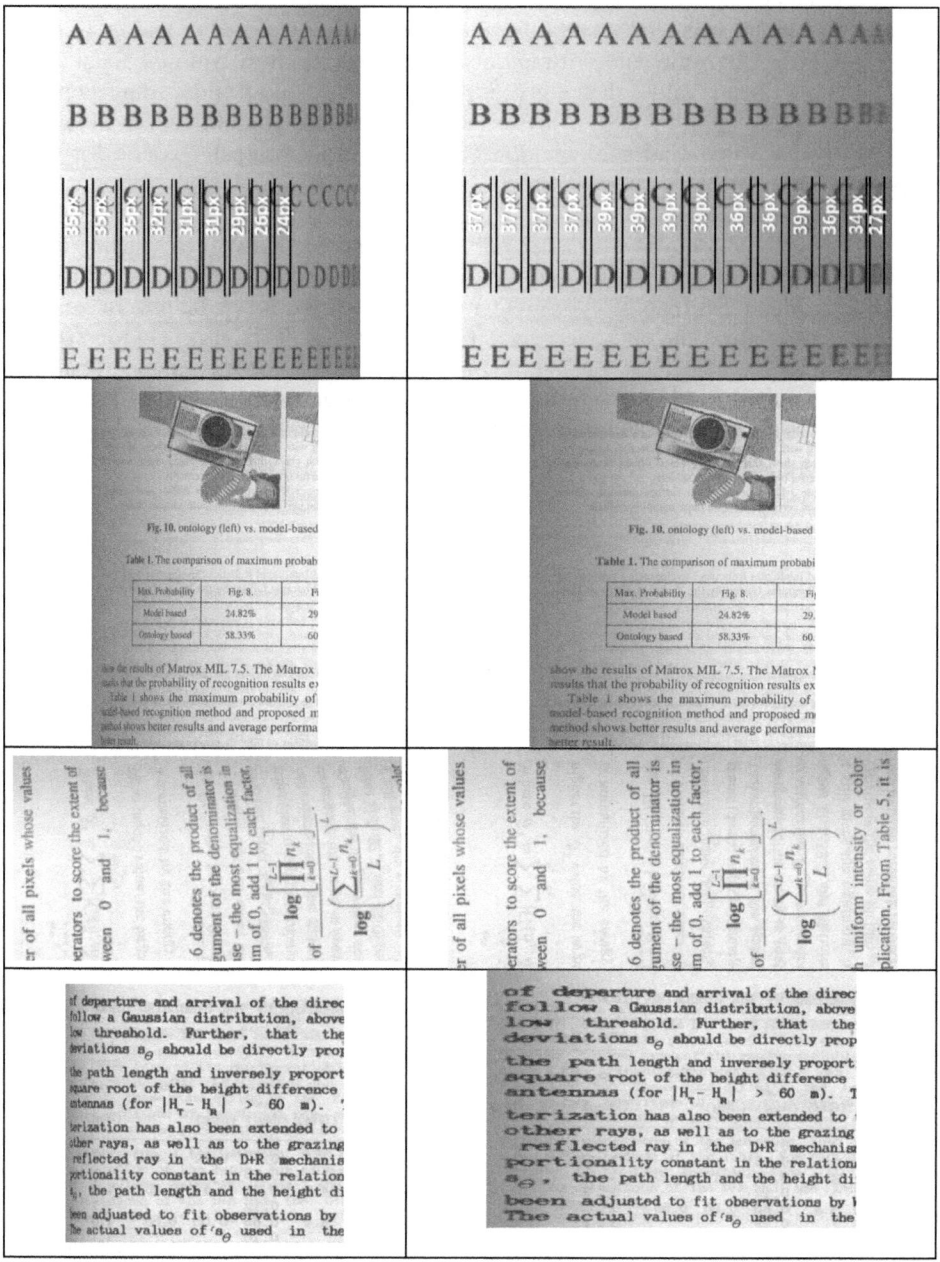

4.1 Pre-processing Improvements

Very often, despite the best efforts of those who scan bound volumes to avoid image skew, the obtained images may not be perfectly parallel to the axis of the scanner flatbed. To correct such distortion, PhotoDoc [16] perspective correction routine is used. An example of such pre-processing is shown in Figure 8, with zoomed results in (d)-(f). The improvement of the new method can be seen in (e) if compared with method [15] in (f).

Text-line segmentation can be also used to detect portrait/landscape orientation assuming that horizontal lines are dominant. As the end-user does not know when to apply shape-from-shading or text-line depth extraction, the flow in Figure 7 is proposed. For all tested images it correctly identified all page orientations.

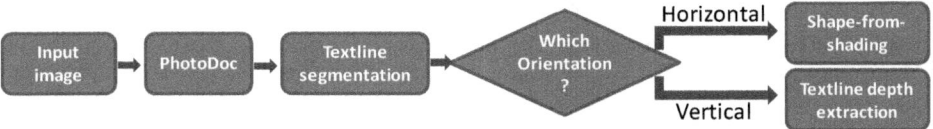

Fig. 7. End-users processing flow

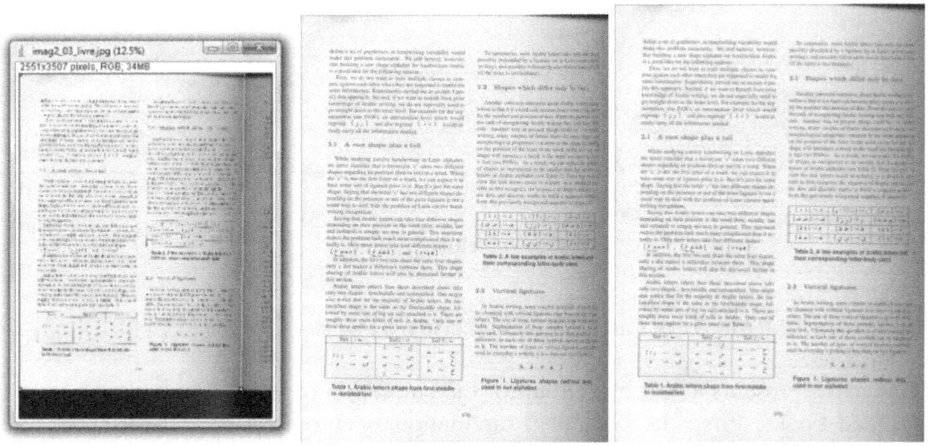

a. Original image	**b.** Perspective corrected image	**c.** De-warped image
d. Zoom into (a) binding	**e.** New method on (d)	**f.** Method [15] on (d)

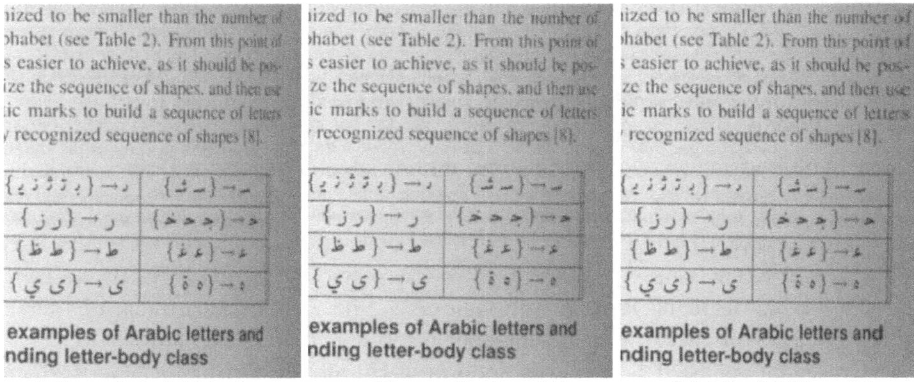

Fig. 8. Example of image with perspective and volume binding warp

364   R.D. Lins et al.

4.2 Post-processing Improvements

The image de-warping algorithm presented herein does not correct the variation in illumination which darkens the region towards the volume binding of the page. Compensating illumination variation is part of the full de-warping procedure. Figure 9 shows the application of two [7] [12][13] shading removal algorithms. Oliveira and Lins [7] searches for blocks with low color variation then tries to find overlapping blocks that belongs to document paper followed by shading removal. Fan's method [12][13] uses a watershed-based color segmentation that works in a wide range of document contents such as those containing large continuous-tone image regions, with a high computational complexity. Figure 9.a shows the application of [7] with inferred paper value; Figure 9.b shows the result forcing the final background to pure white. Figure 9.c presents the result of applying the algorithms in references [12][13].

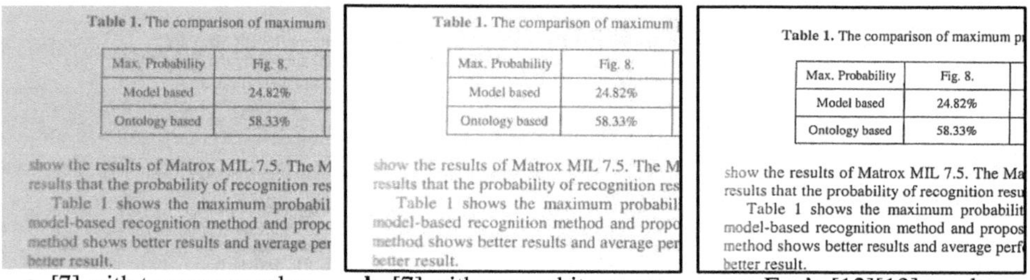

a. [7] with true paper value   b. [7] with pure white paper   c. Fan's [12][13] result

Fig. 9. Shading removal

5   Conclusions and Lines for Further Work

This paper presents a new algorithm for the correction of volume biding distortion in scanned books. First, the document orientation is automatically detected, and then proper depth estimation is chosen followed by exponential fitting and de-warping is performed by using an exponential line integral. For depth estimation, two new methods were proposed. The first one uses shape-from-shading, the other extracts the height by analyzing the envelope of text-lines. Both of them yielded satisfactory results. The application of shading removal algorithms is also recommended.

As some scanned documents with binding distortion may have blurred characters a document de-blurring strategy seems to be appropriate, but is left for further work.

Acknowledgments

The authors like to thank Professor Liliana Gheorghe, of the Department of Mathematics of UFPE for checking the mathematical formalism.

The research reported herein was sponsored by an MCT-Brazilian Government R&D Grant and CNPq funding.

References

[1] Tan, C.L., Zhang, L., Zhang, Z., Xia, T.: Restoring Warped Document Images through 3D Shape Modeling. IEEE Trans. Pattern Analysis and Machine Intelligence 28(2), 195–208 (2006)

[2] Ukida, H., Konishi, K.: 3D Shape Reconstruction Using Three Light Sources in Image Scanner. IEICE Trans. on Inf. & Syst. E84-D(12), 1713–1721 (2001)

[3] Ukida, H., Tanimoto, Y., Sano, T., Yamamoto, H.: 3D Shape Reconstructions Using Image Scanner under Various Number of Illuminations. In: IEEE IST '07, Krakow, Poland, May 5, pp. 1–6 (2007)

[4] Zhang, R., Tsai, P.S., Cryer, J.E., Shah, M.: Shape from shading: A survey. IEEE Transactions on Pattern Analysis and Machine Intelligence 21(8), 690–706 (1999)

[5] Durou, J.-D., Falcone, M., Sagona, M.: Numerical Methods for Shape-from-shading: A New Survey with Benchmarks. Computer Vision and Image Understanding (2007)

[6] Szeliski, R.: Computer Vision: Algorithms and Applications (to be published),
     `http://research.microsoft.com/`
     `en-us/um/people/szeliski/Book/`

[7] Oliveira, D.M., Lins, R.D.: A New Method for Shading Removal and Binarization of Documents Acquired with Portable Digital Cameras. In: CBDAR 2009, pp. 3–10. IAPR Press, Barcelona (2009)

[8] Bukhari, S.S., Shafait, F., Breuel, T.M.: Dewarping of Document Images using Coupled-Snakes. In: CBDAR 2009, pp. 17–24. IAPR Press, Barcelona (2009)

[9] Stamatopoulos, N., Gatos, B., Pratikakis, I., Perantonis, S.J.: A two-step dewarping of camera document images. In: Proceedings 8th IAPR Workshop on Document Analysis Systems, pp. 209–216 (2008)

[10] Pintus, R., Malzbender, T., Wang, O., Bergman, R., Nachlieli, H., Ruckenstein, G.: Photo Repair and 3D Structure from Flatbed Scanners. Computer Vision Theory and Applications, 40–50 (2009)

[11] Schubert, R.: Using a Flatbed Scanner as a Stereoscopic Near-Field Camera. IEEE Computer Graphics and Applications, 38–45 (2000)

[12] Fan, J.: Enhancement of Camera-captured Document Images with Watershed Segmentation. In: CBDAR'07, September 2007, pp. 87–93 (2007)

[13] Fan, J.: Robust Color Image Enhancement of Digitized Books. In: Proceedings of International Conference on Document Analysis and Recognition, pp. 561–565. IEEE Press, Los Alamitos (2009)

[14] Wolfram Resarch. Least Squares Fitting–Exponential,
     `http://mathworld.wolfram.com/`
     `LeastSquaresFittingExponential.html` (accessed October 14, 2009)

[15] Lins, R.D., Oliveira, D.M., Torreão, G., Fan, J., Thielo, M.: A De-Warping Algorithm to Compensate Volume Binding Distortion in Scanned Documents. In: ACM SAC 2010 (March 2010)

[16] Silva, G.F.P., Lins, R.D.: PhotoDoc: A Toolbox for Processing Document Images Acquired Using Portable Digital Cameras. In: CBDAR 2007, Curitiba, pp. 107–115 (2007)

Image-Based Drift and Height Estimation for Helicopter Landings in Brownout

Hans-Ullrich Doehler and Niklas Peinecke

Institute of Flight Guidance, Department of Pilot Assistance,
German Aerospace Center (DLR), Braunschweig, Germany

Abstract. After years of experiences regarding enhanced and synthetic
vision research projects in the fixed wing domain, the DLR's Institute of
Flight Guidance is now addressing helicopter applications as well. The
project ALLFlight is one example. The main objective of this project
is to demonstrate and evaluate the characteristics of different sensors
for helicopter operations within degraded visual environments, such as
brownout or whiteout. Radar, Lidar, IR and TV cameras are part of
this sensor-suite. Although this project aims for the *large solution* of the
brownout problem, there are also simple *small solutions* investigated,
which can be installed in low cost helicopters as well. The following
paper is dealing with such an approach.

We assume a pair of off-the-shelf vertical looking cameras and an
inertial attitude system, which are feeding their data into a low-budget
processing system. The outcome consists of the helicopter's lateral drift
and its altitude above the ground. These data can be shown easily to the
pilot on a cheap display. The paper describes the proposed setup and
methods for drift and height measuring.

1 Introduction

Compared to fixed-wing aircraft, flying a helicopter is still relatively unsafe.
Statistics show that the number of accidents per flight hours is much higher (e.g.,
in 2007: 4.9 per 100,000 flight hours [4]) compared to the number of accidents
of fixed-wing aircraft (1.39 accidents per 1 million flight hours [11]). The main
reasons for this unacceptable number of accidents are pilot errors due to high
workload and bad weather conditions. This is the main motivation for developing
assistant systems to increase the level of safety in flying helicopters. As daily
problems of helicopter operation, like search and rescue (SAR), or helicopter
emergency medical service (HEMS), show, visual assistance for the helicopter's
pilot is even more essential.

During landing on sand, dust or snow, whirled up particles from the ground
can produce a dense cloud around the helicopter so that a visual guidance of
the helicopter becomes impossible. This effect is called *brownout* for landings on
sand, or *whiteout* for landings on snow. During landing the pilot has to ensure
that the lateral drift of the helicopter does not exceed a certain magnitude
directly before touching down. Otherwise a dangerous moment around the roll

A. Campilho and M. Kamel (Eds.): ICIAR 2010, Part II, LNCS 6112, pp. 366–377, 2010.

axis would occur after the first contact of the landing gear or skid with the ground. This turning moment could finally lead to a total roll over of the entire helicopter.

The effect of whirling up dust is caused by the main rotor's down stream. Its strength increases with decreasing flight altitude. As long as the flight path of the helicopter has some forward movement, a horizontal cylinder of dust is formed behind the helicopter. This cylinder becomes a torus (like a "donut") as soon as the altitude and forward speed are falling below certain thresholds (see Fig. 1). After the dust torus has fully developed, the entire external horizontal field-of-view around the helicopter becomes in-transparent. Thus, the pilot is no longer able to acquire visual cues from outside to evaluate the helicopter's lateral drift. Nevertheless, a certain region directly under the helicopter remains free of dust during hovering - at least for a certain amount of time. Through this hole within the "donut" the ground remains recognizable. If the pilot would be able to see a picture of the landing zone below the helicopter (e.g., on a camera display), it would still be difficult for him to interpret this image. This is due to the fact that the lateral shift of the entire image is not only effected by drift, but also by rotations of the helicopter around its roll and pitch axis. Therefore we propose to apply an automatic image analysis system to determine the cross- and along-drift of the helicopter. This system has to be supported by data of the turning rates around the along and transverse axis of the helicopter (pitch and roll axis). Together with a suitable display to show the computed drift rates to the pilot, this system will be able to assist the pilot in controlling the helicopter until touch down. Additionally, to estimate the height above the ground and the possible tilt angles of the landing surface, we apply a stereo camera system.

2 State of the Art

2.1 Optical Position Estimation

Position estimation by optical means is arguably the oldest technique known. Early examples are numerous and cover at most static scene reconstruction [6,3]. However, we will concentrate on the more recent dynamic methods. Application areas included autonomous robot navigation [10,7] and camera pose estimation [2]. Integrated sensor solutions combining optical sensors with Laser measurements will become available in the next years [15,9]. Especially the availability of cheap computer vision hardware integrated into mobile phones has raised some recent interest [16,8].

2.2 Assisted Landing

Different concepts and methods for assisting helicopter pilots during the landing phase under brownout situations have been published within the last years and there are several patent applications available within this field (e.g., [5,14,12]). The proposed methods can be divided into three main categories:

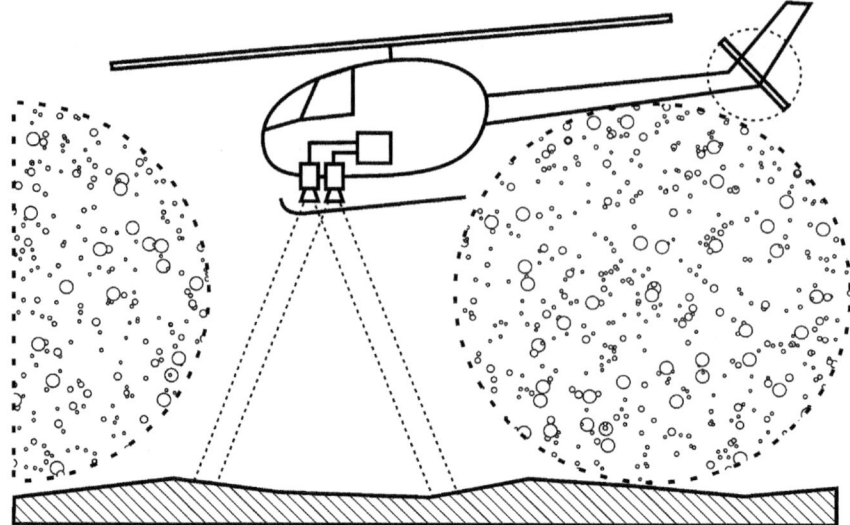

Fig. 1. Imaging situation: A stereo camera is looking downwards in brownout through the remaining hole in the dust below the helicopter's fuselage

- *large solution* - high sophisticated sensor-suite and complex data fusion setup
- *see and remember* - perspective display of real-time acquired 3D terrain data
- *small solution* - downward looking sensors are driving a simple drift and altitude display.

The *large solution* to solve the low visibility problem consists on one hand of a large and complex suite of different imaging sensors, such as millimeter wave Radar systems, optical Lidar systems, and infrared cameras. On the other hand, high accurate data bases and high precision navigation systems are an essential element of this concept. The biggest challenge of this approach is to design intelligent algorithms for data fusion and display generation. The expected advantage is that in every phase of the helicopter landing, at least one sensor is able to look through dust and snow. The main disadvantage is that such a large and complex system will be rather heavy and expensive and therefore is not easy to install [5,14,12,1]. Due to this, such systems will not be affordable for every helicopter.

The second concept, called *see and remember*, means that during the approach phase some imaging sensors are acquiring data from the terrain below as long as possible. From these data a consistent 3D-model of the landing zone is permanently updated in real-time. In combination with precise positioning and attitude data, a perspective view onto this 3D-model is generated and shown to the pilot [14]. After the sensor looses its direct visual contact to the ground, pilots shall still make use of the continuously available perspective display of the 3D-model. Although the perspective presentation of the 3D data is steadily updated with respect to the changing position and attitude of the helicopter, the 3D data itself becomes outdated over time. The main disadvantage of this approach is the scepticism of pilots. Although they are familiar with flight training in simulated environments, they cannot accept to fly a real aircraft based on a (probably) outdated 3D model (although the age of the model data might become rarely older than some 30 seconds until touch-town).

The third approach, the *small solution*, applies only downward looking cameras, which are mounted below the helicopter's fuselage. In [12] a system is presented which makes use of several of so-called PMD sensors [13]. These are solid-state cameras which are able to measure ranging data for each pixel by using some special *range gating* technique. The basic principle is similar to an optical radar system (Lidar), where the whole scene is illuminated by a very short pulse of light. Pfenninger states that such an optical system, mounted below the helicopter's fuselage, would be able to help within the brownout situation. This is true, because during brownout there remains a dust-free zone within the inner part of the "donut-like" cloud. This inner zone allows a visual look-through onto the ground below. We follow this argumentation within our contribution. However, instead PMD cameras (which are more or less in a prototype developing state), we will apply a pair of off-the-shelf standard CMOS-cameras, which are used to built-up a stereo-camera setup.

3 Experimental Setup

We assume the following setup. A pair of cameras are mounted on a common bracket. Additionally, an attitude sensor is also mounted on this bracket for measuring turning rates (Fig. 2).

We purchased two CMOS b/w cameras from IDS, Germany (see table 1). For operation on a Win-XP system, this type of camera has a WDM-driver interface. The implemented software for image acquisition is based on the DirectShow SDK. The camera has a built-in automatic intelligent gain and shutter control. Taking into account the typical levels of light of outdoor scenes, the excellent light sensitivity of the camera allows rather small shutter times (< 5 ms). This is important for operation on a rather dynamic helicopter platform. Otherwise, motion blurring of the acquired images would harm the following feature extraction.

To measure the turning rates around all three axes, we apply the IntertiaCube of Intersense. This system uses a serial interface (RS-232) and delivers its data with a frequency of up to 180 Hz. Such a sensor is needed to build-up a stand-alone demonstrator. Of course, the rate sensor becomes obsolete, when our system is integrated into helicopters with digital interface to the attitude and heading reference system (AHRS), where the turning rates can be read out via any avionics bus (e.g., ARINC-429).

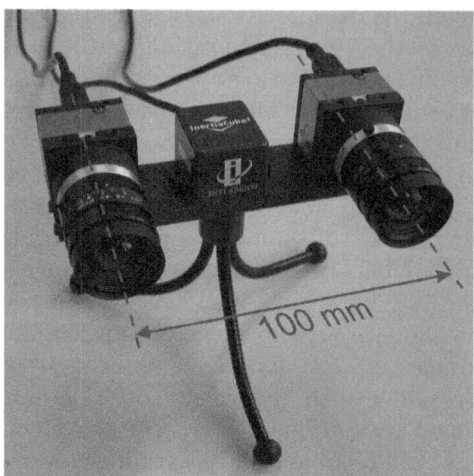

Fig. 2. Applied camera pair (uEye SE1220-M) and attitude sensor (InertiaCube)

Table 1. Applied camera uEye 1220SE-M from IDS

camera type	uEye 1220SE-M
manufacturer	IDS, Obersulm, Germany
sensor type	CMOS - b/w
senor chip	MTV0922 Micron, ID, USA
typical application	automotive industry
resolution	752 × 480 pixel
optical area	4.51 × 2.88 mm
pixel size	6 × 6 micron
ADC resol.	10 bit
frame rate	60 Hz full resolution
global shutter	0.04 ... 5000 ms
lens	Fujinon 1.2/6.0 mm
interface	USB-2.0
Software interface	WDM-driver
size (incl. lens)	32 × 34 × 75 mm
weight (incl. lens)	120 g

4   Image Processing and Feature Extraction

We implemented a feature based image analysis technique. First each input image is converted into a list of features representing the main image content. Feature are extracted by multistage processing. The single stages of processing are:

1. Low pass or median filtering for noise reduction and application of some gradient filter algorithms (e.g., a Sobel filter) for edge enhancement.

2. Converting the gray scale image into a set of binary images (black and white). By applying a number of different predefined binarization thresholds, pixels with values below the threshold become black. Remaining pixels become white. Usually three or four binary images are sufficient for processing images with a reasonable brightness distribution. Although the binarisation levels are currently set manually they can easily be adjusted by applying histogram analysis techniques.

3. Contour extraction and contour following is carried out for each binary layer. The result is a set of closed contours. Each contour is denoted by a starting point and a list of contour points. In case the number of contour points is below a predefined threshold the contour becomes a candidate for a blob feature and is further processed in step 6. All other contours-sets are processed in step 4.

4. Cutting the closed contours into linear sub-sets. Each identified sub-set is fed into an orthogonal regression algorithm, which computes the best fitting line for the sub-set. The result is a list of line features which is further processed in step 5.

5. Post processing of extracted lines: parallel and/or interrupted lines are grouped or combined into one representative line. Each line is denoted by its starting and ending point and its contrast, i.e., the difference of gray values between the left and right side of the line within the original image.

6. Post processing of extracted blobs: different blobs from different binary layers are grouped into one representative blob in case the differences in size and position are below a certain predefined level. Each blob is denoted by its center coordinates, its width and height, and its contrast, i.e., the difference of gray values between inner and outer regions of the blob within the original image.

The implemented program allows adjustment of most predefined processing parameters during run-time. Thus, adaptation to different types of input images can be done easily. Processing time for feature extraction takes something between 20 and 40 ms for a single typical outdoor image (320×240 px, 4 binary layers, Intel Core2, 1.6 GHz).

5   Drift Estimation and Stereo Reconstruction

5.1   Drift Estimation

Drift estimation is based on a shift analysis of the feature lists over time from image to image. This is done in two steps:

1. Each acquired image is converted into a list of features as described above. The list of features from the last `image(t-1)` is denoted as `list(t-1)` and the list from the actual `image(t)` is denoted as `list(t)`. It is assumed that for each feature of the `list(t)` exists exactly one corresponding feature of the same type of the `list(t-1)`. The most probable global shift vector (denoted in x and y-direction) of all features is estimated by means of a 2D

shift histogram. At first, each feature from `list(t)` is assumed to have potentially several corresponding features of `list(t-1)`. However, assuming that feature parameters change only slightly between images the assignment possibilities are substantially reduced. For each established candidate pair, the corresponding entry in the 2D shift histogram is incremented by one. The resolution of histogram bins must not be set too high in order to avoid spreading of similar entries over several histogram bins. To this end we applied a value of two or more pixels. The maximum histogram entry is interpreted as a first coarse estimation of the global 2D shift vector, which is applied as starting value for the next step.

2. We can assume that the local shift vector of each feature in the image has a relatively small deviation from the global coarse shift value, determined in step 1. So, we run again through the feature pairing process, now with a smaller expectation window with a size of a few pixels around the global coarse shift vector. The result is an unambiguous assignment of features of `image(t)` to the features of `image(t-1)`. This assignment is used for computing the fine mean value of the global 2D shift vector.

Considering the time-stamp of each acquired image, we can compute the global 2D shift speed vector (denoted in pixels per second). With regard to the known field of view (FOV) of the camera (in degrees), the 2D angular speed vector (denoted in degrees per second) is obtained. Finally, by adding distance data (computed from stereo reconstruction, or – if available – from the helicopter's radar altimeters) and the turn rates from the rate sensor, the lateral 2D drift vector (denoted in meters per second) results. The resulting drift vector can be displayed to the pilot.

Of course it is true that the image drift will not be constant over the entire image field below some small distance. However, regarding the application of helicopter landing this effect can be neglected since the smallest distance between camera and ground will stay above half a meter for most helicopter models and landings are usually carried out on planar terrain only.

5.2   Height Estimation and Stereo Reconstruction

To estimate the helicopter's height above the ground, a stereo matching process of the images from the left and the right camera has to be carried out. This is done with a similar method as applied for drift estimation. Before starting the actual estimation we need to calibrate the camera setup. We apply a fronto-parallel camera setup (see Fig. 3). For such a setup the distance z from a given 3D point $P(z)$ results from the following equation:

$$z(d) = D\frac{f}{d} \tag{1}$$

where $d = d_R - d_L$ denotes the disparity of the image of $P(z)$ in both images, f denotes the focal length of the cameras and D is the lateral distance of both cameras. For calibration purposes and to adjust some deviation of the principle points between the cameras, as well as for correcting small disalignments between the cameras we apply an additional shift d'. To calibrate the setup some points with known distance z_0 have to be analysed and the value of d' has to be adjusted so that the result is just the known distance z_0.

The contents of `image(L)` and `image(R)` are already converted to `list(L)` and `list(R)`. Here L stands for the left and R for the right camera. Now we apply the following feature matching process:

1. It is assumed that for each feature of the `list(L)` exists exactly one corresponding feature of the same type of the `list(R)`. Again, the most probable global shift vector (denoted in x and y direction) of all features is estimated by means of a 2D shift histogram. By assuming the fact that feature parameters are similar between the left and right image, the assignment possibilities are substantially reduced. For each established candidate pair, the corresponding entry in the 2D shift histogram is incremented by one. Unlike the above method of drift estimation, we assume now that the vertical shift component is small. Instead, the horizontal shift values could take larger values. Again, we obtain a first coarse estimation of the global 2D shift vector from the maximum histogram entry, which is applied as starting value for the next step. In case of acquiring images from 3D objects which are randomly distributed in depth, this method will adapt to some type of a majority disparity. Concerning the application of helicopter landing, this behaviour can be regarded as an advantage. It is true that each helicopter's landing field can be regarded as a more or less tilted plane. Thus, all features from the ground plane are projected at similar disparities.

2. Again, we can assume that the local shift vector of each feature in the image has some deviation from the global coarse shift value. Again we run through the feature pairing process, but now with a much wider expectation window in horizontal direction. Again, the result is an unambiguous assignment of features of `image(L)` to the features of `image(R)`. For each feature we store the individual horizontal shift value which is fed into the stereo reconstruction process.

From the measured disparity list of features between the left and right image we compute a set of 3D points (denoted in relative sensor coordinates). In order to generate an estimated best fitting plane from this list we use a plane regression algorithm. Basically, we implemented two similar approaches: Least square fitting using direct matrix inverse and least square fitting using eigenvalues of the covariance matrix.

In the first case we use the plane equation $ax + by + z + d = 0$ taking advantage of the fact that the face normal will be oriented towards the camera in positive z-direction. Thus, a coefficient for z can be set to 1, the resulting normal will be of the form $(a, b, 1)$. Minimization yields:

$$\begin{pmatrix} \sum x^2 & \sum xy & \sum x \\ \sum xy & \sum y^2 & \sum y \\ \sum x & \sum y & N \end{pmatrix}^{-1} \begin{pmatrix} \sum xz \\ \sum yz \\ \sum z \end{pmatrix} \tag{2}$$

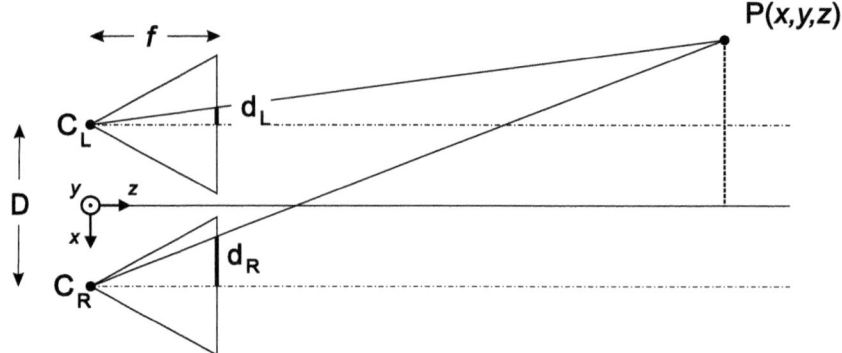

Fig. 3. Geometry setup for stereo camera calibration

with $\sum_{kl} := \sum_{i=1}^{N} k_i l_i$ for $k, l \in \{x, y, z\}$ and $((x_i, y_i, z_i))_{i=1}^{N}$ the list of measured 3D points.

In the second case we first compute the center (x_0, y_0, z_0) of all points. Then we compute the covariance matrix

$$\begin{pmatrix} \sum_{(x-x_0)^2} & \sum_{(x-x_0)(y-y_0)} & \sum_{(x-x_0)(z-z_0)} \\ \sum_{(x-x_0)(y-y_0)} & \sum_{(y-y_0)^2} & \sum_{(y-y_0)(z-z_0)} \\ \sum_{(x-x_0)(z-z_0)} & \sum_{(y-y_0)(z-z_0)} & \sum_{(z-z_0)^2} \end{pmatrix} \tag{3}$$

The normal vector is then the eigenvector of this matrix belonging to the smallest eigenvalue.

In our case both methods perform well with the first method being slightly faster.

6   Experimental Results

We conducted an experiment with a white plate with dark markings on it mounted on a rotateable platform with a highly adjustable angle and the cameras at a fixed distance (see Fig. 4). We recorded a set of measurements for predefined angles between $-20°$ and $+20°$.

The results can be seen in Fig. 5. The maximal, mean and mean square error of the measurements against the exact angles are 2.36, 0.5 and 0.48 respectively, the variance of the error is 0.24. With an image resolution of 352×240 pixels and 19 randomly distributed dark targets on a white plane at a distance of 0.5 m, our method provides an angular accuracy better than 0.5 degrees. The accuracy of the measurements could be improved by using an appropriate camera calibration model. In the present state the method generates a small systematic error due to radial distortion. However, for most applications the method is already adequate ever since angles beyond $10°$ are of no practical importance for the intended helicopter application. The implementation runs at frame rates better than 10 Hz on a typical desktop computer system.

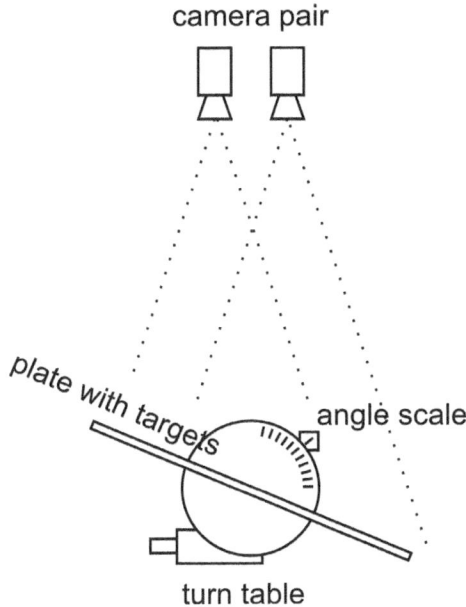

Fig. 4. Experimental setup to test measurement accuracy

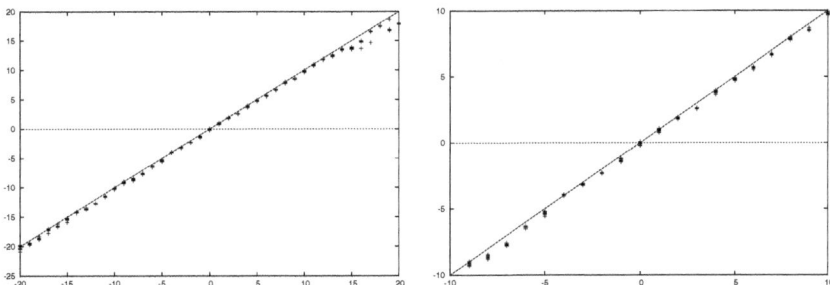

Fig. 5. Results of angular measurements (same data plotted in two resolutions)

7   Conclusion

We presented a contribution to the *small solution* as support of the landing of an helicopter in brownout situations. The proposed system consists of two light-weight downward-looking cameras, one sensor for measuring the pitch and roll rate of the helicopter, and a computer system for image processing, feature matching and display generation. We expect that the system is suitable for easy integration in smaller helicopters, which do not provide any avionics interface. For application in helicopters with digital avionics systems, data from the AHRS (Attitude and Heading Reference System) and the radar altimeter can be fed directly into our system. In that case the separate rate sensor would become obsolete. The radar altimeter data could serve for monitoring the system's integrity.

It should be mentioned, that state-of-the-art inertial navigation systems (INS) can measure drift rates, as well. However, due to their operational principle, which is mainly based on integrating data from accelerometers, the smallest detectable drift value exceeds the demand for controlling the helicopter shortly before touch-down. It is another advantage of our proposed system that it provides increasing drift measuring sensitivity as the height above ground decreases. In fact, it reaches its best drift measuring performance during touch-down.

Flight trials in typical environments are planned to be conducted in the forthcoming months, as part of the project ALLFlight. It is planned that the accuracy of the drift estimation is evaluated during the flight tests comparing the rsults to the highly precise drift estimators in the helicopter's INS.

References

1. Doehler, H.U., Lueken, T., Lantzsch, R.: ALLFlight - a full scale enhanced and synthetic vision sensor suite for helicopter applications. In: Enhanced and Synthetic Vision, SPIE Proceedings, vol. 7328. SPIE, Bellingham (2009)
2. Gilbert, S., Laganiére, R., Roth, G.: Stereo motion from feature matching and tracking. In: IEEE Instrumentation and Measurements Technology Conference, pp. 1246–1250. IEEE, Los Alamitos (2005)
3. Grimson, W.: Computational experiments with a feature based stereo algorithm. IEEE Transactions on Pattern Analysis and Machine Intelligence 7(1), 17–34 (1985)
4. Helicopter Association International: Five-year comparative U.S. civil helicopter safety trends through 4th quarter (January 1- December 31, 2008-2004) (2008), http://www.rotor.com
5. Judge, J.H., Occhiato, J.J., Stiles, L., Sahasrabudhe, V., Macisaac, M.A.: Technical design concepts to improve helicopter obstacle avoidance and operations in brownout conditions. Tech. Rep. US7106217B2, US-Patent application (2004)
6. Kanatani, K.: Detection of surface orientation and motion from texture by a stereological technique. Artificial Intelligence 23(2), 213–237 (1984)
7. Kaszubiak, J., Tornow, M., Kuhn, R.W., Michaelis, B.: Real-time, 3-d-multi object position estimation and tracking. In: International Conference on Pattern Recognition, vol. 1, pp. 785–788 (2004)
8. Keitler, P., Pankratz, F., Schwerdtfeger, B., Pustka, D., Rödiger, W., Klinker, G., Rauch, C., Chathoth, A., Collomosse, J., Song, Y.Z.: Mobile augmented reality based 3d snapshots. In: Proc. Sechster Workshop Virtuelle und Erweiterte Realität der GI-Fachgruppe VR/AR. Braunschweig, Germany (November 2009)
9. Konietschke, R., Busam, A., Bodenmüller, T., Ortmaier, T., Suppa, M., Wiechnik, J., Welzel, T., Eggers, G., Hirzinger, G., Marmulla, R.: Accuracy identification of markerless registration with the dlr handheld 3d-modeller in medical applications. In: Prodeedings of CURAC, vol. 6 (October 2007)
10. Krotkov, E., Hebert, M., Buffa, M., Cozman, F., Robert, L.: Stereo driving and position estimation for autonomous planetary rovers. In: IARP Workshop on Robotics in Space (July 1994)

11. NTSB: Accidents, fatalities, and rates, 1988 through 2007, for U.S. air carriers operating under 14 cfr 121, scheduled and nonscheduled service (airlines) (2008)
12. Pfenninger, T.: Landehilfesystem für senkrecht startende und landende Luftfahrzeuge, insbesondere Helikopter. Tech. Rep. DE102007019808A1, German Patent application (2007)
13. PMD Technologies GmbH: (2009),
    http://www.pmdtec.com/products-services/pmdvisionr-cameras/
    pmdvisionr-camcub%e-20/
14. Scherbarth, S.: Method of pilot support in landing helicopters in visual flight under brownout or whiteout conditions. Tech. Rep. US2006/0087452A1, US-Patent (2005)
15. Suppa, M., Kielhoefer, S., Langwald, J., Hacker, F., Strobl, K.H., Hirzinger, G.: The 3d-modeller: A multi-purpose vision platform. In: Proceedings of International Conference on Robotics and Automation (ICRA), April 2007, vol. 6 (2007)
16. Zhang, L., Rusdorf, S., Brunnett, G.: Echtzeit-Bewegungserfassung mit geringem Marker-Satz und monokularen Videodaten. In: Proceedings, 6. Workshop der GI-Fachgruppe VR/AR, pp. 115–126. Shaker Verlag, Aachen (November 2009)

Can Background Baroque Music Help to Improve the Memorability of Graphical Passwords?

Haichang Gao[1,2], Xiuling Chang[1], Zhongjie Ren[1],
Uwe Aickelin[2], and Liming Wang[1]

[1] Software Engineering Institute, Xidian University, Xi'an, Shaanxi 710071, P.R. China
[2] School of Computer Science, The University of Nottingham, Nottingham, NG8 1BB, U.K.
hchgao@xidian.edu.cn

Abstract. Graphical passwords have been proposed as an alternative to alphanumeric passwords with their advantages in usability and security. However, they still tend to follow predictable patterns that are easier for attackers to exploit, probably due to users' memory limitations. Various literatures show that baroque music has positive effects on human learning and memorizing. To alleviate users' memory burden, we investigate the novel idea of introducing baroque music to graphical password schemes (specifically DAS, PassPoints and Story) and conduct a laboratory study to see whether it is helpful. In a ten minutes short-term recall, we found that participants in all conditions had high recall success rates that were not statistically different from each other. After one week, the music group coped PassPoints passwords significantly better than the group without music. But there was no statistical difference between two groups in recalling DAS passwords or Story passwords. Further more, we found that the music group tended to set significantly more complicated PassPoints passwords but less complicated DAS passwords.

Keywords: Graphical password, Baroque music, Memorability, DAS, Passpoints.

1 Introduction

Graphical passwords have been proposed as an alternative to alphanumeric passwords and the main motivation is the hypothesis that people perform far better when remembering pictures rather than words [1, 2]. Visual objects seem to offer a much larger set of usable passwords. It is conceivable that humans would be able to remember stronger passwords of a graphical nature. However, users still tend to choose passwords that are memorable in some way, which means that the graphical passwords still tend to follow predictable patterns that are easier for attackers to exploit [6, 13, 14].

Various literatures reveal that users are the 'weakest link' in any password authentication mechanism, probably due to their memory limitations [11]. Although human memory capacity is unlikely to increase significantly over the next few years, recent psychological and physiological studies indicate that certain music like baroque music has positive effects of great importance on human memorizing and learning [20, 22].

A. Campilho and M. Kamel (Eds.): ICIAR 2010, Part II, LNCS 6112, pp. 378–387, 2010.

Motivated by these observations, we investigate the novel idea of introducing background baroque music to graphical password schemes with the purpose of alleviating users' memory burden and improving usable security. Based on DAS, PassPoints and Story schemes, we conduct a laboratory study to explore the efficiency of background baroque music on memorizing graphical passwords. We are also interested in whether the background music would enable users to choose more complicated or less predictable passwords, which are usually more resistant to dictionary and other guess attacks.

The following section briefly reviews graphical password schemes and related works. Sections 3 and 4 describe the methodology of our studies and present the results respectively. Section 5 provides several interpretations to the experiment and discusses the experimental results. Conclusion and future work are addressed in Section 6.

2   Related Works

2.1   Graphical Password Schemes

In the open literature to date, the ubiquity of graphical interfaces for applications and input devices, such as the mouse, stylus and touch-screen, has enabled the emergence of graphical authentications. There have been three dominant techniques available which can be defined as: Drawmetrics (DAS [4], Syukri [9], YAGP [21]), Locimetrics (Blonder [3], PassPoints [7]) and Cognometrics (Déjà Vu [5], Story [6], Passfaces [10]) [19].

Drawmetrics systems require users to reproduce a pre-drawn outline drawing on a grid. A well-known scheme in this category is DAS which liberates users from remembering complicate text strings and has the advantage of better security over alphanumerical passwords [4]. Nevertheless, Passdoodle revealed that people are able to remember complete doodle images while less likely to recall the stroke order [8]. Furthermore, Thorpe and Van Oorschot found that users tend to design symmetrical and centered or approximately centered passwords, significantly reducing password space in practice and impacting the security [14]. Gao et al. proposed a modification to DAS where approximately correct drawings can be accepted, based on Levenshtein distance string matching and "trend quadrants" looking at the direction of strokes [21].

Locimetrics systems are based on the method of loci, an old and well-known mnemonic [18]. Originating in Blonder's work, the approach involves users choosing several sequential locations in an image [3]. PassPoints [7] is a representative scheme of this category, where users may choose any place in the image as a password click point. Since it is a cue of great importance for users to recall their passwords, the image should be complex and visually rich enough to have many potentially memorable click points. This scheme was found that although relatively usable, security concerns remains. A primary security problem is hotspots: people tend to select obvious points in the image with high visual salience, leading to a reduced effective password space that facilitates more successful dictionary attacks [12,13].

In the Cognometrics systems, users must recognize the target images embedded amongst a set of distractor images. This category includes Passfaces which relies on face recognition [10], Déjà Vu [5] based on abstract images and Story [6] where users are suggested to create a story and so on. User studies by Valentine have shown that

Passfaces has a high degree of memorability [15, 16], but Davis found that people tended to select faces of their own race and gender [6]. Assigning faces to users arbitrarily may alleviate the problem, whereas it would lead people hard to remember the password. A similar scheme to Passfaces is Story where the password selection is sufficiently free from bias [6]. But, the Story is not as good as Passfaces in memorability, because few people actually choose stories despite the suggestion. In addition, memorability for abstract images in Déjà Vu was found to be only half as good as that for photographic images with a clear central subject [17].

Through the above discussion, we find that most graphical passwords either tend to follow predictable patterns or have a low degree of memorability. The crux of the problem is the users' memory limitations. As human memory capacity is unlikely to increase significantly over the next few years, creating a nice environment for memorizing passwords might alleviate users' burden. There are demonstrations that music can improve memory and in what flows we will illustrate it.

2.2 The Efficiency of Baroque Music

Extensive researches have shown that music has different uses for education and therapy [20]. As our particular interest is to explore the role of music in learning and memorizing graphical passwords, we will briefly review the researches into the effects of music on learning in this subsection.

Georgi Lozanov, a Bulgarian psychologist, made remarkable impact in integrating music into teaching practice. He created a teaching method called 'Suggestopedia', wherein the use of background music, particularly the baroque music with a rate of 50 to 70 beats per minute (BPM), is a cornerstone of accelerated learning techniques. It is stated that the method of Suggestopedia involves three stages where different types of music are used for specific purposes. First, introduce music to relax participants and help them to achieve the optimum state for learning. Second, listen to an "active concert" with music from Mozart, Beethoven and Brahms. Finally, apply a "passive concert" to help participants move the information into the long-term memory. While no details are given as to which exact music is suggested for the first stage, both the concerts in the later two stages result in high memory retention [22]. Further more, Lozanov says that "well organized Suggestopedia accelerates learning 5 times on an average" [22].

Baroque music can help the brain produce alpha waves, and information imbued with music has a greater likelihood of being encoded in the long-term memory by the brain. That is why accelerated learning techniques introduce music into the learning process. For example, 'Mozart Effect' [23] is a phenomenon that music has a positive effect on learning and memory. In the following sections, we bring background baroque music to graphical password schemes, specifically, PassPoints, DAS and Story, and do an investigation to check whether it can improve users' memory or induce users to set stronger passwords.

3   User Study

As mentioned earlier, our evaluation is based on three representative graphical password schemes. For the purpose of collecting and analyzing the success rate, user habits, and login time automatically, we reproduce three schemes which are intentionally very closely modeled after DAS [4], PassPoints [7] and Story [6], respectively. We still adopt the names "DAS", "PassPoints" and "Story" for convenience. In this section, after describing the three schemes deployed in our experiments, we will present our methodology in great detail.

3.1   Brief Introduction of the Reproduced Schemes

DAS is a drawing reproduction based scheme, where a 5×5 grid was deployed for users to draw on. Each grid cell is denoted by rectangular discrete coordinates (x, y) $\in$ [0, 4] $\times$ [0, 4]. A completed drawing is encoded as the ordered sequence of cells that the user crosses whilst constructing the secret, with a distinguished coordinate pair (5, 5) inserted in both ends of each stroke. Two passwords are identical if the encoding is the same. Figure 1 shows how DAS works. Input a graphical password consisting of three strokes, which are colored by black, green and red in sequence. The drawing is mapped to (5,5)(1,2)(1,3)(2,3)(3,3)(3,2)(5,5); (5,5)(2,1)(2,2)(2,3)(5,5); (5,5)(2,1)(5,5).

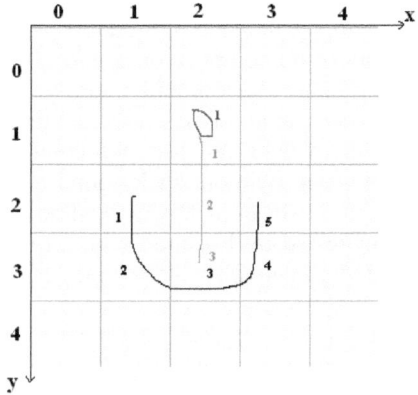

Fig. 1. An example of DAS password with length being 9

In the PassPoints scheme, users are required to select several positions in a single image as their passwords and click close to the chosen points in correct order and within a tolerance distance for authentication. For example, the password in Figure 2 contains five click points orderly labeled by small red rectangle.

In Story, a password is a sequence of k (k≤9) images selected by the user to make a "story". To keep consistent with that in [6], the images used here are also classified into nine categories, which are animals, cars, women, food, children, men, objects, nature, and sports. Images of "men" and "women" are gathered from FordModels.com and the others http://images.google.com. Figure 3 shows the interface of Story, where the man, woman, car and the house are orderly selected and the underlying story is "a gentle man and his girlfriend drive a car to their house".

Fig. 2. Passwords in PassPoints with length being 5   **Fig. 3.** An example of Story password

3.2 Experiments

We conducted a lab study with 28 subjects (16 males and 12 females). All the subjects were university students of computer science and in the age range of 20 to 30. We hypothesized that background music could improve humans memory and then induced people to choose more complex passwords and take less time to log in. This study used a between-subjects design and had two conditions; half of the subjects were assigned to the control group (without background music) and half to the music group. None of them had previously used DAS, PassPoints or Story passwords. We chose the baroque music suggested by Lozanov with a rate of 50 to 70 BMP as the background music and utilized a Lenovo speaker to play it. The volume was set to 30-40 decibels as suggested.

Our study included two lab-based sessions. Session 1 took about two hours. At the beginning of Session 1, each participant was asked to read an instruction document. This provided information of their activities on the experiments and helped them know how DAS, PassPoints and Story work. To make the rules clearer, an example was included in each scheme. Then participants were required to complete the registration and login of DAS, then PassPoints, and finally Story. People were asked to reenter the password to confirm it. After a short delay (about 10 minutes), participants were asked to log in within three attempts. In the end, participants need answer a demographic questionnaire collecting information including age, sex and experience on graphical passwords.

One week later, at Session 2, all the participants returned to the lab and tried to log in each scheme within three attempts using their previously created passwords.

4   Results

We used two types of statistical tests to assess whether differences in the data reflect actual differences between conditions or whether these may have occurred by chance. A t-test (two tails) was used for comparing the means of two groups and Fisher's exact test was used to compare recall success rates. In all cases, we regard a value of P<0.20 as indicating that the groups being tested are different from each other with at least 80% probability, making the result statistically significant. In the tables, "not significant" indicates that the test revealed no statistically significant difference between the two conditions (i.e., P>0.20).

4.1 Success Rates

We first examine success rates as a measure of participants' performance. Table 1 compares the successful recalls in each group.

Table 1. Success rates in each group for DAS, PassPoints and Story

Group	10-minute test		1-week test	
	ratio	Fisher-test	ratio	Fisher-test
DAS (no music)	78.6%	P=0.59	71.4%	P=1
DAS (music)	92.9%		64.3%	
PassPoints (no music)	100%	P=1	35.7%	P=.004
PassPoints (music)	100%		92.9%	
Story (no music)	100%	P=1	92.9%	P=1
Story (music)	100%		92.9%	

In the 10 minutes short term phase, the success rates were high on the whole, indicating that participants' memory was not strongly taxed. In PassPoints and Story, participants under both conditions recalled their passwords. In DAS, the success rate of the music group was 92.9%, higher than that of the control one (78.6%). However, a Fisher's exact test yields a result of P=0.59, indicating that the difference was not statistically significant.

Table 2. Complexity of DAS secrets

Group		DAS (no music)	DAS (music)
Strokes	Avg.	3.36	3.71
	t-test	Not significant	
	S.d.	1.71	2.25
	Max	7	7
	Min	1	1
Password Length	Avg.	13.79	10.43
	t-test	t=1.34, P<0.20	
	S.d.	6.39	6.41
	Max	27	21
	Min	2	1

After one week, the performances of two groups varied in schemes. Both groups in Story had the same success recall rate 92.9%, but differed in DAS and PassPoints. In DAS, only 64.3% of the music group and 71.4% of the control group were able to recall their passwords. It appears that the control group performed better than the music group. It should be noted that it was only a difference of one person in practice. The result of Fisher's exact test showed that there was no statistical difference between two conditions. In PassPoints, we found a significant difference between two groups. The music group was significantly more likely to successfully recall the passwords than the control group. In addition, the success rate of the control group decreased from 100% in the previous phase to 35.7% while the success rate of the music group only decreased by 7.1%. It aligns with psychology research which continues to show that certain music advance the long-term memory.

The results suggest that the background music works differently when it was available in different graphical password schemes. In Drawmetrics and Cognometrics systems, background music seems to have no influence on short-recalls or long-term memory. But in Locimetrics systems, it appears that background music could significantly help people remember passwords in long-term memory.

4.2 Password Complexity

For each scheme, we compare password complexity in both groups. While the password length in PassPoints or Story is easy to understand, it is necessary to explain it in DAS. In DAS, the length of a password yields by adding the lengths of its component strokes wherein the length of a stroke is the number of coordinate pairs it contains exclusive of the distinguished ones(5,5). For example, for the password in Figure2, the length of each stroke is 5, 3 and 1 respectively, producing a password length of 9.

Table 3. Complexity of PassPoints and Story secrets

Group	Password length				
	Avg.	t-test	S.d.	Max	Min
PassPoints (no music)	3.79	t=1.61, P<0.20	1.20	5	1
PassPoints (music)	4.5		1.05	6	3
Story (no music)	3.64	Not significant	0.97	6	2
Story (music)	4.07		0.70	6	3

In DAS (see Table 2), the average password length with music was 10.43 and without, 13.79. The standard deviation of password length with music was 6.41, compared to 6.39 without. A t-test yields a result of t=1.34, P<0.20(two tails), indicating that the password length in the music group was significantly shorter than that in the control group. The background music increased the stroke count of passwords on average, but not to a statistically significant level. The standard deviation with respect to stroke count was higher with music (2.25 vs. 1.71).

While background music reduced the password length in DAS, it increased the password lengths in PassPoints and Story. As shown in Table 3, the average password length with music in PassPoints was 4.5 as opposed to 3.79 without. A t-test yields a result of t=1.61, P<0.20(two tails), indicating that there was statistically significant difference between two conditions. In Story, the password length for two groups differed by 0.43 (4.07 vs. 3.64), which is not statistically significant.

As such, the background music had a negative effect on DAS password length, but encouraged people to choose more complex passwords in PassPoints and Story.

5   Discussion

5.1   Validation of Hypotheses

Based on the results of our study, we now revisit our hypotheses that background music could improve humans' memory and then induced people to choose more complex passwords. This hypothesis was only supported in PassPoints. In PassPoints, people in the music condition not only chose significantly more complicated passwords, but also had significantly higher recall success rates in the long-term test. However, in DAS, the average password length of the music group was much shorter than that of the control group.

5.2   Recall Errors

This subsection will discuss the recall errors in DAS and PassPoints (Few errors occurred in Story and thus be ignored). People committed different types of error shown in Figure 4 (DAS) and Figure 5 (PassPoints). In DAS, there are three types of error: Stroke (i.e., entering more or less strokes), Pwd-Len (i.e., people could recall stroke count but forget the length of password) and Position (others including mixing up the stroke order or crossing incorrect cells). From figure 4, we can see that errors in Stroke and Pwd-Len account for the main proportion of recall errors. At the same time, music group committed more errors than the non-music group and the difference resulted possibly from the long-term recall test.

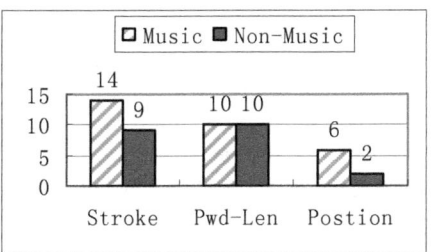   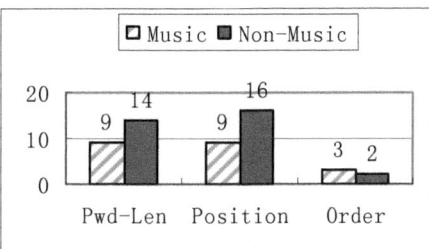

Fig. 4. Recall errors in DAS   **Fig. 5.** Recall errors in PassPoints

There are also three types of error in PassPoints: Pwd-Len (i.e., forgetting the password length), Position (i.e., people can recall the password length but click points outside the tolerance region) and Order (i.e., only mixing up the click-points order) (see Figure 5). In this scheme, the nature of many recall failure was down to either forgetting the password length or clicking points outside the tolerance region. In recall errors and especially in Position errors, music group had a great advantage over non-music group, probably due to its higher success recall rate in the long-term recall test.

5.3 Limitations

Our intent in this study was to examine the effects of background music on the memorability of graphical passwords. We made our study follow the established methods of experimental psychology as much as possible and acknowledged that it did not mirror real-life usage. First, the participants in our study (all of them were university students of computer science) only represented a small part of the whole. It was important to get a good selection of people with various backgrounds in the further studies. Second, users are unlikely to familiar with and create three different graphical passwords one after the other in real life, or be asked to recall in quick succession them after one week (without having used any of them in the intervening time). Third, the participants had no incentive to perform as if protecting or accessing anything of real-life value to them, therefore it was not difficult to understand that many passwords created in both conditions were weak. For example, in Story, the average password length of the control group was less than 4. Furthermore, the effect of the background music volume remains to be discussed when it was embedded into a scheme. Despite these limitations, our controlled laboratory experiment paved the road to numerous further studies.

6 Conclusion

Results of the user study have shown that it is an effective enhancement to introduce baroque music to the PassPoints scheme. Surrounding with music, people not only tended to construct significantly more complicated passwords than their counterparts without the music stimulus, but also performed significantly better in terms of recall success in the long-term tests. This result indicated that the background music improved the memorability of passwords in PassPoints.

In DAS and Story, the introduction of background music has been shown unnecessary for security and usability. The recall of the passwords in both conditions was not statistically different from each other in short-term or long-term test. Further more, the background music significantly impaired the complexity of DAS passwords.

Although results obtained in three representative schemes are not consistent and should be treated with caution, we believe that this work provides a significant extension to the study of security and usability of graphical passwords. The future work includes a larger sale of studies with careful experimental design and Locimetrics systems will be our focus.

Acknowledgments. The authors would like to thank the reviewers for their careful reading of this paper and for their helpful and constructive comments. Project 60903198 supported by National Natural Science Foundation of China.

References

1. Madigan, S.: Picture memory. In: Imagery, Memory, and Cognition, pp. 65–86. Lawrence Erlbaum Associates, Mahwah (1983)
2. Nelson, D.L., Reed, U.S., Walling, J.R.: Picture superiority effect. Journal of Experimental Psychology: Human Learning and Memory 3, 485–497 (1977)

3. Blonder, G.E.: Graphical password. US Patent 5559961, Lucent Technologies, Inc., Murray Hill (August 30, 1995)
4. Jermyn, I., Mayer, A., Monrose, F., Reiter, M., Rubin, A.: The design and analysis of graphical passwords. In: Proceedings of the 8th USENIX Security Symposium (August 1999)
5. Dhamija, R., Perrig, A.: Déjà Vu: A User Study Using Images for Authentication. In: 9th USENIX Security Symposium (2000)
6. Davis, D., Monrose, F., Reiter, M.K.: On user choice in graphical password schemes. In: Proceedings of the 13th Usenix Security Symposium, San Diego, CA (2004)
7. Wiedenbeck, S., Waters, J., Birget, J.C., Brodskiy, A., Memon, N.: Design and longitudinal evaluation of a graphical password system. International J. of Human-Computer Studies 63, 102–127 (2005)
8. Goldberg, J., Hagman, J., Sazawal, V.: Doodling Our Way to Better Authentication. Presented at Proceedings of Human Factors in Computing Systems (CHI), Minneapolis, Minnesota, USA (2002)
9. Syukri, A.F., Okamoto, E., Mambo, M.: A User Identification System Using Signature Written with Mouse. In: Boyd, C., Dawson, E. (eds.) ACISP 1998. LNCS, vol. 1438, pp. 403–441. Springer, Heidelberg (1998)
10. Passfaces, http://www.realuser.com (site accessed January 10, 2010)
11. Notoatmodjo, G.: Exploring the 'Weakest Link': A Study of Personal Password Security. Thesis of Master Degree, The University of Auckland, New Zealand (2007)
12. Dirik, A.E., Memon, N., Birget, J.C.: Modeling user choice in the PassPoints graphical password scheme. In: Symp. on Usable Privacy and Security, SOUPS (2007)
13. Thorpe, J., van Oorschot, P.C.: Human-Seeded Attacks and Exploiting Hot-Spots in Graphical Passwords. In: USENIX Security Symp. 2007 (2007)
14. Nali, D., Thorpe, J.: Analyzing User Choice in Graphical Passwords. Technical Report, School of Information Technology and Engineering, University of Ottawa, Canada (May 27, 2004)
15. Valentine, T.: An evaluation of the Passface personal authentication system. Technical Report, Goldsmiths College, University of London (1998)
16. Valentine, T.: Memory for Passfaces after a Long Delay, Technical Report, Goldsmiths College, University of London (1999)
17. Weinshall, D., Kirkpatrick, A.S.: Passwords you'll never forget, but can't recall. In: Proc. CHI 2004 (2004)
18. Higbee, K.L.: Your Memory: How it Works and How to Improve it, 2nd edn. Prentice-Hall Press, New York (1988)
19. DeAngeli, A., Coventry, L., Johnson, G., Renaud, K.: Is a picture really worth a thousand words? Exploring the feasibility of graphical authentication systems. International Journal of Human-Computer Studies 63, 128–152 (2005)
20. Fassbender, E., Richards, D., Kavakli, M.: Game engineering approach to the effect of music on learning in virtual-immersive environments. In: International Conference on Games Research and Development: CyberGames, Western Australia (2006)
21. Gao, H., Guo, X., Chen, X., Wang, L., Liu, X.: YAGP: Yet Another Graphical Password Strategy. In: ACSAC, California, USA, pp. 121–129 (2008)
22. Lozanov, G.: Suggestology and Suggestopedy, http://lozanov.hit.bg/
23. Rauscher, F.H., Shaw, G.L., Ky, K.N.: Music and spatial task performance. Nature 365(6447), 611 (1993)
24. Wiedenbeck, S., Waters, J., Birget, J.C., Brodskiy, A., Memon, N.: Authentication using graphical passwords: Effects of tolerance and image choice. In: Symposium on Usable Privacy and Security (SOUPS). Carnegie-Mellon University, Pittsburgh (2005)

Color Texture Analysis for Tear Film Classification: A Preliminary Study

D. Calvo[1], A. Mosquera[2], M. Penas[1], C. García-Resúa[3], and B. Remeseiro[1]

[1] VARPA Group, Dept. of Computer Science, Univ. A Coruña, Spain
{dcalvo,mpenas,bremeseiro}@udc.es
[2] Artificial Vision Group, Dept. of Electronics and Computer Science, Univ. Santiago de Compostela, Spain
antonio.mosquera@usc.es
[3] Optometry Group, Dept. Aplied Physics, Univ. Santiago de Compostela, Spain
carlos.garcia.resua@rai.usc.es

Abstract. The tear lipid layer is not homogeneous among the population and its classification depends on its width. Too thin or too thick films can lead to unhealthy eyes as well as create problems when interacting with contact lenses. This work proposes a preliminary methodology to classify the tear lipid layer according to its texture into four main categories. The proposed methodology works on several stages to detect the region of interest, extract the texture descriptors on colour information and classify these descriptors. The method has been tested on several images from each tear type. In some cases, we obtain classification results over the 90%.

Keywords: Tear film, lipid layer, opponent colors, band pass filtering.

1 Introduction

The tear film is a complex, dynamic structure of lipids, proteins, and mucins riding on the hydrophobic surface of the epithelium [1]. Classically, the normal tear film is described as a trilaminar structure comprising a superficial lipid layer, an intermediate aqueous phase and an underlying mucous layer[2]. The tear film provides a smooth optical surface by compensating for the micro irregularities of the corneal epithelium[3]. It also plays an essential role in the maintenance of ocular integrity by removing foreign bodies from the front surface of the eye, supplying antimicrobial and mechanical protection to the corneal epithelium[4,5]. Furthermore, since the corneal surface is avascular, it is highly dependent on the tear film for its nutrition[6].

There are several clinical tests available to evaluate quality or quantity aspects of the tear film. However, some of them are invasive and may disrupt the tear film [7,8,9,10] and others are very expensive for clinical settings[11]. The observation of the superficial lipid layer offers a valuable and non invasive technique that evaluates the tear film quality, since the lipid layer enhances the stability of the tear film by retarding water evaporation from the surface of the open eye[12].

A. Campilho and M. Kamel (Eds.): ICIAR 2010, Part II, LNCS 6112, pp. 388–397, 2010.

Lipid layer thickness can be evaluated by the observation of the interference phenomena [13,14,15], which correlates with tear film quality [16,17,18], since a thinner lipid layer speeds up water evaporation, decreasing tear film stability.[19]

Guillon [3,13] proposed five main categories of lipid interference patterns. In order of increasing thickness and visibility these are: open meshwork, closed meshwork, waves, amorphous and colours. Abnormal appearances and phenomena are also described in [20]. Thicker lipid layers (≥ 90 nm) are readily observed since they result in colour and wave patterns. Thinner lipid layers (≤ 60 nm) are difficult to observe, since the colours and other distinct morphological features are not present. If the lipid layer is ≤ 50 nm, only a gray or white surface, without other features, is observed. However, this technique is affected by the subjective interpretation of the observer and sometimes is difficult to observe, especially with thinner lipid layers that lack distinct features. Some techniques have been designed to objectively calculate the lipid layer thickness where a sophisticated optic system was necessary [21,22], other techniques used an interference camera that evaluates the lipid layer thickness by only analyzing the interference colour[23].

The purpose of this study is to present a novel methodology for the classification of the eye lipid layer into four of the categories defined by Guillon, based on the characterization of both texture and colour patterns. The amorphous category has not been included in this research due to the lack of images from this category in the clinical image dataset available for the validation of our methodology. In a first step the lipid layer patterns have been classified into waves, colours and meshworks; in a second step a refinement of the meshwork patterns into open meshwork and closed meshwork, has been performed.

This paper is organized as follows. Section 2 describes the proposed methodology including the acquisition of the image, the extraction of the region of interest and the color texture analysis. Section 3 shows the experimental results obtained by the system using a set of images provided by opticians. And section 4 briefly exposes and discusses the conclusions of this work.

2   Methodology

The proposed methodology consists of four main stages. The first stage involves the acquisition of the digital image of the lipid tear film. The second stage entails the extraction of the region of interest where the classification takes place. In the third stage, the underlaying texture is analyzed. Finally, the last stage classifies the images into the categories previously presented. In the following sections, all these stages will be explained in detail.

2.1   Acquisition of the Lipid Film Image

The acquisition of the image is the first step towards the tear film classification. The lipid layer was assessed with the TearScope Plus® (Keeler, Windsor, United Kingdom) attached to a Topcon SL-D4 slit lamp. The Tearscope Plus®

(Keeler, Windsor, United Kingdom), designed by Guillon [13,24], is the instrument of choice for rapid evaluation of lipid layer thickness in clinical settings. The Tearscope Plus® projects a cylindrical source of cool white fluorescent light onto the lipid layer illuminating almost all of the corneal surface area and the interference patterns are observed with the magnification of the slit-lamp microscope.

The interference lipid layer images were acquired by a Topcon DV-3 digital camera and stored via Topcon IMAGEnet i-base with a spatial resolution of 1024×768 pixels. Since the tear lipid film is not static between blinks, a video was recorded and analyzed to select the image to be processed. An image is suitable to go through the processing step when the tear lipid film is completely expanded after the eye blink, so only the frame fulfilling this condition is processed to extract the region of interest, as explained in the next section.

2.2   Extraction of the Region of Interest

The extracted image, as depicted in Figure 1(a), contains several areas of the eye, including the sclera, eyelids, eyelashes and other non interesting areas. The experts that analyze the image usually focus on the bottom part of the iris, since this is the area where the tear is shown with better contrast. This fact forces a previous preprocessing step to extract the region where the lipid tear film classification takes place.

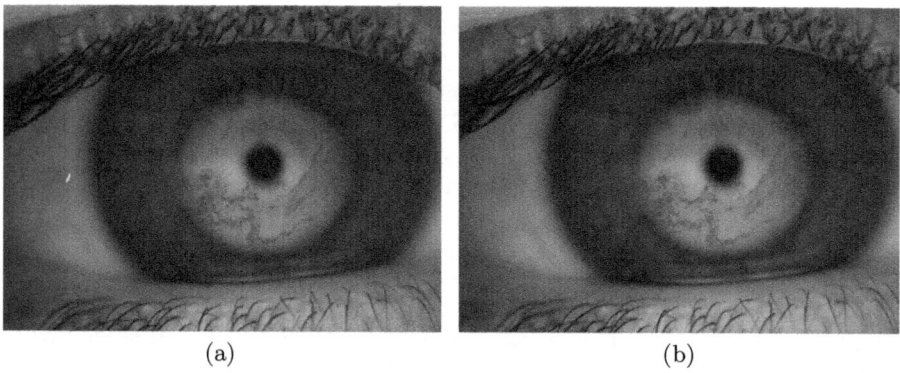

(a)   (b)

Fig. 1. (a)An original image acquired with the *TearscopePlus*. (b) The luminance component in the LAB color space.

The acquisition technique that we use generates a central area in the image, more illuminated than the others, that corresponds to the area used by the experts in the classification step. Thus, to obtain the region of interest we restrict our analysis to the most illuminated area in the image.

Concretely, the proposed method uses a normalized cross-correlation to locate the most illuminated area of the image. In order to restrict our analysis to the illumination, the input image is transformed to the LAB color space and only the luminance component L, depicted in Figure 1(b), is analyzed in this step. The cross-correlation technique matches the luminance component with several templates that cover the various shapes the region of interest can have. Figure 2(a) shows the matching score between one of the templates and the image. The highest peak corresponds to the best match and, therefore, with the position of the area of interest, as depicted in Figure 2(b).

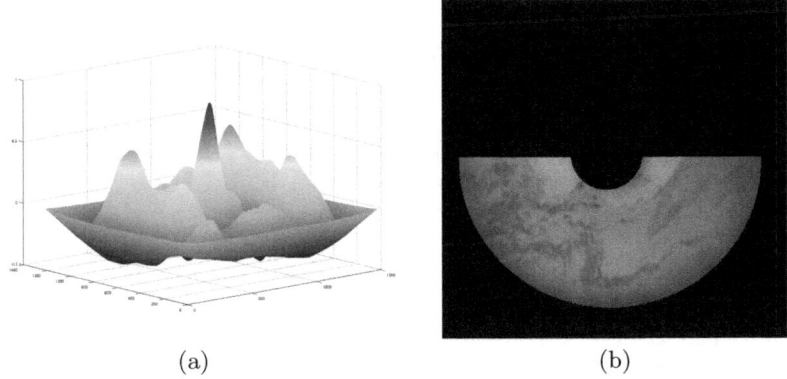

(a)   (b)

Fig. 2. (a) Correlation value for a given mask. (b) Final region of interest.

2.3   Texture Analysis

Our textural features are extracted by applying a bank of filters to the input image and computing the energy of the filter responses. Concretely, we have used the rotationally invariant bank of band-pass filters described below.

Rotational invariant filter bank. The Difference of Gaussians (DoG) is one of the most widely used filters to extract texture features. The difference of two smoothed images using different Gaussian kernels highlights the image features present at different scale ranges. The kernels we have used in this project are defined as,

$$G(x, y, \sigma_1, \sigma_2) = \frac{1}{2\pi\sigma_1^2} e^{-\frac{x^2+y^2}{2\sigma_1^2}} - \frac{1}{2\pi\sigma_2^2} e^{-\frac{x^2+y^2}{2\sigma_2^2}} \qquad (1)$$

Using this non orientation selective filter kernel, the texture features present at particular spatial frequency ranges are extracted by varying the parameters σ_1 and σ_2.

As stated in section 1, tears can be classified into five different categories according to the width of the lipid layer which influences the texture features of the tear. This work is focused on the classification of four of these categories: colour fringes, wave, closed meshwork and open meshwork. The simplified texture pattern for this four tear types is shown in Figure 3. As the pattern for each type of tear is different, a different frequency band is expected to have a higher response. Thus, a wide range of frequencies, covering the whole spectrum of frequencies of interest, have been studied.

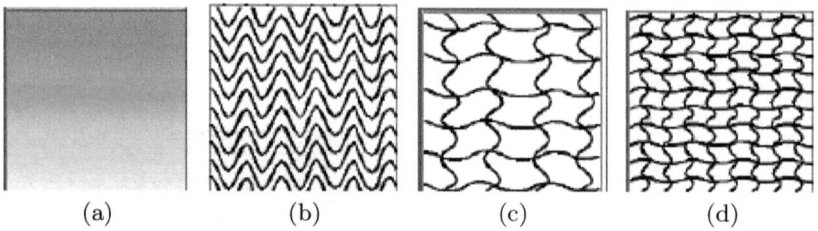

(a)   (b)   (c)   (d)

Fig. 3. Simplified texture pattern of the (a) colour fringe, (b) wave, (c) open meshwork and (d) closed meshwork tear film types

The experimental results, widely discussed in section 3, will confirm our intuition that the four types of tears produce high responses to different and very limited frequency bands, which will allow us to focus our future research in those particular frequency ranges.

As shown in Figure 3, some of the target patterns contain distinctive colour features. For this reason, we have included colour information in the texture analysis. We have extracted the texture in each colour component -R, G and B- and analyzed the results in an opponent colour space, as described in the next section.

2.4   Color Space: Opponent Colours

The opponent process theory of human colour vision was proposed by Hering [25] in the 1800's. Following the experiments of Hurvich and Jameson [26] in 1957, considerable neurophysiological evidence emerged in 1960's supporting the colour opponency [27].

A receptive field is a pattern of photoreceptors in the retina that determines the behaviour of a cell in the visual system. Receptive fields have a centre-surround organization so that, for example, a cell that is excited by a light stimulus in the centre of its receptive field will be inhibited by a light stimulus in the annulus surrounding the excitatory centre. This causes the cell to exhibit spatial antagonism.

The receptive fields for some cells include different classes of photoreceptors, causing the cells to exhibit chromatic antagonism. Single opponent cells are excited (or inhibited) by the response to a class of photoreceptor in the centre field and inhibited (or excited) by the response to a different kind of photoreceptors in the surrounding field.

The existence of these receptive fields was stated by Edward Hering as pairs of colours that are never seen together at the same place at the same time. This pairs of opponent colous, red-green, green-red and blue-yellow; are the differences used in our methodology to analyze the colour of the regions of interest. This three components correspond to a cell with a red, green or blue centre and its opposite surround. More precisely,

$$R_G = R_F - p * G_F$$

$$G_R = G_F - p * R_F \tag{2}$$

$$B_Y = B_F - p * (R_F + G_F)$$

where p is a low pass filter and R_F, G_F and B_F are the filtered RGB components, as explained in the previous section. Given these three new color components - R_G,G_R and B_Y - the texture descriptor is calculated as its probability density function in the region of interest of the input image.

3 Results

The proposed methodology was tested over a set of 91 digital images of the tear lipid layer including 22 colour fringes, 14 wave, 27 open meshwork and 28 closed meshwork images. All the images have a spatial resolution of 1024×768 pixels and have been acquired with the *TearScopePlus*.

The technique chosen for the classification step is the k-nearest neighbour algorithm [28] that, despite being a simple machine learning algorithm, is widely used in pattern recognition and produces good results. In this particular case, we have used an approximation with k=1 nearest neighbour and the euclidean distance as the distance metric.

In order to analyze the generalization of our results to larger data set, a 6-fold cross validation has been performed. The original dataset has been randomly partitioned into six subsets, using five of them for the training step and the remaining one for the validation. The process has been repited six times and the results averaged over these six executions.

Our experiment was divided into two stages. The first stage classifies the tear film image into three main categories - colour fringes, waves and meshworks - and the second stage classifies the images in the last category into open or closed meshwork. This is due to the fact that the similarities between the open and closed meshwork patterns create too many misclassifications when analyzed separately with the other tear types but the accuracy improves when both tear types are agrupated into a category that we have called *meshwork*.

Thus in the first experiment, all the 91 images have been classified into three categories - colour fringe, wave and *meshwork*. Tables 1, 2 and 3 show the confusion matrix for the three frequency ranges corresponding to the frequencies were the best classification result for each type is achieved.

The results obtained confirm our intuition that the three main categories produce high responses to different and very limited frequency bands, which will allow us to focus our future research in those particular frequency bands. Table 4 shows how the different categories are distributed along the frequency spectrum.

Table 1. Best classification for the colour fringe tear film type. Confusion matrix for the filter with $\sigma_1 = 7$ and $\sigma_2 = 1$.

Expert Type / Obtained Type	Colour Fringe	Wave	Meshwork
Colour Fringe	**94.58%**	1.67%	3.75%
Wave	20.75%	73.75%	5.50%
Meshwork	9.64%	11.18%	79.18%

Table 2. Best classification for the wave tear type. Confusion matrix for the filter with $\sigma_1 = 17$ and $\sigma_2 = 7$.

Expert Type / Obtained Type	Colour Fringe	Wave	Meshwork
Colour Fringe	81.42%	1.58%	17.00%
Wave	0.00%	**94.24%**	5.75%
Meshwork	11.60%	11.42%	76.98%

Table 3. Best classification for the *meshwork* category. Confusion matrix for the filter with $\sigma_1 = 3$ and $\sigma_2 = 0.5$.

Expert Type / Obtained Type	Colour Fringe	Wave	Meshwork
Colour Fringe	91.67%	1.75%	3.68%
Wave	21.00%	70.00%	9.00%
Meshwork	8.16%	6.24%	**85.60%**

Table 4. Distribution in the frequency space of the best result for each category

σ_1 \ σ_2	98	41	17	7	3	1
41		***	***	***	***	***
17			***	***	***	***
7			WAVE	***	***	***
3					***	***
1			COLOUR			***
0.5				MESHWORK		

Meshwork classification refinement. As previously explained our *meshwork* category includes the open and closed meshwork tear types. Our next step is to further analyze the images classified in this category in order to assign them to their correct tear types. To this end, the images classified in the *meshwork* category have been analyzed with the same methodology presented in this paper.

Table 5. Best classification for the open meshwork tear type. Confusion matrix for the filter with $\sigma_1 = 17$ and $\sigma_2 = 1$.

Expert Type / Obtained Type	Open Meshwork	Closed Meshwork
Open Meshwork	**86.41%**	13.59%
Closed Meshwork	40.72%	59.28%

Table 6. Best classification for the closed meshwork tear type. Confusion matrix for the filter with $\sigma_1 = 7$ and $\sigma_2 = 0.5$.

Expert Type / Obtained Type	Open Meshwork	Closed Meshwork
Open Meshwork	77.41%	22.59%
Closed Meshwork	37.33%	**62.67%**

4 Conclusions and Future Research

In this paper we have presented a preliminary methodology to classify the tear lipid film into the colour fringe, wave, open meshwork and close meshwork tear types. Tear film classification is necessary to evaluate both quality and quantity aspects from the tear and the automatization of this classification is important ta avoid the burden and subjectivity of the manual task. The results obtained so far show that a colour-based frequency analysis is suitable for the task in mind. Furthermore, this work shows how the target features that allow the correct classification of the tear types are present in limited frequency bands.

The next step in our research is the integration of the results in the frequency bands of interest, the analyis of different texture descriptors and the analysis of different classification techniques among others.

The obtained results show a quite good accuracy in the classification of some of the tear types achieving over a 90% of correct classifications. Despite these resutls, the main contribution of the paper to our future work is the determination of the limited and very defined frequency bands where the response for each type is highest. This contribution will allow us to focus our future work in those frequency bands with the aim to improve the results obtained by the presented methodology.

Acknowledgements

This paper has been partly funded by Ministerio de Ciencia y Tecnología–Instituto de Salud Carlos III (Spain) through the grant PI08/90420 and FEDER funds.

References

1. Holly, F.: Tear film physiology. Int. Ophthalmol. Clin. 27, 2–6 (1987)
2. Holly, F., Lemp, M.: Tear film physiology and dry eyes. Surv. Ophthalmol. 22, 69–87 (1977)
3. Guillon, M., Styles, E., Guillon, J., Maissa, C.: Preocular tear film characteristics of nonwearers and sof contact lens wearers. Optom. Vis. Sci. 74, 273–279 (1997)
4. Fullard, R., Snyder, C.: Proteins level in nonstimulated and stimulated tears of normal human subjects. Invest. Ophtalmol. Vis. Sci. 31, 1119–1126 (1990)
5. Lemp, M., Holly, F., Iwata, S., Dohlman, C.: The precorneal tear film. i. factors in spreading and maintaining a continuous tear film over corneal surface. Arch. Ophtalmol. 83, 89–94 (1970)
6. Craig, J.: Structure and function of the preocular tear film. The Tear Film, structure, function and clinical evaluation. Butterworth-Heinemann, London (2002)
7. Macri, A., Pflugfelder, S.: Correlation of the schirmer 1 and fluorescein clearance tests with the severity of corneal epithelial and eyelid disease. Arch. Ophtalmol. 118, 1632–1638 (2000)
8. Tomlinson, A., Blades, K., Pearce, E.: What does the phenol red thread test actually measure? Optom. Vis. Sci. 78, 142–146 (2001)
9. Balofun, M., Ashaye, A., Ajayi, B., Osuntokun, O.: Tear break-up time in eyes with pterygia and pingueculae in ibadan. West Afr. J. Med. 24, 162–166 (2005)
10. Cho, P., Brown, B.: Review of the tear break-up time and a closer look at the tear break-up time of hong kong chinese. Optom. Vis. Sci. 70, 30–38 (1993)
11. Tomlinson, A., Khanal, S., Ramaesh, K., et al.: Tear film osmolarity: determination of a referent for dry eye diagnosis. Invest. Ophthalmol. Vis. Sci. 47, 4309–4315 (2006)
12. Foulks, G., Bron, A.: meibomian gland dysfunction: a clinical scheme for description, diagnosis, classification, and grading. Ocul. Surf. 1, 107–126 (2003)
13. Guillon, J.: Non-invasive tearscope plus routine for contact lens fitting. cont lens anterior eye. Ocul. Surf. 21(suppl. 1), 31–40 (1998)
14. Giraldez, M., Naroo, S., Garcia-Resua, C.: A preliminary investigation into the relationship between ocular surface temperature and lipid layer thickness. Cont. Lens Anterior Eye 32, 177–180 (2009)
15. Garcia-Resua, C., Lire, M., Yebra-Pimentel, E.: Clinical evaluation of the tears lipid layer in a young university population. Rev. Esp. Contact 12, 37–41 (2005)
16. Craig, J., Tomlinson, A.: Importance of the lipid layer in human tear film stability and evaporation. Optom. Vis. Sci. 74, 8–13 (1997)
17. Foulks, G.: The correlation between the tear film lipid layer and dry eye disease. Surv. Ophthalmol. 52, 369–374 (2007)
18. Isreb, M., Greiner, J., Korb, D.: Correlation of lipid layer thickness measurements with fluorescein tear film break-up time and schirmer's test. Eye 17, 79–83 (2003)
19. King-Smith, P., Hinel, E., Nichols, J.: Application of a novel interferometric method to investigate the relation between lipid layer thickness and tear film thinning. Invest. Ophthalmol. Vis. Sci. (2009)
20. Guillon, J.: Abnormal lipid layers: observation, differential diagnosis and classification. Lacrimal Gland, Tear Film and Dry Eye Síndromes 2. Sullivan, New York (1998)
21. Fogt, N., King-Smith, P., Tuell, G.: Interferometric measurement of tear film thickness by use of spectral oscillations. J. Opt. Soc. Am. A Opt. Image Sci. Vis. 15, 268–275 (1998)

22. King-Smith, P., Fink, B., Fogt, N.: Three interferometric methods for measuring the thickness of layers of the tear film. Optom. Vis. Sci. 76, 19–32 (1999)
23. Goto, E., Dogru, M., Tsubota, K.: Computer-synthesis of an interference color chart of human tear lipid layer, by a colorimetric approach. Invest. Ophthalmol. Vis. Sci. 44, 4693–4697 (2003)
24. Guillon, J.: Use of the tearscope plus and attachments in the routine examination of the marginal dry eye contact lens patient. Adv. Exp. Med. Biol. 438, 859–867 (1998)
25. Morris, C.G., Maisto, A.A.: Understandig Psycology. Prentice Hall, Englewood Cliffs (2000)
26. Huervich, L., Jameson, D.: An opponent-process theory of color vision. Physiol. Rev. 64, 384–404 (1957)
27. Hering, E.: Outlines of a Theory of the Light Sense. Harvard University Press, Cambridge (1964)
28. Bishop, C.: Pattern Recognition and Machine Learning. Springer, Heidelberg (2006)

A New Method for
Text-Line Segmentation for Warped Documents

Daniel M. Oliveira[1], Rafael D. Lins[1], Gabriel Torreão[1],
Jian Fan[2], and Marcelo Thielo[3]

[1] Universidade Federal de Pernambuco, Departamento de Eletrônica e Sistemas.
Av. Prof. Luiz Freire, s/n, Cidade Universitária
50740-540, Recife, PE, Brasil
{rdl,daniel.moliveira,gabriel.dsilva}@ufpe.br
[2] Hewlett-Packard Labs, Palo Alto, USA
jian.fan@hp.com
[3] Hewlett-Packard Labs, Porto Alegre, Brazil
marcelo.resende.thielo@hp.com

Abstract. Bound documents either scanned or captured with digital cameras often present a geometrical warp that makes text-lines curled. The identification of text-lines is one of the steps for document de-warping when only a single image is available. This paper presents a new method for text-line segmentation. It is based on a simple, but effective, skew detector proposed by Ávila-Lins and simplifies the idea of coupled snakes introduced by Bukhari to a moving parallel line regression. The proposed method performed better than the best of the similar algorithms in the literature.

Keywords: Text-line segmentation, document de-warping, layout analysis.

1 Introduction

The digitalization of bound documents, such as books, either performed by flatbed scanners or digital cameras often yields images that exhibit a geometrical distortion in the region close to the book spine. Such distortion not only makes more difficult the document reading for humans, but also degrades OCR performance. Text-line envelope segmentation is one of the pre-processing steps for many algorithms. The segmentation process can be accompanied of a baseline and/or mean line estimation, Figure 1 illustrates these typographic lines and others.

Bukhari, Shafait, and Breuel [9] introduce the concept of baby snakes for extraction of text-lines. Later on, they use coupled *snakelets* [7] for the same purpose. Finally, in references [8][10] they obtain text-lines by ridges detection on grayscale images. Coupled snakes are used for base/mean line estimation. De-warping is done by calculating y-coordinates using upper/lower neighboring lines followed by a perspective correction.

Stamatopoulos and his colleagues [6] detect text-lines and estimate a document 3D model by approximating the border lines with a polynomial of degree three. Fu *et al*

A. Campilho and M. Kamel (Eds.): ICIAR 2010, Part II, LNCS 6112, pp. 398–408, 2010.
© Springer-Verlag Berlin Heidelberg 2010

ascent
ascender height
cap height
mean line
median
baseline
descent
descender height
Sphinx
x-height

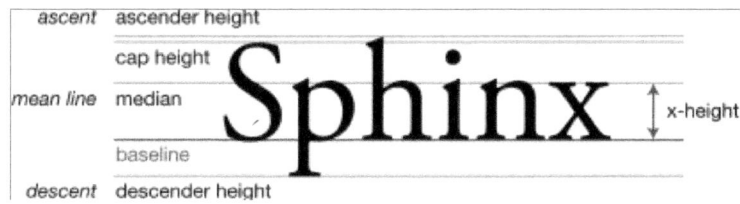

Fig. 1. Font typographic lines following reference [14]

[2] estimate border points fitting them in 3D-cylinder model. Masalovitch and Mestetskiy [1] estimate the spaces between lines; a bezier patch is built and followed by de-warping procedure.

This paper proposes new text-line segmentation method. It borrows ideas from a simple but effective skew detector proposed by Ávila-Lins [3] and coupled snakes introduced by Bukhari, Shafait, and Breuel [7] and can be used regardless document orientation. Reference [4] uses the new method presented herein for de-warping scanned documents. It is organized as follows. Section 2 presents details of the new algorithm. Section 3 shows that the proposed algorithm benchmarked on the CBDAR 2007 de-warping contest test-set achieved an accuracy rate of 91.10% with an under segmentation rate of 1.81%, while the performance of algorithm by Bukhari, Shafait, and Breuel [7], the best algorithm in the literature yields 89.65% and 3.30%, respectively. Section 4 concludes this works.

2 Segmenting Text-Lines

The proposed algorithm improves the Ávila-Lins skew detection scheme [3] for text-line segmentation which is summarized below as illustrated in Figure 2. Black and white images are assumed as input for the algorithm.

1. Component labeling transforming components as enclosing blocks (Figure 2.a);
2. For each unvisited block B do:
   a. Locate the nearest unvisited neighbor block N of block B (Figure 2.b);
   b. Group a text-line starting from blocks B and N (Figure 2.c-d) forming upper/lower or right/left lines;
   c. Detect the skew angle and landscape/portrait orientation of the document;
   d. Detect the up-down orientation of the document;
3. Detect total document rotation;

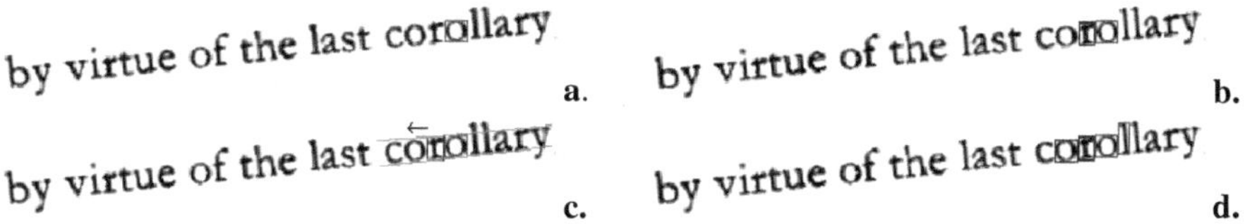

by virtue of the last corollary
a.
by virtue of the last corollary
b.
by virtue of the last corollary
c.
by virtue of the last corollary
d.

Fig. 2. Ávila-Lins skew detection [3]

This (in C language) algorithm is fast. For a 200dpi scanned image, it takes 115ms on a Pentium IV of 2.4GHz and 512MB of RAM with accuracy of 98%. Despite the good results in skew detection, text-line segmentation requires a more robust approach.

2.1 Letter/Line Properties

Let V be the vector formed by the extreme points of a line. The slope of V, called the V angle, is considered to be the angle that the text-line forms with the horizontal line. If the absolute value for the x-component is greater than the y-component of this vector, the line is considered to be horizontal, otherwise it is vertical. The Text-line length is set to V length.

Table 1. Letter properties definition

Letter property Orientation	Horizontal	Vertical
Letter case height	Block height	Block width
Letter case width	Block width	Block height
Letter case top point	Upper middle point	Left middle point
Letter case bottom point	Lower middle point	Right middle point

One may notice that letter properties are subject to orientation, thus term "block" is used for the "letter case", the enclosing box relative to image; term "letter" depends on the document orientation as described in Table 1. The discrimination between the character height and width is useful as the width is less stable than the height due to: variable font width values and character merging caused by digitization and/or binarization (e.g. see "precond<u>iti</u>on" and "wid<u>th</u>" of Fig. 3).

For the steps presented in the next section, some terms are underlined in a high level language with a more precise definition below; the ratio function is defined by (1):

- Small block – Box where width and height have less than 6 pixels;
- Similar size letters – Letters N and M have similar size if the $ratio(N_{height}, M_{height}) \geq 0.6$ and $ratio(N_{width}, M_{width}) \geq 0.1$.
- Parallel lines with offset – an offset of 40% relative to y-axis intercept (i.e. $|b_t - b_d|$) is added to top/down lines;
- Smaller than window mean widths/heights – Letter N and properties window W with $ratio(N_{height}, W_{\overline{height}}) < 0.6$ and $N_{width} \leq W_{\overline{width}}$.
- Maximum distance between letters – 2.50 times window mean height;
- Search for text-line upwards/downwards – search range are limited to 3 times the height of a letter;

$$ratio(a, b) = min(a, b) / max(a, b) \ . \tag{1}$$

2.2 The New Approach

The main idea of the new method proposed here is to group together characters with same properties by "walking" through the document to form a text-line. Instead of coupled snakes [7], moving parallel straight lines are used; Section 2.2.1 explains how to obtain them. As the warp level may distort character sizes, a moving letter window is used while a text-line is formed; A window of length of 7 components was used herein. Listing 1 summarizes the whole procedure.

Listing 1

1. Label components transforming them as enclosing boxes;
2. Remove <u>small blocks</u> or if a block encloses another totally;
3. For each block B do (term "block" is used; orientation is not available):
   a. Locate the nearest neighbor block N of block B;
   b. If N and B have <u>similar sizes</u>, place them in $Q_{NEIGHBORS}$ priority queue, with $priority = ratio(N_{width}, B_{width}) + ratio(N_{height}, B_{height})$;
4. While $Q_{NEIGHBORS}$ is not empty
   a. Pull-out neighbors (B and N) from $Q_{NEIGHBORS}$
   b. If any of B or N was visited go to step 4
   c. Create new text-line TL and add (B and N) neighbors
   d. Search letters between in B to N direction (width/height are orientation dependent):
      i. Create a moving properties window;
      ii. Search for a letter using parallel lines;
      iii. If a letter is found add it to TL if:
         1. It has <u>similar size</u> when compared to moving mean of widths/heights or if it is <u>smaller than window mean widths/heights</u>;
         2. The box center is between <u>parallel lines with offset</u>;
         3. The distance between the last letter and new one is less than the <u>maximum distance between letters</u>;
      iv. Add the letter onto neighbor candidate list on TL if conditions 1-2 above are met and the third is not;
      v. If the letter was added to TL, append it to properties window if it is not <u>smaller than window mean widths/heights</u>; this prevents from adding small components (e.g. accents, punctuation marks) to parallel regression;
   e. Execute previous step for direction N to B;
   f. Place text-line in $Q_{TEXTLINE}$ priority queue, with *priority = textline length*
   g. Mark all letters on the new text-line as visited;
5. Remove text-lines whereas its angle is 90° apart most common text-line angle;
6. While $Q_{TEXTLINE}$ is not empty (process bigger text-lines first)
   a. Pull-out text-line TL from $Q_{TEXTLINE}$
   b. If TL was merged go to step 6
   c. For each letter on TL <u>search for text-line upwards</u> and add it to UTL;
   d. For each letter on TL <u>search for text-line downwards</u> and add it to DTL;

e.   If there are any letter in common between UTL and DTL then
    i.   Mark current TL text-line as an invalid text-line;
    ii. Delete it from text-line list;
    iii. Go back to step 6;
f.   Merge two text-lines in UTL if one have a candidate neighbor of the other and vice-versa;
g.   Add new textlines to $Q_{TEXTLINE}$ with *priority = textline length*
h.   Repeat steps 6.f-g for DTL;
7. Remove text-lines where letter count is less the moving window;
8. Remove text-lines if it contains a letter on 10% of image border (mark them as noise);
9. For each text-line
    a.   Calculate a simple moving linear regression [12] for top/down points
    b.   Compute corresponding point in the line and its deviation error
10. Set top or down points as baseline whether the set with less error;

Figure 3 shows an example of the execution of step 4, where letters with boxes belong to the moving window. Figure 4 shows on upper left finding parallel lines and neighbor candidates between lines on upper right, on bottom shows merging result. Figure 5 shows the execution of step 7-8, with top and down sample points in gray and baseline points in black. The Method described herein can also be used to estimate document orientation using text-line angle histogram.

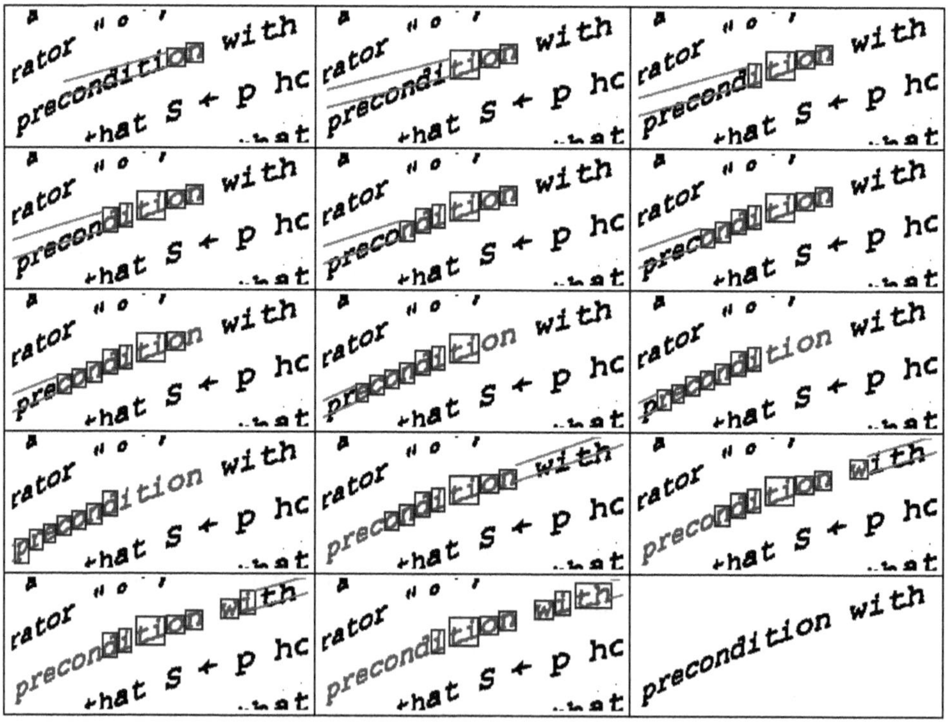

Fig. 3. Example of text-line formation by the new method

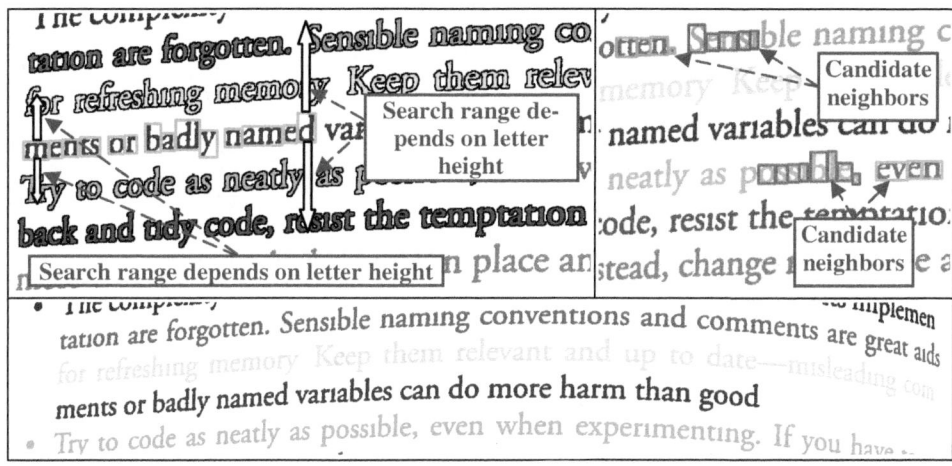

Fig. 4. Text-lines merging procedure: (upper left) parallel lines; (upper right) candidate neighbors; (bottom) merging result

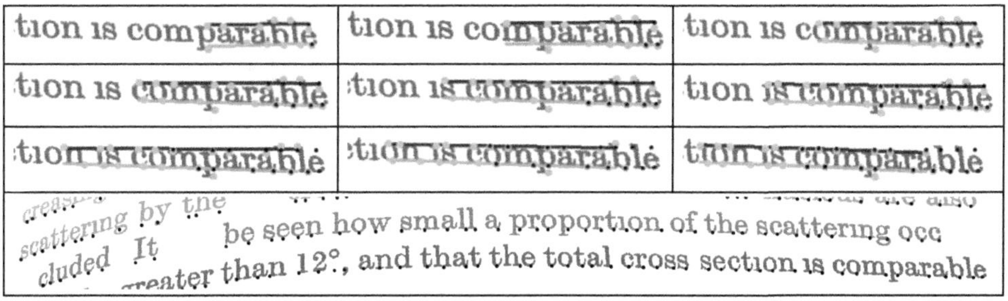

Fig. 5. Baseline estimation with simple moving simple linear regression

2.1.1 Parallel Line Regression

The aim of parallel lines regression is to minimize the error function of equation (2). Where m is the slope which is the same for both lines; b_t and b_d parallel lines intercept for top and bottom lines, respectively; (x_i, y_i) and (x_k, y_k) are the top and bottom samples points, respectively; N is the number of pairs of sample points.

$$E(m, b_t, b_d) = \sum_{i=1}^{N} (m \times x_i + b_t - y_i)^2 + \sum_{k=1}^{N} (m \times x_k + b_d - y_k)^2 \qquad (2)$$

Making $\partial E/\partial m = \partial E/\partial b_t = \partial E/\partial b_d = 0$, results in eqs. (3)-(5).

$$m\left(\sum_{i=1}^{N} x_i^2 + \sum_{k=1}^{N} x_k^2\right) + b_t \sum_{i=1}^{N} x_i + b_d \sum_{k=1}^{N} x_k = \sum_{i=1}^{N} y_i x_i + \sum_{k=1}^{N} y_k x_k \qquad (3)$$

$$m \sum_{i=1}^{N} x_i \qquad + \qquad b_t N \qquad\qquad = \qquad \sum_{i=1}^{N} y_i \qquad (4)$$

$$m \sum_{k=1}^{N} x_k \qquad + \qquad\qquad b_d N \qquad = \qquad \sum_{k=1}^{N} y_k \qquad (5)$$

Using Cramer's rule, m, b_u and b_l values are obtained in eqs. (6)-(9).

$$\Delta = \begin{vmatrix} \left(\sum_{i=1}^{N} x_i^2 + \sum_{k=1}^{N} x_k^2 \right) & \sum_{i=1}^{N} x_i & \sum_{k=1}^{N} x_k \\ \sum_{i=1}^{N} x_i & N & 0 \\ \sum_{k=1}^{N} x_k & 0 & N \end{vmatrix} \qquad (6)$$

$$m = \frac{\Delta m}{\Delta}; \ \Delta m = \begin{vmatrix} \left(\sum_{i=1}^{N} y_i x_i + \sum_{k=1}^{N} y_k x_k \right) & \sum_{i=1}^{N} x_i & \sum_{k=1}^{N} x_k \\ \sum_{i=1}^{N} y_i & N & 0 \\ \sum_{k=1}^{N} y_k & 0 & N \end{vmatrix} \qquad (7)$$

$$b_t = \frac{\Delta b_t}{\Delta}; \ \Delta b_t = \begin{vmatrix} \left(\sum_{i=1}^{N} x_i^2 + \sum_{k=1}^{N} x_k^2 \right) & \left(\sum_{i=1}^{N} y_i x_i + \sum_{k=1}^{N} y_k x_k \right) & \sum_{k=1}^{N} x_k \\ \sum_{i=1}^{N} x_i & \sum_{i=1}^{N} y_i & 0 \\ \sum_{k=1}^{N} x_k & \sum_{k=1}^{N} y_k & N \end{vmatrix} \qquad (8)$$

$$b_d = \frac{\Delta b_d}{\Delta}; \ \Delta b_d = \begin{vmatrix} \left(\sum_{i=1}^{N} x_i^2 + \sum_{k=1}^{N} x_k^2 \right) & \sum_{i=1}^{N} x_i & \left(\sum_{i=1}^{N} y_i x_i + \sum_{k=1}^{N} y_k x_k \right) \\ \sum_{i=1}^{N} x_i & N & \sum_{i=1}^{N} y_i \\ \sum_{k=1}^{N} x_k & 0 & \sum_{k=1}^{N} y_k \end{vmatrix} \qquad (9)$$

3   Results

Reference [11] compares the methods presented in [7][9][10] using CBDAR de-warping dataset [5]. Herein, the same comparison methodology, described in [13], is used. The ground truth and hypothesized (processed) image have each line painted using a different color. Their similarity is compared by a pixel-correspondence graph, where each node represents a text line in both images; the edges are text-line pixels that are shared between them where the weight is the total number of pixels shared. Black and white are not text-line colors; they stand for non-textual (noise) and back-ground pixels, respectively. An incoming edge is significant if $w_i/P \geq T_r$ and $w_i \geq T_a$, where w_i is the weight of the edge; P is the total number of node pixels; T_r and T_a are the relative and absolute thresholds. The following parameters are computed (copied from [11]).

- Number of ground truth lines (N_g) – total number ground truth lines in the whole database.
- Total correct segmentation (N_{o2o}) – the number of one-to-one matches between the ground-truth components and the segmentation components.
- Total over segmentations (N_{oseg}) – the number of significant edges that ground truth lines have, minus the number of ground truth lines.
- Total undersegmentations (N_{useg}): the number of significant edges that segmented lines has minus the number of segmented lines.
- Oversegmented components (N_{ocomp}): the number of ground truth lines having more than one significant edge.
- Undersegmented components (N_{ucomp}): the number of segmented lines having more than one significant edge.
- Missed components (N_{mcomp}): the number of ground truth components that matched the background in the hypothesized segmentation.
- False alarms (N_{falarm}): the number of components in the hypothesize segmentation that did not match any foreground component in the ground-truth segmentation.
- % correct segmentation (P_{o2o}) – N_{o2o}/N_g
- % oversegmented text-lines (P_{ocomp}) – N_{ocomp}/N_g
- % undersegmented text-lines (P_{ucomp}) – N_{ucomp}/N_g
- % missed text-lines (P_{mcomp}) – N_{mcomp}/N_g

Table 2 shows results of new algorithm and [7][9][10] (copied from [11] until writing of this article), where G-ridges and B-ridges stands for [10] the segmentation in grayscale and binary images, respectively. The proposed method has the best perfor-mance for under segmentation, false positives and correct segmentation figures. No parallel line merging was registered, lines where merged if they were aligned but belong to other column such as in Figure 6. Despite the highest missed components among other algorithms, missed lines are suppressed by matched ones when it is used together with a de-warping method. An example of successful (P_{o2o}=100%) processing can be seen in Figure 7 with noisy pixels in black. The proposed algorithm proved also to be fast, running in 8.75s with Java implementation over Windows Vista Business on a Dell D531 3GB.

Table 2. Algorithms comparison metrics

Algorithm	N_g	N_s	P_{o2o}	P_{ocomp}	P_{ucomp}	P_{mcomp}	N_{oseg}	N_{useg}	N_{falarm}
New	3091	2924	**91.10%**	21.71%	**1.81%**	4.43%	682	**57**	**785**
B-Snakes	3091	3371	87.58%	5.79%	2.91%	0%	294	117	13199
Ridges (G)	3091	3045	89.10%	3.53%	3.85%	0.91%	115	131	1186
Ridges (B)	3091	3115	89.65%	4.40%	3.30%	0.29%	144	110	2183
C-Snakes	3091	2799	78.26%	**1.26%**	9.06%	0%	**39**	359	3251

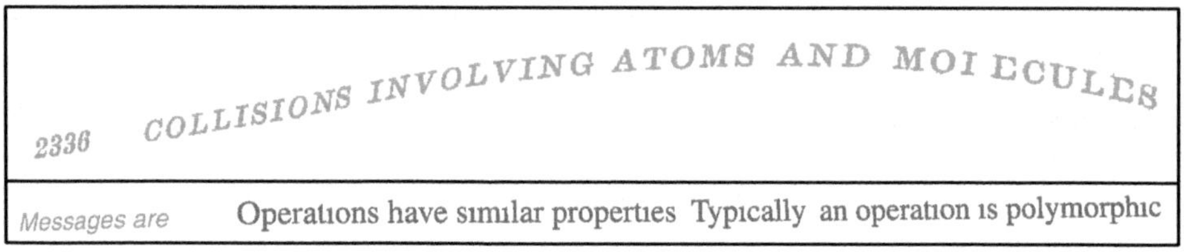

2336 COLLISIONS INVOLVING ATOMS AND MOLECULES

Messages are Operations have similar properties Typically an operation is polymorphic

Fig. 6. Line merging examples

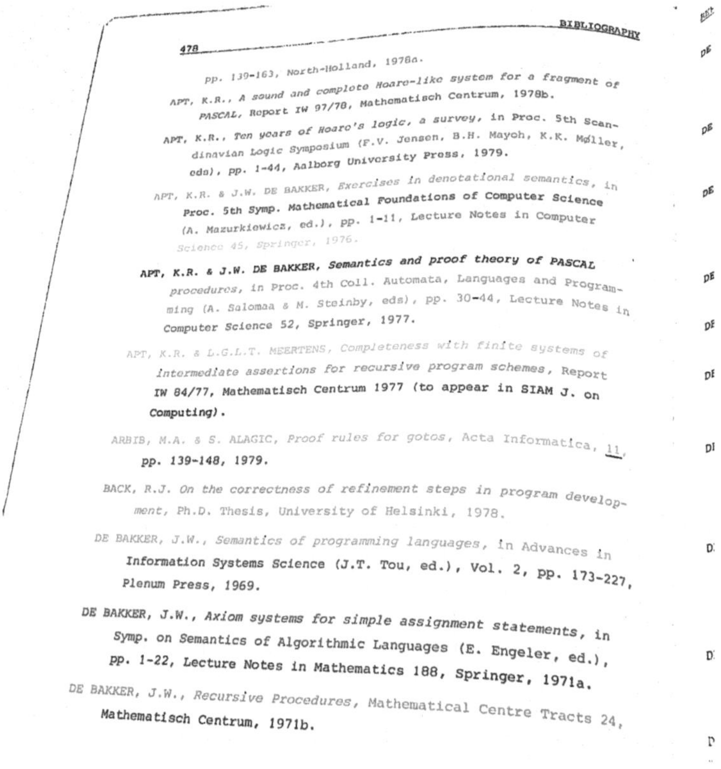

478   BIBLIOGRAPHY

pp. 139-163, North-Holland, 1978a.
APT, K.R., A sound and complete Hoare-like system for a fragment of
PASCAL, Report IW 97/78, Mathematisch Centrum, 1978b.
APT, K.R., Ten years of Hoare's logic, a survey, in Proc. 5th Scan-
dinavian Logic Symposium (F.V. Jensen, B.H. Mayoh, K.K. Møller,
eds), pp. 1-44, Aalborg University Press, 1979.
APT, K.R. & J.W. DE BAKKER, Exercises in denotational semantics, in
Proc. 5th Symp. Mathematical Foundations of Computer Science
(A. Mazurkiewicz, ed.), pp. 1-11, Lecture Notes in Computer
Science 45, Springer, 1976.
APT, K.R. & J.W. DE BAKKER, Semantics and proof theory of PASCAL
procedures, in Proc. 4th Coll. Automata, Languages and Program-
ming (A. Salomaa & M. Steinby, eds), pp. 30-44, Lecture Notes in
Computer Science 52, Springer, 1977.
APT, K.R. & L.G.L.T. MEERTENS, Completeness with finite systems of
intermediate assertions for recursive program schemes, Report
IW 84/77, Mathematisch Centrum 1977 (to appear in SIAM J. on
Computing).
ARBIB, M.A. & S. ALAGIC, Proof rules for gotos, Acta Informatica, 11,
pp. 139-148, 1979.
BACK, R.J. On the correctness of refinement steps in program develop-
ment, Ph.D. Thesis, University of Helsinki, 1978.
DE BAKKER, J.W., Semantics of programming languages, in Advances in
Information Systems Science (J.T. Tou, ed.), Vol. 2, pp. 173-227,
Plenum Press, 1969.
DE BAKKER, J.W., Axiom systems for simple assignment statements, in
Symp. on Semantics of Algorithmic Languages (E. Engeler, ed.),
pp. 1-22, Lecture Notes in Mathematics 188, Springer, 1971a.
DE BAKKER, J.W., Recursive Procedures, Mathematical Centre Tracts 24,
Mathematisch Centrum, 1971b.

Fig. 7. Example of successful processing

4 Conclusions

A new algorithm for text-line segmentation is presented. It outperforms the other state-of-the-art algorithms with 91.10% of accuracy and 1.81% of under segmentation rates. It can automatically detect text baselines with any orientation, proving also to be fast running in 8.75s with Java implementation for CBDAR 2007 images. The new process was used successfully in correcting the binding distortion in scanned books [4] where the document orientation is arbitrary.

Acknowledgments

The authors like to thank Syed Bukhari, Thomas Breuel and Faisal Shafait for providing page segmentation performance evaluation program source code and for discussions on the subject.

The research reported herein was sponsored by a MCT-Brazilian Government R&D Grant and CNPq funding.

References

[1] Masalovitch, A., Mestetskiy, L.: Usage of continuous skeletal image representation for document images de-warping. In: Proceedings of International Workshop on Camera-Based Document Analysis and Recognition, Curitiba, pp. 45–53 (2007)

[2] Fu, B., Wu, M., Li, R., Li, W., Xu, Z.: A model-based book de-warping method using text line detection. In: 2nd Int. Workshop on Camera-Based Document Analysis and Recognition, Curitiba, Brazil (September 2007)

[3] Ávila, B.T., Lins, R.D.: A fast orientation and skew detection algorithm for monochromatic document images. In: Proceedings of the ACM Symposium on Document Engineering, Bristol, UK, pp. 118–126 (2005)

[4] Lins, R.D., Oliveira, D.M., Torreão, G., Fan, J., Thielo, M.: Correcting Book Binding Distortion in Scanned Documents. In: Campilho, A., Kamel, M. (eds.) ICIAR 2010, Part II. LNCS, vol. 6112, pp. 355–365. Springer, Heidelberg (2010)

[5] Shafait, F., Breuel, T.M.: Document Image De-warping Contest. In: 2nd Int. Workshop on Camera-Based Document Analysis and Recognition, CBDAR 2007, Brazil, September 2007, pp. 181–188 (2007)

[6] Stamatopoulos, N., Gatos, B., Pratikakis, I., Perantonis, S.J.: A two-step de-warping of camera document images. In: Proceedings 8th IAPR Workshop on Document Analysis Systems, Nara, Japan, pp. 209–216 (2008)

[7] Bukhari, S.S., Shafait, F., Breuel, T.M.: Coupled snakelet model for curled textline segmentation of camera-captured document images. In: Proceedings 10th International Conference on Document Analysis and Recognition, Barcelona, Spain, pp. 61–65 (2009)

[8] Bukhari, S.S., Shafait, F., Breuel, T.M.: Ridges based curled textline region detection from grayscale camera-captured document images. In: Jiang, X., Petkov, N. (eds.) Computer Analysis of Images and Patterns. LNCS, vol. 5702, pp. 173–180. Springer, Heidelberg (2009)

[9] Bukhari, S.S., Shafait, F., Breuel, T.M.: Segmentation of curled textlines using active contours. In: Proceedings 8th IAPR Workshop on Document Analysis Systems, Nara, Japan, pp. 270–277 (2008)

[10] Bukhari, S.S., Shafait, F., Breuel, T.M.: Textline information extraction from grayscale camera-captured document images. In: Proc. The 13th International Conference on Image Processing, Cairo, Egypt (2009)

[11] Bukhari, S.S.: Technical Report: Performance Evaluation and Benchmarking of Three Curled Textline Segmentation Algorithms. IUPR Techinal Report, Kaiserslautern (2010)

[12] Wolfram Resarch. Least Squares Fitting,
http://mathworld.wolfram.com/LeastSquaresFitting.html
(accessed January 15, 2010)

[13] Shafait, F., Keysers, D., Breuel, T.M.: Performance evaluation and benchmarking of six page segmentation algorithms. IEEE Transactions on Pattern Analysis and Machine Intelligence 30(6), 941–954 (2008)

[14] Naylor, M.: Typographic line terms,
http://en.wikipedia.org/wiki/File:Typography_Line_Terms.svg

HistDoc - A Toolbox for Processing Images of Historical Documents

Gabriel Pereira e Silva, Rafael Dueire Lins, and João Marcelo Silva

Universidade Federal de Pernambuco, Recife, Brazil
gfps.cin@gmail.com, rdl@ufpe.br, joao.mmsilva@ufpe.br

Abstract. HistDoc is a software tool designed to process images of historical documents. It has two operation modes: standalone mode - one can process one image a time; and batch mode - one can process thousands of documents automatically. This tool automatically detects noises present in the document image including back-to-front interference (also called bleeding or show-through) and uses the best techniques to filter it out. Besides that it removes noisy borders and salt-and-pepper degradation introduced during the digitalization process. PhotoDoc also allows document binarization and image compression.

Keywords: Back-to-front interference, bleeding, show-through, historical documents, border removal, binarization, document enhancement.

1 Introduction

Document images - acquired either by scanners or digital cameras - almost always present some kind of noisy artifacts. This statement is particularly true in the case of images of historical documents, in which one often finds back-to-front interference [10] (also known as bleeding [6] or show-through [19]), darkened paper, faded ink, folding marks, stains and damaged or torn off regions. The bequest of the letters of Joaquim Nabuco, a Brazilian statesman, writer, and diplomat, one of the key figures in the campaign for freeing black slaves in Brazil (b.1861-d.1910) is a file of historical documents of paramount importance to understand the formation of the political and social structure of the countries in the Americas and their relationship with other countries. This rich file is kept by the Joaquim Nabuco Foundation [3] (a social science research institute in Recife - Brazil). It encompasses over 6,000 letters of active and passive correspondence. The HistDoc tool presented here was conceived as a way to preserve this important heritage, as the chemical process used in producing paper in the late 19th century used too much beach and the papers are in a fast decomposition process. An example of a document of the Nabuco collection is presented in Figure 1, in which one may observe back-to-front interference, paper darkened, document filing annotation, and writing in different directions, a feature often found in such documents.

HistDoc was conceived as a device independent software tool to run on PCs. Whenever the user unloads the images of the historical documents, he will be able to run the tool prior to storing, printing or sending through networks the document images. HistDoc works in two different ways user driven standalone mode and batch mode. In standalone mode the user chooses which filters to use to enhance the document image.

A. Campilho and M. Kamel (Eds.): ICIAR 2010, Part II, LNCS 6112, pp. 409–419, 2010.

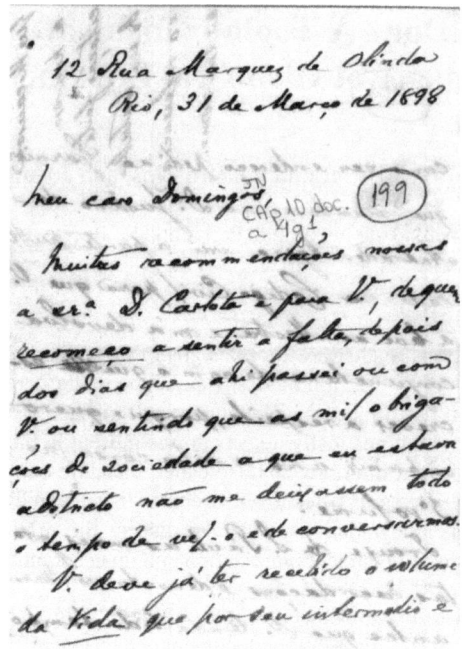

Fig. 1. Letter from Nabuco data base

In batch mode HistDoc uses the noise classifier presented in reference [33] specially tuned for historical documents, which automatically detects which undesirable artifacts are present in each document image and applies the suitable filtering technique. One should observe that such *a priori* noise classification is an important new feature in batch image processing.

HistDoc also encompasses a document compression module which decomposes the document image into paper background and writing. The color distribution and texture of the paper and writing are collected and the monochromatic image of the document is stored. Whenever the document is to be printed the data collected allows to colorize the monochromatic image yielding an image similar to the original one, at the cost of storing (or network transmitting) a compressed monochromatic document.

This paper is organized as follows. Section 2 briefly sketches the automatic noise classifier. The image filters implemented and the user interface for the standalone operation mode is presented in Section 3. The document image compression scheme is presented in Section 4. Conclusions and lines for further work are drawn at the final section.

2   The HistDoc Noise Classifier

Each document exhibits different noises and in general batch processing applies filters blindly and this may even cause document image degradation. In this section, the HistDoc Noise Classifier is outlined.

A number of features are extracted from each image to allow classification and training set as specified in [33].The noise classifier used is Random Forest [31] which was implemented in Weka [32], an open source tool for statistical analysis developed at the University of Waikato, New Zealand. The noise detection architecture is formed by parallel classifiers that detect framing border noises, skew, orientation and back-to-front interference. The first three classifiers detect noises with almost 100 % accuracy, while the last one due to its complex nature claimed for a more sophisticated noise detection and classification strategy as explained below.

2.1 The Back-to-Front Interference Classifier

Researchers [29] [30] have pointed out that no algorithm in the literature is good enough to remove bleeding noise in all sorts of documents. Depending on the strength of the noise, some algorithms may perform better than others. Unfortunately, the back-to-front noise appears more often in the digitalization of documents than one may assume to start with. The test set of documents with show-though used was formed by 2,027 real-world documents (no synthetic ones) which were obtained either from historical files (such as the one shown in Figure 1 from Nabuco bequest) or from the scanning of printed proceedings of technical events. Images were hand labeled according to four levels of interference as: strong, medium, light and none. The classifiers for this noise were cascaded. The architecture of the cascaded classifier to handle the spotting of the back-to-front noise is shown in Figure 2.

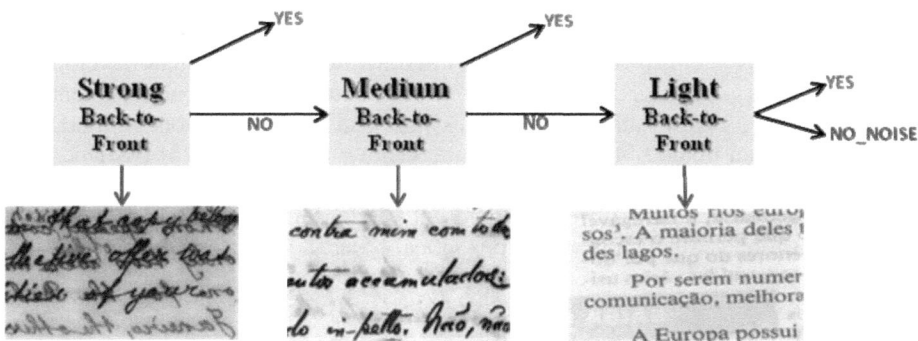

Fig. 2. Cascaded back-to-front noise detection architecture

The strong-classifier was trained with the images human tagged as strong in the training set, against all the remaining images (Medium-Light-None) from the training set. Similarly, the medium-classifier was trained with the images labeled as medium, against the others with a lighter or no interference. The classification results obtained are shown in Table 1.

The analysis of the data obtained shows that the classifier was able to detect the back-to-front noise in 90.97% of the noisy images and also to classify 93.90% of the noise-less images correctly. It is also worth mentioning that the misclassification of the images without noise was in the direction that they had a light back-to front interference. If one takes into account that such images were in JPEG format and that the

Table 1. Confusion matrix of the back-to-front noise classifier with sub-sampled images

Back-to-front	Strong	Medium	Light	None	Accuracy %
Strong	1,073	65	3	1	93.95
Medium	91	638	15	19	83.61
Light	5	9	96	12	76.22
None	24	53	106	2,817	93.90

background of many documents was not solid white, but also encompassed other noises due to aging, stains, etc, the results obtained are quite reasonable.

One should also note that the noisy documents, whenever misclassified, tend to be placed in the group immediately below. For instance 91 of the documents labeled as having strong bleeding noise were classified as having a medium noise, an acceptable result as the tagging followed no quantitative criteria. The adoption of synthetic noisy images could be of some help in solving the aforementioned problems, but their generation is far from being a simple task as it involves not only the overlapping of two images, one of which is faded. The image in the background also presents some degree of blur and this scenario gets complicated further in the case of the simulation of aged documents, a situation very often found whenever dealing with historical documents.

3   HistDoc Filters

This section explains the filters implemented in HistDoc and presents the user interface for operating them in standalone user driven filtering. The filters developed in HistDoc are able to process the kinds of noise often found in historical document. The same filters are used in batch mode processing. The current version of HistDoc is implemented as an ImageJ [4] plug-in. Figure 3 shows a screen shot of HistDoc being activated from ImageJ.

As one may observe in Figure 3, the present version of the HistDoc plug-in offers five different filters, which appear in alphabetical order:

1. Back-to-front interference removal
2. Binarization
3. Border removal
4. Document Enhancement
5. Compression

The fact that HistDoc is now in ImageJ also allows the user to experiment with the different filters and other plug-ins already present in ImageJ.

ImageJ, as an open code library, allows the developer to extract from it only the needed functionality in such a way that the developer may provide to ordinary user a HistDoc interface that looks independent from ImageJ. At present, the authors of this paper consider such possibility premature. Such tool particularization seems to be more adequate if the processing tool becomes embedded into a particular device, which allows also a better tuning of the algorithms implemented in HistDoc developed for such a device. In what follows the HistDoc filter operations are described.

Fig. 3. HistDoc plugin in ImageJ

3.1 Border Removal

Very often document binarization either performed with scanners or cameras yield an image framed with some background which served of physical support to the document, an instance of which may be found in Figure 4 left. There are obvious drawbacks in keeping such frame: larger space and network bandwidth are needed for storage and transmission, respectively; The visualization area in a device such as a CRT is wasted in exhibiting pixels that convey no information and ink or toner are used in printing such border noise. Besides that, the digitalization border has a serious

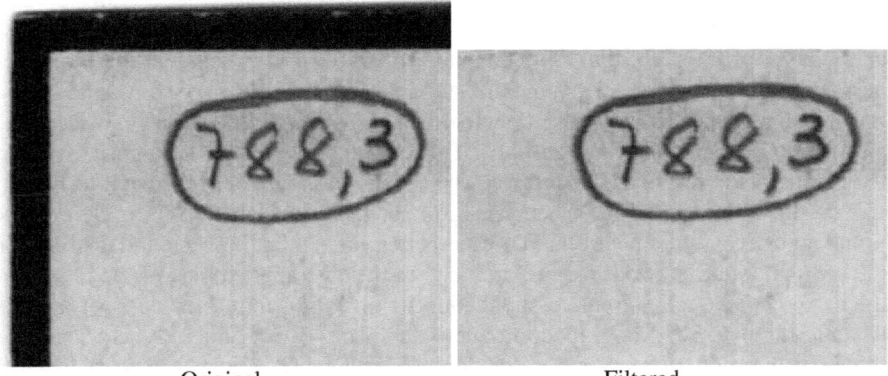

Original   Filtered

Fig. 4. Images: original and border removed

deleterious impact in the quality of the image subjected to palette reduction. This brings important implication as most automatic transcription tools (OCR and ICR) pre-process their input images into grayscale or binary before character recognition. The very first step to perform in processing a document image in HistDoc is to detect the actual physical limits of the original document [3]. Reference [3] reports on the binarization of documents. HistDoc offers to the user 16 thresholding techniques suitable for this sort of document, as it is detailed in the next section. Global and even local binarization algorithms take into account a statistical analysis of the document image, thus the presence of such border mislays the binarization process.

The algorithm presented in reference [20] was used in the development of the Hist-Doc (see Figure 4).

3.2 Back-to-Front Interference Removal

The HistDoc document processing environment offers three different strategies for filtering out the back-to-front noise [11], [21], [25] (see Figure 5). Whenever HistDoc is used in the user driven mode the user may select the most suitable algorithm for removing the back-to-front noise present in the document. If operated in batch mode the noise classifier will automatically choose the filter to be applied based on the strength of the interfering artifact.

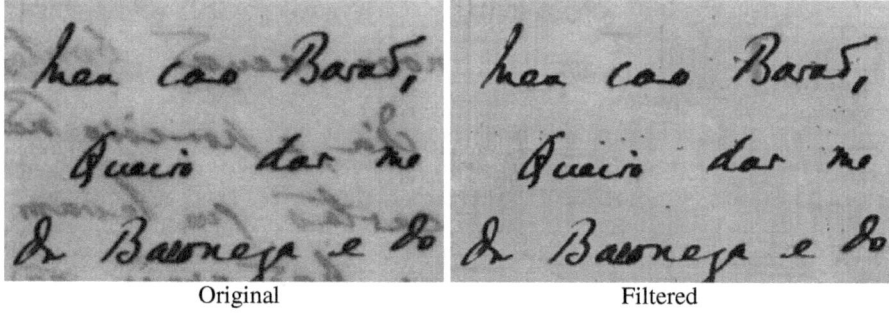

Original   Filtered

Fig. 5. Zoom into parts with back-to-front interference

The basic idea of the algorithm presented in reference [25], the most sophisticated and general of the algorithms implemented in HistDoc is to segment the three components of the document (background, ink and interference). Figure 6 shows the segmentation of the components of the document. The scheme used applies twice a global entropy-based thresholding algorithm. The first application of the algorithm separates the text from the rest of the document. The second pass separates the back-to-front interference from the rest of the paper background. Different loss factors α, an empirically found adjustment parameter that allows a better adjustment between the distributions of the original and binarized images, are used in the two applications of the algorithm. In the case of the batch automatic application of this algorithm three pairs of are used to suitably remove the strong, medium and weak back-to-front interference. The result of the application of such scheme to the document shown in Figure 1 appears in Figure 6. This scheme is also of central importance in the parametric image compression strategy presented in Section 4 below, also implemented in HistDoc.

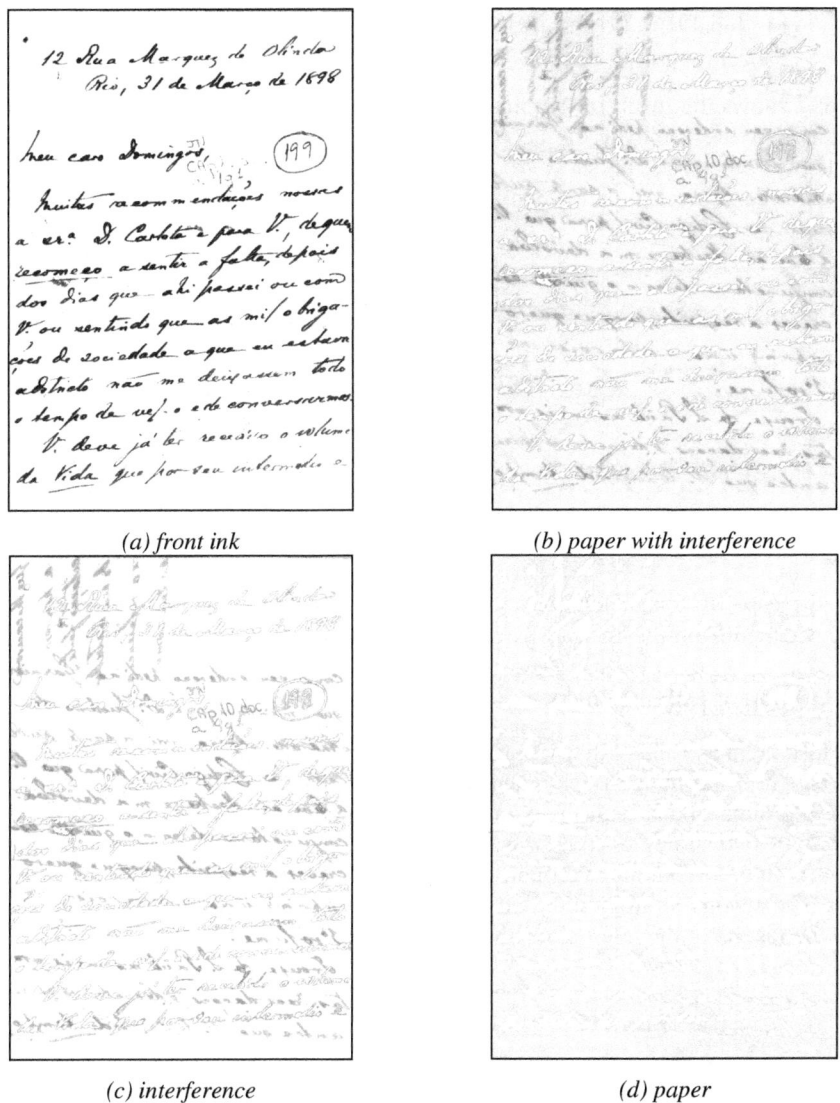

(a) front ink   *(b) paper with interference*

(c) interference   *(d) paper*

Fig. 6. Image segments of a document with back-to-front interference

3.3 Binarization

Document binarization is an important operation not only because a binary image is much smaller than its color counterpart but also due to most automatic transcription tools (OCR and ICR) pre-process their input images into grayscale or binary before character recognition. Reference [34] presents a survey of the most important binarization techniques applied to documents. HistDoc in user driven mode offers to the user 16 thresholding techniques suitable for this sort of document:

- 11 global ([5], [7], [8], [9], [13], [15], [17], [23], [24], [27], [28]) and,
- 5 local ([1], [14], [16], [18], [26]).

Figure 7 shows the result of the binarization of the image in Figure 3 and provides an account that the removal of the back-to-front interference prior to binarization is mandatory; otherwise the show-trough noise irrecoverably degrades document information in the monochromatic version.

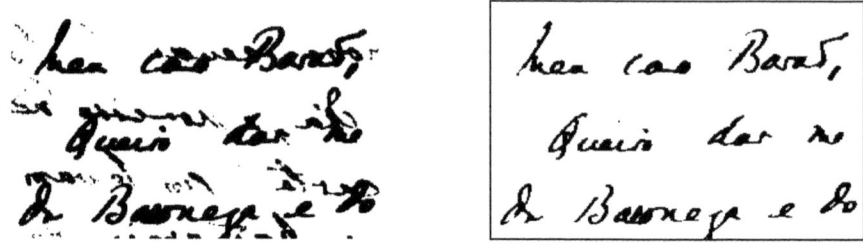

Fig. 7. Direct binarization (left) and after back-to-front interference removed (right)

In the case of using HistDoc in the automatic batch mode the binarization algorithm is called from the back-to-front noise detector.

3.4 Document Enhancement

This task creates a mask that identifies the pixels of the foreground and background objects. The final image is obtained through keeping the object pixels and replacing the background pixels with the average of the colors of the pixels in that class. Hist-Doc brings two strategies to do this. The first is the proposed by Castro and Pinto [2], that uses the Sauvola and Pietikainen's binarization algorithm [18] to determine the mask. The second strategy is based on the algorithm in reference [23]. Figure 8 presents the results of the latter algorithm.

Original   Filtered

Fig. 8. Images: original and enhanced (filtered)

4 HistDoc Compression Module

If the user wants to obtain an image that resembles the color original image, but is very efficiently compressed, HistDoc offers the compression scheme described in reference [22], in which the image is decomposed and stored as a compressed mono-chromatic image together with the colors and texture of the different graphical elements in the document (paper, printing, signature, etc.). The basic principle adopted in this compression scheme is shown in Figure 9.

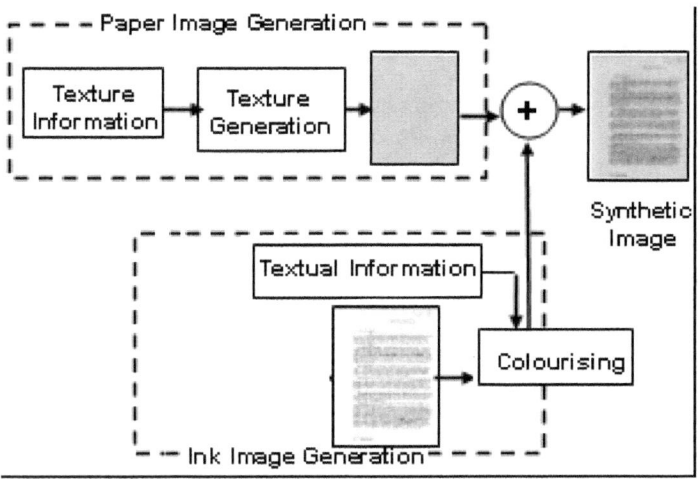

Fig. 9. Parametric generation of synthetic color document images

The user may also save images with the several file formats available in ImageJ (jpg, jpeg2000, png, tiff, etc), with and without losses.

5 Conclusions and Lines for Further Work

HistDoc is a user friendly tool for processing images of historical documents. It works in two different modes: user driven and automatic batch filtering mode. The batch mode makes use of a noise detecting tool that automatically detects and removes noisy framing borders, skew, orientation and back-to-front interference. The output may be either a binary image, a color image in the same file format of the input image or a parametrically compressed image which closely resembles the original one but is far more efficiently compressed.

The user driven operating mode of HistDoc provides a wide range of filters to enhance the document image at will. The first version was developed using the MAT-LAB [12] environment. It can be used as a MATLAB Tool, but a standalone version is also available. Aiming to speed-up the document processing phase, some of the algorithms are implemented in C.

The current version of HistDoc was developed as a plug-in in ImageJ, an open source portable Java library freely available. HistDoc runs on the users' PC and has the advantage of the great portability of Java. The executable code of HistDoc is freely available and may be obtained by requesting to the authors of this paper.

Several lines may be followed to provide further improvements to HistDoc filters and environment. Some of them are: being able to easily erase marks and stains from the digital version of the document, incorporate screens in which the user may provide annotations, interface with an OCR to automatically transcribe or find keywords in documents. The interfacing of Tesseract [35] with HistDoc is on progress. Incorporating into HistDoc some of the functionalities of Gamera [36] another free platform of similar purpose is also a possibility. Preliminary tests performed with Gamera showed that although its OCR mechanism presents a much lower recognition performance than Tesseract, it allows the user to train the OCR recognizer with new font types, for instance, which may be of interest in some files of historical documents in which all documents were typed using a particular machine. Gamera is implemented in Python and C++ and is slightly faster than the current version HistDoc which is implemented as an ImageJ plugin in Java.

Acknowledgments

Research reported herein was partly sponsored by CNPq – Conselho Nacional de Pesquisas e Desenvolvimento Tecnológico and CAPES – Coordenação de Aperfeiçoamento de Pessoal de Nível Superior, Brazilian Government.

The authors also express their gratitude to the Fundação Joaquim Nabuco, for granting the permission to use the images from Nabuco bequest.

References

1. Lins, R.D., et al.: An Environment for Processing Images of Historical Documents. In: Microprocessing & Microprogramming, pp. 111–121. North-Holland, Amsterdam (1994)
2. Kasturi, R., ÓGorman, L., Govindaraju, V.: Document image analysis: A primer. Sadhana (27), 3–22 (2002)
3. Sharma, G.: Show-through cancellation in scans of duplex printed documents. IEEE Trans. Image Processing 10(5), 736–754 (2001)
4. FUNDAJ, http://www.fundaj.gov.br (accessed on 20/03/2010)
5. Lins, R.D., Silva, G.F.P., Banergee, S., Kuchibhotla, A., Thielo, M.: Automatically Detecting and Classifying Noises in Document Images. In: ACM-SAC 2010, ACM, New York (2010)
6. Breiman, L.: Random Forests. Machine Learning 45(1), 5–32 (2001)
7. Weka 3: Data Mining Software in Java, http://www.cs.waikato.ac.nz/ml/weka/
8. Lins, R.D., Silva, J.M.M., Martins, F.M.J.: Detailing a Quantitative Method for Assessing Algorithms to Remove Back-to-Front Interference in Documents. Journal of Universal Computer Science 14, 299–313 (2008)
9. Stathis, P., Kavallieratou, E., Papamarkos, N.: An Evaluation Technique for Binarization Algorithms. Journal of Universal Computer Science 14, 3011–3030 (2008)
10. IMAGEJ, http://rsbweb.nih.gov/ij/ (accessed on 20/03/2010)
11. e Silva, A.R.G., Lins, R.D.: Background removal of document images acquired using portable digital cameras. In: Kamel, M.S., Campilho, A.C. (eds.) ICIAR 2005. LNCS, vol. 3656, pp. 278–285. Springer, Heidelberg (2005)
12. Lins, R.D., Netto, I.G.: Uma Nova Estratégia para Filtrar a Interferência Frente-Verso em Documentos Históricos. In: XXV SBrT, Recife, Brazil (2007)
13. da Silva, J.M.M., Lins, R.D.: Um Novo Método de Filtragem de Interferência Frente-Verso em Documentos Coloridos. In: XXV SBrT, Recife, Brazil (2007)

14. da Silva, J.M.M., Lins, R.D., Silva, G.F.P.: Enhancing the quality of color documents with back-to-front interference. In: Kamel, M., Campilho, A. (eds.) ICIAR 2009. LNCS, vol. 5627, pp. 875–885. Springer, Heidelberg (2009)
15. Sezgin, M., Sankur, B.: Survey over Image Thresholding Techniques and Quantitative Performance Evaluation. Journal of Eletronic Imaging 13(1), 145–165 (2004)
16. Bernsen, J.: Dynamic thresholding of gray level images. In: ICPR'86: Proc. Intl. Conf. Patt. Recog., pp. 1251–1255 (1986)
17. Castro, P., Pinto, J.R.C.: Methods for Written Ancient Music Restoration. In: Kamel, M.S., Campilho, A. (eds.) ICIAR 2007. LNCS, vol. 4633, pp. 1194–1205. Springer, Heidelberg (2007)
18. Kapur, J.N., Sahoo, P.K., Wong, A.K.C.: A new method for gray-level picture thresholding using the entropy of the histogram. G. Models I. Process. 29, 273–285 (1985)
19. Kavallieratou, E., Antonopoulou, H.: Cleaning and Enhancing Historical Document Images. In: Blanc-Talon, J., Philips, W., Popescu, D.C., Scheunders, P. (eds.) ACIVS 2005. LNCS, vol. 3708, pp. 681–688. Springer, Heidelberg (2005)
20. Khashman, A., Sekeroglu, B.: A Novel Thresholding Method for Text Separation and Document Enhancement. In: 11th Panhellenic Conf. on Informatics, Greece, May 18–20 (2007)
21. Kittler, J., Illingworth, J.: Minimum error thresholding. Patt. Recog. 19, 41–47 (1986)
22. Mello, C.A.B., Lins, R.D.: Generation of images of historical documents by composition. In: ACM Document Engineering 2002, McLean, VA, USA (2002)
23. Niblack, W.: An Introduction to Image Processing, pp. 115–116. Prentice-Hall, Englewood Cliffs (1986)
24. Otsu, N.: A threshold selection method from gray level histograms. IEEE Trans. Syst. Man Cybernetics 9, 62–66 (1979)
25. Palumbo, P.W., Swaminathan, P., Srihari, S.N.: Document image binarization: Evaluation of algorithms. In: Proc. SPIE, vol. 697, pp. 278–286 (1986)
26. Ridler, T.W., Calvard, S.: Picture thresholding using an iterative selection method. IEEE Trans. Syst. Man Cybern. SMC-8, 630–632 (1978)
27. Sauvola, J., Pietaksinen, M.: Adaptive document image binarization. Pattern Recogn. 33, 225–236 (2000)
28. da Silva, J.M.M., Lins, R.D., Martins, F.M.J., Wachenchauzer, R.: A New and Efficient Algorithm to Binarize Document Images Removing Back-to-Front Interference. Journal of Universal Computer Science 14, 299–313 (2008)
29. da Silva, J.M.M., Lins, R.D., da Rocha Jr., V.C.: Binarizing and Filtering Historical Documents with Back-to-Front Interference. In: ACM-SAC'06, ACM Press, New York (2006)
30. White, J.M., Rohrer, G.D.: Image thresholding for optical character recognition and other applications requiring char. image extraction. IBM J. Res. Dev. 27(4), 400–411 (1983)
31. Wu, L.U., Songde, M.A., Hanqing, L.U.: An effective entropic thresholding for ultrasonic imaging. In: ICPR'98: Intl. Conference Pattern Recognition, pp. 1522–1524 (1998)
32. Yen, J.C., Chang, F.J., Chang, S.: A new criterion for automatic multilevel thresholding. IEEE Trans. Image Process. IP-4, 370–378 (1995)
33. da Silva, J.M.M., Lins, R.D.: Color Document Synthesis as a Compression Strategy. In: ICDAR 2007, vol. 1, pp. 466–470. IEEE Press, Los Alamitos (2007)
34. MATHWORKS, http://www.mathworks.com/
35. Tesseract OCR, http://code.google.com/p/tesseract-ocr/ (accessed on 20/03/2010)
36. Gamera Project, http://gamera.informatik.hsnr.de/ (accessed on 20/03/2010)

Urban Road Extraction from High-Resolution Optical Satellite Images

Mohamed Naouai[1,2], Atef Hamouda[1], and Christiane Weber[2]

[1] Faculty of Science of Tunis, University campus el Manar DSI
2092 Tunis Belvédaire-Tunisia
Research unit URPAH
naouai@polytech.unice.fr, atef_hammouda@yahoo.fr
[2] Laboratory Image Ville Environnement ERL 7230-CNRS-University Strasbourg
3rue de l'Argonne F-67000 Strasbourg
naouai@polytech.unice.fr, christiane.weber@live-cnrs.unistra.fr

Abstract. Road extraction research has always been an active research on automatic identification of remote sensing images. With the availability of high spatial resolution images from new generation commercial sensors, how to extract roads quickly, accurately and automatically has been a cutting-edge problem in remote sensing related fields. In this paper, we present a novel road extraction approach which uses a scale space segmentation and two measures of the shape index to filter all regions from the result of the segmentation. The approach makes full use of spectral and geometric properties of roads in the imagery, and proposes a new algorithm named "Road Segments joint Algorithm" to ensure the continuity of roads.

Keywords: Object recognition, road extraction, shape index, scale-space segmentation, skeletonization, remote sensing, Feature Extraction.

1 Introduction

At first, roads are among the most important objects that are extracted from high resolution images; they are necessary for many applications, for example navigation systems or spatial planning. Since roads are subject to frequent changes, it is necessary to check road databases frequently to eliminate errors and to add new road objects. Many approaches for road extraction have been developed; some of them are summarized in [1]. However, only few approaches work in urban or suburban areas due to the highly complex structure found in urban scenes which complicates the task of automatic road extraction[35][36]. In [4][27] , the road network is expected to be a more or less regular grad but this constraint is not suitable for many urban areas. Another approach can use a very sophisticated road and context model and is based on grouping small extracted entities to lanes, carriageways and road networks [28]. It employs a large set of parameters that must be carefully adapted for different scenes. In recent works, color properties are exploited, for example in [26]: the authors perform a pixel-based multispectral classification and use shape descriptors to reduce the number of misclassifications. But they still only have a completeness and correctness

A. Campilho and M. Kamel (Eds.): ICIAR 2010, Part II, LNCS 6112, pp. 420–433, 2010.

rate of approximately 50%. In our opinion, this is due to the fact that the multispectral classification does not take into account the spatial relations of the pixels and that color and shape properties are treated separately.

From the above mentioned works we can deduce that a proper segmentation algorithm is essential for the extraction of roads in urban areas and that it is important to combine several features for the segmentation. A simple line based road model as used in many road extraction approaches for rural areas is not applicable.

In this part, this paper deals with road extraction in urban scenes with a focus on segmentation. Initially we use scale space segmentation, for a set of objects, these ones represent different structures that exist in the image. After the segmentation, the filtering operation could be necessary to remove objects which do not have geometric properties similar to road. In fact, our algorithm uses two shape indexes to make this step. However, the main goal of the algorithm is to output a set of segments representing the entire road network in the segmented image. For this, we apply the technique skeletonization, which is a technique of mathematical morphology on the region list obtained after a filtering operation. Finally, we propose a new algorithm named "Road Segments joint Algorithm" to ensure the continuity of roads.

2 Methodological Framework

In this work we introduce a road model in high spatial resolution remotely sensed images. This model is based on several properties with geometric and radiometric characteristics, these had the following properties:

Stability of spectral property: The spectral properties of uncovered roads are stable to a certain degree. Because urban roads are mainly constructed by asphalt or cement, especially asphalt dominates a large part; spectral properties of roads are limited to a fixed range which corresponds to the spectral range of road materials. However, in the imagery, objects on roadsides like zebra crossings, cars and people cause noises due to the huge spectral difference to roads.

Continuity of roads: Normally roads in reality are continuous and regular in geometry, while in the imagery, trees and shadows of high buildings interrupt the continuity of roads to a large degree. But on the whole, roads in the imagery still have impressive connectivity and regularity.

Straightness: On high spatial resolution images, urban roads are straight and smooth with no small wiggles thus can be recognized as combinations of straight road segmentation.

Topological property: Road segments are always connected with each other constituting road networks, and impossible to be broken suddenly.

Since this, in our case, the geometric and radiometric characteristics appear together, it is possible to apply a combination of these characteristics by representing a segmentation algorithm. We assume that each road segment is represented as an elongate rectangle has constant width and length, and they branches from often wide angles. The flow diagram of automatic road extraction process is shown in (fig1).

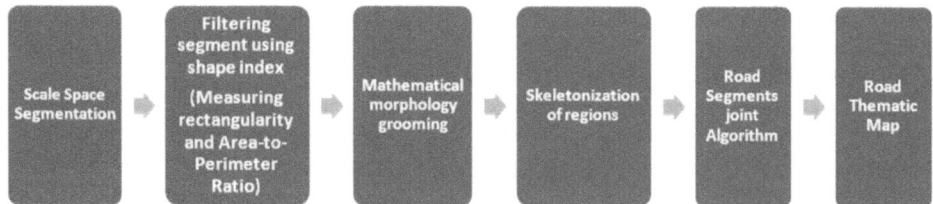

Fig. 1. Automatic road extraction process

3   Automatic Road Extraction Approach

Automatic road extraction can be concentrated on road model, which embody the global features and local features of the road. So achieve to road detection, the key problem is correct description and understanding of the road and the establishment of appropriate road model. In this section, an innovative method (see fig. 1) is presented to guide the road extraction in an urban scene starting from a single complex high-resolution image.

3.1   Scale-Space Segmentation

The basis of scale-space segmentation is the extraction of the hierarchical structure of the image. In (fig. 2) the description represent the general algorithm that we can see. First, the Scale-Space representation is generated. Right after, the structure analysis is realized building up the tree-like hierarchy (fig.3). From this, a set of segments is obtained. Those correspond to all the pixels hanging down the selected roots from the hierarchy. In the end, just a morphological filtering on the encountered regions mask the performance to erase little spots or regions corresponding to mistakes occurred during the phase of structure analysis. In this work, the multi-resolution segmentation algorithm by Vincken [2] is taken as a starting point.

Fig. 2. Scale Space Segmentation Diagram

3.1.1   Scale-Space *Generation*
Assumptions made by Lindeberg [20] are based on the idea of using successive con-volutions to generate the scale-space. Koendering first realized [3] about which should be the basis for the structure of images analysis. Under several constraints, he defined the diffusion equation, given by (1), as the generator of its scale-space.

$$\frac{\partial I\,(\vec{u}\,,t)}{\partial t} = \Delta I\,(\vec{u}\,,t) \tag{1}$$

Where I stand for the luminance of the image which depends on position $\vec{u} = (x, y)$ and t, scale. From (1) and from the constraint of using convolution to generate the subsequent scale levels, one finds that the unique kernel that satisfies both is the Gaussian.

3.1.2 Linking Up through Space

The algorithm for the construction of the structure, on a simple approach [2], is based on the tracking of the iso-intensity paths through scale. Other algorithms where proposed relying on extrema [23], [22], [24], [21], [25], but we considered that could be more consistent and generic to search for the iso-intensity paths.

This is because image pixels can not be fully described by maxima and minima. The algorithm sets up the structure establishing relations between pixels of consecutive levels. On the finest scale (the original image) all the pixels are related to the pixel from the first blurred image on the scale direction. At this level, not all pixels will receive a link from a pixel or from the level below. That because due to blurring, the image contains less information, and so a pixel from the upper level (bigger scale), will be related to a bigger number than one pixel from the level below (finer scale). Consequently, Pixels provided from the finer scale level will represent the details lost by blurring in the upper level. This linking up is performed between all the scale levels. (Fig. 3) shows a simple schema of the idea. Levels are linked in a tree like structure. These links converge through scale according to the reduction of information imposed by the low-pass filtering.

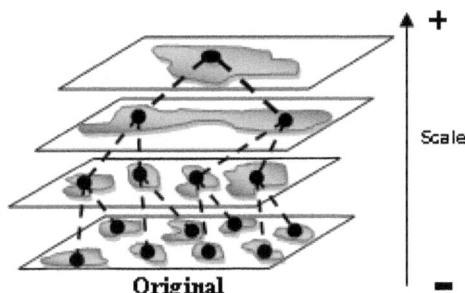

Fig. 3. Hierarchical analysis of the image structure linking pixels through levels

In addition to the base criteria of gray level difference, some others where added in order to help the convergence [2]. Those are relying on different features like for example volume of pixels hanging from the selected parent pixel. That can influence by the way that a pixel having many children is very likely to have more. Another feature would be the average gray level of the hanging pixels. Such a characteristic is quite advantageous when segmenting regions with a uniform gray level, for example medical images. Factors are represented by:

$$C_I = 1 - \frac{\left| I_p - I_c \right|}{\Delta I_{max}} \qquad C_G = \frac{SG_p}{SG_{max}} \qquad (2)$$

424   M. Naouai, A. Hamouda, and C. Weber

Where C_I is the driving feature that relies on the intensity of pixels (I_P) parent and (I_C) Children.

C_G represents the accessory feature that favorizes big segments (SG_P) represents the number of pixels associated to a parent pixel, and (SG_{max}) the maximum value associated to a parent pixel.

$$C_M = 1 - \frac{\left| M_p - M_c \right|}{\Delta I_{max}} \qquad (3)$$

Where C_M is the feature associated to the mean gray value of segments.

3.1.3 Reconstructing Segments

There for, the image structure has been estimated, the obtention of segments is evident. To carry out the segmentation it is necessary to select the scale of analysis. From this, all the nodes at that scale level will define a segment separately. The segment will be all the pixels connected through the hierarchical tree to the upper selected node (fig4).

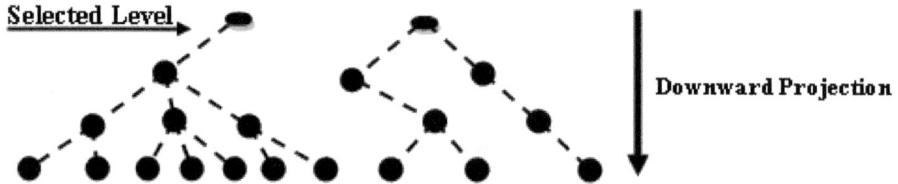

Fig. 4. Scanning of the image structure to obtain the segments

So we can see, selection of the upper nodes that define the final number of segments can be done in different ways (fig. 5). The most simple is the selection of scale level and from there takes all the segments emerging from the hierarchical tree.

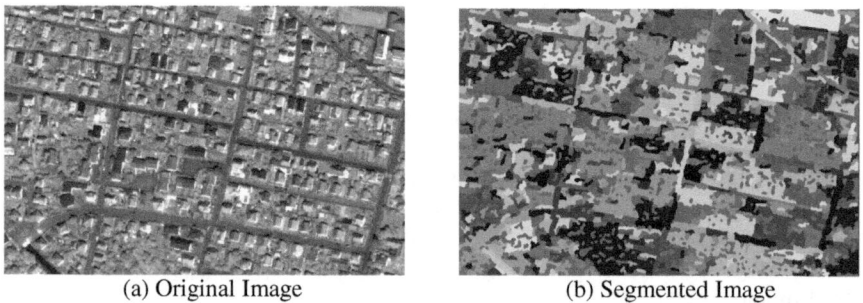

(a) Original Image   (b) Segmented Image

Fig. 5. A QuickBird satellite image covering the City of Strasbourg in 2008

3.1.4 Cleaning Up Regions

The problem of little regions miss-segmented (fig.5.(b)) due to linking errors intro-duces quite a high number of little segments of few pixels. It is clear that they do not belong to the selected scale level. A way to remove them is to delete regions smaller (fig.6) than certain area proportional to scale and re-assign those pixels to the big neighbords segments on the basis of some criteria, like average gray level, or big existing regions can be grown using geodesy with some morphological operators.

Fig. 6. Result of remove small region

3.2 Filtering

The segmentation algorithm can provide as output a set of regions or a set of objects where we will have several different geometry ways, for this reason, a geometric filtering step is necessary to keep only the objects that have properties similar to the geometric road parameters. The shape index offers a natural and invariant description of pure 2nd order image structure. Therefore the filtering step uses a shape index to keep only the structure having a rectangular form.

3.2.1 Area-to-Perimeter Ratio of a Region

In general, the width of a road is almost constant, or piecewise constant. However, it is difficult to estimate the width of a road for two reasons. One is that some of the extracted vertices are not on road, and in another way the vertices at T intersections has multiple directions at the road cross section. Instead of directly calculating the width of road at a vertex v, we consider an area-to-perimeter ratio in that region, called A/P ratio and that is denoted by d(v). Fig. 7(a) shows an ideal two-directional rectangular region of vertex v. The lengths of the shorter edge and the longer edge are denoted by w and w + 2r, respectively, where w is determined by the width of road, and r is dependent on the length of the spoke. In this case, the A/P ratio is:

$$d(v) = (\frac{w}{2}) \times (1 - (\frac{w}{2 \times r + 2 \times w})) \qquad (6)$$

When the spokes are selected so that their lengths are much longer than the width of road, then r >> w; therefore,

$$d\ (v)\ \approx\ \frac{w}{2} \tag{7}$$

Similarly, for the T-shaped region [see Fig.7(b)], the A/P ratio is:

$$d\,(v) = (\frac{w}{2}) \times (1 - (\frac{w}{3 \times r + 2 \times w})) \approx \frac{w}{2} \tag{8}$$

For the X-shaped region [see Fig.7(c)], the A/P ratio as presented in equation (9). We conclude that the A/P ratio of a region is close to half the road width and is independent of the number of toes in a region. In the next section, we prune the superfluous paths using A/P ratio d(v) over the set of extracted vertices.

$$d\,(v) = (\frac{w}{2}) \times (1 - (\frac{w}{4 \times r + 2 \times w})) \approx \frac{w}{2} \tag{9}$$

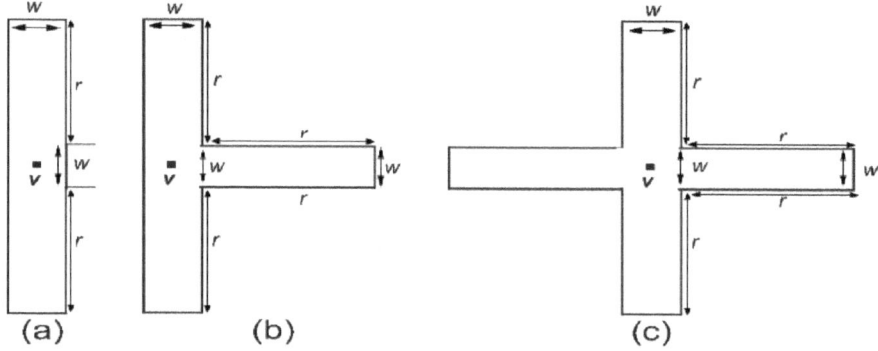

Fig. 7. (a) Ideal rectangular region. (b) Ideal T-shaped region. (c) Ideal X-shapedregion.

3.2.2 Rectangle Fitting

Our approach for rectangle fitting is based on [34] [37], where the author propose one shape attribute Q → [0,1] called Rectangularity, which is obtained by the ratio between one object area and its bounding box area. However, due to rotation this

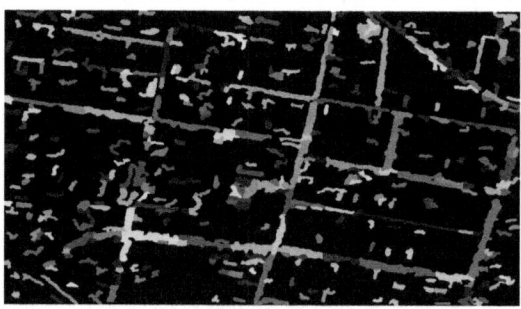

Fig. 8. Filtering using shape index

measure can not be correctly represent the object rectangularity, unless a pre-processing step is performed to transform Rectangularity invariant to rotation.

Given an object and its internal points coordinates, the eigenvectors are calculated. The first eigenvector shows the object's main angle. Than a new object is created by rotating it, in relation to this main angle. Afterward, the unbiased Q is obtained by dividing the object area and the area of its rotated bounding box. This value is used for inspecting each alternative for merging regions (see Fig. 8).

3.3 Mathematical Morphology Grooming

The rude result image deriving from the last procedure is groomed using mathematical morphology in this stage. The grooming stage relies on four basic steps: connecting, smoothing, thinning and linking. The connecting joins discrete road segments using morphological dilation. The smoothing, which combines morphological opening and closing operator, reduces the roughness of road edges significantly. The image is split into equal sized regions and in each region, so morphological thinning operators are selected automatically according to local road width information.

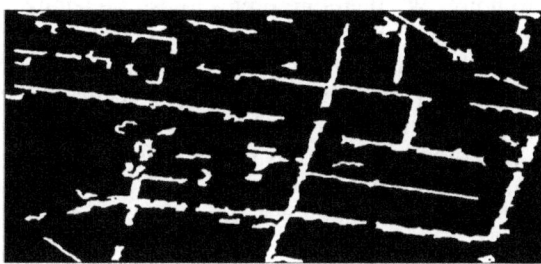

Fig. 9. Result of mathematical morphology grooming

The linking, the last step of grooming stage, concentrates on correct connection of one-pixel wide road segments and final elimination of non-road information from the image. Geometrical features such as size, connectivity and distance between road segments are considered to achieve the purpose. Single or too short segments would be eliminated from the image. After this final step, we acquire the result image (fig.9.) which contains road network information extracted from original remotely sensed imageries.

3.4 Skeleton-Based Methods

The main family of methods for finding the lines is to compute the skeleton. There are two well-known paradigms for skeletonization methods: The first is that of "peeling an onion", i.e. iterative thinning of the original image until no pixel can be removed without altering the topological and morphological properties of the shape [32]. These methods require only a small number of lines in an image buffer at any time, which can be an advantage when dealing with large images. But on the other side, multiple passes are necessary before reaching the final result, so that computation times may become a high quality. The second definition used for a skeleton is that of the ridge lines formed by the centers of all maximal disks included the original shape, is to preserve the connectivity. This leads directly is using on distance transforms or similar measures [29, 30, 33], which can be computed in only two passes on the image. In our group, we have been testing both approaches.

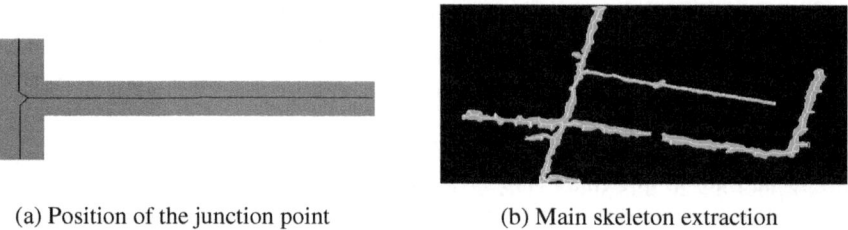

(a) Position of the junction point   (b) Main skeleton extraction
with a skeleton-based method

Fig. 10.

The iterative thinning algorithm is straightforward and can give good results, but it's very sensitive to noise. We therefore prefer to use skeletons computed from distance transforms, to guarantee the precision of the skeleton, we advocate the use of chamfer distances, which come closer to approximation :the Euclidean distance. A good compromise between precision and simplicity seems to be the 3–4 chamfer distance transform (see reference [31] for details.), for which a good skeletonization algorithm has been proposed by Sanniti di Baja [31](see Fig.10(b)). A single threshold on the significance of a branch enables correct removal of the smallest barbs. The correct positioning of junctions is often very important in graphics recognition applications. All skeleton based methods are weak with respect to the correct restitution of the junction at the location the draftsman wanted to be. This is a direct consequence of the fact that the skeleton follows the centers of the maximal discs of the pattern, whereas the position of the junction as envisioned by the draftsman is not on these centers (see Fig. 10 (a)).

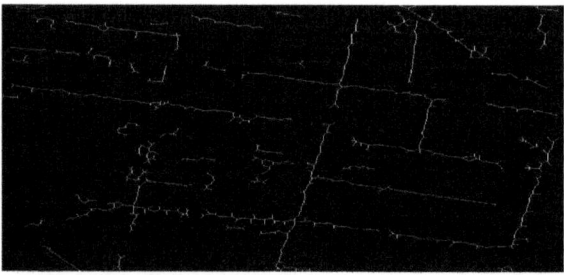

Fig. 11. Result after skeletonization algorithm

4   Road Segments Joint Algorithm

In order to extract roadside in spite of heavy shadows on roads, we bring forward a new algorithm named "road connection algorithm". This is the key procedure of the whole frame work, in which spectral and geometric features of roads are represented in two rules: (1) spectral feature: for each pixel, only when its three radiometric properties are similar to those of roads, it can be identified as road candidates. (2) geometric feature: on high resolution image, roads are usually smooth with no small wiggles. Thus in an appropriate distance, road segments, including smooth curves, can be recognized as straight line segments.

According to road geometrical properties mentioned above, for each pixel on the image, search in specific direction in a straight line at fixed step. If the proportion of road candidate pixels (refers to pixels whose spectral property satisfies the road) to all pixels on the line exceeds a threshold value S that we give in advance, all pixels on the line L could be identified as pixels of roads . After that, we convert original non-road pixels to road pixels. However, we assume the image after skeletonization as IS, and the result after road connection algorithm as RC, and the value of number 1 marks road. Then the pseudo codes are as follows:

```
W = image width, H = image height
RC is an image of dimension [W,H] and whose values equal 0
for  i = 1 to H
        for i = 1 to W
                for φ = 0 to 180 by 5
                        X= the pixel positions on the line segments
                           starting from (i,j) in φ direction with
                           length L.
                        RN= number of pixels in X in image
                        If (RN ≥ 1)
                                N=number of pixels in X in image IS
                                  whose value equal 1
                                N1=number of pixels in X in image IS
                                If (N1/N) ≤ S then
                                    RC (X) = 0
                                End if
                                If (N1/N) ≥ S then
                                    RC (X) = 1
                                End if
                        End if
                End for
        End for
End for
```

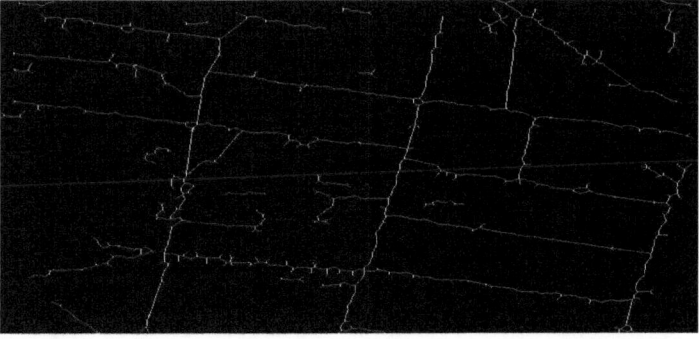

Fig. 12. Final Road Network using "Road Segments joint Algorithm"

5   Evaluation

In this section we discuss some results of our algorithm. In order, to simplify the evaluation of the result, we define main road as those with the width larger than 10 pixels or the length more than 300 pixels, and the sub-road as those with the width less than 10 pixels and the length between 100 and 300 pixels. Road segment which is also defined to evaluate the result refers to the segment between intersections of roads of the same level. Accuracy is given at last in table 1. As results turn out, 93.4% main road segments and 54.2% sub-road segments are extracted correctly, while 5.0% main road segments and 10% sub-road segments are recognized partially. Only 2.44% main road segments are failed to be extracted, while that proportion of sub road segments is 15.8%.

Table1. Road information extraction accuracy assessment

	Main roads	*Sub-roads*
Complete	93.4	54.2%
Incomplete	5%	10%
Missing	2.44%	15.8%
Manual Ref	45	32
Result	45	38
Wrong	0	15.7%

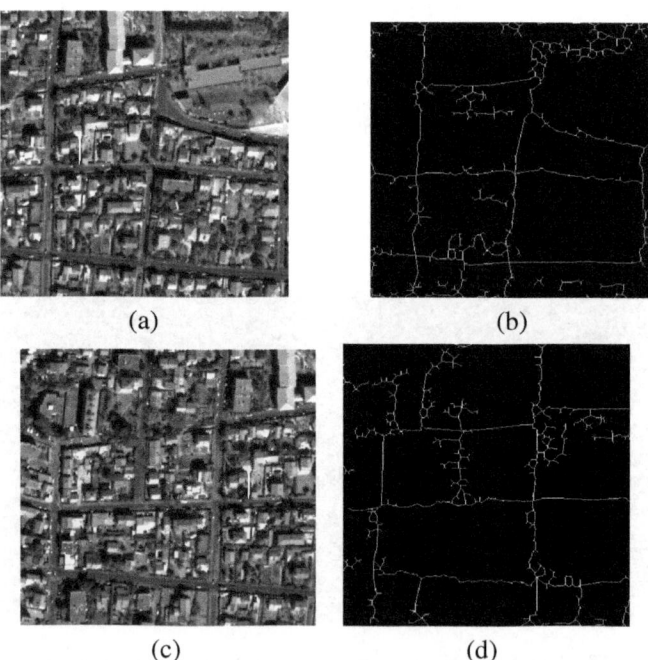

(a)   (b)

(c)   (d)

Fig. 13. Results of road line extraction ((a),(c) Original Image (c), (d) Road line Extraction)

6 Conclusion

The proposed approach of automatic road extraction from high spatial resolution images can improve the accuracy of road extraction and reduce the effects of occlusions on roads such as shadows. Through tests, the method proves to be simple, accurate, and highly automatic, applying well to road extraction from high spatial resolution images of huge volumes. We use scale space segmentation, for a set of objects, these ones represent different structures that exist in the image. After the segmentation, the filtering operation could be necessary to remove objects which do not have geometric properties similar to road. In fact, our algorithm uses two shape indexes to make this step.

We process the rude result image through morphological operators to connect discrete line segments and smooth the lines. So, the subjects of morphological process are roads, excluding non-road objects, and will not cause errors on non-road region, which acquires higher accuracy compared to simply morphological process in the whole image. Finally we use our algorithm '**Road Segments joint**' to eliminate the discontinuity of roads segments. This procedure significantly ensures the continuity of roads, reducing the road occlusions caused by other unrelated objects. Besides, if the pixel on a direction has been searched, the next pixel will be searched immediately, which avoids the repetitive search and improves the efficiency significantly.

After connecting and smoothing, we introduce the concept of region to erode roads according to local road width information to get the skeleton of roads, and the result is one-pixel road width image. Considering the topological and geometric properties of roads, we can eliminate single and too short line segment further. To sum up, the approach in this paper could extract urban road network from high spatial resolution image accurately and automatically and is of satisfactory practical value.

References

1. Mayer, H., Reznik, S.: MCMC linked with implicit shape models and plane sweeping for 3D building façade interpretation in image sequences. In: International Archives of Photogrammetry and Remote Sensing, vol. XXXVI(3), pp. 130–135 (2006)
2. Vincken, K.: Probabilistic Multi-Scale Image Segmentation by the Hyperstack, Ph.D. thesis, Utrecht University (1995)
3. Lindeberg, T.: Scale-Space Theory in Computer Vision. Kluwer Academic Publishers, Dordrecht (1994)
4. Price, K.: Road grid extraction and verification. In: International Archives of Photogrammetry and Remote Sensing, part 3-2W5, vol. 32, pp. 101–106 (1999)
5. Quam, L.H.: Road Tracking and Anomaly Detection. In: DARPA Image Understanding Workshop, May 1978, pp. 51–55 (1978)
6. Nevatia, R., Babu, K.R.: Linear feature extraction. In: DARPA'78, pp. 73–78 (1978)
7. Bolles, R.C., Quam, L.H., Fischler, M.A., Wolf, H.C.: Automatic determination of image to database correspondence. In: IJCAI'79, pp. 73–78 (1979)
8. Fischler, M.A., Tenenbaum, J.M., Wolf, H.C.: Detection of roads and linear structures in low- resolution aerial imagery using a multisource knowledge integration technique. In: CGIP, March 1981, vol. 15(3), pp. 201–223 (1981)

9. Gong, P., Li, X., Xu, B.: Interpretation Theory and Application Method Development for Information Extraction from High Resolution Remotely Sensed Data. Journal of remote sensing 10(1), 1–5 (2006)
10. Shi, W.Z., Zhu, C.C., Wang, Y.: Road Feature Extraction from Remotely Sensed Image: Review and Prospects. Acta geodaetica et cartographica sinica 30(3) (2001)
11. Mena, J.B.: State of the art on automatic road extraction for GIS update: a novel classification. Pattern Recognition Letters 24, 3037–3058 (2003)
12. Gruen, A., Li, H.: Road Extraction from Aerial and Satellite Images by Dynamic Programming. ISPRS Journal of Photogrammetry and Remote Sensing 50(4), 11–20 (1995)
13. Trinder, J.C.: Semi-automatic Feature Extraction by Snakes. In: Automatic Extraction of Manmade Objects from Aerial and Space Images, Birkhacuser Verlag, Basel (1995)
14. Merlet, N., Zérubia, J.: New prospects in line detection by dynamic programming. IEEE Transactions on Pattern Analysis and Machine Intelligence 18(4)
15. Bhattacharya, U., Parui, S.K.: An improved back propagation neural network for detection of road-like features in satellite imagery. International Journal of Remote Sensing 18(16), 3379–3394 (1997)
16. Baumgartner, A., Steger, C., Mayer, H., Eckstein, W., Heinrich, E.: Automatic road extraction based on multiscale, grouping, and context. Photogrammetric Engineering and Remote Sensing 65(7), 777–785 (1999)
17. Couloigner, I., Ranchin, T.: Mapping of urban areas: A multiresolution modeling approach for semiautomatic extraction of streets. Photogrammetric Engineering and Remote Sensing 66(7), 867–874 (2000)
18. Zhu, C.Q., Wang, Y.G., Ma, Q.H.: Road Extraction from High-resolution Remotely Sensed Image Based on Morphological Segmentation. Acta geodaetica et cartographica sinica 33(4), 347–351 (2004)
19. An, R., Feng, X.Z., Wang, H.L.: Road Feature Extraction form Remote Sensing Classified Imagery Based on Mathematical Morphology and Analysis of Road Networks. Journal of Image and Graphic 8(7), 798–804 (2003)
20. Lindeberg, T.: Scale-space for discrete signals. IEEE Transactions on Pattern Analysis and Machine Intelligence 12 (March 1990)
21. Lifshitz, L.M., Pizer, S.M.: A multi-resolution hierarchical approach to image segmentation based on intensity extrema. IEEE Transactions on Pattern Analysis and Machine Intelligence 12(6) (June 1990)
22. Henkel, R.D.: Segmentation with synchronising neural oscillators. Tech. Rep., Zentrum fr Kognitionswissenschaften, Universitt auf Bremen (1994)
23. Henkel, R.D.: Segmentation in scale-space. In: Hlaváč, V., Šára, R. (eds.) CAIP 1995. LNCS, vol. 970. Springer, Heidelberg (1995)
24. Florack, L.M., ter Haar Romeny, B.M., Koenderink, J.J., Viergever, M.A.: Linear scale-space. Journal of Mathematical Imaging and Vision 4 (1994)
25. Lindeberg, T., Eklundh, J.O.: Scale detection and region extraction from a scale-space primal sketch. In: Third International Conference on Computer Vision (1990)
26. Zhang, Q., Couloigner, I.: Automated Road Network Extraction from High Resolution Multi-Spectral Imagery. In: ASPRS 2006 Annual Conference, Reno, Nevada (2006a)
27. Youn, J., Bethel, J.S.: Adaptive snakes for urban road extraction. In: International Archives of Photogrammetry and Remote Sensing, part B3, vol. 35, pp. 465–470 (2004)
28. Hinz, S.: Automatic road extraction in urban scenes and beyond. In: International Archives of Photogrammetry and Remote Sensing, part B3, vol. 35, pp. 349–355 (2004)
29. Borgefors, G.: Distance Transforms in Digital Images. Computer Vision, Graphics and Image Processing 34, 344–371 (1986)

30. Chiang, J.Y., Tue, S.C., Leu, Y.C.: A New Algorithm for Line Image Vectorization. Pattern Recognition 31(10), 1541–1549 (1998)
31. di Baja, G.S.: Well-Shaped, Stable, and Reversible Skeletons from the (3,4)-Distance Transform. Journal of VisualCommunication and Image Representation (1994)
32. Lam, L., Lee, S.-W., Suen, C.Y.: Thinning Methodologies A Comprehensive Survey. IEEE Transactions on PAMI 14(9), 869–885 (1992)
33. Niblack, C.W., Gibbons, P.B., Capson, D.W.: Generating Skeletons and Centerlines from the Distance Transform. CVGIP: Graphical Models and Image Processing 54(5), 420–437 (1992)
34. Korting, T.S., Fonseca, L.M.G., Dutra, L.V., Silva, F.C.: Image Re-Segmentation – A New Approach Applied to Urban Imagery, pp. 467–472 (2008)
35. Naouai, M., Hamouda, A., Weber, C.: Detection of road in a high resolution image using a multicriteria directional. In: IADIS Computer Graphics, Visualization, Computer Vision and Image Processing (CGVCVIP) 2009 conference, Algarve, Portugal, June 20–22 (2009)
36. Naouai, M.: Atef Hamouda et Christiane Weber, Extraction de route dans une image à très haute résolution spatiale, Taima'09, Mai 4-9, Hammamet Tunisie (2009)
37. Rosin, P.L.: Measuring rectangularity. Mach. Vis. Appl. 11(4) (December 1999)

Geometrical Characterization of Various Shaped 3D-Aggregates of Primary Spherical Particules by Radial Distribution Functions

Marthe Lagarrigue[1], Johan Debayle[2], Sandra Jacquier[1],
Frédéric Gruy[1], and Jean-Charles Pinoli[2]

[1] SPIN/LPMG, UMR CNRS 5148
[2] CIS/LPMG, UMR CNRS 5148
Ecole Nationale Supérieure des Mines de Saint-Etienne,
158 cours Fauriel, 42023 Saint-Etienne cedex 2, France

Abstract. Multi-scale aggregates are composed of particles which results themselves of agglomeration of other primary particles. If particles are modeled by their centers, the geometrical characterization of aggregates refers to point pattern analysis. Radial distribution and function of pairs allow a description of the point pattern to be performed. They describe how points are radially packed around each other. In this paper, the characterization of different simulated aggregates are computed and compared.

1 Introduction

In precipitation process, the final product is often obtained in the form of aggregates of particles, which themselves consist of assembling of smaller crystals.

The purpose of this research work is to characterize 3D solid aggregates by a morphological method. Ultimately, this characterization will be related with an optical method which consists in analysing the scattering parameter of an aggregate under an incident light beam. Indeed, the scattering parameter particularly depends on the internal and external geometry of the aggregates e.g. the chord length distribution (see, for instance Jacquier and Gruy [1]). The final aim of this study is to find a link between the underlying optical and morphological parameters.

This paper is focused on the geometrical characterization of aggregates. Two methods of morphological characterization of the internal and external geometry are proposed: the radial distribution function, and the function of pairs.

Several experimental studies are then performed with computationally simulated aggregates. A comparison of the two proposed methods is carried out for aggregates constituted by different shape of convex hulls, different ratios of filling, and different geometrical shape ratios.

A. Campilho and M. Kamel (Eds.): ICIAR 2010, Part II, LNCS 6112, pp. 434–443, 2010.

2 3D Aggregates Modelling

To study the aggregates morphology, it is necessary to simulate them in order to understand the influence of several geometrical parameters.

First of all, an aggregate is defined by its scale number. In this paper only the case of aggregates with two scales is presented.

1. The smaller scale level consists in spherical particles (imposed by the optical model [2]). The centers of these primary particles are distributed along the close-packed hexagonal mesh [4], selected for its compactness. The radius is chosen equal to 10 nm because it is the usual order of magnitude for primary particles of the first scale level in the optical domain.
2. The second scale level is defined by geometrical shapes: sphere, cube, cylinder, spheroïds (oblate and prolate). The cylindrical convex hull is defined by its base diameter and its height which is k-proportional to the base diameter, with $k \in \{1; 2; 8; 20\}$.

   The geometry of the two spheroïdal convex hulls (oblate and prolate) are defined by the axis a, b and c, with an equality between two axis lengths ($a = b$ for example). The third parameter, c, is proportional to the first one by a factor k, $k > 1$ for the prolate and $0 < k < 1$ for the oblate, respectively. In this paper will be used $k \in \{2; 20\}$ for the prolate case and $k \in \{1/2; 1/20\}$ for the oblate case.

   Some examples of aggregates are shown in the figure 1.

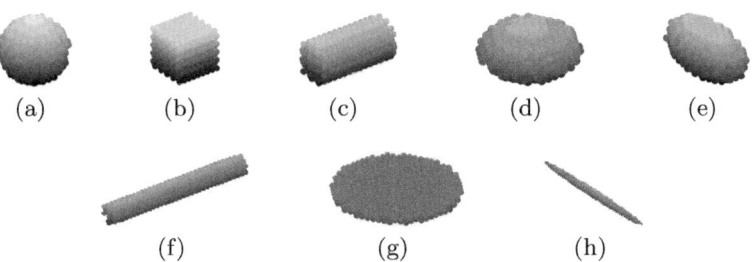

(a)   (b)   (c)   (d)   (e)

(f)   (g)   (h)

Fig. 1. Representation of different aggregates with a ful convex hull: (a) spherical, (b) cubic, (c) cylindrical with $k = 2$, (d) oblate with $k = 1/2$, (e) prolate with $k = 2$, (f) cylindrical with $k = 20$, (g) oblate with $k = 1/20$, (h) prolate with $k = 20$

Moreover, in order to compare the aggregates, the volume of their convex hull is the same value for all of them. This volume is fixed equal to that of a sphere with a 300 nm diameter, because this size is an usual order of magnitude of the second scale length in the optical domain.

The last studied parameter is the filling ratio of the convex hull by spherical primary particles: 100%, 75%, 50% and hollow aggregates.

The 100% filling aggregate is composed of particles whose center is inside the convex hull. This convex hull is placed so that it would be as fulfilled as possible. The method is shown in the figure 2).

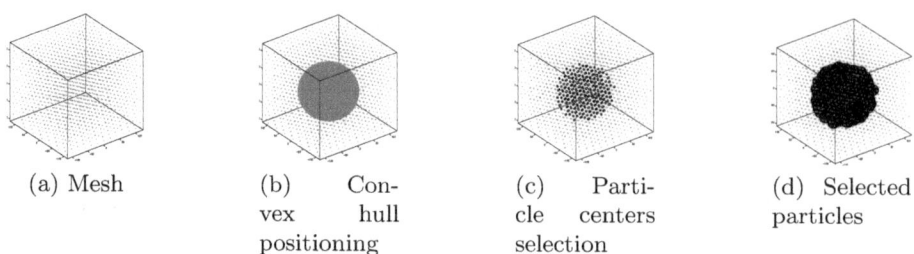

(a) Mesh   (b)   Con-   (c)   Parti-   (d) Selected
                  vex   hull   cle   centers   particles
                  positioning   selection

Fig. 2. Building of the 100% filling spherical convex hull

The 75% et 50% random filling correspond to a random choice (standard uniform law) of, respectively, 75% and 50% of the particles selected in the 100% filling case.

Concerning the full aggregate, each primary particles has 12 adjacent primary particles, implied by the closed-packed hexagonal mesh, except the ones located on the aggregate's surface. So, the particles constituting the hollow aggregates are those of the corresponding aggregates with a full convex hull, which doesn't have their 12 neighbours.

The figure 3 illustrate the different filling ratios for the spherical convex hull.

(a)   (b)   (c)   (d)

Fig. 3. Representation of aggregates with a spherical convex hull and different filling ratio: (a) full filling, (b) hollow filling, (c) random filling at 50%, (d) random filling at 75%

After aggregates simulation, the study of their geometrical characterization using two methods is performed in the next section. The particles are modeled by their center. As a consequence the aggregate is analysed such as a distribution of points (point pattern analysis).

3   Geometrical Characterization

Firstly, for each method, some results are presented to compare the different filling ratios (explained below): this is done for only one type of convex hull (the spherical convex hull) because the comments done for one are similar for the other ones. Next, analogies and differences between aggregates with quasi-similar convex hull are analyzed: spherical, cubic, cylindrical with $k = 2$, oblate with $k = 1/2$ and prolate with $k = 2$. Lastly, the cylindrical convex hull with several k-parameter values are compared.

3.1   Radial Distribution (RD)

The radial distribution (RD) method uses a sphere S, the center of which is chosen within the aggregate, and the radius r of which is variable. The value of r starts from 0 and then increases until the sphere totally incircles the aggregate. For each r value, the number of particle centers included in S is calculated. The same process could be done with the particle volume (quantity of matter), included in S as shown in Fig.4 with an aggregate constituted of non-connected particles. The study is focused on the distribution of the particle centers. Therefore, the cumulative radial distribution function (CRDF) can be extracted with regard to the parameter r. In this paper, the center of S is the geometrical center of the aggregate. Concerning the discretization of the r value, the step between two r values is fixed to 20 nm, because it is the smallest distance between two particle centers, the radius of one particle being equal to 10 nm.

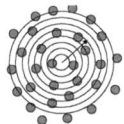

Fig. 4. Process of radial distribution function with an increasing radius r

Mathematically, the formula for the CRDF is defined by:

$$CRDF(r) = \frac{Number\ of\ particle\ centers\ at\ a\ distance \leq r}{Total\ number\ of\ particles\ in\ the\ aggregate} \quad (1)$$

Since the aggregates have similar volume and are built along the same mesh, the focus has been placed on the particle mean number, normalized or not by the total number of particle within the aggregate.

Characterization of the filling ratios of one convex hull
Fig.5 shows the CRDF for the spherical convex hull aggregates with different filling ratios.

In Fig.5(a), it can be noticed that random filled aggregates are uniformely filled, because for each value of r, the ratios 50% and 75% are conserved between the concerned curves, until the r-value equal to the convex hull radius.

In Fig.5(b), for aggregates with full convex hull, or filled at 75% or at 50% respectively, the normalized number of inclusion follows the same curve. This

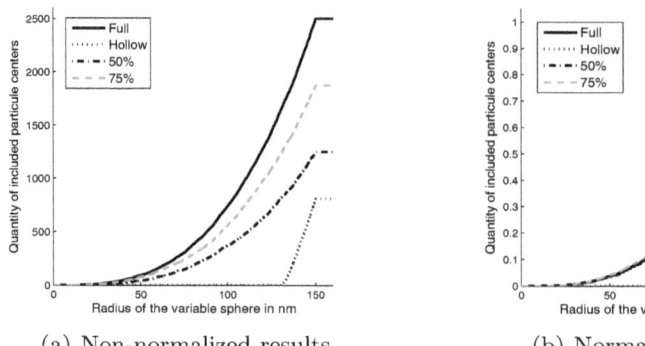

(a) Non-normalized results   (b) Normalized results

Fig. 5. Radial cumulative distribution function for the aggregates with a spherical convex hull and different filling ratios. The graph (a) is non-normalized. In (b) the number of centers included in S is normalized (CRDF(r)) by the total number of particles within each aggregate.

curve's equation is $f(r) = \frac{r^3}{R^3}$, where r and R are the radius of the sphere S, and the radius of the spherical hull, respectively. This equation comes from the fact that the distribution of the centers is uniform (standard uniform law). Consequently, the normalized radial cumulative distribution function with the aggregate is calculated as the volume of the sphere S, normalized by the volume of the convex hull of the aggregate.

However, this is different for the aggregate with a hollow convex hull. Indeed, the particles are along the convex hull so that the centers are included at the same step. It is the reason why, the curve for the hollow spherical convex hull is (theorically) a Heaviside function (see Fig. 5(b)).

Characterization of aggregates of quasi-similar convex hulls
Spherical, cubic, cylindrical, prolate with $k = 2$ and oblate $k = 1/2$ (Fig.7) convex hull aggregates are now studied.

The CRDF curves corresponding to each type of convex hull are different. All the curves have a common part: there, S is included within the aggregate. The equation of the radial cumulative distribution function is linked to the volumic fraction of the variable sphere S (Fig.5(b)). It is the reason why four phases in the curve corresponding to the cubic convex hull aggregate can be observed as in Fig.6(b) and Fig.7.

The first graph (Fig. 7(a)), is identical for all the type of convex hull: this is the phase where the variable sphere is totally included in the aggregate.

The second phase (Fig. 7(b)), is when the sphere overflows the aggregate forming spherical caps. The form of the cap basis depends on the aggregate convex hull: it is plane for the cubic convex hull, but, for example, curved for the oblate.

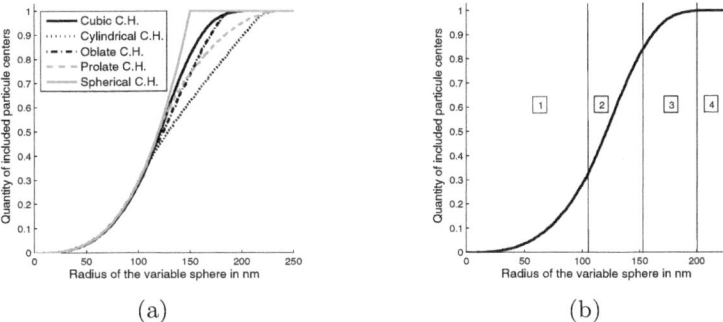

(a)   (b)

Fig. 6. (a) CRDF for quasi-similar and full convex hull aggregates. (b) CRDF for only the cubic convex hull.

(a)   (b)   (c)   (d)

Fig. 7. For the cubic convex hull, visualisation of the different phases CRDF curve

In some cases, there may be a third phase (Fig. 7(c)), where the caps begin to join, even if the aggregate is not totally incircled. For example, concerning the cubic convex hull, caps join before that the corners would be inside the variable sphere S. The ultimate phase (Fig. 7(d)) starts when the aggregate is totally incircled.

Characterization of aggregates with the same convex hull and several aspect ratios

The figure 8 shows the results for a cylindrical convex hull with several values for the parameter k (shape ratio).

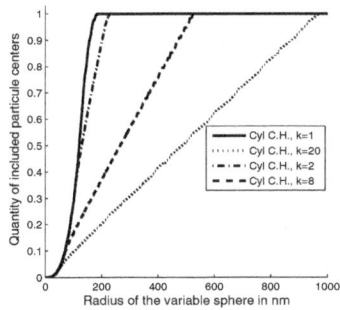

Fig. 8. CRDF for cylindrical and full convex hull aggregates with several shape ratio k

As mentioned for the previous graph, four phases of the cumulative radial distribution function curve can be seen (Fig.9), especially for the case with $k = 1$. These four stages can be also observed for $k \neq 1$ cases. As all the convex hulls have the same volume, the larger k is, the shorter the base diameter is (and longer the heigth is). Thus the more large is k, the longer the second phase is, contrary to the others phases.

(a)   (b)   (c)   (d)

Fig. 9. Four phases for CRDF of the cylindrical convex hull aggregates

To conclude on this first quantification method, the radial distribution method allows to differenciate the external structure of an aggregate.

3.2   Function of Pairs

The functions of pairs are morphological functions developped by means of integral geometry in Santalo [5]. They act as radial distribution functions, but are applied to each center of the particles constituting an aggregate. A pair designates the distance between a couple of particle centers. This function is closed to Ripley's function exposed in [6]. In the works of Gruy [7] are expressed the analytical pairs distribution functions of a spheroïd, oblate and prolate. In this paper, a simulated cumulative distribution of inter-center distances (averaged over the total number of pairs) is then computed.

The mathematical formula of the cumulative pair distribution function (CPDF) is:

$$CPDF(r) = \frac{Number\ of\ pairs \leq r}{Total\ number\ of\ pairs\ in\ the\ aggregate} \tag{2}$$

Characterization of the filling ratios of one convex hull

The results for the spherical convex hull aggregates ar shown in Fig.10.

As in the CRDF, the CPDF does not distinguish the filling ratio. Indeed, after normalization by the total number of inter-center distances of each aggregate respectively, the curves of 100%, 75% and 50% exactly coincide. Besides, the curve of the hollow convex hull aggregate remains isolated.

In Fig.10, especially in Fig.10(b), a inflection point of the curves can be noticed. It means that, for the spherical convex hull aggregate, there is a particular inter-center distance (about 150 nm), which is the same for full, 75% and 50% filling ratios, and another particular inter-center distance for the hollow spherical convex hull which is equal to 250 nm.

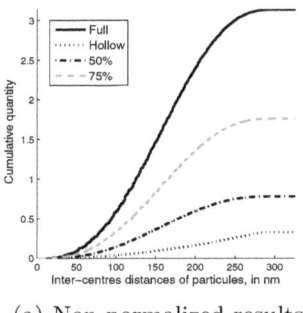

(a) Non-normalized results

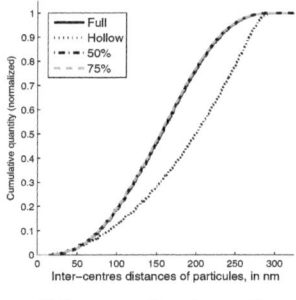

(b) normalized results

Fig. 10. Inter-center distances distribution for the aggregates of spherical convex hull, with different filling ratios. (a) non-normalized case. (b) Distribution normalized by the total number of inter-centers distances of each aggregate respectively, i.e. CPDF.

Characterization of aggregates of quasi-similar convex hulls

The results obtained for quasi-similar full convex hull aggregates are compared (Fig.11).

The CPDF for pherical, cubic, cylindrical, prolate with $k = 2$ and oblate with $k = 1/2$ convex hull aggregates are calculate and shown in the figure 11.

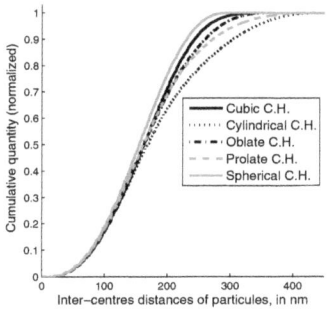

Fig. 11. CPDF for quasi-similar convex hull and full aggregates (spherical, cubic, cylindric with k=2, oblate with k=1/2, prolate with k=2=)

As in the figure 10, an inflection point is also observed at some inter-center distance values in the figure 11. These values are the same for the different convex hulls presented, and corresponds to a statistical mode (a class of the distribution having the maximum of elements).

Characterization of aggregates with the same convex hull type and several shape ratios

A similar inflection point can be remarked in the figure 12, which represents the results for aggregates with a cylindrical of convex hull, and different values for the k-parameter (1, 2, 8, 20).

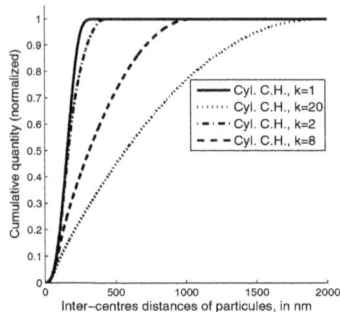

Fig. 12. CPDF for full and cylindrical convex hull aggregates, with different shape ratio k

Firstly, a proportionality between the largest inter-center distance values of each cylindrical convex hull aggregate and the k-parameter can be confirmed by the figure 12.

Further, the inflection point can be located for smaller inter-distance value while k-parameter increases.

The inter-center distance corresponding to the inflection point, and the maximal inter-center distances of each CPDF characterize the isotropy of the aggregate shape. These relation between these two elements characterize if the aggregate is hunched up (cubic, oblate convex hulls) or if the hull presents extensions (cylindric convex hull with k=8 or 20...), or anisotropies.

4   Conclusion and Perspectives

This article deals with two statistical methods for the morphological characterization of an aggregate of spherical particles. In a first time, cumulative radial distribution function allows an external analysis of the convex hull aggregate to be performed. In addition, this function is linked with the volumic fraction of the sphere S, normalized by the volume of the convex hull of the aggregate. It would be interesting to find analytically the equation of this function. On a second time, the analysis of the cumulative distribution function of pairs, i.e. distances between all of the particle centers of an aggregate, is performed. This analysis has shown that two elements seem to be important in the distribution of inter-centers distances: the inflection point and the spreading of all the distances. These two parameters characterize the isotropy of the aggregates shape. A good discrimination between the different convex hulls is reached with the two methods. However, they don't allow distinguishing the internal structure of the aggregates. For a better discrimination ofa ggregates, the authors are currently working on their geometrical characterization using more specific tools of point pattern analysis.

References

1. Jacquier, S., Gruy, F.: Anomalous diffraction approximation for light scattering cross section: Case of random clusters of non-absorbent spheres. J. Quant. Spectrosc. Radiat. Transfer 109(5), 789–810 (2008)
2. Xu, Y.-l.: Electromagnetic scattering by an aggregate of spheres. Applied Optics 34(21), 4573–4588 (1995)
3. Lagarrigue, M.: PhD thesis: Caractérisation optique et géométrique des agrégats multi-échelles. École nationale Supérieure des Mines de Saint-Étienne (in progress)
4. Mecke, K., Stoyan, D.: Finite packings and parametric density. In: Mecke, K.R., Stoyan, D. (eds.) Statistical Physics And Spatial Statistics. LNP, vol. 554, pp. 332–845. Springer, Heidelberg (2000)
5. Santalo, L.: Integral geometry and geometric probability. Cambridge University Press, Cambridge (2004)
6. Ripley, B.D.: Modelling spatial patterns. J. Roy. Statistical Society B 39, 172–212 (1977)
7. Gruy, F.: Light-scattering cross section as a function of pair distribution density J. Quant. Spectrosc. Radiat. Transfer 110(3), 240–246 (2009)

Author Index

GPSR Compliance

The European Union's (EU) General Product Safety Regulation (GPSR) is a set of rules that requires consumer products to be safe and our obligations to ensure this.

If you have any concerns about our products, you can contact us on ProductSafety@springernature.com

In case Publisher is established outside the EU, the EU authorized representative is:

Springer Nature Customer Service Center GmbH
Europaplatz 3
69115 Heidelberg, Germany

Batch number: 09473985

Printed by Printforce, the Netherlands